abe 5201

abe 5201

Annelids in Modern Biology

Annelids in Modern Biology

Edited by

Daniel H. Shain

WILEY-BLACKWELL

A John Wiley & Sons, Inc., Publication

Wiley-Blackwell is an imprint of John Wiley & Sons, formed by the merger of Wiley's global Scientific, Technical, and Medical business with Blackwell Publishing.

Published by John Wiley & Sons, Inc., Hoboken, New Jersey.
Published simultaneously in Canada.

For general information on our other products and services or for technical support, please contact our Customer Care Department within the United States at (800) 762-2974, outside the United States at (317) 572-3993 or fax (317) 572-4002.

Wiley also publishes its books in a variety of electronic formats. Some content that appears in print may not be available in electronic formats. For information about Wiley products, visit our web site at www.wiley.com.

Library of Congress Cataloging-in-Publication Data:

Annelids in modern biology / edited by Daniel H. Shain.
    p. cm.
    Includes index.
    ISBN 978-0-470-34421-7 (cloth)
  1. Annelida. I. Shain, Daniel H.
    QL391.A6.A56 2009
    592′.6–dc22
                            2008042925

Printed in the United States of America.

10 9 8 7 6 5 4 3 2 1

In Memoriam

Charlie D. Drewes (1946–2005)

On Monday, July 4, 2005, those of us interested in annelids and their use as models in research and teaching lost a good friend and enthusiastic colleague. Charlie Drewes, Professor of Ecology, Evolution and Organismal Biology at Iowa State University, died in Ames, Iowa.

Charlie was born and raised in Minnesota and received a bachelor's degree in biology from Augustana College in South Dakota. As a National Science Foundation Fellow in the early 1970s, he completed both master's and doctoral degrees in zoology at Michigan State University. He was recruited as an assistant professor to Iowa State University where he attained the rank of professor and, in 2003, was honored with the title of University Professor. As a physiologist at Iowa State University for 31 years, Charlie taught hundreds of students in neurobiology, invertebrate biology, advance vertebrate physiology and science ethics. Charlie mentored 19 graduate students in neurobiology research, many of them now mentoring students of their own at universities in California, Illinois, Texas, Vermont and Georgia, among others. He published over 50 refereed research articles and numerous book chapters, invited reviews and education-related publications. Many of his most important contributions followed his development of noninvasive, recording devices for detecting giant fiber action potentials in freely moving worms. He applied this approach to studies of escape reflex physiology, annelid nerve regeneration and the sublethal effects of environmental toxicants. In the last decade or more, Charlie invested considerable effort in developing *Lumbriculus variegatus* as a tool for biological science education (see Chapter 10). With groups now interested in sequencing the genome of *L. variegatus* and with its expansion as a model in teaching and research laboratories, Charlie's legacy might be linked to his keen enthusiasm for the California mudworm as a model for discovery.

Charlie Drewes focused the last years of his life on educational outreach to high school and college science teachers. He conducted an annual "total immersion" teaching workshop at Iowa Lakeside Laboratory in Lake Okoboji, Iowa. A legion of high school and community college biology teachers from 16 states participated in these workshops over the years. He conducted numerous invited workshops and presentations at local and national platforms, including a guest instructorship at the Woodrow Wilson National Fellowship Foundation and recurring visiting instructorships at Advanced Placement Biology workshops for State of Texas biology teachers at Texas A&M University. Thus, Charlie's impact on biologists, and especially the group of individuals he fondly called wormologists, was far broader than the borders of Iowa. Charlie was a recipient of several distinguished teaching awards from Iowa State University, the Iowa Academy of Science and the National Association of Biology Teachers (NABT). In 2004, the NABT named him the National College Biology Teacher of the Year. Along with his research publications, Charlie produced a long list of useful documents, informational and procedural, on annelids and other invertebrates as tools for science education. This can be found online at a site (http://www.eeob.iastate.edu/faculty/DrewesC/htdocs/) maintained by the Department of Ecology, Evolution, and Organismal Biology, Iowa State University.

Charlie Drewes was willing to answer e-mails from wondering high school students at almost any time of the day or night. He would talk for hours at a time in person or by phone with colleagues seeking his knowledge. We will miss Charlie's inexhaustible excitement toward biology and annelids in particular. His passing has created a void in our scientific community and in many of our hearts. Charlie enjoyed life and the lives of worms. Let us continue the joy.

Contents

Preface   xi

Contributors   xiii

Part I   Annelids as Model Systems
in Biology

1. Developing Models for
   Lophotrochozoan and Annelid
   Biology   3

   Kenneth M. Halanych and
   Elizabeth Borda

   1.1   Introduction   3
   1.2   Phylogenetic Considerations   4
   1.3   Genetic and Developmental
          Tools   5
   1.4   Annelid Model Organisms   6
   1.5   Other Potential Annelid
          Models   8

2. Annelid Phylogeny—Molecular
   Analysis with an Emphasis on
   Model Annelids   13

   Christoph Bleidorn

   2.1   Introduction   13
   2.2   Genes   14
   2.3   Molecular Annelid
          Phylogeny   16
   2.4   Choosing Model Organisms   24
   2.5   Branch Lengths   24
   2.6   Problems in Inferring Annelid
          Phylogeny   24
   2.7   Conclusions   25

3. Cryptic Speciation in Clitellate
   Model Organisms   31

   Christer Erséus and Daniel Gustafsson

   3.1   Introduction   31

   3.2   Sources and Kinds
          of Variation   32
   3.3   Examples of Clitellate Model
          Organisms   34
   3.4   Cryptic Speciation   41
   3.5   Conclusions and
          Recommendations   41

4. Annelid Life Cycle Cultures   47

   Donald J. Reish and Bruno Pernet

   4.1   Introduction   47
   4.2   Criteria for the Selection of
          Species   48
   4.3   Summary of Culture
          Techniques   48
   4.4   Life Cycle Cultures of
          Polychaeta   50
   4.5   Life Cycle Cultures of
          Oligochaeta   56
   4.6   Life Cycle Cultures of Hirudinea
          (Leeches)   58

Part II   Evolution and Development

5. Annelids in Evolutionary
   Developmental Biology   65

   Dian-Han Kuo

   5.1   Introduction   65
   5.2   Evo-Devo Today   66
   5.3   Evo-Devo as Comparative
          Biology   66
   5.4   Why Annelid Development Is
          Interesting for Metazoan
          Evo-Devo Biologists   67
   5.5   Case Study 1:
          Segmentation   68
   5.6   Case Study 2: Spiral Cleavage and
          Axis Specification   74

5.7   Tools for Analyzing Molecular Mechanisms of Development   79

5.8   The Future of the Annelid Model Systems for Evo-Devo   81

6. Evolution, Development and Ecology of *Capitella* sp. I: A Waxing Model for Polychaete Studies   88

Susan D. Hill and Robert M. Savage

6.1   Introduction   88
6.2   Speciation Studies   89
6.3   *Capitella* Sp. 1 Morphology   91
6.4   Replacement of Lost Segments and Reproductive Trade-Offs   92
6.5   Metatrochophores, Ciliary Bands and Musculature   93
6.6   Gene Expression during the Specification and Differentiation of Germ Layers   95
6.7   Sex among the Vermes   100
6.8   Annelids and the Segmentation Debate   101
6.9   A-P Polarity—*Hox* and *ParaHox* Genes   106
6.10   Annelid Genomics: Draft Genome Sequence   108
6.11   The Future—Where Is This Going?   109

7. Stem Cell Genesis and Differentiation in Leech   116

Shirley A. Lang and Daniel H. Shain

7.1   Introduction   116
7.2   Stem Cell Genesis and Development   117
7.3   Factors Affecting Stem Cell Genesis   121
7.4   Stem Cell Differentiation   124
7.5   Gene Expression   128
7.6   Conclusion   130

Part III   Neurobiology and Regeneration

8. Cellular and Behavioral Properties of Learning in Leech and Other Annelids   135

Kevin M. Crisp and Brian D. Burrell

8.1   Introduction   135
8.2   Learning in the Leech Whole-Body Shortening Reflex and Role of the S Interneuron   137
8.3   Role of the S Interneuron: Modulation of Excitability   139
8.4   Learning in the Leech Swim Circuit   144
8.5   Using the Leech to Study Intrinsic Forms of Sensitization   146
8.6   Synaptic Plasticity in Leech CNS   147
8.7   Conclusions   150

9. Development, Regeneration and Immune Responses of the Leech Nervous System   156

Michel Salzet and Eduardo Macagno

9.1   Introduction   156
9.2   Background   157
9.3   Recent Work on the Development of the Nervous System   157
9.4   Neuronal Regeneration and Repair   169
9.5   Neuroimmune Responses   173
9.6   Cellular and Humoral Immune Mechanisms: A Leech Innate Immune Response   176
9.7   Conclusions and Future Directions   179

10. *Lumbriculus variegatus* and the Need for Speed: A Model System for Rapid Escape, Regeneration and Asexual Reproduction   185

Mark J. Zoran and Veronica G. Martinez

10.1   Introduction   185
10.2   Neural Regeneration in Oligochaetes   187

10.3 *Lumbriculus variegatus*, a Model System for Regeneration and Asexual Reproduction 191

10.4 Neural Morphallaxis 192

10.5 Accessible Model for Life Science Education 198

Part IV Environmental and Ecological Studies

11. Polychaetes in Environmental Studies 205

Victoria Díaz-Castañeda and Donald J. Reish

11.1 Introduction 205

11.2 Estuarine Occurrence 206

11.3 Intertidal Occurrence 207

11.4 Mussel Beds 207

11.5 Sea Grasses 208

11.6 Sabellarid and Serpulid Reefs 209

11.7 Benthic Community Structure 209

11.8 Unusual Benthic Habitats 210

11.9 Feeding Guilds 213

11.10 Algal "Gardening" Behavior 214

11.11 Polychaetes as Environmental Indicators and Remediators 214

11.12 Biomonitoring 217

11.13 Toxicological Tests 219

11.14 Economic Importance of Polychaetes 221

11.15 Conclusions 221

12. Oligochaete Worms for Ecotoxicological Assessment of Soils and Sediments 228

Jörg Römbke and Philipp Egeler

12.1 Introduction 228

12.2 Principles of Environmental Risk Assessment 229

12.3 Soil Tests with Lumbricidae 230

12.4 Soil Tests with Enchytraeidae 232

12.5 Sediment Tests with Lumbriculidae and Tubificidae 235

12.6 Oligochaetes in Ecotoxicology 237

12.7 Conclusions 238

13. Evolution and Ecology of *Ophryotrocha* (Dorvilleidae, Eunicida) 242

Daniel J. Thornhill, Thomas G. Dahlgren, and Kenneth M. Halanych

13.1 Introduction 242

13.2 General Morphology 242

13.3 Taxonomic and Phylogenetic Considerations 243

13.4 Reproductive Biology 247

13.5 Ecology 250

13.6 Future Research 252

14. Cosmopolitan Earthworms—A Global and Historical Perspective 257

Robert J. Blakemore

14.1 Introduction 257

14.2 Number of Earthworm Species 267

14.3 Characteristics and Origins of Cosmopolitan Earthworms 268

14.4 Overview of Results 268

14.5 Discussion 269

14.6 Regional Species Totals and Proportions of Exotics 271

14.7 Earthworms, Archaeology and Human History 273

14.8 Benefits and Risks of Earthworm Transportations 277

14.9 Conclusions 278

Part V   Extreme Environments and
Biological Novelties

15. Hydrothermal Vent Annelids   287

Florence Pradillon and Françoise Gaill

15.1   Introduction   287
15.2   *Alvinella pompejana*: a Symbiotic
       System   289
15.3   Temperature Adaptation   290
15.4   Temperature Adaptation at a
       Molecular Level   290
15.5   *Alvinella* Tubes   291
15.6   Collagens   292
15.7   Temperature Adaptation at a
       Cellular Level: the Case of
       Developing Embryos   295
15.8   Behavioral Adaptation to a
       High-Temperature
       Environment   297
15.9   Future Development of Thermal
       Adaptation Studies   297
15.10 Perspectives   298

16. Glacier Ice Worms   301

Paula L. Hartzell and Daniel H. Shain

16.1   Introduction   301
16.2   Natural History   303
16.3   Classification and Phylogenetic
       Relationships   305
16.4   Origins   306
16.5   Clades   307
16.6   Physiology   309
16.7   Conservation Status   311

17. Sperm Ultrastructure in Assessing Phylogenetic Relationships among Clitellate Annelids   314

Roberto Marotta and Marco Ferraguti

17.1   Introduction   314
17.2   The Spermatozoon of *Propappus
       volki* (Michaelsen 1916)   317
17.3   Sperm Ultrastructure in
       Branchiobdellids, *Acanthobdella
       peledina*, and Hirudineans   320
17.4   Sperm Ultrastructure inside
       Tubificidae   320
17.5   Plesiomorphic Spermatozoon of
       Clitellates and Spermatological
       Apomorphic Trends   324
17.6   Patterns of Spermatological
       Characters among
       Clitellates   324

18. Clitellate Cocoons and Their Secretion   328

Jon'elle Coleman and Daniel H. Shain

18.1   Introduction   328
18.2   Reproductive Biology   330
18.3   Clitellum and CGCs   330
18.4   Cocoon Production   331
18.5   Brooding Behavior within
       Glossiphoniidae   334
18.6   Cocoon Structure: Surface
       Topology and Ultrastructural
       Properties   335
18.7   Evolution of Clitellate Cocoons
       and Their Secretion   341
18.8   Biomaterials Applications   342

Index   347

Preface

Annelids (segmented worms) are among the most ecologically diverse animal groups, occupying habitats that range from hydrothermal vents to glacier ice. This diversity, along with their simple body plan and broad experimental accessibility, has made them desirable subjects of scientific curiosity over the past century. Unfortunately, trends in modern biology have clustered mainstream researchers around a few model systems that support contemporary molecular techniques, creating a relatively narrow scope in the scientific community and sometimes overcrowded and undercollaborative fields. A consortium of scientists is eager to increase the diversity of research subjects in the sciences, as a means of enhancing our ability to dissect fundamental biological questions as well as to promote the discovery of biological novelties. Annelids are particularly suited and well poised for these challenges, as they occupy a strategic evolutionary position in animal phylogeny (i.e. Lophotrochozoa) and have an experimental infrastructure that supports state-of-the-art methodologies.

In my own somewhat eclectic research interests, which include neurodevelopment, embryonic stem cell specification, cold temperature physiology and biomaterials science, I have found that annelids of various species offer unexpectedly versatile experimental systems. For instance, stem cells in glossiphoniid leeches are among the largest in the Metazoa (up to 300 μm in some species), which facilitates their homogenous isolation and subsequent characterization; clitellate cocoons comprise thin, flexible biomembranes sealed with underwater glues, and display remarkable resiliency to heat and denaturing chemicals; glacier ice worms represent the largest and most sophisticated cold-adapted ectotherm. Others have employed the experimental accessibility of annelids to examine fundamental questions in neurobiology (for which hirudinid leeches are the classical preparation), learning and memory, immunology, development, ecology, life at extreme environments, and for assessing environmental quality and toxicology—all topics that are addressed in this book.

This compilation of annelid works is by no means intended to be comprehensive; rather, it provides a sampling of the types of questions that can be investigated using annelids as research subjects. I am committed to promoting and fostering the use of annelid worms in science, as they remain a relatively unexplored animal group that are potential gold mines in disparate scientific disciplines. My hope is to inform and encourage members of the scientific community, particularly emerging young scientists, to consider the use annelids in their future studies, with the long-term and necessary goal of increasing diversity in scientific research.

DANIEL H. SHAIN

Contributors

Robert J. Blakemore, COE Soil Ecology Group, Graduate School of Environment and Information Sciences, Yokohama National University, Tokiwadai, Yokohama 240-8501 Japan; Email: robblakemore@bigpond.com

Christoph Bleidorn, Unit of Evolutionary Biology/Systematic Zoology, Institute of Biochemistry and Biology, University of Potsdam, Karl-Liebknecht-Strasse 24-25, Haus 26, D-14476 Potsdam-Golm, Germany; Email: bleidorn@uni-potsdam.de

Elizabeth Borda, Department of Biological Sciences, Auburn University, 101 Rouse, Auburn, AL

Brian D. Burrell, Neuroscience Group, Division of Basic Biomedical Sciences, Sanford School of Medicine at the University of South Dakota, Vermillion, SD

Jon'elle Coleman, Biology Department, Rutgers The State University of New Jersey, Camden, NJ

Kevin M. Crisp, Department of Biology and Neuroscience Program, St. Olaf College, Northfield, MN; Email: crisp@stolaf.edu

Victoria Díaz-Castañeda, Departamento de Ecología, CICESE. Km 107 Carr. Tijuana-Ensenada, C. P. 22860 Baja California, Mexico; Email: vidiaz@cicese.mx

Thomas G. Dahlgren, Göteborgs Universitet, Zoologiska Institutionen, Systematik och Biodiversitet, Box 463, 405 30 Göteborg, Sweden

Philipp Egeler, ECT Oekotoxikologie GmbH, Böttgerstr. 2-14, D-65439 Flörsheim, Germany

Christer Erséus, Department of Zoology, University of Gothenburg, Box 463, SE-405 30 Göteborg, Sweden; Email: christer.erseus@zool.gu.se

Marco Ferraguti, Dipartimento di Biologia, Università di Milano, Italy

Françoise Gaill, UMR CNRS 7138 Systématique, Adaptation et Evolution, Université Pierre et Marie Curie, 7 quai Saint-Bernard, 75252 Paris cedex 05, France; Email: francoise.gaill@snv.jussieu.fr

Daniel Gustafsson, Department of Zoology, University of Gothenburg, Box 463, SE-405 30 Göteborg, Sweden

Kenneth M. Halanych, Department of Biological Sciences, Auburn University, 101 Rouse, Auburn, AL; Email: Ken@auburn.edu

Paula L. Hartzell, Mountain Research Institute, POB 2154, Kealakekua, Hawaii

Susan D. Hill, Department of Zoology, Michigan State University, East Lansing, MI; Email: hillss@msu.edu

Dian-Han Kuo, Department of Molecular and Cell Biology, University of California, Berkeley, Berkeley, CA; Email: dhkuo@berkeley.edu

Shirley A. Lang, Biology Department, Rutgers, The State University of New Jersey, Camden, NJ

Eduardo Macagno, Division of Biological Sciences, University of California, San Diego, La Jolla, CA; Email: emacagno@ucsd.edu

Veronica G. Martinez, Department of Biology, University of the Incarnate Word, San Antonio, TX

Roberto Marotta, Dipartimento di Biologia, Università di Milano, Italy; Email: roberto.marotta@unimi.it

Bruno Pernet, Department of Biological Sciences, California State University, Long Beach, Long Beach, CA 90840

Florence Pradillon, UMR CNRS 7138 Systématique, Adaptation et Evolution, Université Pierre et Marie Curie, 7 quai Saint-Bernard, 75252 Paris cedex 05, France

Donald J. Reish, Department of Biological Sciences, California State University, Long Beach, Long Beach, CA 90840; Email: djReish@aol.com

Jörg Römbke, ECT Oekotoxikologie GmbH, Böttgerstr. 2-14, D-65439 Flörsheim, Germany; Email: j-roembke@ect.de

Michel Salzet, Laboratoire de Neuroimmunologie des Annélides, FRE-CNRS 2933, IFR 147, Université des Sciences et Technologies de Lille, 59655 Villeneuve d'Ascq, France

Robert M. Savage, Biology Department, Williams College, Williamstown, MA

Daniel H. Shain, Biology Department, Rutgers, The State University of New Jersey, Camden, NJ; Email: dshain@camden.rutgers.edu

Daniel J. Thornhill, Department of Biological Sciences, Auburn University, 101 Rouse, Auburn, AL

Mark J. Zoran, Department of Biology, Texas A&M University, College Station, TX; Email: zoran@mail.bio.tamu.edu

Part I

Annelids as Model Systems in Biology

Chapter 1

Developing Models for Lophotrochozoan and Annelid Biology

Kenneth M. Halanych and Elizabeth Borda

Department of Biological Sciences, Auburn University, Auburn, AL

1.1 INTRODUCTION

Most investigators using model organisms assume that, for the question being asked, the model acts in a similar (or at least predictably different) way compared to the organism of interest (e.g. human, an insect pest, ecologically important taxon). Given the processes of evolutionary change, we can expect that similarity in processes or mechanisms and thus the ability to extrapolate information are often correlated with phylogenetic relatedness. Thus, one might expect that, as a community of scientists, we would try to judiciously choose model organisms to span the diversity of organismal lineages. However, phylogenetic considerations are only one of several factors that determine the suitability of an organism as a model. For example, the ease and cost of propagating lineages in the laboratory, generation time and ability to genetically manipulate are also important considerations.

Nonetheless, model species have historically been heavily skewed toward only two of the three bilaterian lineages (i.e. Ecdysozoa, Deuterostoma). Species including the ecdysozoans *Drosophila melanogaster* and *Caenorhabditis elegans* and the deuterostomes *Mus musculus* and *Danio rerio* dominate the genetic, cellular biology, physiological and developmental literature because they are inexpensive to maintain; rearing and breeding in a laboratory setting is easy; they exhibit short generation times, and tools for genetic manipulation have been well described. Of significance, many of these species were subject to large-scale early mutational screens, thereby making them of interest to a wide range of scientific disciplines. In contrast, these types of screens are lacking for lophotrochozoans, the third bilateran lineage that has been largely ignored for holistic (i.e. multidisciplinary) model organisms.

Because model system choice has been closely tied to the ease of experimentation in a laboratory setting, considerations of evolutionary history have been neglected, resulting in the most morphologically diverse branch of Bilateria being overlooked (i.e. Lophotrochozoa). This situation has considerable implications for our understanding of animal evolution and diversity. In particular, studies on the evolution of developmental mechanisms have been particularly informative because early studies on ecdysozoan and deuterostome models, and later studies on cnidarians, have suggested conservation of homologous developmental mechanisms, such as axial patterning systems, endoderm formation and neuronal formation (Martindale 2005). However, some genes (e.g. genes related to segmentation) appear to have different roles in the three major bilaterian lineages (Seaver 2003). Without the inclusion of lophotrochozoan data, we cannot fully understand the evolution of these important mechanisms in the context of animal evolution.

Admittedly, choosing a holistic model organism within Lophotrochozoa is not easy. The challenge of selecting model species that will be used by multiple disciplines to represent all of Lophotrochozoa is exacerbated by several factors. Two of these factors are (1) an unresolved phylogeny for Lophotrochozoa, and (2) genetic and development protocols for laboratory experimentation have not been adequately developed. Fortunately, scientists are making considerable advances on both of these issues.

1.2 PHYLOGENETIC CONSIDERATIONS

As mentioned, in order for a model organism to be useful, the information must be extrapolated to other taxa. Thus, a comparative framework is an implicit, but often forgotten, component of studies using model organisms. In biological studies, the comparative framework is phylogeny. Thus, better understanding of evolutionary history allows more refined and accurate comparisons to be made. We now have a reasonable understanding of most major animal lineages. The most basal lineages are likely sponges and placozoans followed by the dipoblastic ctenophores and cnidarians. Within bilaterian animals are three major clades: deuterostomes, ecdysozoan and lophotrochozoans. The deuterostomes consist of Chordata, Hemichordata, Echinodermata and *Xenoturbella*. Major ecdysozoan lineages include Arthopoda, Nematoda, Priapulida, Kinorhyncha and their kin. Sister to ecdysozoans are lophotrochozoans, which include Annelida, Mollusca, Nemertea, Bryozoa, Brachiopoda and their allies. An apparent subgroup of Lophotrochozoa, i.e. Platyzoa, includes Playthelminthes (but not Acoelomorpha) and several meiofaunal taxa, such a gastrotrichs, gnathostomulids and possibly entoprocts. Our current understanding of animal relationships has been recently reviewed elsewhere (Halanych 2004; Giribet et al. 2007; and see Dunn et al. 2008), and thus we will only focus on issues of annelid affinities that are germane to this volume.

The early diversification of Lophotrochozoa presumably started in the late Proterozoic and very early Cambrian (which started 543 mya). Historically, relationships within Lophotrochozoa have been difficult to resolve in part due to the short time between diversification events, thus allowing little time for the accumulation of phylogentic signal, relative to the time since such events (Halanych 1998; Rokas et al. 2005). Despite this limitation, recent molecular studies have made progress in unraveling evolutionary relationships within Lophtrochozoa. Large ribosomal subunit (LSU) data and combined small subunit (SSU) + LSU data suggest that platyzoans are among the most derived lineages of lophotrochozoans, with annelids occupying a more basal or central position, respectively (Passamaneck and Halanych 2006). Unfortunately, support for internal nodes within these data is weak. In contrast, express sequence tag (EST) data (cDNA libraries generated from expressed messenger RNA that is then sequenced) are becoming a more widely used tool and hold promise for helping provide some resolution (Telford et al. 2008). Although several interesting EST studies have included lophotrochozoans (e.g. Philippe et al. 2005; Struck and Fisse 2007), the work of Dunn et al. (2008) includes the best taxon sampling of Lophotrochozoa to date (note that all of these studies have large proportions of missing data and the influence of these missing characters on topology reconstruction and robustness of results is not clear). In their analyses, Dunn et al. (2008) found annelids as sister to a brachiopod/phoronid/nemertean clade, which in turn is sister to mollusks. This general cluster of taxa has been loosely observed previously but without nodal support (see Halanych 2004). The close placement of mollusks, annelids and nemerteans is in good agreement with morphological and developmental information based on larval form, cleavage patterns and cell fate (Nielsen 2004; Martindale 2005). The placement of brachiopods and phoronids close to these taxa is puzzling based on morphological or embryological grounds, but note that both annelids and brachiopods have similar chaetal structures (Orrhage 1973).

Traditionally, several presumed phyla have been allied with annelids, including sipunulids, echiurids and siboglinids (formerly pogonophorans; Halanych et al. 2002). A considerable body of evidence shows that all three of these worm groups are within Annelida (McHugh 1997; Rousset et al. 2006; Struck et al. 2007; Dunn et al. 2008). Similarly, myzostomids, an unusual parasitic group of worms that have been variably placed outside annelids, are within the clade (Bleidorn et al. 2007). Several molecular studies (e.g. McHugh 1997; Bleidorn et al. 2003; Rousset et al. 2006; Struck et al. 2007) have placed the Clitellata (Oligochaeta + Huridinea) well within polychaetes rendering the node that defined "Polychaeta" as the same node as Annelida. The following chapter by C. Bleidorn more thoroughly covers the advances in annelid relationships, and therefore this discussion will be forgone here.

1.3 GENETIC AND DEVELOPMENTAL TOOLS

Traditionally, the field of genetics has largely driven the development of "classical" model systems using organisms for which a considerable literature and resources existed. Advances in molecular genetic technology have the potential to change how model organisms can be established. Namely, with the power of genome sequencing and the availability of functional genomic tools, an organism that has received rather limited attention can be very quickly developed into a useful multidisciplinary resource. For example, the sea squirt *Ciona intestinalis* quickly reached prominence as a deuterostome model organism once EST projects began and the genome was sequenced (Satoh et al. 2003).

In the case of lophotrochozoans, such efforts are being made, but as the scientific community interested in such resources tends to be fewer in number, they unfortunately have a smaller collective voice. Poignant illustrations of this fact are recent decisions by the US National Institutes of Health (NIH) to fund major genomic projects on all orders of mammals and tens of species of *Drosophila* before much investment has been made on annelids or mollusks. Nonetheless, the situation is improving as the gastropod limpet *Lottia gigantea*, the leech *Hellobdella robusta*, the capetillid annelid *Capitella* sp. I and the nereidid annelid *Platynereis dumerilii* have been the subject of major genomic efforts. All of these animals are emerging model systems, but additional resources need to be developed. Also, sequencing of the behavioral and neural model gastropod *Aplysia californica*, which has a large genome, is under way. Major genome initiatives also have taken (or are taking) place for some parasitic platyhelminthes (e.g. *Schmidtea mediterranea,*

Taenia solium, Schistosoma mansoni). However, these genomic efforts aim to understand the biology of these parasitic organisms versus developing the resources to be used as a biological model.

Development of these new model systems usually takes into serious consideration the ability of genomic information to be used by disciplines such as evolution of developmental mechanisms or neurobiology. Given that genome projects are getting cheaper and faster, the availability of large amounts of genomic data will not be sufficient to develop a model system. Other resources need to be developed. In the case of lophotrochozoans, several organisms have been used as models within particular disciplines. For example, the squid *Loligo pealei* and nudibranch *Tritonia diomedea* have been used heavily in neurobiology for years (see symposium by McPhie and Miller 2006). Also, the bivalve *Mya arenaria* has served as a model for leukemia; snail species of *Illyanassa* have been used in developmental biology, and *Capitella* species have long been bioindicators in ecotoxicology experiments. However, for all of these models, limitations in developing interdisciplinary resources (e.g. undeveloped methods for surveying gene expression), life history issues (e.g. long generation times) or rearing issues (e.g. hard to maintain in the lab) have hindered the interest in these organisms by other disciplines.

Below we highlight some of the "up-and-coming" annelid systems that may serve as models not only for annelids but for Lophotrochozoa as well. We provide only a brief description of each, as many of these taxa are discussed in more detail in the chapters that follow. For any of these organisms to reach the celebrity status of "model organism," the community must continue to develop resources that have interdisciplinary appeal.

1.4   ANNELID MODEL ORGANISMS

Historically, annelid species have been models for specific biological disciplines, and developing a more interdisciplinary or "holistic" approach has been underway and continues to improve. The combination of remarkable diversity in life history adaptations, form and function has promoted several annelid species that have the potential to become a general model organism. Improved understanding of annelid phylogeny has provided more clues toward identifying models that are broadly representative of other lineages. Although major efforts are being made toward identifying an appropriate model, currently, several annelid lineages hold promise as model systems. The following annelid taxa have already been used extensively as models in molecular and developmental biology (Weisblat and Shankland 1985; Irvine and Seaver 2006; references therein).

1.4.1   "The Leech"

One of the most widely used annelids in developmental biology is the leech. The leech is a conglomeration of several species representing two families: leeches within Glossiphoniidae (i.e. *Helobdella robusta, Helobdella triserialis, Theromyzon tessulatum* and *Haementeria ghilianii*) and the European medicinal leech *Hirudo medicinalis* (Hirudinidae) (Sawyer 1986; Irvine and Seaver 2006). Studies of annelid segmentation and development initially came from leeches (e.g. Sawyer 1986; Shankland 1991). The leech embryo is relatively large and exhibits a stereotypical cleavage program, which has enabled the description of segment formation during development.

Of all leeches, *H. medicinalis* is the most renowned member of Hirudinida. *Hirudo medicinalis* was heavily exploited for its sanguinivorous capacity during the late 19th century and is one of few invertebrates approved by the U.S. Food and Drug Administration (FDA) as a medical device (Rados 2004). *Hirudo medicinalis* has also been extensively studied with respect to its neurophysiology (e.g. Hagadorn 1966; Brodfuehrer et al. 2008), anticoagulant biochemical properties (e.g. Harvey et al. 1986; Mao et al. 1988), developmental genetics (e.g. Baker and Macagno 2001; Venkitaramani et al. 2004), behavior (e.g. French et al. 2005) and morphology (e.g. Sawyer 1986; Orevi et al. 2000). The lesser-known species of Glossiphoniidae have particularly been amenable to developmental work (Weisblat and Huang 2001). For example, the non-blood-feeding *Helobdella* species have been used for understanding embryogenesis (e.g. Kang et al. 2003), body plan development (Bruce and Shankland 1998; Kuo and Shankland 2004) and segmentation (Seaver 2003). Both *H. medicinalis* and *H. robusta* are subjects of genomic undertakings (ESTs and ESTs and whole genome sequencing, respectively).

Surprisingly, recent application of molecular tools has revealed significant issues with species identification among leeches. Specifically, DNA bar coding and microsatellite markers indicates that the FDA-approved and commercially available *H. medicinalis* is a different species altogether, *Hirudo verbana* (Trontelj and Utevsky 2005; Siddall et al. 2007). Furthermore, the two widely used *Helobdella* species (*H. robusta* and *H. triserialis*) each consist of a species complex (Bely and Weisblat 2006). These identification issues are likely confounded in the literature (see Chapter 3).

1.4.2   The Earthworm

Lumbricus terrestris, the common earthworm, is likely the best-known annelid because it is an easily recognized oligochaete and is widely used as an education tool for invertebrate zoology. Like many annelids, *Lumbricus* is a strong biological indicator due to its sensitivity to soil toxicities and bioaccumulation (e.g. Wright and Stringer 1974; Mahmoud 2007). Although important in agricultural systems, *L. terrestris* is considered a pest of soil communities due to human-assisted introductions that compete with native oligochaete species (Eisenhauer et al. 2007; Holdsworth et al. 2007). In terms of genomics, two species are dominating as model oligochaetes. *Lumbricus terrestris* has been the subject of EST studies (see GenBank) and has a completed mtDNA genome (Boore and Brown 2000), and in 2003, Sturzenbaum and colleagues initiated an EST project for the hummus earthworm, *Lumbricus rubellus*.

1.4.3   *Capitella*

Capitella capitata is an excellent biological indicator (Méndez et al. 1998, 2001) that was thought to be a cosmopolitan species dominating polluted marine environments. Allozyme and life history data reveal that *C. capitata* comprises a species complex with little morphological variation (Grassle and Grassle 1976, 1977; Méndez et al. 2000), suggesting that years of environmental data need to be reevaluated. The work by Grassle and Grassle (1976) identified at least two sibling species, designated *Capitella* sp. I and II as ideal laboratory animals for comparative biology. As a result, speciation (e.g. Grassle 1984; Du et al. 2007), ecology (e.g. Tsutsumi et al. 2005; Martin and Bastida 2006), environmental impact (e.g. Cardoso et al. 2007) and toxicology (e.g. Méndez and Green-Ruiz 2006) of *Capitella* species continue to be of great interest. However, in the recent literature, *Capitella* sp. I is one of the best-developed annelid models being used to understanding body plan formation and segmentation (Seaver et al. 2005; Seaver and Kaneshige 2006).

Capitella sp. I was one of the first lophotrochozoan representatives for which whole genomic sequence was obtained (completed by the Joint Genome Institute).

1.4.4 *Platynereis dumerilii*

Dumeril's clam worm, *P. dumerilii*, is a marine annelid that has been the subject of considerable study. In general, when most think of a polychaete worm, they typically envision members of the family Nereididae (e.g. *Platynereis, Nereis*), despite the amazing diversity of forms across annelid lineages. *Platynereis dumerilii* was chosen as another organism well suited for comparative studies in development (Fischer and Dorresteijn 2004). The study of genes responsible in eye development and photoreception (Arendt et al. 2002) as well as segmentation (e.g. Prud'homme et al. 2003; Kulakova et al. 2007) in *P. dumerilii* has expanded our knowledge of developmental mechanism across bilatarians. Furthermore, *P. dumerilii* has been amendable to culturing in the lab, helping it emerge as a leading annelid model system, especially among European researchers. Arguments that this worm serves as an ancestral or basal model within annelids (Tessmar-Raible and Arendt 2003) are problematic as this position is not supported by molecular (Rousset et al. 2007; Struck et al. 2007) or morphological (Rouse and Pleijel 2001) data. To the contrary, nereidids seem highly derived within Phyllodocid worms.

1.4.5 *Ophryotrocha* sp.

Ophryotrocha species (Dorvilleidae) have been a laboratory model in comparative biology since the late 19th century. Our understanding of *Ophryotrocha* biology has resulted from comparative studies of several different species. Of the 40 described species, *Ophryotrocha labronica, Ophryotrocha puerilis* and *Ophryotrocha diadema* have been heavily studied, particularly with respect to reproduction. *Ophrytrocha* exhibits a diversity of reproductive modes including gonochorism, and both simultaneous and sequential hermaphroditism, which have facilitated comparative studies of reproductive biology (Massamba N'Siala et al. 2006; Prevedelli et al. 2006). *Ophyotrocha* species have also been important in ecology (e.g. Åkesson 1976a, b; Cassai and Prevedelli 1999), behavior (e.g. Schleicherova et al. 2006; Lorenzi et al. 2006), development (e.g. Åkesson 1973; Jacobsohn 1999) and toxicology (e.g. Åkesson 1970; Lee et al. 2006). With respect to the phylogenetic relationships of *Ophryotrocha* species, recent work based on mitochondrial 16S (Dahlgren et al. 2001) in combination with cytochrome c oxidase subunit I (Heggoy et al. 2007) remains incongruent with morphology (Pleijel and Eide 1996) and warrants further study.

1.5 OTHER POTENTIAL ANNELID MODELS

Several other annelids have also been the subject of considerable study and are worthy of mention here. However, a goal of this chapter is not to provide an exhaustive list, but rather to provide insight as to why some species may be emerging as models. Three additional species have been the subject of considerable evolutionary developmental work. These include the serpulid *Hydroides elegans* (Seaver et al. 2005; Seaver and Kaneshige 2006; Arenas-Mena 2007), the chaetopterid *Chaetopterus variopedatus* (Irvine et al. 1997; Irvine and Martindale 2001; Potenza et al. 2003) and the oligocheates *Tubifex tubifex*

(Oyama and Shimizu 2007) and *Paranais litoralis* (Bely 1999). Taxonomic issues concerning identification are certainly problematic for *H. elegans* and *C. variopedatus*, the latter of which has clear tagmosis of the body. *Paranais litoralis* is particularly interesting because of its regenerative properties. *Tubifex tubifex*, a freshwater oligochaete, has been used largely as a food source for cultured animals and as a bioindicator.

Annelids have been playing important roles in comparative biology, and although a single model system for Annelida may not be possible, the emergence of multiple annelid species as models provides a greater advantage in establishing a comparative evolutionary framework for a comprehensive understanding of annelid and lophotrochozoan evolution.

ACKNOWLEDGMENTS

The efforts of Dan Shain to bring this volume to fruition are greatly appreciated. This work was made possible by National Science Foundation grants EAR-0120646 and OCE-0425060 to KMH and DBI-0706856, a Minority Postdoctoral Fellowship, to EB. This work is Auburn University Marine Biology Program contribution #38.

REFERENCES

ÅKESSON, B. 1970. *Ophryotrocha labronica* as test animal for the study of marine pollution. Helgolander Wiss. Meeresunters. 20, 293–303.

ÅKESSON, B. 1973. Reproduction and larval morphology of five *Ophryotrocha* species (Polychaeta, Dorvilleidae). Zool. Scr. 2, 145–155.

ÅKESSON, B. 1976a. Temperature and life-cycle in *Ophryotrocha labronica* (Polychaeta, Dorvilleidae). Ophelia 15, 37–47.

ÅKESSON, B. 1976b. Morphology and life-cycle of *Ophryotrocha diadema*, a new polychaete species from California. Ophelia 15, 23–35.

ARENAS-MENA, C. 2007. Sinistral equal-size spiral cleavage of the indirectly developing polychaete *Hydroides elegans*. Dev. Dyn. 236, 1611–1622.

ARENDT D, TESSMAR K, MEDEIROS DE CAMPOS-BAPTISTA, M., DORRESTEIJN, A., WITTBRODT, J. 2002. Development of pigment-cup eyes in the polychaete *Platynereis dumerilii* and evolutionary conservation of larval eyes in Bilateria. Development 129, 1143–1154.

BAKER, M.W., MACAGNO, E.R. 2001. Neuronal growth and target recognition: lessons from the leech. Can. J. Zool. 79, 204–217.

BELY, A.E. 1999. Decoupling of fission and regenerative capabilities in an asexual oligochaete. Hydrobiologia 406, 243–251.

BELY, A.E., WEISBLAT, D.A. 2006. Lessons from leeches: a call for DNA barcoding in the lab. Evol. Dev. 8, 491–501.

BLEIDORN, C., VOGT, L., BARTOLOMAEUS, T. 2003. New insights into polychaete phylogeny (Annelida) inferred from 18S rDNA sequences. Mol. Phylogenet. Evol. 29, 279–288.

BLEIDORN, C., EECKHAUT, I., PODSIADLOWSKI, L., SCHULT, N., MCHUGH, D., HALANYCH, K.M., MILINKOVITCH, M.C., TIEDEMANN, R. 2007. Mitochondrial genome and nuclear sequence data support Myzostomida as part of the annelid radiation. Mol. Biol. Evol. 24, 1690–1701.

BOORE, J.L., BROWN, W.M. 2000. Sequence and gene arrangement comparisons indicate that Pogonophora is not a phylum and Annelida and Arthropoda are not sister taxa. Mol. Biol. Evol. 17, 87–106.

BRODFUEHRER, P.D., MCCORMICK, K., TAPYRIK, L., ALBANO, A.M., GRAYBEAL, C. 2008. Activation of two forms of locomotion by a previously identified trigger interneuron for swimming in the medicinal leech. Invert. Neurosci. 8, 31–39.

BRUCE, A.E.E., SHANKLAND, M. 1998. Expression of the head gene Lox22-Otx in the leech *Helobdella* and the origin of the bilateral body plan. Dev. Biol. 201, 101–112.

CARDOSO, P.G., BANKOVIC, M., RAFFAELLI, D., PARDAL, M.A. 2007. Polychaete assemblages as indicators of habitat recovery in a temperate estuary under eutrophication. Estuar. Coast. Shelf Sci. 71, 301–308.

CASSAI, C., PREVEDELLI, D. 1999. Fecundity and reproductive effort in *Ophryotrocha labronica* (Polychaeta: Dorvilleidae). Mar. Biol. 133, 489–494.

DAHLGREN, T.G., ÅKESSON, B., SCHANDER, C., HALANYCH, K.M., SUNDBERG, P. 2001. Molecular phylogeny of the model annelid *Ophryotrocha*. Biol. Bull. 201, 193–203.

DU, H., HAN. J., LIN. K., QU. X., WANG. W. 2007. Characterization of 11 microsatellite loci derived from genomic sequences of polychaete *Capitella capitata* complex. Mol. Ecol. Notes 7, 1144–1146.

DUNN, C.W., HEJNOL, A., MATUS, D.Q., PANG, K., BROWNE, W.E., SMITH, S.A., SEAVER, E.C., ROUSE, G.W., OBST, M., EDGECOMBE, G.D., SØRENSEN, M.V., HADDOCK, S.H.D., SCHMIDT-RHAESA, A., OKUSU, A., KRISTENSEN, R.M., WHEELER, W.C., MARTINDALE, M.Q., GIRIBET, G. 2008. Broad phylogenomic sampling improves resolution of the animal tree of life. Nature 452, 745–749.

EISENHAUER, N., PARTSCH, S., PARKINSON, D., SCHEU, S. 2007. Invasion of a deciduous forest by earthworms: changes in soil chemistry, microflora, microarthropods and vegetation. Soil Biol. Biochem. 39, 1099–1110.

FISCHER, A., DORRESTEIJN, A. 2004. The polychaete *Platynereis dumerilii* (Annelida): a laboratory animal with spiralian cleavage, lifelong segment proliferation and a mixed benthic/pelagic life cycle. Bioessays 26, 314–325.

FRENCH, K.A., CHANG, J., REYNOLDS, S., GONZALEZ, R., KRISTAN, W.B., KRISTAN, W.B. 2005. Development of swimming in the medicinal leech, the gradual acquisition of a behavior. J. Comp. Phys. A Sens. Neur. Behav. Physiol. 191, 813–821.

GIRIBET, G., DUNN, C.W., EDGECOMBE, G.D., ROUSE, G.W. 2007. A modern look at the animal tree of life. Zootaxa 1668, 61–79.

GRASSLE, J.P., GRASSLE, J.F. 1976. Sibling species in the marine pollution indicator *Capitella* (Polychaeta). Science 4239, 567–569.

GRASSLE, J.F., GRASSLE, J.P. 1977. *Temporal Adaptations in Sibling Species of* Capitella in *Ecology of Marine Benthos*. Belle W. Baruch Library in Marine Science, vol. 6, edited by B.C. Coull. University of South Carolina Press, Columbia, South Carolina, pp. 177–189.

GRASSLE, J.P. 1984. Speciation in the genus *Capitella* (Polychaeta, Capitellidae). Prog. Zool. 29, 293–298.

HAGADORN, I.R. 1966. The histochemistry of the neurosecretory system in *Hirudo medicinalis*. Gen. Comp. Endocrin. 6, 288–294.

HALANYCH, K.M. 1998. Considerations for reconstructing metazoan history: signal, resolution, and hypothesis testing. Am. Zool. 38, 929–941.

HALANYCH, K.M. 2004. The new view of animal phylogeny. Ann. Rev. Ecol. Evol. Sys. 35, 229–256.

HALANYCH, K.M., DAHLGREN, T.G., McHUGH, D. 2002. Unsegmented annelids? Possible origins of four lophotrochozoan worm taxa. Int. Comp. Biol. 42, 678–684.

HARVEY, R.P., DEGRYSE, E., STEFANI, L., SCHAMBER, F., CAZENAVE, J.P., COURTNEY, M., TOLSTOSHEV, P., LECOCQ, J.P. 1986. Cloning and expression of a cDNA coding for the anticoagulant from the bloodsucking leech, *Hirudo medicinalis*. Proc. Natl. Acad. Sci. Biol. Sci. 83, 1084–1088.

HEGGOY, K.K., SCHANDER, C., AKESSON, B. 2007. The phylogeny of the annelid genus *Ophryotrocha* (Dorvilleidae). Mar. Biol. Res. 3, 412–420.

HOLDSWORTH, A.R., FRELICH, L.E., REICH, P.B. 2007. Effects of earthworm invasion on plant species richness in northern hardwood forests. Cons. Biol. 21, 997–1008.

IRVINE, S.Q., MARTINDALE, M.Q. 2001. Comparative analysis of *Hox* gene expression in the polychaete Chaetop-terus: implications for the evolution of body plan regionalization. Am. Zool. 41, 640–651.

IRVINE, S.Q., SEAVER, E.C. 2006. Early annelid development, a molecular perspective. In *Reproductive Biology and Phylogeny of Annelida*, vol. 4, edited by G.W. Rouse and F. Pleijel. Enfield, New Hampshire, Science Publishers, pp. 93–140.

IRVINE, S.Q., WARINNER, S.A., HUNTER, J.D., MARTINDALE, M.Q. 1997. A survey of homeobox genes in *Chaetopterus variopedatus* and analysis of polychaete homeodomains. Mol. Phylogenet. Evol. 7, 331–345.

JACOBSOHN, S. 1999. Characterization of novel F-actin envelopes surrounding nuclei during cleavage of a polychaete worm. Int. J. Dev. Biol. 43, 19–26.

KANG, D., HUANG, F., LI, D., SHANKLAND, M., GAFFIELD, W., WEISBLAT, D.A. 2003. A hedgehog homolog regulates gut formation in leech (*Helobdella*). Development 130, 1645–1657.

KULAKOVA, M., BAKALENKO, N., NOVIKOVA, E., COOK, C.E., ELISEEVA, E., STEINMETZ, P.R.H., KOSTYUCHENKO, R.P., DONDUA, A., ARENDT, D., AKAM, M., ANDREEVA, T. 2007. *Hox* gene expression in larval development of the polychaetes *Nereis virens* and *Platynereis dumerilii* (Annelida, Lophotrochozoa). Dev. Genes Evol. 217, 39–54.

KUO, D., SHANKLAND, M. 2004. Evolutionary diversification of specification mechanisms within the O/P equivalence group of the leech genus *Helobdella*. Development 131, 5859–5869.

LEE, H.W., BAILEY-BROCK, J.H., McGURR, M.M. 2006. Temporal changes in the polychaete infaunal community surrounding a Hawaiian mariculture operation. Mar. Ecol. Prog. Ser. 307, 175–185.

LORENZI, M.C., SCHLEICHEROVA, D., SELLA, G. 2006. Life history and sex allocation in the simultaneously hermaphroditic polychaete worm *Ophryotrocha diadema*: the role of sperm competition. Int. Comp. Biol. 46, 381–389.

McHUGH, D. 1997. Molecular evidence that echiurans and pogonophorans are derived annelids. Proc. Natl. Acad. Sci. U.S.A. 94, 8006–8009.

McPHIE, D.L., MILLER, M.W. 2006. Virtual symposium: marine invertebrate models of learning and memory. Biol. Bull. 210, 170–173.

MAHMOUD, H.M. 2007. Earthworm (*Lumbricus terrestris*) as indicator of heavy metals in soils. J. Vet. Res. 11, 23–37.

MAO, S.J.T., YATES, M.T., OWEN, T.J., KRSTENANSKY, J.L. 1988. Interaction of *hirudin* with *thrombin*: identification of a minimal binding domain of *hirudin* that inhibits clotting activity. Biochemistry 27, 8170–8173.

MARTIN, J.P., BASTIDA, R. 2006. Life history and production of *Capitella capitata* (Capitellidae: Polychaeta) in Rio de la Plata estuary (Argentina). Thalassas 22, 25–38.

MARTINDALE, M.Q. 2005. The evolution of metazoan axial properties. Nature Rev. Gen. 6, 917–927.

MASSAMBA N'SIALA, G., PREVEDELLI, D., SIMONINI, R. 2006. Gonochorism versus hermaphroditism: relationship between sexuality and fitness in the genus *Ophryotrocha*

(Polychaeta). Biologia Marina Mediterranea 13, 1104–1108.

MÉNDEZ, N., GREEN-RUIZ, C. 2006. Cadmium and copper effects on larval development and mortality of the polychaete *Capitella* sp. Y from Estero del Yugo, Mazatlan, Mexico. Water Air Soil Pollut. 171, 291–299.

MÉNDEZ, N., FLOS, J., ROMERO, J. 1998. Littoral soft-bottom polychaete communities in a pollution gradient in front of Barcelona (western Mediterranean, Spain). Bull. Mar. Sci. 63, 167–178.

MÉNDEZ, N., LINKE-GAMENICK, I., FORBES, V.E. 2000. Variability in reproductive mode and larval development within the *Capitella capitata* species complex. Invert. Reprod. Dev. 38, 131–142.

MÉNDEZ, N., LINKE-GAMENICK, I., FORBES, V.E., BAIRD, D.J. 2001. Sediment processing in *Capitella* spp. (Polychaeta: Capitellidae): strain-specific differences and effects of the organic toxicant fluoranthene. Mar. Biol. 138, 311–319.

NIELSEN, C. 2004. Trochophora larvae: cell-lineages, ciliary bands, and body regions. 1. Annelida and Mollusca. J. Exp. Zool. 302B, 35–68.

OREVI, M., ELDOR, A., GIGUZIN, I., RIGBI, M. 2000. Jaw anatomy of the blood-sucking leeches, Hirudinea *Limnatis nilotica* and *Hirudo medicinalis*, and its relationship to their feeding habits. J. Zool. 250, 121–127.

ORRHAGE, L. 1973. Light and electron microscope studies of some brachiopod and pogonophoran setae; with a discussion of the 'annelid seta' as a phylogenetic-systematic character. Zeitschrift Morph. Tiere 74, 253–270.

OYAMA, A., SHIMIZU, T. 2007. Transient occurrence of vasa-expressing cells in nongenital segments during embryonic development in the oligochaete annelid *Tubifex tubifex*. Dev. Genes Evol. 217, 675–690.

PASSAMANECK, Y., HALANYCH, K.M. 2006. Lophotrochozoan phylogeny assessed with LSU and SSU data: evidence of lophophorate polyphyly. Mol. Phylogenet. Evol. 40, 20–28.

PHILIPPE, H., LARTILLOT, N., BRINKMANN, H. 2005. Multigene analyses of bilaterian animals corroborate the monophyly of Ecdysozoa, Lophotrochozoa, and Protostomia. Mol. Biol. Evol. 22, 1246–1253.

PLEIJEL, F., EIDE, R. 1996. The phylogeny of *Ophryotrocha* (Dorvilleidae: Eunicida: Polychaeta). J. Nat. Hist. 30, 647–659.

POTENZA, N., DEL GAUDIO, R., CHIUSANO, M.L., RUSSO, G.M.R., GERACI, G. 2003. Specificity of cellular expression of *C. variopedatus* polychaete innexin in the developing embryo: evolutionary aspects of innexins' heterogeneous gene structures. J. Mol. Evol. 57, S165–S173.

PREVEDELLI, D., N'SIALA, G.M., SIMONINI, R. 2006. Gonochorism vs. hermaphroditism: relationship between life history and fitness in three species of *Ophryotrocha* (Polychaeta: Dorvilleidae) with different forms of sexuality. J. An. Ecol. 75, 203–212.

PRUD'HOMME, B., DE ROSA, R., ARENDT, D., JULIEN, J., PAJAZITI, R., DORRESTEIJN, A.W.C., ADOUTTE, A., WITTBRODT, J., BALAVOINE, G. 2003. Arthropod-like expression patterns of engrailed and wingless in the annelid

Platynereis dumerilii suggest a role in segment formation. Curr. Biol. 13, 1876–1881.

RADOS, C. 2004. Beyond bloodletting: FDA gives leeches a medical makeover. FDA Consumer Magazine, p. 38.

ROKAS, A., KRUEGER, D., CARROLL, S.B. 2005. Animal evolution and the molecular signature of radiations compressed in time. Science 310, 1933–1938.

ROUSE, G.W., PLEIJEL, F. 2001. *Polychaetes*. Oxford: University Press.

ROUSSET, V., PLEIJEL, F., ROUSE, G.W., ERSEUS, C., SIDDALL, M.E. 2006. A molecular phylogeny of annelids. Cladistics 23, 41–63.

SATOH, N., SATOU, Y., DAVIDSON, B., LEVINE, M. 2003. *Ciona intestinalis*: an emerging model for whole-genome analyses. Trends Genet. 19, 376–381.

SAWYER, R.T. 1986. *Leech Biology and Behaviour*. Oxford: Clarendon Press, pp. 1–418.

SCHLEICHEROVA, D., LORENZI, M.C., SELLA, G. 2006. How outcrossing hermaphrodites sense the presence of conspecifics and suppress female allocation. Behav. Ecol. 17, 1–5.

SEAVER, E.C. 2003. Segmentation: mono- or polyphyletic? Int. J. Dev. Biol. 47, 583–595.

SEAVER, E.C., KANESHIGE, L.M. 2006. Expression of segmentation genes during larval and juvenile development in the polychaetes *Capitella* sp. I and *H. elegans*. Dev. Biol. 289, 179–194.

SEAVER, E.C., THAMM, K., HILL, S.D. 2005. Growth patterns during segmentation in the two polychaete annelids, *Capitella* sp. I, *Hydroides elegans*: comparisons at distinct life history stages. Evol. Dev. 7, 312–326.

SHANKLAND, M. 1991. Leech segmentation: cell lineage and the formation of complex body patterns. Evol. Dev. 144, 221–231.

SIDDALL, M.E., TRONTELJ, P., UTEVSKY, S.Y., NKAMANY, M., MACDONALD, K.S. 2007. Diverse molecular data demonstrate that commercially available medicinal leeches are not *Hirudo medicinalis*. Proc. Royal Soc. Biol. Sci. Ser. B. 274, 1481–1487.

STRUCK, T.H., FISSE, F. 2007. Phylogenetic position of nemertea derived from phylogenomic data. Mol. Biol. Evol. 25, 728–736.

STRUCK, T.H., SCHULT, N., KUSEN, T., HICKMAN, E., BLEIDORN, C., McHUGH, D., HALANYCH, K.M. 2007. Annelid phylogeny and the status of Sipuncula and Echiura. BMC Evol. Biol. 7, 1–11.

STURZENBAUM, S.R., PARKINSON, J., BLAXTER, M., MORGAN, A.J., KILLE, P., GEORGIEV, O. 2003. The earthworm Expressed Sequence Tag project. Pedobiologia 47, 447–451.

TELFORD, M.J., BOURLAT, S.J., ECONOMOU, A., PAPILLON, D., ROTA-STABELLI, O. 2008. The evolution of the Ecdysozoa. Philos. Trans. R. Soc. Lond. Ser. B. 363, 1529–1537.

TESSMAR-RAIBLE, K., ARENDT, D. 2003. Emerging systems: between vertebrates and arthropods, the Lophotrochozoa. Curr. Op. Gen. Dev. 13, 331–340.

TRONTELJ, P., UTEVSKY, S.Y. 2005. Celebrity with a neglected taxonomy: molecular systematics of the medicinal leech (genus *Hirudo*). Mol. Phylogenet. Evol. 34, 616–624.

TSUTSUMI, H., TANIGUCHI, A., SAKAMOTO, N. 2005. Feeding and burrowing behaviors of a deposit-feeding capitelid polychaete, *Capitella* sp. I. Benthos Res. 60, 51–58.

VENKITARAMANI, D.V., WANG, D., JI, Y., XU, Y., PONGUTA, L., BOCK, K., ZIPSER, B., JELLIES, J., JOHANSEN, K.M., JOHANSEN, J. 2004. Leech *filamin* and *tractin*: markers for muscle development and nerve formation. J. Neurobiol. 60, 369–380.

WEISBLAT, D.A., HUANG, F.Z. 2001. An overview of glossiphoniid leech development. Can. J. Zool. 79, 218–232.

WEISBLAT, D.A., SHANKLAND, M. 1985. Cell lineage and segmentation in the leech. Philos. Trans. R. Soc. Lond. Ser. B. 312, 39–56.

WRIGHT, M.A., STRINGER, A. 1974. The toxicity of thiabendazole, benomyl, methyl benzimidazol-2-yl carbamate and thiophanate-methyl to the earthworm, *Lumbricus terrestris*. Pesticide Sci. 4, 431–432.

Chapter 2

Annelid Phylogeny—Molecular Analysis with an Emphasis on Model Annelids

Christoph Bleidorn

Unit of Evolutionary Biology/Systematic Zoology, Institute of Biochemistry and Biology, University of Potsdam, Germany

2.1 INTRODUCTION

Analyses of arthropod and nematode species have traditionally dominated the genomics of invertebrates. The model animals *Caenorhabditis elegans* and *Drosophila melanogaster* are assumed to be unrelated, and thus extrapolating these results to all protostomes seems reasonable. In the last decade, however, the view of bilaterian phylogeny has changed (see Halanych 2004 for review) and current evidence indicates that *C. elegans* and *D. melanogaster*, which represent members of the Ecdysozoa, are more closely related than previously thought. Lophotrochozoan model taxa (e.g. spiralians including annelids, molluscs) are underrepresented in these comparisons and therefore efforts are being made to gain complete genome sequences of representative species within this taxon.

Among annelids, *Platynereis dumerilii, Capitella* sp. 1 and *Helobdella robusta* are the most common model systems for phylogenomic approaches and developmental investigations (e.g. Seaver and Shankland 2000; Raible and Arendt 2004; Raible et al. 2005; Rokas et al. 2005; Seaver et al. 2005; Seaver and Kaneshige 2006), and draft genomes of *Capitella* sp. I and *H. robusta* are available. First analyses of the genomic data have revealed, for example, the presence of toll-like receptors, which are an important part of an innate immunity system (Davidson et al. 2008). In addition to the aforementioned species, *Chaetopterus variopedatus, Hirudo medicinalis, Ophryotrocha* species and *Lumbricus terrestris* are used as model annelids for questions regarding evolutionary developmental biology (e.g. Wysocka-Diller et al. 1995; Irvine and Martindale 2000), behavioral ecology (e.g. Sella and Ramella 1999) and molecular evolution (e.g. Schatz et al. 1995).

The phylogeny of annelids remains controversial (Bartolomaeus et al. 2005; Bleidorn 2007). Several predictions can be derived from cladistic morphological analyses (Rouse

& Fauchald 1997); for example, the unsegmented Echiura and Sipuncula fall outside of the annelid clade, within annelids Clitellata and Polychaeta represent the most highly ranked sister taxa, and within polychaetes the palpless Scolecida are sister to Palpata, which comprise the Aciculata (former Errantia) and Canalipalpata. This view of annelid phylogeny has been challenged by several molecular analyses in recent years. In this chapter, I will present an overview of these new insights into annelid phylogeny as derived from molecular data, and summarize which genes and subtaxa have been analyzed, placing an emphasis on model annelid species.

2.2 GENES

2.2.1 Nuclear Genes

The gene for which the most comprehensive taxon sampling is available is nuclear ribosomal 18S rRNA (*18S*). This gene has been sequenced in representatives of nearly all polychaete and clitellate families, and several annelid phylogenies based on 18S rRNA have been published (Apakupakul et al. 1999; Martin et al. 2000; Martin 2001; Rota et al. 2001; Struck et al. 2002a; Bleidorn et al. 2003a, b; Erséus and Källersjö 2004; Hall et al. 2004; Struck and Purschke 2005). Partial and complete sequences of the large ribosomal subunit (28S) are also available (e.g. Joerdens et al. 2004; Struck et al. 2005a, 2006, 2007), but among the nuclear protein coding genes, only elongation factor-1α (EF-1α) (Kojima et al. 1993; McHugh 1997, 2000; Kojima 1998; Eeckhaut et al. 2000; Struck et al. 2007) and histone H3 (Brown et al. 1999; Rousset et al. 2007) have been used for annelid systematics.

2.2.2 Mitochondrial Genes

Complete or nearly complete mitochondrial genomes have been sequenced for three clitellates, *L. terrestris* (Boore and Brown 1995), *H. robusta* (Boore and Brown 2000) and *Perionyx excavatus* (Kim et al. unpublished, but available in NCBI GenBank), and 10 polychaetes, *Clymenella torquata* (Jennings and Halanych 2005), *P. dumerilii* (Boore and Brown 2000), *Nephtys* sp. (Vallés et al. 2008), the two orbiniids *Orbinia latreillii* (Bleidorn et al. 2006b) and *Scoloplos armiger* (Bleidorn et al. 2006a), two siboglinids *Riftia pachyptila* (Jennings and Halanych 2005) and *Galathealinum brachiosum* (Boore and Brown 2000), and three members of the Terebellida, namely, *Pista cristata, Eclysippe vanelli* and *Terebellides stroemi* (Zhong et al. 2008). Dozens of other annelid mitochondrial sequencing projects are currently in progress (K. Halanych, pers. comm.), and mitochondrial genome data are available for the echiurid *Urechis caupo* (Boore 2004), the sipunculid *Phascolopsis gouldii* (Boore and Staton 2002) and the myzostomid *Myzostoma seymourcollegiorum* (Bleidorn et al. 2007). Among mitochondrial genes, mt16S rRNA (*16S*) and cytochrome c oxidase subunit I (*cox1*) have been used in several phylogenetic studies (Table 2.1), whereas cytochrome b (*cytb*) has been used less in annelids (Burnette et al. 2005). The NADH dehydrogenase subunit I gene (*nad1*) has been used in some molecular phylogenetic studies of clitellate taxa (e.g. Siddall and Borda 2003). A gene fragment containing the 3′ end of cytochrome oxidase subunit III (*cox3*), the complete tRNA for glutamine (*trnQ*) and most of NADH dehydrogenase subunit VI (*nad6*) was recently used by Bleidorn et al. (2006a).

Mitochondrial genomes have proven useful in phylogenetic analyses because of their relatively rapid divergence rates, and are widely used in animal systematics (Boore 1999).

Table 2.1 Representative studies address annelid subtaxa relationships

Taxon	Authors	Genes
Arenicolidae (Polychaeta)	Bleidorn et al. 2005	*16S, 18S, 28S*
Arhynchobdellida (Clitellata)	Borda and Siddall 2004	*12S, 18S, 28S, cox1*
Autolytinae (Syllidae, Polychaeta)	Nygren and Sundberg 2003	*16S, 18S*
Branchiobdellidae (Clitellata)	Gelder and Siddall 2001	*18S, cox1*
Chaetopteridae (Polychaeta)	Martin et al. 2008	*18S, cox1*
Erpobdelliformes (Clitellata)	Oceguera-Figueroa et al. 2005	*12S, 18S, cox1*
Euhirudinea (Clitellata)	Trontelj et al. 1999	*12S, 18S*
Eunicida (Polychaeta)	Struck et al. 2002b, 2005a, 2006	*16S, 18S, 28S, cox1*
Flabelligeridae (Polychaeta)	Burnette et al. 2005	*18S, cytB*
Glossiphoniidae (Clitellata)	Siddall et al. 2005	*18S, cox1, nad1*
Helobdella (Clitellata, Glossiphoniidae)	Bely and Weisblat 2006	*cox1*
Hesionidae (Polychaeta)	Ruta et al. 2007	*16S, 18S, 28S, cox1*
Hirudo (Clitellata, Hirudinidae)	Trontelj and Utevsky 2005	*12S*
Lumbricidae (Clitellata)	Pop et al. 2007	*16S, 18S, cox1*
Macrobdellidae (Clitellata)	Phillips and Siddall 2005	*12S, 18S, 28S*
Megascolecidae & Crassiclitellata (Clitellata)	Jamieson et al. 2002	*12S, 16S, 28S*
Myzostomida	Lanterbecq et al. 2006	*16S, 18S, cox1*
Naididae (Clitellata)	Bely and Wray 2004	*cox1*
Nereidiformia (Polychaeta)	Dahlgren et al. 2000	*cox1*
Nerillidae (Polychaeta)	Worsaae 2005	*18S*
Ophryotrocha (Polychaeta)	Dahlgren et al. 2001	*16S*
Orbiniidae (Polychaeta)	Bleidorn 2005	*16S, 18S*
Palola (Polychaeta, Eunicidae)	Schulze 2006	*16S, cox1*
Phallodrilinae (Clitellata)	Nylander et al. 1999	*cox1*
Phyllodocidae (Polychaeta)	Eklöf et al. 2007	*16S, 18S, 28S, cox1*
Piscicolidae (Clitellata)	Utevsky et al. 2007	*12S, 28S, cox1, nad1*
Sabellidae (Polychaeta)	Kupriyanova and Rouse 2008	*18S, 28S, histone H3*
Scalibregmatidae (Polychaeta)	Persson and Pleijel 2005	*18S, 28S*
Scoloplos (Polychaeta, Orbiniidae)	Bleidorn et al. 2006a	*cox3-trnQ-nad6*
Serpulidae (Polychaeta)	Kupriyanova et al. 2006	*18S, 28S*
Siboglinidae (Polychaeta)	Halanych et al. 2001	*16S, 18S, cox1*
Syllidae (Polychaeta)	Aguado et al. 2007	*16S, 18S, cox1*
Terebelliformia (Polychaeta)	Rousset et al. 2003	*28S*
Tubificidae (Clitellata)	Envall et al. 2006	*12S, 16S, 18S*

In animals, the mitochondrial genome is typically a single circular duplex, about 16 kB in length containing 13 protein coding genes, two ribosomal RNA genes, 22 transfer RNA genes and one AT-rich, noncoding control region (Wolstenholme 1992). Nucleotide and amino acid sequences, mitochondrial gene order, transfer RNA secondary structure and mitochondrial deviations from the universal genetic code are all used as characters for phylogenetic analyses.

Hassanin et al. (2005) have shown strand asymmetry in metazoan mitochondrial genomes, with one strand biased toward A and C, and the other toward G and T. Further, taxa characterized by reverse strand bias (i.e. change of the transcribed strand) tend to cluster due to long-branch attraction (LBA). This problem might not apply for annelids, however, as all mitochondrial genomes investigated thus far are transcribed from the same strand.

Jennings and Halanych (2005) found a conserved mitochondrial gene order among annelid taxa, suggesting that gene order data are of limited utility in Annelida. I have contested this view based on the available data (Bleidorn et al. 2006b) and suggest that gene translocations within Annelida seem to be less frequent than in Mollusca, but are more frequent than previously realized. This result was supported by the study of Zhong et al. (2008), who found further evidence for tRNA translocations within mitochondrial genomes of annelid taxa.

2.3 MOLECULAR ANNELID PHYLOGENY

2.3.1 Relationships among Annelids and Allies

The most comprehensive phylogenetic dataset available is that of the 18S rRNA gene. Analysis of this dataset, which includes representatives of available polychaete families and "model annelids," is presented in Fig. 2.1 (see Table 2.2 for included taxa). The predicted evolutionary relationships are largely congruent with many analyses of annelid phylogeny previously conducted with the 18S rRNA gene (e.g. Rota et al. 2001; Struck et al. 2002a; Bleidorn et al. 2003a, b; Hall et al. 2004; Struck and Purschke 2005). In contrast to phylogenetic studies using morphological data, neither the monophyly of Annelida nor of Polychaeta is recovered. Interestingly, the unsegmented Sipuncula and Echiura clades are nested within the segmented annelids; in the case of Echiura, a sister group relationship with capitellids is well supported by bootstrap values (Fig. 2.1). This result received further support in analyses based on an 18S + 28S + EF-1α dataset (Struck et al. 2007; see Fig. 2.2), and by an expressed sequence tag (EST) data-based analysis including more than 140 genes (Dunn et al. 2008).

The monophyly of most annelid taxa traditionally assigned as families is corroborated (e.g. Arenicolidae, Chaetopteridae, Lumbrineridae, Sabellidae, Serpulidae, Syllidae), but the relationships between these taxa remain ambiguous. None of the clades (Scolecida, Aciculata, Canalipalpata) proposed by morphological analysis (Rouse and Fauchald 1997) are recovered. The monophyly of Clitellata is well supported and, according to this analysis, they group with annelids. However, other analyses group clitellates with other polychaete taxa (e.g. Hall et al. 2004) or with sipunculids (e.g. Bleidorn et al. 2003b). However, none of the proposed sister group relationships for clitellates have gained valuable support. The Siboglinidae (former Pogonophora + Vestimentifera) are resolved as derived polychaetes, also confirming results from morphological analyses (Bartolomaeus 1995, Rouse and Fauchald 1997). The monophyly of Archiannelida (Hermans 1969), a taxon comprising the morphologically simple meiofaunal taxa Polygordiidae, Protodrilidae, Dinophilidae, Saccocirridae and Nerillidae is rejected, as it was also by all previous molecular (e.g. Struck and Purschke 2005) and cladistic morphological analyses (Rouse and Fauchald 1997).

Some remarkable results of this analysis are the close relationship of *Questa* with Orbiniidae (see also Bleidorn et al. 2003a), the inclusion of *Ctenodrilus* within Cirratulidae (as in Bleidorn et al. 2003b) and the exclusion of Dinophilidae from Eunicida (supporting Struck et al. 2002b, 2005a). The scaleless *Pisione* has been placed within scale worms (as previously shown in Struck et al. 2005b; Wiklund et al. 2005), and an evolutionary origin of holopelagic *Poebius* is assumed to be within benthic flabelligerids (corroborating

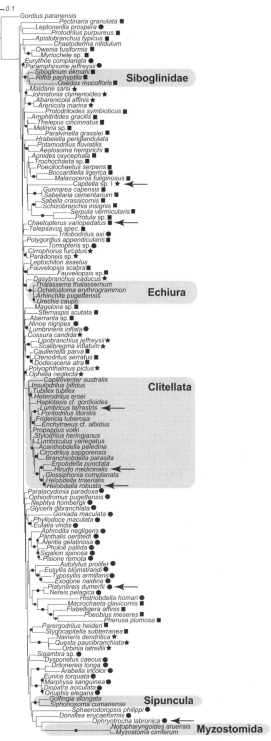

Figure 2.1 Maximum likelihood tree of the 18S rRNA gene dataset including representatives of all available annelid families and model annelids using the GTR + I + Γ model. Clades with bootstrap support >70% are marked with a circle on the branch leading to it (•). Arrows identify model annelid taxa discussed in the text. Taxa assigned as Aciculata are marked with a circle (•), Canalipalpata with a square (■) and Scolecida with a star (⋆). In contrast to morphological phylogenetic analyses, none of these taxa are supported as monophyletic. The sister group relationship of Echiura and Capitellidae is well supported and suggests an inclusion of Echiura within Annelida. The enigmatic Myzostomida as well as the annelid model organisms *Capitella* sp. I and *Ophryotrocha* show long branches.

Table 2.2   List of taxa analyzed with 18S rRNA (higher-ranked taxa without available 18S rRNA gene data are indicated)

Taxon	Higher-Ranked Taxon	GenBank Accesion No.
	Polychaeta	
	Scolecida	
Abarenicola affinis	Arenicolidae	AY569661
Arenicola marina	Arenicolidae	AF508116
Capitella sp. I	Capitellidae	http://genome.jgi-psf.org/euk_cur1.html
Dasybranchus caducus	Capitellidae	AF448153
Cossura candida	Cossuridae	AY532350
Johnstonia clymenoides	Maldanidae	AY569656
Maldane sarsi	Maldanidae	AY612617
Ophelia neglecta	Opheliidae	AF448156
Polyophthalmus pictus	Opheliidae	AF519236
Naineris dendritica	Orbiniidae	AY532358
Orbinia latreillii	Orbiniidae	AY532355
Paradoneis sp.	Paraonidae	AY612618
Cirrophorus furcatus	Paraonidae	AY532349
Questa paucibranchiata	Questidae	AF209464
Lipobranchus jeffreysii	Scalibregmatidae	AF508120
Scalibregma inflatum	Scalibregmatidae	AF448163
	Aciculata	
Panthalis oerstedi	Acoetidae	AY839572
Aphrodita negligens	Aphroditidae	AY894294
Alentia gelatinosa	Polynoidae	AY525630
Pholoe pallida	Pholoidae	AY894302
Sigalion spinosa	Sigalionidae	AY894304
Pisione remota	Pisionidae	AY525628
Disponetus caecus	Chrysopetalidae	AY839568
Ophiodromus pugettensis	Hesionidae	DQ790086
No sequence available	**Nautilinidae**	
Sigambra sp.	Pilargidae	AY340444
Autolytus prolifer	Syllidae	AF474295
Exogone naidina	Syllidae	AF474290
Eusyllis blomstrandi	Syllidae	AF474281
Typosyllis armillaris	Syllidae	AF474292
Platynereis dumerilii	Nereididae	AY894303
Nereis pelagica	Nereididae	AY340438
Glycera dibranchiata	Glyceriformia	AY995208
Goniada maculata	Glyceriformia	AY176288
No sequence available	**Lopadorhynchidae**	
Myzostoma cirriferum	Myzostomida	AF260585
Notpharyngoides aruensis	Myzostomida	AF260587
Nephtys hombergii	Nephtyidae	AY527054
Paralacydonia paradoxa	Paralacydonidae	DQ790088
Eulalia viridis	Phyllodocidae	AY340428
Phyllodoce maculata	Phyllodocidae	AY176302
Sphaerodoropsis phillipi	Sphaerodoridae	AY176307
Tomopteris sp.	Tomopteridae	DQ790095
Eurythoe complanata	Amphinomida	AY364851
Paramphinome jefreysii	Amphinomida	AY176299
Dorvillea erucaeformis	Dorvilleidae	AY838846

(Continued)

Table 2.2   (*Continued*)

Taxon	Higher-Ranked Taxon	GenBank Accesion No.
Ophryotrocha labronica	Dorvilleidae	AY838855
Eunice torquata	Eunicidae	AY838851
Marphysa sanguinea	Eunicidae	AY525621
No sequence available	**Hartmaniellidae**	
Histriobdella homari	Histriobdellidae	AY527053
Ninoe nigripes	Lumbrineridae	AY838852
Lumbrineris inflata	Lumbrineridae	AY525622
Drilonereis longa	Oenonidae	AY838847
Arabella iricolor	Oenonidae	AY525624
Onuphis elegans	Onuphidae	AY838854
Diopatra aciculata	Onuphidae	AY838845
Trilobodrilus axi	Dinophilidae	AF412806
Aberranta sp.	Aberrantidae	AY834760
Leptonerilla prospera	Nerillidae	AY834758
No sequence available	**Spintheridae**	
	Canalipalpata	
Owenia fusiformis	Oweniidae	AF448160
Myriochele sp.	Oweniidae	AY340437
Sabellaria cementarium	Sabellariidae	AY732223
Gunnerea capensis	Sabellariidae	AY577892
Schizobrnachia insignis	Sabellidae	AY732222
Sabella crassicornis	Sabellidae	AY527059
Serpula vermicularis	Serpulidae	AY732224
Protula sp.	Serpulidae	AY611453
Osedax mucofloris	Siboglinidae	AY941263
Riftia pachyptila	Siboglinidae	AF168745
Siboglinum ekmani	Siboglinidae	AF315062
Macrochaeta clavicornis	Acrocirridae	AY612620
Dodecaceria atra	Cirratulidae	AF448154
Caulleriella parva	Cirratulidae	AF448151
Ctenodrilus serratus	Ctenodrilidae	AF508119
Fauveliopsis sp.	Fauveliopsidae	AY340429
Fauveliopsis scabra	Fauveliopsidae	AY708537
Flabelligera affinis	Flabelligeridae	AY708532
Pherus plumosa	Flabelligeridae	AY708528
Poeobius meseres	Poeobiidae	AY708526
Sternaspis scutata	Sternaspis	AY532353
Paralvinella grasslei	Alvinellidae	AY577886
Melinna sp.	Ampharetidae	AY611459
Pectinaria granulata	Pectinariidae	AY577890
Amphitritides gracilis	Terebellidae	AF508115
Thelepus cincinnatus	Terebellidae	AY611462
Apistobranchus typicus	Apistobranchidae	AF448150
Chaetopterus variopedatus	Chaetopteridae	U67324
Telepsavus sp.	Chaetopteridae	AF448165
Magelona sp.	Magelonidae	AY611454
No sequence available	**Heterospionidae**	
Poecilochaetus serpens	Poecilochaetidae	AY569652
Boccardiella ligerica	Spionidae	AY527061
Aonides oxycephala	Spionidae	AF448149

(*Continued*)

Table 2.2 (*Continued*)

Taxon	Higher-Ranked Taxon	GenBank Accesion No.
Malacoceros fuligenosus	Spionidae	AY525632
Trochochaeta sp.	Trochochaetidae	DQ790097
No sequence available	**Uncispionidae**	
Polygordius appendicularis	Polygordiidae	AY525629
Protodrilus purpureus	Protodrilida	AY527057
Protodriloides symbioticus	Protodrilida	AF508125
Aeolosoma hemprichi	Aeolosomatidae	AJ310500
Stygocapitella subterreanea	Parergodrilidae	AJ310505
Parergodrilus heideri	Parergodrilidae	AJ310504
No sequence available	**Psammodrilidae**	
	Polychaeta incertae sedis	
Potamodrilus fluviatilis	Polychaeta incertae sedis	AY527056
Hrabeiella periglandulata	Polychaeta incertae sedis	AJ310501
	Clitellata	
Capilloventer australis	Capilloventridae	AY365455
Insulodrilus bifidus	Phreodrilidae	AF411906
Tubifex tubifex	Tubificidae	AF397152
Heterodrilus ersei	Tubificidae	AY885576
Proppapus volki	Propappidae	AY365457
Haplotaxis cf. *gordioides*	Haplotaxidae	AY365456
Fridericia tuberosa	Enchytraeidae	AF209453
Enchytraeus cf. *albidulus*	Enchytraeidae	AY040683
Lumbricus terrestris	Crassiclitellata	AJ272183
Pontodrilus litoralis	Crassiclitellata	AY365462
Acanthobdella peledina	Acanthobdellida	AY040680
Branchiobdella parasita	Branchiobdellida	AF310690
Cirrodrilus sapporensis	Branchiobdellida	AF310698
Helobdella triserialis	Hirudinida	AY962435
Helobdella robusta	Hirudinida	http://genome.jgi-psf.org/euk_cur1.html
Hirudo medicinalis	Hirudinida	AY786464
Glossiphonia complanata	Hirudinida	AF099943
Erpobdella punctata	Hirudinida	AF116002
Lumbriculus variegatus	Lumbriculidae	AY521551
Stylodrilus heringianus	Lumbriculidae	AF411907
	Allies	
Golfingia elongata	Sipuncula	AY340431
Siphonosoma cumanense	Sipuncula	AF519241
Thalassema thalassemum	Echiura	AY532354
Arhinchite pugettensis	Echiura	AY210441
Urechis caupo	Echiura	AF342805
Ochethostoma erythrogrammon	Echiura	X79875
	Out-groups	
Chaetoderma nitidulum	Mollusca	AY275894
Leptochiton asselus	Mollusca	AY145382
Gordius paranensis	Nematomorpha	AF421766

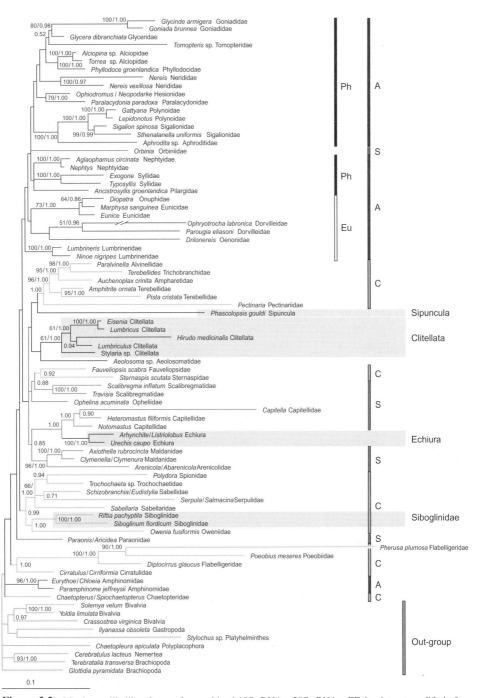

Figure 2.2   Maximum likelihood tree of a combined 18S rRNA + 28S rRNA + EF-1α dataset modified after Struck et al. (2007). BS values above 50 are shown at the branches on the left; posterior probabilities are on the right or alone. Abbreviations: A = Aciculata; C = Canalipalpata; S = Scolecidae; Eu = Eunicida; Ph = Phyllodocida.

Burnette et al. 2005). A well-supported sister group relationship between Flabelligeridae and Acrocirridae is also suggested by this analysis.

The taxon sampling of other genes used for analyzing annelid phylogeny is not as extensive as for the 18S rRNA dataset alone. McHugh (1997, 2000) used EF-1α sequences to examine annelid phylogeny and proposed the inclusion of pogonophorans (=siboglinids) and echiurans within annelids. An analysis using EF-1α supports the exclusion of myzostomids from annelids, instead favoring a flatworm affinity (Eeckhaut et al. 2000, but see Littlewood et al. 2001). This result is contradicted by a study using complete 28S rRNA sequences (Passamaneck and Halanych 2006) in which myzostomids were included in a bryozoan clade. However, Bleidorn et al. (2007) demonstrated that the 28S rRNA data analysis was most likely misled by a long-branch artifact. Instead, the hypothesis of an annelid origin of myzostomids received strong support by analyzing mitochondrial gene order and sequence data.

Brown et al. (1999) assessed the usefulness of histone H3, U2 snRNA and fragments of the 28S rRNA gene for polychaete systematics, but none of the relationships were supported through bootstrapping. Colgan et al. (2006) conducted a combined analysis using histone H3, U2 snRNA, two fragments of the 28S rRNA gene (D1 and D9–10 expansion regions), the 18S rRNA gene and *cox1*. Although they provide twice as many aligned nucleotide positions than previous 18S single-gene studies, the support for basal annelid splittings remains poor. In contrast with previous studies, their analyses do not unequivocally support an echiurid-capitellid clade. A recent study by Rousset et al. (2007) used the largest taxon sampling to analyze polychaete relationships, combining data from 18S rRNA, 28S rRNA (D1 and D9–10 expansion regions), mt16S rRNA and histone H3 sequences. Their results mainly corroborate those of 18S rRNA analysis alone. Contrary to the expectations of Rousset et al. (2007), a dense taxon sampling did not significantly improve the resolution of basal annelid relationships. In a study using a 2-kb fragment of 28S and complete 18S rRNA sequences, Joerdens et al. (2004) found a close relationship between Parergodrillidae and orbiniid worms. Struck et al. (2007) analyzed a combined dataset including a 2-kb fragment of the 28S, complete 18S and EF-1α data (Fig. 2.2). This dataset corroborated the monophyly of annelids including allied taxa (Echiura, Sipuncula) and, as analyses of 18S alone, supported the monophyly of most families. A clade including orbiniids and all aciculate taxa besides amphinomids has been recovered, but has received no bootstrap support. Interestingly, chaetopterids and amphinomids are found at the base of the annelid tree.

Complete mitochondrial genomes are only available for a few annelids. Phylogenetic analyses using this sequence data recover well-supported clades including polychaetes, clitellates, echiurans and sipunculans (Jennings and Halanych 2005; Bleidorn et al. 2006b; Zhong et al. in press), and some support exists for including unsegmented sipunculans within annelids, suggesting that sipunculans have secondarily lost their segmentation (Bleidorn et al. 2006b). Inclusion of sipunculids within annelids finds further support in phylogenomic analyses using large EST-based datasets (Hausdorf et al. 2007; Dunn et al. 2008). All these studies support reciprocally monophyletic Mollusca and Annelida phyla (the latter including Sipuncula and Echiura).

Rousset et al. (2004) and Worsaae et al. (2005) conducted combined analyses using ribosomal sequence data and morphological data matrices to infer the relationship of annelid subtaxa. In contradiction to morphological analyses, which favor the inclusion of siboglinids within Sabellida, Rousset et al. (2004) propose a sister group relationship between Siboglinidae and Oweniidae.

2.3.2 Clitellate Phylogeny

Whereas the relationships between most polychaete families remain unresolved, phylogenetic relationships within clitellates are better understood (see Erséus 2005 for review). As in the analysis presented here, the enigmatic, parapodia-bearing Capilloventridae are regarded as a sister taxon of all remaining clitellates (Erséus 2005). The monophyly of all leeches with *Acanthobdella* as sister group to Branchiobdellida + Hirudinida (true leeches) is recognized by 18S rRNA datasets (Fig. 2.1; Siddall et al. 2001). Lumbriculidae are the closest relatives of leeches (Fig. 2.1; Erséus and Källersjö 2004), with Enchytraeidae + Crassiclitellata (Erséus & Källersjö 2004) branching off next. In contrast, the analysis presented here suggests that Propappidae branch between Lumbriculidae and Enchytraeidae + Crassiclitellata (Fig. 2.1). The phylogeny of remaining clitellate families remains unclear in hitherto published analyses (Erséus 2005). However, analyses show that all members of the Naididae are nested within Tubificidae (see Erséus et al. 2008).

2.3.3 Annelid Subtaxa

In recent years, many molecular studies on the phylogeny of annelid subtaxa have been published (see Table 2.1 for a representative selection). Usually, 18S rRNA gene sequence data serve as a backbone in these analyses, and mitochondrial genes such as mt16S rRNA and cytochrome c oxidase subunit 1 are used as less conserved genes to resolve younger nodes. Clearly, more phylogenetic analyses of clitellate taxa than polychaete taxa are available, although the taxon Siboglinidae is well studied (Table 2.1). Note that four molecular analyses of sipunculid ingroup relationships challenge the traditional morphological view of their evolution (Maxmen et al. 2003; Staton 2003; Schulze et al. 2005, 2007).

2.3.4 Model Annelids

Whereas the phylogenetic relationships of model organisms like *D. melanogaster* (e.g. DeSalle 1992) and *C. elegans* (Kiontke et al. 2004) are well studied, the phylogeny of most model annelids has not been investigated using molecular markers. Capitellid and chaetopterid relationships are poorly understood (but see Osborn et al. 2007 and Martin et al. 2008 for initial studies of chaetopterid relationships), and only morphological studies are available for Nereididae (Rouse and Pleijel 2001). Exceptions are *Ophrytrocha, Helobdella* and *Hirudo*, for which molecular phylogenetic studies have been conducted (Dahlgren et al. 2001; Siddall and Borda 2003; Trontelj and Utevsky 2005; Bely and Weisblat 2006). An interesting point was raised in the study on *Hirudo* by Trontelj and Utevsky (2005) and later by DeSalle et al. (2005) and Siddall et al. (2007). Their results show that individuals distributed by commercial suppliers and shipped as "*H. medicinalis*" group with a closely related Asian species and not the true *H. medicinalis*. Using *cox1* bar coding, Bely and Weisblat (2006) found that laboratory isolates of *Helobdella* probably represent five distinct species belonging to two different major clades. Whereas other commercially bred model organisms are usually taxonomically well defined, this seems not to be the case for annelids. Grassle and Grassle (1976) showed in their landmark study on *Capitella capitata* that this species represents a complex of many sibling species, of which most are morphologically indistinguishable. The laboratory strain *Capitella* sp. I is currently being used for whole genome sequencing. The taxonomic status of *C. variopedatus* is also unclear, and Petersen (1984) proposed that this widely distributed species comprises many similar species. Clearly, future investigations into the relationships of model annelids are desirable.

2.4 CHOOSING MODEL ORGANISMS

Typically, practical features such as ease of rearing in the laboratory, availability or genome size are considered when choosing model organisms; however, the phylogenetic position of a model organism is also crucial. If organisms that are too closely related are chosen, extrapolating the results for a broader range of taxa becomes problematic. Judging from current data, species chosen as annelid model organisms seem not too closely related (Fig. 2.1). In the case of *Capitella*, we have the promising situation of finding a segmented annelid species closely related to the unsegmented echiurids. Therefore, segmentation must be assumed to have been lost secondarily in echiurids (e.g. Bleidorn 2007). A comparative investigation of "segmentation genes" and their expression in both of these taxa could provide further insights into the evolution of segmentation. Note that Seaver and Kaneshige (2006) have recently started to investigate "segmentation genes" in *Capitella*.

2.5 BRANCH LENGTHS

The length of a branch leading to an operational taxonomic unit (OTU) is proportional to the amount of inferred character change. Usually, for trees calculated with model-based inference methods such as maximum likelihood, Bayesian inference or distance methods, the number of substitutions per site is indicated. Branch lengths derived from a phylogenetic analysis can be compared to each other and thus "rapidly evolving" taxa can be identified. Comparing branch lengths of bilaterian model animals using amino acid data from 442 genes, Raible et al. (2005) have shown that the polychaete *Platynereis* can be regarded as a "slowly evolving" protostome, in contrast with *D. melanogaster* and *C. elegans*, which show higher substitution rates.

LBA, i.e. the erroneous grouping of long-branched taxa due to methodological artifacts, is a major problem in phylogenetic inferences (see Bergsten 2005 for review). Thus, a comparison of branch lengths of model annelids with other annelids is worth considering. For example, the *P. dumerilii* branch is approximately as long as branches of other annelids, which applies to species of *Helobdella*, *Hirudo*, *Lumbricus* and *Chaetopterus* as well. In contrast, *Capitella* sp. I and *Ophryotrocha labronica* display a distinctly longer branch (at least twice as long as the mean of branches of all polychaete OTUs, see Fig. 2.1) than most other annelids, and a higher substitution rate can be assumed for these taxa.

2.6 PROBLEMS IN INFERRING ANNELID PHYLOGENY

Considering the molecular phylogenetic analyses reviewed above, we have two possible lines of arguments. First, the lack of resolution in the annelid tree is due to a lack of phylogenetic information in the chosen gene(s), and second, the lack of resolution reflects a rapid radiation of annelid worms from each other that has left few traces. In the first case, we can hope that the choice of other genes and/or gene combinations will increase resolution and unravel annelid relationships. Momentarily, however, neither 28S rRNA nor EF-1α genes appear to significantly increase the resolution of annelid phylogeny. Considering the large number of annelid families described, which reflects the morphological diversity of this group, much time and effort will be needed to establish a multigene dataset representing this variety. However, construction of EST libraries for annelid taxa is already in progress (e.g. Hausdorf et al. 2007; Dunn et al. 2008) and will certainly expedite this process.

With respect to the second argument, the time frame of annelid radiation is of special interest. The oldest unequivocal annelid fossils stem from the Cambrian and are ~530 million years old (Conway Morris and Peel 2008). This corresponds with the earliest records of sipunculid fossils (Huang et al. 2004), which are morphologically very similar to modern sipunculids. Given that sipunculids are included in the annelid radiation, one can conclude that a radiation of at least 530 million years must be reconstructed. If this radiation occurred rapidly—at least in evolutionary time—the assumption that little phylogenetic information is left to reconstruct annelid phylogeny is reasonable. In a comparative study and using simulation studies, Rokas et al. (2005) have shown that the lack of resolution in deep metazoan trees is a signature of a closely spaced series of cladogenetic events (but see Baurain et al. 2007). Aside from this, the root of the annelid tree remains largely unknown (Rouse and Pleijel 2003; McHugh 2005). In 18S rRNA gene datasets, the basal splittings are not supported (e.g. Fig. 2.1). Furthermore, annelids are not supported as a monophyletic group in these analyses, and other lophotrochozoan taxa (e.g. molluscs) are scattered between them. The combined 18S rRNA + 28S rRNA + EF-1α analysis by Struck et al. (2007) (see Fig. 2.2) recovers monophyletic annelids, and chaetopterids and amphinomids are found at the base. However, bootstrap analyses do not support any of the basal annelid splittings in this tree. Thus, at present, the molecules give no answers about which taxa constitute the basal annelids. McHugh (2005) has outlined how to find the root of the annelid tree using the analysis of different gene copies, but no such dataset is currently available.

2.7 CONCLUSIONS

Summarizing data from molecular (this chapter) as well as morphological investigations (see Bartolomaeus et al. 2005 for review) shows that no uncontroversial system of annelid phylogeny is available. Nevertheless, progress has been observed in the field of molecular annelid systematics. Most analyses agree that clitellates represent derived polychaetes (McHugh 2005). Additionally, pogonophorans (a.k.a. Siboglinidae) seem to be polychaete annelids (reviewed in Halanych 2005), and the same is evidently true for Echiura (McHugh 1997; Bleidorn et al. 2003a, b; Struck et al. 2007; Dunn et al. 2008) and Sipuncula (Boore and Staton 2002; Jennings and Halanych 2005; Bleidorn et al. 2006b; Hausdorf et al. 2007; Struck et al. 2007; Dunn et al. 2008). These studies have elucidated the plasticity of morphological characters; i.e. characters and character complexes including segmentation (lost in Echiura and Sipuncula, reduced in Siboglinidae) or the presence of coelomic cavities (lacking in some interstitial polychaete species) are not as evolutionarily conserved as previously thought. Those analyses with a broad taxon sampling show that many taxa assigned as polychaete families are monophyletic (e.g. Bleidorn et al. 2003b; Hall et al. 2004; Rousset et al. 2007), whereas the relationships between these taxa remain unresolved. Clearly, molecules are powerful tools for analyzing phylogenetic relationships of morphologically simple taxa in cases where only few morphological characters are available (e.g. Bleidorn 2005; Envall et al. 2006).

Nevertheless, phylogenetic studies have to explain the evolution of both molecular and morphological characters. Apart from the need to gain more genetic data for all annelid taxa (EST library sequencing is a fruitful attempt for generating data for nonmodel organisms), a synthesis is needed to integrate currently available data in phylogenetic studies. The development of sophisticated analysis methods is urgently needed for this purpose. Attempts to develop simple models for morphology and integrating morphological data into Bayesian analyses have been initiated (Lewis 2001; Nylander et al. 2004).

The availability of complete genomes of model annelids will greatly improve molecular systematic studies, enabling investigations into the phylogenetic utility of genomic markers such as gene arrangements, the distribution of introns or absence/presence pattern of transposable elements [e.g. SINE, LINE—even though this type of marker seems to be restricted to resolving relatively young (50 mya and younger) divergences (Shedlock and Okada 2000)]. Future investigations will clarify whether current problems in unravelling annelid phylogeny are based on lack of data or lack of phylogenetic signal due to rapid radiation.

ACKNOWLEDGMENTS

Sincere thanks to Laura Epp (University of Potsdam, Germany) and Sonia Krutzke (Rutgers University, NJ, USA) for proofreading the manuscript. Thanks to Dan Shain for editing and giving me the opportunity to participate in this book project.

REFERENCES

AGUADO, M.T., NYGREN, A., SIDDALL, M.E. 2007. Phylogeny of Syllidae (Polychaeta) based on combined molecular analysis of nuclear and mitochondrial genes. Cladistics 23, 552–564.

APAKUPAKUL, K., SIDDALL, M.E., BURRESON, E. 1999. Higher level relationships of leeches (Annelida: Clitellata: Euhirudinea) based on morphology and gene sequences. Mol. Phylogenet. Evol. 12, 350–359.

BARTOLOMAEUS, T. 1995. Structure and formation of the uncini in *Pectinaria koreni, Pectinaria auricoma* (Terebellida) and *Spirorbis spirorbis* (Sabellida): implications for annelid phylogeny and the position of the Pogonophora. Zoomorphology 115, 161–177.

BARTOLOMAEUS, T., PURSCHKE, G., HAUSEN, H. 2005. Polychaete phylogeny based on morphological data—a comparison of current attempts. Hydrobiologia 535/536, 341–356.

BAURAIN, D., BRINKMANN, H., PHILIPPE, H. 2007. Lack of resolution in the animal phylogeny: closely spaced cladogeneses or undetected systematic errors? Mol. Biol. Evol. 24, 6–9.

BELY, A.E., WEISBLAT, D.A. 2006. Lessons from leeches: a call for DNA barcoding in the lab. Evol. Dev. 8, 491–501.

BELY, A.E., WRAY, G.A. 2004. Molecular phylogeny of naidid worms (Annelida: Clitellata) based on cytochrome oxidase I. Mol. Phylogenet. Evol. 30, 50–63.

BERGSTEN, J. 2005. A review of long-branch attraction. Cladistics 21, 163–193.

BLEIDORN, C. 2005. Phylogenetic relationships and evolution of Orbiniidae (Annelida, Polychaeta) based on molecular data. Zool. J. Linn. Soc. 144, 59–73.

BLEIDORN, C. 2007. The role of character loss in phylogenetic reconstruction as exemplified for the Annelida. J. Zool. Syst. Evol. Res. 45, 299–307.

BLEIDORN, C., VOGT, L., BARTOLOMAEUS, T. 2003a. A contribution to sedentary polychaete phylogeny using 18S rRNA sequence data. J. Zool. Syst. Evol. Res. 41, 186–195.

BLEIDORN, C., VOGT, L., BARTOLOMAEUS, T. 2003b. New insights into polychaete phylogeny (Annelida) inferred by 18S rDNA sequences. Mol. Phylogenet. Evol. 29, 279–288.

BLEIDORN, C., VOGT, L., BARTOLOMAEUS, T. 2005. Molecular phylogeny of lugworms (Annelida, Arenicolidae) inferred from three genes. Mol. Phylogenet. Evol. 34, 673–679.

BLEIDORN, C., KRUSE, I., ALBRECHT, S., BARTOLOMAEUS, T. 2006a. Mitochondrial sequence data expose the putative cosmopolitan polychaete *Scoloplos armiger* (Annelida, Orbiniidae) as a species complex. BMC Evol. Biol. 6, 47.

BLEIDORN, C., PODSIADLOWSKI, L., BARTOLOMAEUS, T. 2006b. The complete mitochondrial genome of the orbiniid polychaete *Orbinia latreillii* (Annelida, Orbiniidae)—a novel gene order for Annelida and implications for annelid phylogeny. Gene 370, 96–103.

BLEIDORN, C., EECKHAUT, I., PODSIADLOWSKI, L., SCHULT, N., MCHUGH, D., HALANYCH, K.M., MILINKOVITCH, M.C., TIEDEMANN, R. 2007. Mitochondrial genome and nuclear sequence data support Myzostomida as part of the annelid radiation. Mol. Biol. Evol. 24, 1690–1701.

BOORE, J.L. 1999. Animal mitochondrial genomes. Nucleic Acids Res. 27, 1767–1780.

BOORE, J.L. 2004. Complete mitochondrial genome sequence of *Urechis caupo*, a representative of the phylum Echiura. BMC Genomics 5, 67.

BOORE, J.L., BROWN, W.M. 1995. Complete sequence of the mitochondrial DNA of the annelid worm *Lumbricus terrestris*. Genetics 141, 305–319.

BOORE, J.L., BROWN, W.M. 2000. Mitochondrial genomes of *Galathealinum, Helobdella*, and *Platynereis*: sequence and gene arrangement comparisons indicate that Pogonophora is not a phylum and Annelida and Arthropoda are not sister taxa. Mol. Biol. Evol. 17, 87–106.

BOORE, J.L., STATON, J.L. 2002. The mitochondrial genome of the Sipunculid *Phascolopsis gouldii* supports its association with Annelida rather than Mollusca. Mol. Biol. Evol. 19, 127–137.

BORDA, E., SIDDALL, M.E. 2004. Arhynchobdellida (Annelida: Oligochaeta: Hirudinida): Phylogenetic relationships and evolution. Mol. Phylogenet. Evol. 30, 213–225.

BROWN, S., ROUSE, G.W., HUTCHINGS, G.W., COLGAN, D. 1999. Assesing the usefulness of histone H3, U2 snRNA and 28S rDNA in analyses of polychaete relationships. Austral. J. Zool. 47, 499–516.

BURNETTE, A.B., STRUCK, T.H., HALANYCH, K.M. 2005. Holopelagic *Poeobius meseres* ("Poeobiidae", Annelida) is derived from benthic flabelliegerid worms. Biol. Bull. 208, 213–220.

COLGAN, D.J., HUTCHINGS, P.A., BRAUNE, M. 2006. A multigene framework for polychaete phylogenetic studies. Org. Div. Evol. 6, 220–235.

Conway MORRIS, S., PEEL, J.S. 2008. The earliest annelids: Lower Cambrian polychaetes from the Sirius passet lagerstätte, Peary Land, North Greenland. Acta Palaeontol. Pol. 53, 137–148.

DAHLGREN, T.G., LUNDBERG, J., PLEIJEL, F., SUNDBERG, P. 2000. Morphological and molecular evidence of the phylogeny of nereidiform polychaetes (Annelida). J. Zool. Syst. Evol. Res. 38, 249–253.

DAHLGREN, T.G., AKESSON, B., SCHANDER, C., HALANYCH, K.M., SUNDBERG, P. 2001. Molecular phylogeny of the model annelid *Ophryotrocha*. Biol. Bull. 201, 193–203.

DAVIDSON, C.R., BEST, N.M., FRANCIS, J.W., COOPER, E.L., WOOD, T.C. 2008. Toll-like receptor genes (TLRs) from *Capitella capitata* and *Helobdella robusta* (Annelida). Dev. Comp. Immun. 32, 608–612.

DESALLE, R. 1992. The phylogenetic relationships of flies in the family Drosophilidae deduced from mtDNA sequences. Mol. Phylogenet. Evol. 1, 31–40.

DESALLE, R., EGAN, M.G., SIDDALL, M.E. 2005. The unholy trinity: taxonomy, species delimitation and DNA barcoding. Phil. Trans. R. Soc. 360, 1905–1916.

DUNN, C.W., HEJNOL, A., MATUS, D.Q., PANG, K., BROWNE, W.E., SMITH, S.A., SEAVER, E., ROUSE, G.W., OBST, M., EDGECOMBE, G.D., SÖRENSEN, M.V., HADDOCK, S.H.D., SCHMIDT-RHAESA, A., OKUSU, A., KRISTENSEN, R.M., WHEELER, W.C., MARTINDALE, M.Q., GIRIBET, G. 2008. Broad phylogenomic sampling improves resolution of the animal tree of life. Nature 452, 745–749.

EECKHAUT, I., MCHUGH, D., MARDULYN, P., TIEDEMANN, R., MONTEYNE, D., JANGOUX, M., MILINKOVITCH, M.C. 2000. Myzostomida: a link between trochozoans and flatworms? Proc. R. Soc. Lond. B 267, 1383–1392.

EKLÖF, J., PLEIJEL, F., SUNDBERG, F. 2007. Phylogeny of benthic Phyllodocidae (Polychaeta) based on morphological and molecular data. Mol. Phylogenet. Evol. 45, 261–271.

ENVALL, I., KÄLLERSJÖ, M., ERSÉUS, C. 2006. Molecular evidence for the non-monophyletic status of Naidinae (Annelida, Clitellata, Tubificidae). Mol. Phylogenet. Evol. 40, 570–584.

ERSÉUS, C. 2005. Phylogeny of oligochaetous Clitellata. Hydrobiologia 535/536, 357–372.

ERSÉUS, C., KÄLLERSJÖ, M. 2004. 18S rDNA phylogeny of Clitellata (Annelida). Zool. Scr. 33, 187–196.

ERSÉUS, C., WETZEL, M.J., GUSTAVSSON, L. 2008. ICZN rules—a farewell to Tubificidae (Annelida, Clitellata). Zootaxa 1744, 66–68.

GELDER, S.R., SIDDALL, M.E. 2001. Phylogenetic assessment of the Branchiobdellidae (Annelida: Clitellata) using 18S rDNA and mitochondrial cytochrome c oxidase subunit I characters. Zool. Scr. 30, 215–222.

GRASSLE, J., GRASSLE, J.F. 1976. Sibling species in the marine pollution indicator *Capitella* (Polychaeta). Science 192, 567–569.

HALANYCH, K.M. 2004. The new view of animal phylogeny. Ann. Rev. Ecol. Syst. 35, 229–256.

HALANYCH, K.M. 2005. Molecular phylogeny of siboglinid annelids (a.k.a. pogonophorans): a review. Hydrobiologia 535/536, 27–307.

HALANYCH, K.M., FELDMAN, R.A., VRIJENHOEK, R.C. 2001. Molecular evidence that *Sclerolinum brattstromi* is closely related to vestimentiferans, not frenulate pogonophorans (Siboglinidae, Annelida). Biol. Bull. 201, 65–75.

HALL, K.A., HUTCHINGS, P., COLGAN, D. 2004. Further phylogenetic studies of the Polychaeta using 18S rDNA sequence data. J. Mar. Biolog. Assoc. U.K. 84, 949–960.

HASSANIN, A., LÉGER, N., DEUTSCH, J. 2005. Evidence for multiple reversals of asymmetric mutational constraints during the evolution of the mitochondrial genome of Metazoa, and consequences for phylogenetic inferences. Syst. Biol. 54, 277–298.

HAUSDORF, B., HELMKAMPF, M., MEYER, A., WITEK, A., HERLYN, H., BRUCHHAUS, I., HANKELN, T., STRUCK, T.H., LIEB, B. 2007. Spiralian phylogenomics supports the resurrection of Bryozoa comprising Ectoprocta and Entoprocta. Mol. Biol. Evol. 24, 2723–2729.

HERMANS, C.O. 1969. The systematic position of the Archiannelida. Syst. Zool. 18, 85–102.

HUANG, D.Y., CHEN, J.Y., VANNIER, J., Saiz SALINAS, J.I. 2004. Early Cambrian sipunculan worms from southwest China. Proc. R. Soc. Lond. B 271, 1671–1676.

IRVINE, S.Q., MARTINDALE, M.Q. 2000. Expression patterns of anterior hox genes in the polychaete *Chaetopterus*: correlation with morphological boundaries. Dev. Biol. 217, 333–351.

JAMIESON, B.G.M., TILLIER, S., TILLIER, A., JUSTINE, J.L., LING, E., JAMES, S., MCDONALD, K., HUGALL, A.F. 2002. Phylogeny of the Megascolecidae and Crassiclitellata (Annelida, Oligochaeta):combined versus partitioned analysis using nuclear (28S) and mitochondrial (12S, 16S) rDNA. Zoosystema 24, 707–734.

JENNINGS, R.M., HALANYCH, K.M. 2005. Mitochondrial genomes of *Clymenella torquata* (Maldanidae) and *Riftia pachyptila* (Siboglinidae). Evidence for conserved gene order in Annelida. Mol. Biol. Evol. 22, 210–222.

JOERDENS, J., STRUCK, T., PURSCHKE, G. 2004. Phylogenetic inference regarding Parergodrilidae and *Hrabeiella periglandulata* ("Polychaeta", Annelida) based on 18S rDNA, 28S rDNA and COI sequences. J. Zool. Syst. Evol. Res. 42, 270–289.

KIONTKE, K., GAVIN, N.P., RAYNES, Y., ROEHRIG, C., PIANO, F., FITCH, D.H.A. 2004. *Caenorhabditis* phylogeny predicts convergence of hermaphroditism and extensive intron loss. Proc. Natl. Acad. Sci. U.S.A. 101, 9003–9008.

KOJIMA, S. 1998. Paraphyletic status of Polychaeta suggested by phylogenetic analysis based on the amino acid sequences of elongation factor-1alpha. Mol. Phylogenet. Evol. 9, 255–261.

KOJIMA, S., HASHIMOTO, T., HASEGAWA, M., MURATA, S., OHTA, S., SEKI, H., OKADA, N. 1993. Close phylogenetic relationship between Vestimentifera (Tube worms) and Annelida revealed by the amino acid sequence of elongation factor-1alpha. J. Mol. Evol. 37, 66–70.

KUPRIYANOVA, E.K., ROUSE, G.W. 2008. Yet another example of paraphyly in Annelida: molecular evidence that Sabellidae contains Serpulidae (2008). Mol. Phylogenet. Evol. 46, 1174–1181.

KUPRIYANOVA, E.K., MADONALD, T.A., ROUSE, G.W. 2006. Phylogenetic relationships within Serpulidae (Sabellida, Annelida) inferred from molecular and morphological data. Zool. Scr. 35, 421–39.

LANTERBECQ, D., ROUSE, G.W., MILIKOVITCH, M.C., EECKHAUT, I. 2006. Molecular phylogenetic analyses indicate multiple independent emergences of parasitism in Myzostomida (Protostomia). Syst. Biol. 55, 208–227.

LEWIS, P.O. 2001. A likelihood approach to estimating phylogeny from discrete morphological character data. Syst. Biol. 50, 913–925.

LITTLEWOOD, D.T., OLSON, P.D., TELFORD, M.J., HERNIOU, E.A., RIUTORT, M. 2001. Elongation factor 1-alpha sequences alone do not assist in resolving the position of the Acoela within the Metazoa. Mol. Biol. Evol. 18, 437–442.

MCHUGH, D. 1997. Molecular evidence that echiurans and pogonophorans are derived annelids. Proc. Natl. Acad. Sci. U.S.A. 94, 8006–8009.

MCHUGH, D. 2000. Molecular phylogeny of the Annelida. Can. J. Zool. 78, 1873–1884.

MCHUGH, D. 2005. Molecular systematics of polychaetes (Annelida). Hydrobiologia 535/536, 309–318.

MARTIN, D., GIL, J., CARRERAS-CARBONELL, J., BHAUD, M. 2008. Description of a new species of *Mesochaetopterus* (Annelida, Polychaeta, Chaetopteridae), with redescription of *Mesochaetopterus xerecus* and an approach to the phylogeny of the family. Zool. J. Linn. Soc. 152, 201–225.

MARTIN, P. 2001. On the origin of the Hirudinea and the demise of the Oligochaeta. Proc. R. Soc. Lond. B 268, 1089–1098.

MARTIN, P., KAYGORODOVA, I., SHERBAKOV, D.Y., VERHEYEN, E. 2000. Rapidly evolving lineages impede the resolution of phylogenetic relationships among Clitellata (Annelida). Mol. Phylogenet. Evol. 15, 355–368.

MAXMEN, A.B., KING, B.F., CUTLER, E.B., GIRIBET, G. 2003. Evolutionary relationships within the protostome phylum Sipuncula: a molecular analysis of ribosomal genes and histone H3 sequence data. Mol. Phylogenet. Evol. 27, 489–503.

NYGREN, A., SUNDBERG, P. 2003. Phylogeny and evolution of reproductive modes in Autolytinae (Syllidae, Annelida). Mol. Phylogenet. Evol. 29, 235–249.

NYLANDER, J.A.A., ERSÉUS, C., KÄLLERSJÖ, M. 1999. A test of monophyly of the gutless Phallodrilinae (Oligochaeta, Tubificidae) and the use of a 573-bp region of the mitochondrial cytochrome oxidase I gene in analysis of annelid phylogeny. Zool. Scr. 28, 305–313.

NYLANDER, J.A.A., RONQUIST, F.R., HUELSENBECK, J.P., NIEVES-ALDREY, J. 2004. Bayesian phylogenetic analysis of combined data. Syst. Biol. 53, 47–67.

OCEGUERA-FIGUEROA, A., LÉON-RÈGAGNON, V., SIDDALL, M.E. 2005. Phylogeny and revision of Erpobdelliformes (Annelida, Arhynchobdellida) from Mexico based on nuclear and mitochondrial gene sequence. Rev. Mex. Biodiversidad 76, 191–198.

OSBORN, K.J., ROUSE, G.W., GOFFREDI, S.K., ROBISON, B.H. 2007. Description and relationships of *Chaetopterus pugaporcinus*, an unusual pelagic polychaete (Annelida, Chaetopteridae). Biol. Bull. 212, 40–54.

PASSAMANECK, Y., HALANYCH, K.M. 2006. Lophotrochozoan phylogeny assessed with LSU and SSU data: evidence of lophophorate polyphyly. Mol. Phylogenet. Evol. 40, 20–28.

PERSSON, J., PLEIJEL, F. 2005. On the phylogenetic relationships of *Axiokebuita, Travisia* and Scalibregmatidae. Zootaxa 998, 1–14.

PETERSEN, M.E. 1984. *Chaetopterus variopedatus* (Renier) (Annelida: Polychaeta: Chaetopteridae): a species complex. What species are being used at MBL? Biol. Bull. 167, 513.

PHILLIPS, A.J., SIDDALL, M.E. 2005. Phylogeny of the New World medicinal leech family Macrobdellidae (Oligochaeta: Hirudinida: Arhynchobdellida). Zool. Scr. 34, 559–564.

POP, A.A., CECH, G., WINK, M., CSUZDI, C., POP, V.V. 2007. Application of 16S, 16S rDNA and COI sequences in the molecular systematics of the earthworm family Lumbricidae (Annelida, Oligochaeta). Eur. J. Soil Biol. 43, Supp. 1, S43–S52.

RAIBLE, F., ARENDT, D. 2004. Metazoan evolution: some animals are more equal than others. Curr. Biol. 14, R106–R108.

RAIBLE, F., TESSMAR-RAIBLE, K., OSOEGAWA, K., WINKER, P., JUBIN, C., BALAVOINE, G., FERRIER, D., BENES, V., de JONG, P., WEISSENBACH, J., BORK, P., ARENDT, D. 2005. Vertebrate-type intron-rich genes in the marine annelid *Platynereis dumerilii*. Science 310, 1325–1326.

ROKAS, A., KRÜGER, D., CARROLL, S.B. 2005. Animal evolution and the molecular signature of radiations compressed in time. Science 310, 1933–1938.

ROTA, E., MARTIN, P., ERSÉUS, C. 2001. Soil-dwelling polychaetes: enigmatic as ever? Some hints on their phylogenetic relationship as suggested by a maximum parsimony analysis of 18S rRNA gene sequences. Contrib. Zool. 70, 127–138.

ROUSE, G.W., FAUCHALD, K. 1997. Cladistics and polychaetes. Zool. Scr. 26, 139–204.

ROUSE, G.W., PLEIJEL, F. 2001. *Polychaetes*. Oxford: University Press.

ROUSE, G.W., PLEIJEL, F. 2003. Problems in polychaete phylogeny. Hydrobiologia 496, 175–189.

ROUSSET, V., ROUSE, G.W., FERAL, J.P., DESBRUYERES, D., PLEIJEL, F. 2003. Molecular and morphological evidence of Alvinellidae relationships (Terebelliformia, Polychaeta, Annelida). Zool. Scr. 32, 185–197.

ROUSSET, V., ROUSE, G.W., SIDDALL, M.E., TILLIER, A., PLEIJEL, F. 2004. The phylogenetic position of Siboglinidae (Annelida) inferred from 18S rRNA, 28S rRNA and morphological data. Cladistics 20, 518–533.

ROUSSET, V., PLEIJEL, F., ROUSE, G.W., ERSÉUS, C., SIDDALL, M.E. 2007. A molecular phylogeny of annelids. Cladistics 23, 41–63.

RUTA, C., NYGREN, A., ROUSSET, V., SUNDBERG, P., TILLIER, A., WIKLUND, H., PLEIJEL, F. 2007. Phylogeny of Hesionidae (Aciculata, Polychaeta) assessed from morphology, 18S rDNA, 28S rDNA, 16S rDNA and COI. Zool. Scr. 36, 99–107.

SCHATZ, M., ORLOVA, E.V., DUBE, P., JAGER, J., van HEEL, M. 1995. Structure of *Lumbricus terrestris* haemoglobin at 30 A resolution determined using angular reconstitution. J. Struct. Biol. 114, 28–40.

SCHULZE, A. 2006. Phylogeny and genetic diversity of Palolo worms (*Palola*, Eunicidae) from the tropical North Pacific and the Caribbean. Biol. Bull. 210, 25–37.

SCHULZE, A., CUTLER, E.B., GIRIBET, G. 2005. Reconstructing the phylogeny of the Sipuncula. Hydrobiologia 535/536, 277–296.

SCHULZE, A., CUTLER, E.B., GIRIBET, G. 2007. Phylogeny of sipunculan worms: a combined analysis of four gene regions and morphology. Mol. Phylogenet. Evol. 42, 171–192.

SEAVER, E.C., KANESHIGE, L.M. 2006. Expression of "segmentation" genes during larval and juvenile development in the polychaetes *Capitella* sp. I and *H. elegans*. Dev. Biol. 289, 179–194.

SEAVER, E.C., SHANKLAND, M. 2000. Leech segmental repeats develop normally in the absence of signals from either anterior or posterior segments. Dev. Biol. 224, 339–353.

SEAVER, E.C., THAMM, K., HILL, S.D. 2005. Growth patterns during segmentation in the two polychaete annelids, *Capitella* sp. I and *Hydroides elegans*: comparisons at distinct life history stages. Evol. Dev. 7, 312–326.

SELLA, G.M., RAMELLA, L.M. 1999. Sexual conflict and mating systems in the dorvilleid genus *Ophryotrocha* and the dinophilid genus *Dinophilus*. Hydrobiologia 402, 203–213.

SHEDLOCK, A.M., OKADA, N. 2000. SINE insertions: powerful tools for molecular systematics. Bioessays 22, 148–160.

SIDDALL, M.E., BORDA, E. 2003. Phylogeny and revision of the leech genus *Helobdella* (Glossiphonidae) based on mitochondrial gene sequences and morphological data and a special consideration of the *triserialis* complex. Zool. Scr. 32, 23–33.

SIDDALL, M.E., APAKUPAKUL, K., BURRESON, E., COATES, K.A., ERSÉUS, C., GELDER, S.R., KÄLLERSJÖ, M., TRAPIDO-ROSENTHAL, H. 2001. Validating Livanow: molecular data agree that leeches, branchiobdellidans, and *Acanthobdella peledina* form a monophyletic group of oligochaetes. Mol. Phylogenet. Evol. 21, 346–351.

SIDDALL, M.E., BUDINOFF, R.B., BORDA, E. 2005. Phylogenetic evaluation of systematics and biogeography of the leech family Glossiphoniidae. Inv. Syst. 19, 105–112.

SIDDALL, M.E., TRONTELJ, P., UTEVSKY, S.Y., NKAMANY, M., MACDONALD, K.S., III 2007. Diverse molecular data demonstrate that commercially available medicinal leeches are not *Hirudo medicinalis*. Proc. R. Soc. B 274, 1481–1487.

STATON, J.L. 2003. Phylogenetic analysis of the mitochondrial cytochrome c oxidase subunit 1 gene from 13 sipunculan genera: intra- and interphylum relationships. Inv. Biol. 122, 252–264.

STRUCK, T.H., PURSCHKE, G. 2005. The sister group relationship of Aelosomatidae and Potamodrilidae (Annelida: "Polychaeta")—a molecular phylogenetic approach based on 18S rDNA and cytochrome oxidase I. Zool. Anz. 243, 281–293.

STRUCK, T.H., HESSLING, R., PURSCHKE, G. 2002a. The phylogenetic position of the Aelosomatidae and Paergodrilidae, two enigmatic oligochaete-like taxa of the "Polychaeta", based on molecular data from 18S rDNA sequences. J. Zool. Syst. Evol. Res. 40, 155–163.

STRUCK, T.H., WESTHEIDE, W., PURSCHKE, G. 2002b. Progenesis in Eunicida ("Polychaeta", Annelida)—separate evolutionary events? Evidence from molecular data. Mol. Phylogenet. Evol. 25, 190–199.

STRUCK, T.H., HALANYCH, K.M., PURSCHKE, G. 2005a. Dinophilidae (Annelida) is most likely not a progenetic Eunicida: evidence from 18S and 28S rDNA. Mol. Phylogenet. Evol. 37, 619–623.

STRUCK, T.H., PURSCHKE, G., HALANYCH, K.M. 2005b. A scaleless scale worm: Molecular evidence for the phylogenetic placement of *Pisione remota* (Pisionidae, Annelida). Mar. Biol. Res. 1, 243–253.

STRUCK, T.H., PURSCHKE, G., HALANYCH, K.M. 2006. Phylogeny of Eunicida (Annelida) and exploring data congruence using a partition addition bootstrap alteration (PABA) approach. Syst. Biol. 55, 1–20.

STRUCK, T.H., SCHULT, N., KUSEN, T., HICKMAN, E., BLEIDORN, C., MCHUGH, D., HALANYCH, K.M. 2007. Annelid phylogeny and the status of Sipuncula and Echiura. BMC Evol. Biol. 7, 57.

TRONTELJ, P., UTEVSKY, S.Y. 2005. Celebrity with a neglected taxonomy: molecular systematics of the medicinal leech (genus *Hirudo*). Mol. Phylogenet. Evol. 34, 616–624.

TRONTELJ, P., SKET, B., STEINBRÜCK, G. 1999. Molecular phylogeny of leeches: congruence of nuclear and mitochondrial rDNA data sets and the origin of bloodsucking. J. Zool. Syst. Evol. Res. 37, 141–147.

UTEVSKY, S.Y., UTEVSKY, A.Y., SCHIAPARELLI, S., TRONTELJ, P. 2007. Molecular phylogeny of pontobdelline leeches and their place in the decent of fish leeches (Hirudinea, Piscicolidae). Zool. Scr. 36, 271–280.

VALLÉS, Y., HALANYCH, K.M., BOORE, J.L. 2008. Group II introns break new boundaries: presence in a Bilaterian's genome. PLoS ONE 3, e1488.

WIKLUND, H., NYGREN, A., PLEIJEL, F., SUNDBERG, P. 2005. Phylogeny of Aphroditiformia (Polychaeta) based on molecular and morphological data. Mol. Phylogenet. Evol. 37, 459–502.

WOLSTENHOLME, D.R. 1992. Animal mitochondrial DNA: structure and evolution. Int. Rev. Cytol. 141, 173–216.

WORSAAE, K. 2005. Phylogeny of Nerillidae (Polychaeta, Annelida) as inferred from combined 18S rDNA and morphological data. Cladistics 21, 143–162.

WORSAAE, K., NYGREN, A., ROUSE, G.W., GIRIBET, G., PERSSON, J., SUNDBERG, P., PLEIJEL, F. 2005. Phylogenetic position of Nerillidae and *Aberranta* (Polychaeta, Annelida), analysed by direct optimization of combined molecular and morphological data. Zool. Scr. 34, 313–328.

WYSOCKA-DILLER, J., AISEMBERG, G.O., MACAGNO, E.R. 1995. A novel homeobox cluster expressed in repeated structures in the midgut. Dev. Biol. 171, 439–447.

ZHONG, M., STRUCK, T.H., HALANYCH, K.M. 2008. Three mitochondrial genomes of Terebelliformia (Annelida) worms and duplication of the methionine tRNA. Gene. 416, 11–21.

Chapter 3

Cryptic Speciation in Clitellate Model Organisms

Christer Erséus and Daniel Gustafsson

Department of Zoology, University of Gothenburg, Göteborg, Sweden

3.1 INTRODUCTION

Several annelid species are easy to breed and maintain in culture and are often used for various kinds of biological experiments. Further, many annelids are benthic and deposit feeders, and are therefore suitable for studies of uptake and bioaccumulation of pollutants in sediments. Members of Clitellata (oligochaetes including leeches; Martin 2001; Siddall et al. 2001) are particularly popular, probably because so many of them dwell in freshwater or soils, making the laboratory independent of infrastructure and protocols for seawater, which are needed for polychaetes. Thus, scientific literature contains a multitude of articles concerning basic and applied studies that have used clitellates as model or test organisms (Table 3.1).

As in all scientific work, the use of model organisms presupposes repeatability as well as predictability and generalization. For instance, if a specimen or population of *Lumbriculus variegatus* reacts to a pollutant in a certain way, another specimen (or population) of the same species will be expected to react in a similar way and, with regard to predictability and generalization, so will other lumbriculids, other clitellates or even other annelids, etc. By the same token, a single worm model can be used to evaluate differences in the effects of different pollutants. Other examples can be given from studies on physiology, developmental biology, genetic regulation, immunology, reproduction, behavior, etc. (see other chapters in this book). However, a prerequisite for this logic is that the taxonomic identity and phylogenetic context of the model organism are securely established and consistent across all research institutions in the world, and especially so when networks of scientists share standardized experimental designs.

Recent advancements in molecular systematics have provided new tools for the improvement of species delimitation and identification. In particular, gene sequences of the mitochondrial genome (mtDNA) have been used in attempts to establish the borderline between intraspecific and interspecific variation. A part of the cytochrome oxidase I (COI) gene has even been suggested as a molecular bar-code marker for animal species (Hebert

Annelids in Modern Biology, Edited by Daniel H. Shain

Table 3.1 Number of articles found in an internet search (May 2007) using Google Scholar (http://scholar.google.com), ISI Web of Knowledge/Web of Science (http://isiknowledge.com/) and PubMed/Medline (http://ncbi.nlm.nih.gov/)

Clitellate Species or Complex	Number of Hits		
	Google Scholar	Web of Science	PubMed
Lumbricus terrestris	4,870	1,029	2,249
Eisenia fetida/Eisenia foetida/Eisenia andrei	4,720	1,959	2,919
Tubifex tubifex	4,570	420	219
Hirudo medicinalis	4,050	725	690
Lumbriculus variegatus	1,280	213	89
Limnodrilus hoffmeisteri	1,110	101	34
Enchytraeus albidus/crypticus	712	125	61
Theromyzon tessulatum	375	102	53
Helobdella triserialis	328	82	37

et al. 2003a, b), and a few empirical studies can now also be used as references for clitellate worms (e.g. Pérez-Losada et al. 2005; Trontelj and Utevsky 2005; Bely and Weisblat 2006; Erséus and Kvist 2007; Siddall et al. 2007). This chapter aims to review the recent evidence for considerable genetic variation in some widely distributed and commonly used model "species" of Clitellata, and also to outline what future research is needed to establish the true diversity, and to improve the taxonomy, of these taxa.

3.2 SOURCES AND KINDS OF VARIATION

3.2.1 Morphology

Historically, zoological taxonomy has focused on the patterns of morphological variation in animals, but a classical problem is that in closely related organisms, intraspecific variation may be difficult to distinguish from interspecific differences. This has resulted in the polarity between "splitters" and "lumpers," i.e. the tendency among taxonomists either to overestimate or to underestimate the number of species within a given higher taxon. During the 19th century and the beginning of the 20th century, numerous new oligochaete species were formally named, but they were often briefly described and/or separated from each other on the basis of rather subtle morphological differences. After a more thorough scrutiny, much of this diversity was subsequently interpreted as continua within a lower number of species, and several old names were regarded as junior synonyms or "species dubiae." The monograph on the aquatic oligochaetes of the world by Brinkhurst and Jamieson (1971) gives many examples of this.

More recently, however, a part of this lumping has been challenged; for example, Holmquist (1983, 1985) referred to historical and new morphological evidence in a plead for separation of different species-level taxa within the *Tubifex tubifex* complex, and work by Erséus and coworkers has provided several examples of small but consistent morphological differences between closely related and apparently radiating species of marine Tubificidae (e.g. Erséus 1981, 1988, 1997, 2003; Wang and Erséus 2004; Erséus and Bergfeldt 2007). Moreover, in recent years, an astonishingly high species diversity has been revealed within *Grania*, a morphologically homogeneous genus of marine Enchytraeidae (e.g. Coates 1990; Coates and Stacey 1993, 1997; Rota and Erséus 1996, 2003; De Wit and Erséus 2007; Rota et al. 2007). Thus, numerous indications suggest that, for aquatic groups in particular, oligochaete diversity is considerably higher than once thought.

Additionally, Timm (2005) refers to a tendency to misuse some species names in oligochaete research. For instance, in many cases, "*T. tubifex*" has been used for "any pink tubificid equipped with dorsal hair chaetae, and living in aquatic mud" in polluted streams, the profundal of lakes or even in brackish waters (Timm 2005, p. 55), while several other tubificids (in particular, species of *Rhyacodrilus, Psammoryctides, Potamothrix* and *Ilyodrilus*), which may be more common than *T. tubifex* in these habitats, are easily mistaken for the latter. As also discussed by Timm (2005), the well-known enchytraeid, *Enchytraeus albidus*, may have been identified in a similar way (just by assumption) from habitats and sites where it does not exist.

3.2.2 Polyploidy

Polyploidy is common in plants, where estimates of incidence range from 30 to 80% of the described species (Otto and Whitton 2000). It appears to be less frequent in animals, but may be underestimated, particularly among invertebrates. Chromosome counts have been performed for several common clitellates (see Christensen 1980), and polyploidy has been found in taxa widely used in the laboratory, such as *L. variegatus* (chromosome numbers reported: 34, 68, c. 85, 136, c. 185) and *T. tubifex* (c. 24, 75, 100, 125, 150). Polyploid oligochaetes have rarely been described as separate species, but they have sometimes been recognized as (trinominal) subspecies taxa or as "varieties," occasionally even before their actual karyotype number has been established (Viktorov 1997).

Polyploidization may disturb the formation of genital elements, causing physiological or morphological incompatibility of the polyploid individual with its ancestral stocks as well as other polyploids, and if mating does occur, the progeny may be sterile or may die before reaching reproductive age. In some cases, the polyploidization event disables sexual reproduction entirely, replacing it with (asexual) fragmenting of genetically identical individuals. Such clones may become firmly established and may even outcompete their closest, sexually reproducing, relatives, but as they lack the possibility of gene recombination, polyploidization leading to obligate asexual reproduction has often been regarded as an evolutionary dead end in the longer term. If, on the other hand, the polyploid organism can switch between sexual and asexual modes of reproduction, it can break free of this constraint, resulting in polyploid populations that evolve independently of their diploid ancestors. In animals, however, polyploid forms appear to be predominant among asexually reproducing forms (Otto and Whitton 2000), and Christensen's (1980) review gives many examples of this, particularly within Enchytraeidae and some tubificid genera (e.g. *Tubifex* and *Limnodrilus*). However, in Naidinae (Tubificidae), a clitellate group well known for alternations between sexual and asexual reproduction (e.g. see Ladle and Todd 2006), polyploidy appears to be uncommon (Christensen 1980).

The existence of multiple copies of the same original chromosome in an individual or a lineage has important consequences, as duplicated genes allow an organism to "experiment" with extra copies not needed for survival (Ohno 1970). When such duplication occurs, some gene copies may diverge from the original ones, and changes may result in morphological, physiological or ecological effects that are selectively advantageous or, as one alternative, may lead to nonfunctional pseudogenes. As shown by Song et al. (1995), genetic changes in connection to polyploidization may occur rapidly, at least in the case of allopolyploidization, i.e. hybridization. Polyploids may be larger than diploids of the same "species," and they often occur in unstable or harsh environments (Stebbins 1950; Bell 1982). They also tend to be more widely distributed and peregrine than their diploid relatives (discussed by Viktorov 1997; Otto and Whitton 2000). As often as not, however, polyploid and diploid individuals of the same taxon are morphologically indistinguishable and can be separated only through genetic information or biological differences (e.g. ecological preferences, physiology, parasite tolerance). If these differences are not already the result of speciation, they may ultimately lead to such a process.

3.2.3 DNA and Barcoding

For at least 30–40 years, assessing genetic variation at various enzyme loci using gel electrophoresis has been possible, and such studies, pertaining to some important clitellate models, are reviewed below. Nevertheless, the rapid development of methods to sequence and analyze DNA has revolutionized the study of genetic variation in organisms. Various genes have been explored, some of which are conserved enough to be suitable for assessment of high-level relationships (e.g. 18S and 28S rDNA), while others evolve at higher rates, revealing differences and unique similarities among closely related taxa and within species. The most popular of the latter category are the mitochondrial COI and 16S genes, and the internal transcriber spacers, ITS1 and ITS2, of the nuclear genome; ITS1-2 comprise, together with 5.8S rDNA, the "ITS region," between the 18S and 28S rDNA genes.

The aim of DNA barcoding, specifically using part of the COI gene, as first suggested by Hebert et al. (2003a, b), is to create a sequence database of most or all known species, in order to aid in the assignment of a given sample to the correct species, and to facilitate the discovery of new taxa (Moritz and Cicero 2004), particularly from organism groups for which morphological characters are either vague, ambiguous, complex or lacking the desired degree of distinctiveness. This idea has its critics on both economical and philosophical grounds (e.g. Will and Rubinoff 2004; Will et al. 2005; Cameron et al. 2006; Fitzhugh 2006), and growing evidence suggests that one of its basic tenets—that intraspecific variation is sufficiently lower than interspecific variation (the "barcoding gap") to allow more or less clear-cut lines between species—does not always hold true for more extensive sampling, especially not when true sister taxa are considered (e.g. Johnson and Cicero 2004; Moritz and Cicero 2004; Meyer and Paulay 2005). However, as exemplified below, barcoding will probably be useful as a tool to routinely determine the degree to which one laboratory's worm isolate is comparable with that of other laboratories. Here, the aim would be to quickly ascertain phylogenetic affinity and to estimate comparability with other studies on the supposedly same nominal species, rather than to confirm or dismiss previously established species boundaries.

3.3 EXAMPLES OF CLITELLATE MODEL ORGANISMS

3.3.1 Tubificidae

Over the last few decades, several common laboratory worms have been recognized as containing great genetic variation, and are thus essentially unsuitable for laboratory studies unless proper care is first taken to identify the lineage of the studied population. The most well-known example is *Tubifex tubifex*, which is now known to comprise several different lineages with different biological features.

Polymorphisms and different ecological forms of *T. tubifex* have been long known (Brinkhurst and Jamieson 1971; Poddubnaja and Timm 1980; Holmquist 1983; Milbrink 1983). Sometimes, the morphological differences have been considered as phenotypic plasticity (e.g. Chapman and Brinkhurst 1987). However, through the use of a segment of the 16S rDNA sequence from *T. tubifex* sampled from some rivers in Eastern and Central Europe, Sturmbauer et al. (1999) established that *T. tubifex* representing five different lineages coexist sympatrically in various combinations. The distinct lineages show differences in their resistance to cadmium, suggesting that considerable physiological and ecological differentiation is associated with the divergence observed in the 16S rDNA pattern.

These findings reinforced what other researchers had previously found. For instance, Reynoldson et al. (1996) found differences in growth, reproduction and tolerance to pollutants between lab cultures from Spain and Canada. Anlauf (1994) differentiated three "types" of *T. tubifex* by enzyme electrophoresis (of glucose phosphate isomerase and isocitrate dehydrogenase), and these types showed differences in adult fresh weight, fecundity and tolerance to anoxia. Sturmbauer et al. (1999) studied the same enzymes and were able to correlate the differences in enzyme patterns to their five lineages.

Beauchamp et al. (2001, 2002) examined 16S rDNA from several North American *T. tubifex* populations, recovering some lineages of Sturmbauer et al. (1999) as well as a sixth lineage unknown in Europe (see Fig. 3.1), and found that susceptibility to infection by *Myxobolus cerebralis*, the myxozoan parasite that spreads salmonid whirling disease, ranges from complete resistance to high susceptibility in these various forms. To this can be added the older reports of varying chromosome numbers in *T. tubifex* (see Christensen 1980), although these different karyotypes have still not been matched with any of the six *T. tubifex* lineages recognized today.

Thus, while *T. tubifex* has long been known as variable, strong evidence now indicates that it cannot be regarded as a single species. The 16S rDNA sequences of the lineages known to date (Fig. 3.1) differ from each other by 5.8–13.0% (Sturmbauer et al. 1999; Beauchamp et al. 2001), which is comparable to distances commonly found between closely related species [e.g. pholcid spiders, 15.0–34.1% (Astrin et al. 2006); *Calopteryx* (Odonata), 0.6–21.6% (Misof et al. 2000)]. Although morphological traits remain diagnostically unreliable, further genetic studies may provide final resolution to the species taxonomy of the *T. tubifex* complex; of course, this should include a correlation with the data of Anlauf (1994), Anlauf and Neumann (1997), Sturmbauer et al. (1999), Beauchamp et al. (2001, 2002) and others. Clearly, routine assignment of cultured or wild-caught *T. tubifex* to that taxon without further genetic characterization is no longer recommended.

Another commonly studied tubificid is *Limnodrilus hoffmeisteri*, for which we have incorporated GenBank data in our 16S rDNA tree (Fig. 3.1). Kennedy (1969), Brinkhurst and Jamieson (1971), and Kathman and Brinkhurst (1998) recognized large morphological variability in this taxon, particularly with regard to the shape of chaetae and cuticular penis sheaths. A 16S rDNA study by Sturmbauer et al. (1999) involved two *L. hoffmeisteri* individuals, and the distance between them was greater than most of the distances between the *T. tubifex* lineages. Beauchamp et al. (2001) analyzed other specimens of *L. hoffmeisteri*, which had sufficient genetic variation to suggest that this taxon also harbors cryptic species. The same authors again pointed out that the morphological characters used to

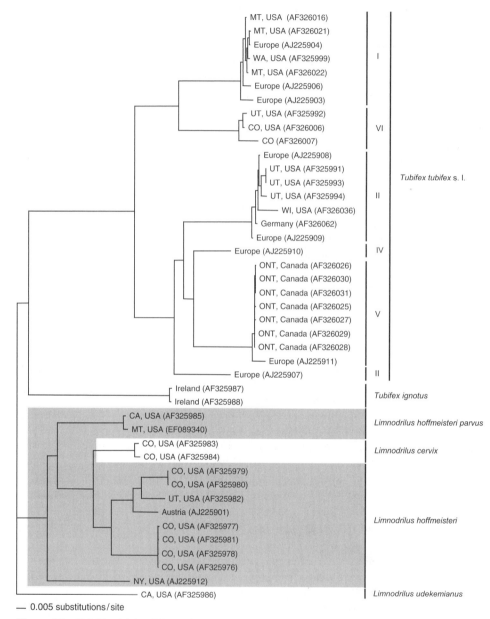

— 0.005 substitutions/site

Figure 3.1 Neighbor-joining (NJ) tree based on a selection of 16S rDNA sequences of the six *Tubifex tubifex* lineages as reported by Sturmbauer et al. (1999) and Beauchamp et al. (2001, 2002), as well as corresponding data of *Limnodrilus hoffmeisteri* from these authors. The shaded area represents *L. hoffmeisteri* s. lat., which may be a paraphyletic group, but note that the NJ tree is a phenetic rather than a phylogenetic tree. All sequences were downloaded from GenBank (March 2007)

distinguish species of *Limnodrilus* differ too much between localities to be reliable. Thus, *L. hoffmeisteri*, as well as some of its congeners, is evidently in great need of a taxonomic revision, and—as in *T. tubifex*—the number of species is greater than the number of names in use today.

3.3.2 Lumbriculidae

Lumbriculus variegatus, commonly known in North America as the "Californian black-worm," is often used as a laboratory worm. It has several desirable features for studies on bioaccumulation and toxicity (e.g. see Phipps et al. 1993; Veltz et al. 1996), as well as regeneration (e.g. Martinez et al. 2005, 2006). These features include (1) its wide geographic distribution and prevalence, and the ease with which it can be obtained and maintained in culture (see Leppänen and Kukkonen 1998a, b), (2) its short generation time and its ability to reproduce asexually by fragmentation and to regenerate lost ends, (3) its burrowing behavior, which exposes it to sediment-bound pollutants and (4) the ease with which it can be distinguished from other clitellates (Dermott and Munawar 1992). A majority of the articles scored for *L. variegatus* in Table 3.1 deal with the effect of pollutants and toxins on this species, which frequently appears in recommendations for suitable model organisms (e.g. American Society for Testing and Materials, 1995; Ducrot et al. 2005).

Lumbriculus variegatus is presumed to be common across the entire northern hemisphere, and has secondarily been introduced to Australia, New Zealand and South Africa (e.g. Cook 1971; Brinkhurst 1989). In a recent molecular study on *L. variegatus* from Europe, Japan and the USA (Gustafsson et al., in press), however, we have shown that this taxon is divided into at least three clades, two of which occur in Europe and North America, in some cases sympatrically. These groups, while likely to be morphologically indistinguishable, can easily be distinguished at mitochondrial COI and 16S loci, as well as the nuclear ITS region, and the genetic distances are great enough to suggest that the three clades represent different species. Our study also provides evidence that the different ploidy levels reported for *L. variegatus* (Christensen 1980) may be clade dependent. At present, how the genetic data translate into ecological, physiological and developmental differences in *L. variegatus* is unknown, but as maturation seems to become inhibited as the ploidy level increases (Christensen 1980), at least some populations may be entirely asexual.

These findings are not entirely surprising. For more than 100 years, occasional sexually mature specimens or even sexually reproducing populations of *L. variegatus* (or *Lumbriculus* sp.) have been reported, mainly from Europe (references in Gustafsson et al., in press). In these reports, the great variation in the position and number of genital elements (e.g. Mrázek 1907), and the season when maturity occurs is striking, even within samples from the same locality. This has usually been interpreted as within-species variation; however, *Lumbriculus* as a whole has always been a problematic genus. Faunal lists for various regions often list only "*Lumbriculus* sp.," which has caused uncertainty about both the circumscription and the actual distribution of *L. variegatus*. Thus, to fully resolve the taxonomy and phylogeny of the *Lumbriculus* complex, a comprehensive reassessment of the morphological and genetic patterns, using DNA technology and a large number of worms from a wide geographical range, will be needed.

3.3.3 Enchytraeidae

Enchytraeids, and particularly members of the genus *Enchytraeus*, have been recommended as suitable terrestrial test organisms to complement data from the lumbricid *Eisenia fetida* (Westheide et al. 1987; Römbke 1989; Römbke and Knacker 1989; Westheide and Bethke-Beilfuss 1991). However, while *Enchytraeus* species have many of the characteristics of a good model organism—easy to culture, short generation times, ecological relevance (Westheide and Bethke-Beilfuss 1991)—they are often difficult to separate on morphological characters alone, and they are suspected to contain cryptic species (Brockmeyer 1991). Westheide and Brockmeyer (1992) even suggest the use of a protein pattern index for species differentiation within *Enchytraeus* and other enchytraeids.

Westheide and Graefe (1992) described a new species of *Enchytraeus, Enchytraeus crypticus*, that could not be differentiated from what was presumed to be its closest relative (*Enchytraeus variatus*) by any "conventional morphological features." Nevertheless, while specimens from cultures of the two species were unable to interbreed, they could be distinguished by DNA restriction fragment patterns (Schlegel et al. 1991), isoenzyme and general protein patterns (Brockmeyer 1991), and sperm ultrastructure (Westheide et al. 1991). In a later study (Schirmacher et al. 1998), *E. crypticus* and *E. variatus* showed slight but consistent genetic distances (using the random amplification of polymorphic DNA [RAPD] technique), sufficient to make them reproductively isolated.

Römbke (1989) suggested that the common "pot worm," *Enchytraeus albidus*, would be suitable as a terrestrial model organism in ecotoxicology. In a study of the COI variation in this taxon (Table 3.2; Erséus, Rota and Gustafsson in preparation), however, we found evidence that this species contains forms that are different enough to qualify for species status, but which may be hard or impossible to distinguish morphologically. As shown in Fig. 3.2, 10 specimens all identified as *E. albidus* represented three main lineages (clades A, B and C in Fig. 3.2). Clade A contains two specimens obtained from a seaweed compost in Galicia, Spain, as well as two worms from a seashore site at Torslanda, Sweden; within this clade, the pairwise distances are 0.9–2.6%, with the larger distances (2.1–2.6%) between the two sites. Clade B is represented by three individuals, one from a second Swedish seashore locality (at Stenungsund) plus two from a German laboratory culture; here the within-clade distances are only 0.2–0.9%, and one of the laboratory worms was more similar to the Stenungsund worm than to the other laboratory specimen. Clade C comprises three individuals from the Spanish compost, i.e. they are sympatric with a part of the clade A material; distances within C range from 2.0 to 2.6%. Between-clade variation is 10.0–16.3%, with a minimum of 10.0–10.6% for A–B comparisons, and a maximum of 15.0–16.3% for A–C comparisons. Thus, the distances between the three

Table 3.2 List of specimens used for analysis of COI mtDNA variation in *Enchytraeus albidus* (see Fig. 3.2), with *E. buchholzi, E. bulbosus* and *E. christenseni* selected as out-groups

Ref. #	Specimen ID	Locality	Source
CE521	*E. albidus*	Seashore, Stenungsund, Sweden	C. Erséus
CE965	*E. albidus*	Seashore, Torslanda, Sweden	A. Ansebo
CE972	*E. albidus*	Seashore, Torslanda, Sweden	A. Ansebo
CE1684	*E. albidus*	Seaweed compost, Galicia, Spain	B. Reboreda Rivera
CE1685	*E. albidus*	Seaweed compost, Galicia, Spain	B. Reboreda Rivera
CE1686	*E. albidus*	Seaweed compost, Galicia, Spain	B. Reboreda Rivera
CE1689	*E. albidus*	Seaweed compost, Galicia, Spain	B. Reboreda Rivera
CE1693	*E. albidus*	Seaweed compost, Galicia, Spain	B. Reboreda Rivera
CE2169	*E. albidus*	Lab culture, Germany	J. Römbke
CE2170	*E. albidus*	Lab culture, Germany	J. Römbke
CE724	*E. buchholzi*	Culture, from Krasnoyarsk, Siberia	E. Rota
CE798	*E. bulbosus*	Götene, Sweden	E. Rota and C. Erséus
CE805	*E. christenseni*	Alingsås, Sweden	E. Rota and C. Erséus

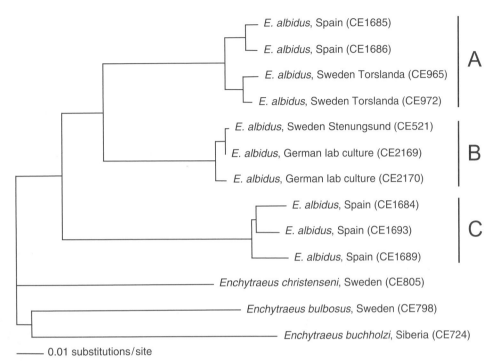

E. albidus, Spain (CE1685)

E. albidus, Spain (CE1686)

E. albidus, Sweden Torslanda (CE965)

E. albidus, Sweden Torslanda (CE972)

A

E. albidus, Sweden Stenungsund (CE521)

E. albidus, German lab culture (CE2169)

E. albidus, German lab culture (CE2170)

B

E. albidus, Spain (CE1684)

E. albidus, Spain (CE1693)

E. albidus, Spain (CE1689)

C

Enchytraeus christenseni, Sweden (CE805)

Enchytraeus bulbosus, Sweden (CE798)

Enchytraeus buchholzi, Siberia (CE724)

—— 0.01 substitutions/site

Figure 3.2 NJ tree of preliminary COI mtDNA analysis of 10 *Enchytraeus albidus* specimens from four European localities shows a division of at least three different genotypes; see text and Table 3.2 for further explanation.

clades of *E. albidus* are only marginally shorter than those (14.7–20.2%) between the individual *E. albidus* worms and any of the three out-groups (*E. christenseni, E. bulbosus* and *E. buchholzi*); distances among the out-groups are 15.6–17.4%.

Hartzell et al. (2005) also noted considerable COI variation in a large material of the North American glacier ice worm, *Mesenchytraeus solifugus*, revealing two geographically distinct clades, an Alaskan lineage being separated from a southern one (including populations from British Columbia, Washington and Oregon) by a COI distance of about 10%. Schmelz (2003), in his monograph on the taxonomy of the species-rich enchytraeid genus *Fridericia*, used protein and isoenzyme patterns to confirm species diagnoses and distinctions. These and other studies on terrestrial species (e.g. Rota 1995; Rota and Healy 1999) suggest that the family Enchytraeidae may contain more variability and more species than traditional morphological studies have been able to establish.

3.3.4   Lumbricidae

A well-known example of diversity in popular laboratory worms is that of the *Eisenia fetida*/*E. andrei* complex, commonly used in ecotoxicology (e.g. Edwards and Coulson 1992; Kokta 1992; Reinecke 1992). These two taxa, first regarded as variants of *E. fetida* (André 1963; but recognized as a heterogeneous group already by Avel 1929), and with *E. andrei*, formally named as a separate subspecies by Bouché (1972), are now generally recognized as different species. [The spelling "*foetida,*" incorrectly proposed by Michaelsen (1900), is still seen in the literature (e.g. Kohlerova et al. 2004; Yuan et al. 2006); see

Easton 1983; Sims 1983.]. Over the last few decades, the interspecific differences between *E. fetida* and *E. andrei* have been confirmed using metabolic profiling (Bundy et al. 2002), life history studies (Reinecke and Viljoen 1991; Domínguez et al. 2005), allozyme electrophoresis (Jaenike 1982; Øien and Stenersen 1984; Bouché 1992; Henry 1999; McElroy and Diehl 2001), and mitochondrial and nuclear DNA (Pérez-Losada et al. 2005).

Robotti (1983; see also Bouché 1992) showed that *E. andrei* was genetically distinct from *E. fetida*, and although *E. andrei* was fairly homogeneous, populations of *E. fetida* from France and Italy differed substantially. Pérez-Losada et al. (2005) showed that the two species are separated genetically, but also that Irish specimens of *E. fetida* were distinct from Spanish ones (COI distances ~18%), indicating that even this taxon contains cryptic species. Thus, while *E. fetida* and *E. andrei* are easily distinguished on morphological characters, they may still include forms deserving status as sibling species. Further, suspicion has been raised (e.g. Westheide and Bethke-Beilfuss 1991) that what is commonly called *E. fetida* or *Eisenia* sp. in the literature may actually be a mixture of *E. fetida* and *E. andrei*. This confusion calls for caution and genetic verification of specimens when using worms of this complex as model organisms.

By and large, genetic variation and cryptic speciation (but referred to as "polyploidy," "polymorphism," "different morphs," etc.) appear to be commonplace in several widespread lumbricid earthworms, such as *Allolobophora chlorotica, Aporrectodea caliginosa, Aporrectodea rosea, Dendrobaena octaedra* and *Dendrodrilus rubidus* (see Sims and Gerard 1985).

3.3.5 Leeches (Hirudinida)

One of the most commonly studied clitellate is the European medicinal leech, *Hirudo medicinalis* (Table 3.1), and in recent years, the genetic variation of this worm has been scrutinized. Using RAPD-DNA (Trontelj et al. 2004), both nuclear and mitochondrial genes (ITS2 + 5.8S, 12S and COI; Trontelj and Utevsky 2005), and the COI gene and microsatellite data (Siddall et al. 2007), these authors have recently reconsidered a long overlooked diversity in *H. medicinalis* s. lat. (*medicinalis, verbana* and a new species once called "variety orientalis"), prompting them to call for molecular characterization of laboratory cultures. These three species, one of which (*Hirudo verbana*) was originally described as early as 1820, are by no means cryptic, as they can be differentiated by their body pigmentation (Nesemann and Neubert 1999). Nevertheless, overeager lumping of *Hirudo* species has caused confusion of the lineages, and, as showed by Siddall et al. (2007), many (if not all) commercial and laboratory stocks of *H. medicinalis* are actually *H. verbana*.

The risk of imprecise taxonomy due to assumption has also been demonstrated for *Helobdella*, a genus of leeches used in developmental research for over 20 years. Recognizing that the current stocks used in the USA originally came from different isolates, Bely and Weisblat (2006) collected *Helobdella* specimens from several laboratory cultures along with wild-type populations and studied their variation in the COI gene. The resulting phylogenetic tree showed that the laboratory worms belong to five different lineages of *Helobdella*, each of which is more closely related to another nominal species than to each other. Moreover, lab cultures labeled *Helobdella robusta* belonged to three different clades, two of which occur sympatrically in Sacramento, CA. Because of this, Bely and Weisblat (2006) emphasized the need for barcoding laboratory isolates to avoid incomparable or unrepeatable data.

3.4 CRYPTIC SPECIATION

Although growing evidence indicates that many well-known and widely distributed "species" of Clitellata exhibit considerable genetic variation, and that some genotypes also have different characteristics in reproduction, physiology, host–parasite relationships, pollution tolerance, etc., whether different lineages should be regarded as separate nominal taxa or just as locally adapted variants of the same biological species may still be disputed. To a great extent, the discussion depends on what kind of species concept one adheres to, of which a plethora exists (e.g. Mayden 1997). In the case of asexual clones, which may be prevalent in complexes such as *L. variegatus* and among naidines (Tubificidae), a biological species concept based on interfertility can hardly be applied and has commonly been replaced by a pragmatic approach referring to diagnostically distinguishable morphospecies. Even with evidence of sexual reproduction, however, reproductive isolation may be difficult to establish without extensive genetic investigation across many populations. Besides, for allopatric forms, predicting whether or not they will be interfertile if they meet in a future event may be impossible. A phylogenetic species concept, regarding a species as the least inclusive clade of specimens sharing a recent ancestor, may be easier to conceive using genetic methods, but it will have problems dealing with ancient lineage sorting and paraphyletic parental stocks.

Regardless of whether or not the clitellates reviewed here will be taxonomically split into several formal species-level taxa, a speciation process is likely to be at work in complexes such as *T. tubifex, L. hoffmeisteri, L. variegatus, E. albidus* and *E. fetida.* Moreover, as long as coherent evidence of long separation in the mitochondrial and (in the case of *L. variegatus*) nuclear genomes of different lineages in these examples, with genetic distances comparable to those known between other closely related species of the same genera, exists, recognition of these lineages as separate species seems reasonable.

What remains to be completed is a more comprehensive assessment of the long-noted morphological variants in these clitellate complexes, to determine in which way they match the lineages recognized by genetic data. If extensive enough and successful, future work will lead to formal nomenclatural revisions, probably including the resurrection of some older names available in the literature, but which are currently on hold as junior synonyms to the more inclusive names used today. Alternatively, or as a supplement, a more strictly DNA-based taxonomy may be seen in the future, i.e. a taxonomy less dependent on diagnostic morphological features, as has been advocated by Vogler and Monaghan (2007).

3.5 CONCLUSIONS AND RECOMMENDATIONS

Both asexuality and polyploidy appear to be more common in clitellates than in many other invertebrate groups. At the same time, morphological characters are often ambiguous, insufficient or too few for establishing stable phylogenies of these worms. Therefore, we have reason to suspect that the cryptic speciation observed in, e.g. *L. variegatus* and *T. tubifex* is prevalent also in many other clitellate taxa. That is, clitellates may be a much more speciose group than previously thought.

As reviewed here, common laboratory species of Clitellata have been found to be either nonmonophyletic (*Helobdella*) or, while probably monophyletic, sufficiently variable to warrant special attention for putative cryptic speciation. In some of these cases (most notably *T. tubifex*, but also in the *Eisenia* complex; Reinecke & Viljoen 1991), physiological or ecological differences are already known, but in others, only the genetic differences are known. In effect, this means that results from one study using a particular clitellate model may not be fully comparable to those from another study of the same model, unless the phylogenetic affinities of the stocks are firmly established. At the same time, one should keep in perspective that model organisms are often used for the very purpose of studying fundamental biological processes such as development, neurobiology, etc., that have proven to be strikingly similar even between animal phyla.

How, then, do we efficiently compare isolates or populations of the worms used? Certainly, a combination of morphology, DNA data, isoenzyme data, karyology and other types of data discussed above would be ideal, but this would be a costly and unwieldy procedure, especially for large sample sizes or smaller laboratories. DNA barcoding using the COI gene may give an estimate of how closely related two isolates are, and may prove to be one important marker for identification and future reference. However, it may be insufficient as the single tool for distinction of "proper" cryptic species from great intra-specific variation, as we still know neither enough about the relationship between COI variation and species delimitation, nor enough about how this variation is reflected in the biological properties of the organisms, i.e. factors that may affect the response of the organism to various experimental situations. Thus, further research, with regard to the variation as such, and also the mechanisms of speciation, involving also its possible elements of polyploidization and hybridization, is greatly needed. Further, the conceptual basis of "a species" must be better considered and stated in all such work. Finally, a search for additional genetic markers, particularly those that will increase phylogenetic resolution, should be encouraged.

ACKNOWLEDGMENTS

We are indebted to Dr. Emilia Rota (Università di Siena, Italy), Dr. Jörg Römbke (ECT Ökotoxikologie GmbH, Flörsheim, Germany) and Ms. Belén Reboreda Rivera (Reboreda s.l., Sandidae Vexetal, Galicia, Spain) for providing specimens of *Enchytraeus* spp.; to Ms. Anna Ansebo (Department of Zoology, University of Gothenburg) for assistance in the field collection and molecular lab work; to Dr. Katherine Beauchamp (National Fish Health Laboratory, U.S. Geological Survey, Kearneysville, WV, USA) for assistance with literature; to the Swedish Research Council (grant # 621-2004-2397 to C.E.), The Swedish Research Council for Environment, Agricultural Sciences and Spatial Planning, FORMAS (grant # 217-2004-1136 to C.E.) and the Swedish Taxonomy Initiative (grant # dha 141/05 1.4 to C.E.) for financial support; and to Sonia Krutzke and Dr. Daniel Shain (both Rutgers University) for excellent comments on the original version of the manuscript.

REFERENCES

AMERICAN SOCIETY FOR TESTING AND MATERIALS. 1995. Standard guide for determination of the bioaccumulation of sediment-associated contaminants by benthic invertebrates. E 1688–95. In *Annual Book of ASTM Standards*, vol. 11.05. Philadelphia: ASTM, pp. 1140–1189.

ANDRÉ, F. 1963. Contribution a l'analyse experimental de la reproduction des lombriciens. Bull. Biol. Fr. Belg. 97, 1–101.

ANLAUF, A. 1994. Some characteristics of genetic variants of *Tubifex tubifex* (Müller, 1774) (Oligochaeta: Tubificidae) in laboratory cultures. Hydrobiologia 278, 1–6.

ANLAUF, A., NEUMANN, D. 1997. The genetic variability of *Tubifex tubifex* (Müller) in 20 populations and its relation to habitat type. Arch. Hydrobiol. 139, 145–162.

ASTRIN, J.J., HUBER, B.A., MISOF, B., KLÜTSCH, C.F.C. 2006. Molecular taxonomy in pholcid spiders (Pholcidae,

Areneae): evaluation of species identification methods using CO1 and 16S rRNA. Zool. Scr. 35, 441–457.

AVEL, M. 1929. Recherches expérimentales sur les caractères sexuels somatiques des Lombriciens. Bull. biol. Fr. Belg. 63, 149–318.

BEAUCHAMP, K.A., KATHMAN, R.D., McDOWELL, T.S., HEDRICK, R.P. 2001. Molecular phylogeny of tubificid oligochaetes with special emphasis on *Tubifex tubifex* (Tubificidae). Mol. Phylogenet. Evol. 19, 216–224.

BEAUCHAMP, K.A., GAY, M., KELLEY, G.O., EL-MATBOULI, M., KATHMAN, R.D., NEHRING, R.B., HEDRICK, R.P. 2002. Prevalence and susceptibility of infection to *Myxobolus cerebralis*, and genetic differences among populations of *Tubifex tubifex*. Diseases Aquat. Org. 51, 113–121.

BELL, G. 1982. *The Masterpiece of Nature: the Evolution and Genetics of Sexuality.* Berkeley, CA: University of California Press.

BELY, A.E., WEISBLAT, D.A. 2006. Lessons from leeches: a call for DNA barcoding in the Lab. Evol. Dev. 8, 491–501.

BOUCHÉ, M.B. 1972. Lombriciens de France, écologie et systematique. INRA. Publ. Ann. Zool. Ecol. Anim. (no hors-series) 72, 1–671.

BOUCHÉ, M.B. 1992. Earthworm species and ekotoxicological studies. In *Ecotoxicology of Earthworms*, edited by P.W. Greig-Smith, H. Becker, P.J. Edwards and F. Heimbach. Andover: Intercept, pp. 20–35.

BRINKHURST, R.O. 1989. A phylogenetic analysis of Lumbriculidae (Annelida, Oligochaeta). Can. J. Zool. 67, 2731–2739.

BRINKHURST, R.O., JAMIESON, B.G.M. 1971. *Aquatic Oligochaeta of the world.* Edinburgh: Oliver and Boyd.

BROCKMEYER, V. 1991. Isozymes and general protein patterns for use in discrimination and identification of *Enchytraeus* species (Annelida, Oligochaeta). Z. Zool. Syst. Evolut.-forsch. 29, 343–361.

BUNDY, J.G., SPURGEON, D.J., SVENDSEN, C., HANKARD, P.K., OSBORN, D., LINDON, J.C., NICHOLSON, J.K. 2002. Earthworm species of the genus *Eisenia* can be phenotypically differentiated by metabolic profiling. FEBS Lett. 521, 115–120.

CAMERON, S., RUBINOFF, D., WILL, K. 2006. Who will actually use DNA barcoding and what will it cost? Syst. Biol. 55, 844–847.

CHAPMAN, P.M., BRINKHURST, R.O. 1987 Hair today, gone tomorrow: induced chaetal changes in tubificid oligochaetes. Hydrobiologia 155, 45–55.

CHRISTENSEN, B. 1980. *Animal Cytogenetics 2: Annelida.* Berlin-Stuttgart: Gebrüder Borntraeger.

COATES, K.A. 1990. Marine Enchytraeidae (Oligochaeta, Annelida) of the Albany area, Western Australia. In *Proceedings of the Third International Marine Biological Workshop: the Marine Flora and Fauna of Albany, Western Australia*, edited by F. Wells, D. Walker, H. Kirkman and R. Lethbridge. Perth: Western Australian Museum, pp. 13–41.

COATES, K.A., STACEY, D. 1993. The marine Enchytraeidae (Oligochaeta, Annelida) of Rottnest Island, Western Australia. In *Proceedings of the Fifth International Marine Biological Workshop: the Marine Flora and Fauna of Rottnest Island, Western Australia*, edited by F. Wells, D. Walker, H. Kirkman and R. Lethbridge. Perth: Western Australian Museum, pp. 391–414.

COATES, K.A., STACEY, D. 1997. Enchytraeidae (Oligochaeta: Annelida) of the lower shore and shallow subtidal of Darwin Harbour, Northern Territory, Australia. In *Proceedings of the Sixth International Marine Biological Workshop: the Marine Flora and Fauna of Darwin Harbour, Northern Territory*, edited by J.R. Hanley, G. Caswell, D. Megirian and H. Larson. Darwin: Museums and Art Galleries of the Northern Territory and the Australian Marine Sciences Association, pp. 67–79.

COOK, D.G. 1971. Family Lumbriculidae. In *Aquatic Oligochaeta of the World*, edited by R.O. Brinkhurst and B. G.M. Jamieson. Edinburgh: Oliver and Boyd, pp. 200–285.

DERMOTT, R., MUNAWAR, M. 1992. A simple and sensitive assay for evaluation of sediment toxicity using *Lumbriculus variegatus* (Müller). Hydrobiologia 235/236, 407–414.

DE WIT, P., ERSÉUS, C. 2007. Seven new species of *Grania* (Annelida: Clitellata: Enchytraeidae) from New Caledonia, South Pacific Ocean. Zootaxa 1426, 27–50.

DOMÍNGUEZ, J., VELANDO, A., FERREIRO, A. 2005. Are *Eisenia fetida* (Savigny 1826) and *Eisenia andrei* Bouché (1972) (Oligochaeta, Lumbricidae) different biological species? Pedobiologia 49, 81–87.

DUCROT, V., USSEGLIO-POLATERA, P., PÉRY, A.R.R., MOUTHON, J., LAFONT, M., ROGER, M.-C., GARRIC, J., FÉRARD, J.-F. 2005. Using aquatic macroinvertebrate species traits to build test batteries for sediment toxicity assessment: accounting for the diversity of potential biological responses to toxicants. Environ. Toxicol. Chem. 24, 2306–2315.

EASTON, E.G. 1983. A guide to the valid names of Lumbricidae (Oligochaeta). In *Earthworm Ecology from Darwin to Vermiculture*, edited by J.E. Satchell. London: Chapman & Hall, pp. 475–485.

EDWARDS, P.J., COULSON, J.M. 1992. Choice of earthworm species for laboratory tests. In *Ecotoxicology of Earthworms*, edited by P.W. Greig-Smith, H. Becker, P.J. Edwards and F. Heimbach. Andover: Intercept, pp. 36–43.

ERSÉUS, C. 1981. Taxonomic revision of the marine genus *Heterodrilus* Pierantoni (Oligochaeta, Tubificidae). Zool. Scr. 10, 111–132.

ERSÉUS, C. 1988. Taxonomic revision of the *Phallodrilus rectisetosus* complex (Oligochaeta: Tubificidae). Proc. Biol. Soc. Wash. 101, 784–793.

ERSÉUS, C. 1997. Marine Tubificidae (Oligochaeta) from the Montebello and Houtman Abrolhos Islands, Western Australia, with descriptions of twenty-three new species. In *The Marine Flora and Fauna of the Houtman*

Abrolhos, Western Australia, edited by F.E. Wells. Perth: Western Australian Museum, pp. 389–458.

ERSÉUS, C. 2003. The gutless Tubificidae (Annelida: Oligochaeta) of the Bahamas. Meiofauna Marina 12, 59–84.

ERSÉUS, C., BERGFELDT, U. 2007. Six new species of the gutless genus *Olavius* (Annelida, Clitellata, Tubificidae) from New Caledonia. Zootaxa 1400, 45–58.

ERSÉUS, C., KVIST, S. 2007. COI variation in Scandinavian marine species of *Tubificoides* (Annelida, Clitellata, Tubificidae). J. Mar. Biolog. Assoc. U.K. 87, 1121–1126.

FITZHUGH, K. 2006. DNA barcoding: an instance of technology-driven science? BioScience 56, 462–463.

GUSTAFSSON, D.R., PRICE, D.A., ERSÉUS, C. In press. Genetic variation in the popular lab worm *Lumbriculus variegatus* (Annelida: Clitellata: Lumbriculidae) reveals cryptic speciation. Mol. Phylogenet. Evol.

HARTZELL, P.L., NGHIEM, J.V., RICHIO, K.J., SHAIN, D.H. 2005. Distribution and phylogeny of glacier ice worms (*Mesenchytraeus solifugus* and *Mesenchytraeus solifugus rainierensis*). Can. J. Zool. 83, 1206–1213.

HEBERT, P.D.N., CYWINSKA, A., BALL, S.L., de WAARD, J.R. 2003a. Biological identifications through DNA barcodes. Proc. R. Soc. Lond. B 270, 313–321.

HEBERT, P.D.N., RATNASINGHAM, S., deWAARD, J.R. 2003b. Barcoding animal life: cytochrome *c* oxidase subunit 1 divergences among closely related species. Proc. R. Soc. Lond. B 270, 96–99.

HENRY, W.B. 1999. Differentiation of allozyme loci to distinguish between two species of Eisenia. MSc Thesis, Mississippi State University.

HOLMQUIST, C. 1983. What is *Tubifex tubifex* (O. F. Müller) (Oligochaeta, Tubificidae)? Zool. Scr. 12, 187–201.

HOLMQUIST, C. 1985. A revision of the genera *Tubifex* Lamarck, *Ilyodrilus* Eisen, and *Potamothrix* Vejdovský & Mrázek (Oligochaeta, Tubificidae), with extensions to some connected genera. Zool. Jb. Syst. 112, 311–366.

JAENIKE, J. 1982. "*Eisenia foetida*" is two biological species. Megadrilogica 4, 6–8.

JOHNSON, N.K., CICERO, C. 2004. New mitochondrial DNA data affirm the importance of Pleistocene speciation in North American birds. Evolution 58, 1122–1130.

KATHMAN, R.D., BRINKHURST, R.O. 1998. *Guide to the Freshwater Oligochaetes of North America*. College Grove, TN: Aquatic Resources Center, pp. 1–264.

KENNEDY, C.R. 1969. The variability of some characters used for species recognition in the genus *Limnodrilus* Claparède (Oligochaeta: Tubificidae). J. Nat. Hist. 3, 53–60.

KOHLEROVA, P., BESCHIN, A., SILEROVA, M., De BAETSELIER, P., BILEJ, M. 2004. Effect of experimental microbial challange on the expression of defense molecules in *Eisenia foetida* earthworm. Dev. Comp. Immunol. 28, 701–711.

KOKTA, C. 1992. A laboratory test on sublethal effects of pesticides on *Eisenia fetida*. In *Ecotoxicology of Earthworms*, edited by P.W. Greig-Smith, H. Becker, P.J. Edwards and F. Heimbach. Andover: Intercept, pp. 213–216.

LADLE, R.J., TODD, P.A. 2006. Sex and sanctuary: how do asexual worms survive the winter? Hydrobiologia 559, 395–399.

LEPPÄNEN, M.T., KUKKONEN, J.V.K. 1998a. Factors affecting feeding rate, reproduction, and growth of an oligochaete *Lumbriculus variegatus* (Müller). Hydrobiologia 377, 183–194.

LEPPÄNEN, M.T., KUKKONEN, J.V.K. 1998b. Relationships between reproduction, sediment type and feeding activity of *Lumbriculus variegatus* (Müller): implications for sediment toxicity testing. Environ. Toxicol. Chem. 17, 2196–2202.

MARTIN, P. 2001. On the origin of the Hirudinea and the demise of the Oligochaeta. Proc. R. Soc. Lond. B 268, 1089–1098.

MARTINEZ, V.G., MENGER, G.J., III, ZORAN, M.J. 2005. Regeneration and asexual reproduction share common molecular changes: upregulation of a neural glycoepitope during morphallaxis in *Lumbriculus*. Mech. Dev. 122, 721–732.

MARTINEZ, V.G., REDDY, P.K., ZORAN, M.J. 2006. Asexual reproduction and segmental regeneration, but not morphallaxis, are inhibited by boric acid in *Lumbriculus variegatus* (Annelida: Clitellata: Lumbriculidae). Hydrobiologia 564, 73–86.

MAYDEN, R.L. 1997. A hierarchy of species concepts: the denouement in the saga of the species problem. In *Species: the Units of Biodiversity*, edited by M.F. Claridge, H.A. Dawah and M.R. Wilson. New York: Chapman & Hall, pp. 381–424.

McELROY, T.C., DIEHL, W.J. 2001. Heterosis in two closely related species of earthworm (*Eisenia fetida* and *E. andrei*). Heredity 87, 598–608.

MEYER, C.P., PAULAY, G. 2005. DNA barcoding: error rates based on comprehensive sampling. PLoS Biology, 3, 2229–2238.

MICHAELSEN, W. 1900. Oligochaeta. Tierreich 10, 1–575.

MILBRINK, G. 1983. An improved environmental index based on the relative abundance of oligochaete species. Hydrobiologia 102, 89–97.

MISOF, B., ANDERSON, C.L., HADRYS, H. 2000. A phylogeny of the damselfly genus *Calopteryx* (Odonata) using mitochondrial 16S rDNA markers. Mol. Phylogenet. Evol. 15, 5–14.

MORITZ, C., CICERO, C. 2004. DNA barcoding: promise and pitfalls. PLoS Biology, 2, 1529–1531.

MRÁZEK, A. 1907. Die Geschlechtsverhältnisse und die Geschlechtsorgane von *Lumbriculus variegatus* Gr. Zool. Jb. Anat. 23, 381–462.

NESEMANN, H., NEUBERT, E. 1999. Annelida, Clitellata: Branchiobdellida, Acanthobdellea, Hirudinea. In *Süsswasserfauna von Mitteleuropa*, 6/2, edited by J. Schwoerbel and P. Zwick. Heidelberg, Berlin: Spektrum Akademischer Verlag, pp. 1–18.

OHNO, S. 1970. *Evolution by Gene Duplication*. New York: Springer.

ØIEN, N., STENERSEN, J. 1984. Esterases of earthworms—III. Electrophoresis reveals that *Eisenia fetida* (Savigny) is two species. Comp. Biochem. Phys. C 78, 277–282.

OTTO, S.P., WHITTON, J. 2000. Polyploid incidence and evolution. Ann. Rev. Genet. 34, 401–437.

PÉREZ-LOSADA, M., EIROA, J., MATO, S., DOMÍNGUEZ, J. 2005. Phylogenetic species delimitation of the earthworms *Eisenia fetida* (Savigny, 1826) and *Eisenia andrei* Bouché, 1972 (Oligochaeta, Lumbricidae) based on mitochondrial and nuclear DNA sequences. Pedobiologia 49, 317–324.

PHIPPS, G.L., ANKLEY, G.T., BENOIT, D.A., MATTSON, V.R. 1993. Use of the aquatic oligochaete *Lumbriculus variegatus* for assessing the toxicity and bioaccumulation of sediment-associated contamitants. Environ. Toxicol. Chem. 12, 269–279.

PODDUBNAJA, T.L., TIMM, T. 1980. Morfo-biologičeskie osobennosti *Tubifex tubifex* (Müll.) (Oligochaeta, Tubificidae) v različnyh presnovodnyh biocenozah [Morphobiological peculiarities of *Tubifex tubifex* (Müll.) (Oligochaeta, Tubificidae) in different freshwater biocenoses]. In *Biologija, morfologija i sistematika vodnyh bezpozvonočnyh*, vol. 41 (44). Leningrad: Nauka, Institut biologii vnutrennih vod, Trudy, pp. 3–16.

REINECKE, A.J. 1992. A review of ecotoxicological test methods using earthworms. In *Ecotoxicology of Earthworms*, edited by P.W. Greig-Smith, H. Becker, P.J. Edwards and F. Heimbach. Andover: Intercept, pp. 7–19.

REINECKE, A.J., VILJOEN, S.A. 1991. A comparison of the biology of *Eisenia fetida* and *Eisenia andrei* (Oligochaeta). Biol. Fertil. Soils 11, 295–300.

REYNOLDSON, T.B., RODRIGUEZ, P., MARTINEZ MADRID, M. 1996. A comparison of reproduction, growth and acute toxicity in two populations of *Tubifex tubifex* (Müller, 1774) from the North American Great Lakes and Northern Spain. Hydrobiologia 334, 199–206.

ROBOTTI, C.A. 1983. Genetic distances among European populations of *Eisenia fetida andrei* and *Eisenia fetida fetida*. Att. Assoc. Genet. Ital. 29, 207–208.

RÖMBKE, J. 1989. *Enchytraeus albidus* (Enchytraeidae, Oligochaeta) as a test organism in terrestrial laboratory systems. Arch. Toxicol. Suppl. 13, 402–405.

RÖMBKE, J., KNACKER, T. 1989. Aquatic toxicity test for enchytraeids. Hydrobiologia 180, 235–242.

ROTA, E. 1995. Italian Enchytraeidae (Oligochaeta). I. Boll. Zool. 62, 183–231.

ROTA, E., ERSÉUS, C. 1996. Six new species of *Grania* (Oligochaeta, Enchytraeidae) from the Ross Sea, Antarctica. Antarct. Sci. 8, 169–183.

ROTA, E., ERSÉUS, C. 2003. New records of *Grania* (Clitellata, Enchytraeidae) in the Northeast Atlantic (from Tromsø to the Canary Islands), with descriptions of seven new species. Sarsia 88, 210–243.

ROTA, E., HEALY, B. 1999. A taxonomic study of some Swedish Enchytraeidae (Oligochaeta), with descriptions of four new species and notes on the genus *Fridericia*. J Nat. Hist. 33, 29–64.

ROTA, E., WANG, H., ERSÉUS, C. 2007. The diverse *Grania* fauna (Clitellata: Enchytraeidae) of the Esperance area, Western Australia, with descriptions of two new species. J. Nat. Hist. 41, 999–1023.

SCHIRMACHER, A., SCHMIDT, H., WESTHEIDE, W. 1998. RADP-PCR investigations on sibling species of terrestrial *Enchytraeus* (Annelida: Oligochaeta). Biochem. Syst. Ecol. 26, 35–44.

SCHLEGEL, M., STEINBRÜCK, G., KRAMER, M., BROCK-MEYER, V. 1991. Restriction fragment patterns as molecular markers for species identification and phylogenetic analysis in the genus *Enchytraeus*. J. Zool. Syst. Evolut.-forsch. 29, 362–372.

SCHMELZ, R.M. 2003. Taxonomy of *Fridericia* (Oligochaeta, Enchytraeidae). Revision of species with morphological and biochemical methods. Abh. Naturwiss. Ver. Hamb. (N. F.) 38, 1–415.

SIDDALL, M.E., APAKUPAKUL, K., BURRESON, E.M., COATES, K.A., ERSÉUS, C., KÄLLERSJÖ, M., GELDER, S.R., TRAPIDO-ROSENTHAL, H. 2001 Validating Livanow's hypothesis: molecular data agree that leeches, branchiobdellidans and *Acanthobdella peledina* are a monophyletic group of oligochaetes. Mol. Phylogenet. Evol. 21, 346–351.

SIDDALL, M.E., TRONTELJ, P., UTEVSKY, S.Y., NKAMANY, M., MACDONALD, K.S., SIDDALL, M.E., III 2007. Diverse molecular data demonstrate that commercially available medicinal leeches are not *Hirudo medicinalis*. Proc. Biol. Sci. 274, 1481–1487.

SIMS, R.W. 1983. The scientific names of earthworms. In *Earthworm Ecology from Darwin to Vermiculture*, edited by J.E. Satchell. London: Chapman & Hall, pp. 467–474.

SIMS, R.W., GERARD, B.M. 1985. Earthworms. In *Synopses of the British Fauna (new series)*, vol. 31, edited by D.M. Kermack and R.S.K. Barnes. London: The Linnean Society of London and The Estuarine and Brackish-water Sciences Association, E.J. Brill / Dr. W. Backhuys, pp. 1–171.

SONG, K., LU, P., TANG, K., OSBORN, T.C. 1995. Rapid genome change in synthetic polyploids of *Brassica* and its implications for polyploid evolution. Proc. Natl. Acad. Sci. U.S.A. 92, 7719–7723.

STEBBINS, G.L. 1950. *Variation and Evolution in Plants*. New York: Columbia University Press.

STURMBAUER, C., OPADIYA, G.B., NIEDERSTÄTTER, H., RIEDMANN, A., DALLINGER, R. 1999. Mitochondrial DNA reveals cryptic oligochaete species differing in cadmium resistance. Mol. Biol. Evol. 16, 967–974.

TIMM, T. 2005. Some misused names in aquatic Oligochaeta. In *Advances in Earthworm Taxonomy II (Annelida: Oligochaeta)*, Proceedings of the 2nd International Oligochaeta Taxonomy Meeting dedicated to Victor Pop, Cluj-Napoca, Romania, September 4–8, 2005, edited by V.V. Pop, A.A. Pop. Cluj-Napoca: Cluj University Press, pp. 53–60.

TRONTELJ, P., UTEVSKY, S.Y. 2005. Celebrity with a neglected taxonomy: molecular systematics of the medicinal leech (genus *Hirudo*). Mol. Phylogenet. Evol. 34, 616–624.

TRONTELJ, P., SOTLER, M., VEROVNIK, R. 2004. Genetic differentiation between two species of the medicinal leech, *Hirudo medicinalis* and the neglected *H. verbana*, based on random-amplified polymorphic DNA. Parasitol. Res. 94, 118–124.

VELTZ, I., ARSAC, F., BIAGIANTI-RISBOURG, S., HABETS, F., LECHENAULT, H., VERNET, G. 1996. Effects of platinum (Pt^{4+}) on *Lumbriculus variegatus* Müller (Annelida, Oligochaeta): acute toxicity and bioaccumulation. Arch. Environ. Contam. Toxicol. 31, 63–67.

VIKTOROV, A.G. 1997. Diversity of polyploid races in the family Lumbricidae. Soil Biol. Biochem. 29, 217–221.

VOGLER, A.P., MONAGHAN, M.T. 2007. Recent advances in DNA taxonomy. J. Zool. Syst. Evol. Res. 1, 1–10.

WANG, H.Z., ERSÉUS, C. 2004. New species of *Doliodrilus* and other Limnodriloidinae (Oligochaeta, Tubificidae) from Hainan and other parts of the northwest Pacific Ocean. J. Nat. Hist. 38, 269–299.

WESTHEIDE, W., BETHKE-BEILFUSS, D. 1991. The sublethal enchytraeid test system: guidelines and some results. In *Modern Ecology: Basic and Applied Aspects*, edited by G. Esser and D. Overdieck. Amsterdam: Elsevier.

WESTHEIDE, W., BROCKMEYER, V. 1992. Suggestions for an index of enchytraeid species (Oligochaeta) based on general protein patterns. Z. Zool. Syst. Evol.-forsch. 30, 89–99.

WESTHEIDE, W., GRAEFE, U. 1992. Two new terrestrial *Enchytraeus* species (Oligochaeta, Annelida). J. Nat. Hist. 26, 479–488.

WESTHEIDE, W., BETHKE-BEILFUSS, D., HAGENS, M., BROCKMEYER, V. 1987. Enchytraeiden als Testorganismen–Voraussetzungen für ein terrestrisches Testverfahren und Testergebnisse. Verh. Ges. Ökol. 17, 793–798.

WESTHEIDE, W., PURSCHKE, G., MIDDENDORF, K. 1991. Spermatozoal ultrastructure of the taxon *Enchytraeus* (Annelida, Oligochaeta) and its significance for species discrimination and identification. Z. Zool. Syst. Evol.-forsch. 29, 323–342.

WILL, K.W., RUBINOFF, D. 2004. Myth of the molecule: DNA barcodes for species cannot replace morphology for identification and classification. Cladistics 20, 47–55.

WILL, K.W., MISHLER, B.D., WHEELER, Q.D. 2005. The perils of DNA barcoding and the need for integrative taxonomy. Syst. Biol. 54, 844–851.

YUAN, X., CAO, C., SHAN, Y., ZHAO, Z., CHEN, J., CONG, Y. 2006. Expression and characterization of earthworm *Eisenia foetida* Lumbrokinase-3 in *Pichia pastoris*. Prep. Biochem. Biotechnol. 36, 273–279.

Chapter 4

Annelid Life Cycle Cultures

Donald J. Reish and Bruno Pernet

Department of Biological Sciences, California State University, Long Beach, CA

4.1 INTRODUCTION

Animals are extraordinarily diverse in their morphology, physiology, reproduction, development and life cycles. Most research on fundamental biological processes in animals has focused on a few "model" species, which together comprise a fraction of the ~1.2 million described animal species (Bolker 1995; Gest 1995). As noted by physiologist August Krogh, most general problems in biology have one or several ideal study systems (Krogh 1929). Undoubtedly, the current small library of model species does not include all, or even most, of these ideal systems. Additionally, with methodological advances over the past few decades—especially in phylogenetics, developmental biology, genomics and proteomics—biology has become more explicitly comparative and integrative. Data on patterns and processes in a broad range of nonmodel system taxa are required to address questions in such a framework.

One problem constraining the use of novel species as model systems, or even obtaining certain kinds of data from nonmodel organisms, is a lack of information on how to raise them in the laboratory. Some species of interest are difficult to obtain from the field for a variety of logistical reasons; others are rare and cannot be sustainably harvested and thus must be cultured in the laboratory for long-term study. Studies of many biological processes (e.g. development) often require continuous laboratory cultures of the target species. However, comparative biologists are often uninterested or unable to spend time developing culture techniques for nonmodel species.

In this review, we provide information for culturing members of the diverse array of segmented, coelomate worms that belong to the phylum Annelida. Annelids are members of the large clade Lophotrochozoa that includes flatworms, rotifers, mollusks and several other phyla (Adoutte et al. 2001). Two major lineages comprise the annelids, the polychaetes and the clitellates (i.e. oligochaetes including leeches; Martin 2001; Siddall et al. 2001). Several clitellate annelids are already well-known as model systems in development (*Helobdella* spp.; Weisblat and Huang 2001; Bely and Weisblat 2006) and neurobiology (*Hirudo medicinalis*; Kristan et al. 2005), and several oligochaetes are important models in aquatic toxicology, development and regeneration (e.g. Bely and Wray 2001). Polychaetes, however, have only recently been appreciated as important potential model

Annelids in Modern Biology, Edited by Daniel H. Shain
Copyright © 2009 John Wiley & Sons, Inc.

systems for understanding such phenomena as the evolution of early development and segmentation (e.g. Fischer 1999; Tessmar-Raible and Arendt 2003; Seaver and Kaneshige 2006).

We provide a summary of annelid species that have been successfully cultured through at least one complete life cycle under laboratory conditions, and provide references to literature describing culture techniques for those species. Many of these techniques can easily be adapted for culturing previously uncultured species. Our aim is to facilitate the use of additional annelid species in basic research by making information on their life cycles and culture requirements easily accessible.

4.2   CRITERIA FOR THE SELECTION OF SPECIES

Most species included in this chapter have been cultured in a laboratory through at least one complete sexual life cycle, usually from adult to the maturity of the F_1 adult, resulting in a net increase in the laboratory population. Many species have been successfully fertilized in the laboratory but are excluded because they have not been reared through a complete life cycle (e.g. *Spirobranchus corniculatus*, *Galeolaria caespitosa*, *Harmothoë imbricata*, *Glycera dibranchiata*). Data on many such polychaete species have been summarized by Schroeder and Hermans (1975), Fischer and Pfannenstiel (1984), Wilson (1991) and Giangrande (1997). We also include a few species in which increases in laboratory populations are more usually derived from asexual reproductive processes. Indeed, clonal propagation has some useful properties (e.g. genetic identity) that can be advantageous for particular questions.

Below, we summarize basic biological data and culturing information on annelid species known to meet the above criteria. Species are organized by taxonomic families [polychaete family usage is as presented in Beesley et al. (2000) and in Rouse and Pleijel (2001)], and families are listed alphabetically within annelid classes. If two or more scientific names have been used for a taxon, the common accepted name is given with the other names enclosed in parentheses.

4.3   SUMMARY OF CULTURE TECHNIQUES

Methods used to culture annelids depend on the biology of the target species (especially body size, habitat use and feeding and reproductive biology). Methods developed for one species can often be adapted for species with similar lifestyles and life histories. In this section, we present a few general rules for the maintenance of aquatic annelid cultures (and other aquatic invertebrates). More extensive discussions of culture techniques can be found in a variety of sources, including Galtsoff et al. (1937), Smith and Chanley (1975), Harrison and Cowden (1982), the National Research Council Committee on Marine Invertebrates (1981) and Strathmann (1987).

4.3.1   Containers

Container size depends upon the size of the worms in question, their behavior, type of substratum they require and the density of the laboratory population. Most species do best with a good deal of "free" space in the culture container. Aquaria of various sizes, finger bowls, petri dishes and tissue culture plates have all been used successfully for culturing annelids. Tube-dwelling species (e.g. spionids) can be removed from their natural tubes and introduced into short pieces of glass or plastic tubing to facilitate microscopic examination of reproductive condition.

4.3.2 Water and Aeration (for Aquatic Species)

Aquatic annelids should be reared in water that approximates salinities in the natural habitat. For estuarine or marine species, clean or filtered (0.45 μm) natural seawater (diluted to appropriate salinities using distilled water if needed) usually works well. Artificial seawater can be used if natural seawater is not available, or in cases where the chemical composition of the medium must be controlled. Several commercially available artificial seawaters (e.g. Instant Ocean) are available, and recipes for artificial seawater can be found in the literature (e.g. Bidwell and Spotte 1985; Strathmann 1987). For freshwater species, filtered (0.22 μm) pond water often works well. If the biomass of the culture is small relative to the water surface area, aeration is often not necessary; in many cases, however, supplementary aeration using aquarium air pumps is generally helpful. Water can be constantly recirculated through filtration units, or simply changed every few days or weeks. Evaporation can be slowed with a tank cover.

4.3.3 Food

Appropriate food sources for larval or adult aquatic annelids obviously depend on their feeding biology. A wide variety of natural and commercially available foods have been used to feed annelids in laboratory culture, and many more inexpensive and convenient potential annelid foods can be purchased in a grocery store or pet shop (e.g. rabbit chow). Using dead, dried or prepared food eliminates the need to culture or purchase one species to feed another; however, care must be taken not to overfeed because it may lead to water fouling as excess food decomposes. As a general rule, underfeeding cultured annelids is better than overfeeding them.

Suspension-feeding larvae and adults often do well on diets of various easily cultured phytoplankton (e.g. in marine systems, these might include *Dunaliella* sp., *Isochrysis* sp. and *Rhodomonas* sp.; see Strathmann 1987 for sources and culture techniques). Some suspension feeders may do better on larger prey, e.g. rotifers or brine shrimp larvae. Herbivorous or omnivorous adults have been cultured using fresh, frozen or dried and subsequently rehydrated seaweeds (e.g. *Enteromorpha* sp., *Ulva* sp. or finely branched red algae). Many species will accept and thrive on foods that they never encounter naturally in the field. For example, frozen spinach can be scraped or ground into small particles and used as a seaweed substitute for herbivores or omnivores (e.g. dorvilleids in the genus *Ophyrotrocha*), or as a food for deposit feeders. Deposit feeders (e.g. many spionids) have frequently been fed powdered fish food (e.g. TetraMin®). Rabbit food pellets are convenient for feeding a variety of species of omnivores, especially nereidids. Various oligochaetes have been reared successfully on foods that include cow manure, other animal wastes, sludge, shredded paper and plant material.

Some species have fairly specific diets, and these are often more difficult to culture. For example, members of the syllid genus *Autolytus* (and perhaps related species in the subfamily Autolytinae) require living hydrozoans as food, such as members of the genus *Eirene* (Schiedges 1979). Some species of leeches feed on such prey as aquatic snails (e.g.

Helobdella sp.), oligochaetes or insect larvae, but adults of some species (e.g. *H. medicinalis*) require vertebrate blood as food.

4.4 LIFE CYCLE CULTURES OF POLYCHAETA

4.4.1 Amphinomidae

Eurythoe complanata: common in warmer seas of the world and often found under rocks or rubble at low tide; never been reared through a complete sexual cycle, but adults easily regenerate both anterior and posterior ends. Raising genetically identical adults by cutting source animals into many pieces is relatively easy; each fragment will subsequently form a complete worm. **Maximum adult size**: 20 cm; **sexual mode**: separate sexes; **spawning method**: eggs released into water column; **adult fate**: survive, repeat spawning; **ovum size**: flattened disks ~90 μm in diameter; **number of ova**: >10^4; **larval habitat**: planktonic; **larval nutrition**: plankton, larvae have not been reared to metamorphosis; **time to maturity**: unknown; **adult nutrition**: omnivore, has been fed *Enteromorpha* sp.; **adult longevity in culture**: unknown; **location of culture**: California; **use of culture**: reproductive biology, regeneration, toxicity testing; **references**: Kudenov (1974), Reish et al. (1989).

4.4.2 Capitellidae

Capitella *sp. I*: The genus *Capitella* comprises multiple species that are morphologically similar as adults, but are differentiated reproductively (e.g. in ovum size, sperm characteristics, spawning habits and development). These taxa are often lumped together as "*Capitella capitata*" in ecological studies. Taxa whose larvae have short planktonic periods tend to be easier to culture than those with longer larval planktonic periods. **Maximum adult size**: 10 cm; **sexual mode**: separate sexes; **spawning method**: eggs fertilized within the maternal tube; **adult fate**: survive, repeat spawning; **ovum size**: 180×260 μm; **number of ova**: 30–400; **larval habitat**: maternal tube, followed by brief planktonic period; **larval nutrition**: yolk; **time to maturity**: 30–40 days; **adult nutrition**: deposit feeder; has been fed *Enteromorpha* sp., powdered fish flakes, other foods; **adult longevity in culture**: ~6 months; **number of laboratory life cycles**: >20; **location of culture**: New Jersey, California, elsewhere; **use of culture**: feeding biology, development, toxicity testing; **references**: Grassle and Grassle (1976), Reish (1980), Eckelbarger and Grassle (1987), Qian and Chia (1992).

4.4.3 Cirratulidae

Cirriformia spirobrancha: Populations can be collected from intertidal mussel beds on the west coast of the USA, especially where sediment accumulates. Worms should be maintained in fine sand to prevent branchia from becoming entangled. **Maximum adult size**: 16 cm; **sexual mode**: separate sexes; **spawning method**: gametes released on surface of sediment; **adult fate**: survive, repeat spawning; **ovum size**: 100–120 μm; **number of ova**: 10^5; **larval habitat**: plankton; **larval nutrition**: yolk; **time to maturity**: ~1 year; **adult nutrition**: deposit feeder, has been fed *Enteromorpha* sp. and Biorell$^©$ powder at 4:1 ratio (mixed with sufficient seawater to form a fluid mixture); **adult longevity in culture**: 17

months; **number of laboratory life cycles**: 1; **location of culture**: California; **use of culture**: toxicity testing; **reference**: Reish (1980).

4.4.4 Ctenodrilidae

Ctenodrilus serratus: cosmopolitan species collected from fouling communities growing on floating docks, or from material accumulated in containers suspended in bays or harbors (Reish 1961); reproduces both asexually and sexually; asexual propagation by fission or budding can lead to rapid increases in laboratory population size. **Maximum adult size**: 3 mm; **sexual mode**: hermaphroditic; **spawning method**: viviparous; **adult fate**: survive, repeat reproduction; **ovum size**: unknown; **number of ova**: unknown, but very few; **larval habitat**: no larval stage; **larval nutrition**: no larval stage; **time to maturity**: 14–21 days; **adult nutrition**: deposit feeder, has been fed powdered dried algae and various commercial animal foods as a powder; **adult longevity in culture**: unknown; **number of laboratory life cycles**: >10; **location of culture**: California; **use of culture**: toxicity testing; **reference**: Reish (1980).

4.4.5 Dorvilleidae

Dinophilus gyrociliatus: This minute species is cosmopolitan in distribution and is easily cultured. Curiously, females produce eggs of two sizes: small eggs develop into males, and large eggs into females. *D. gyrociliatus* can be collected by placing material scraped from harbor fouling communities in a jar of unaerated seawater; the next day, worms will have migrated to the edge of the jar just below the air–seawater interface. **Maximum adult size**: 1 mm; **sexual mode**: separate sexes; **spawning method**: deposit egg capsules; **adult fate**: females survive, repeat reproduction, but males die after first reproduction; **ovum size**: 40–80 μm; **number of ova**: three to five per capsule (two to four female eggs, one to two male eggs; ratios vary); **larval habitat**: no larval stage; **larval nutrition**: no larval stage; **time to maturity**: 5–7 days; **adult nutrition**: omnivore; has been fed frozen spinach particles and powdered fish food; **adult longevity in culture**: unknown; **number of laboratory life cycles**: >20; **location of culture**: California; **use of culture**: toxicity testing; **reference**: Reish (1980).

Dorvillea (Schistomeringos) longicornis [=Stauronereis ruldolphi; Dorvillea articulata]: widely distributed species known by several names. Richards (1967) collected specimens from mussel beds attached to floating boat docks. **Maximum adult size**: 5 cm; **sexual mode**: separate sexes; **spawning method**: swarming and release gametes in water column; **adult fate**: survive, repeat spawning; **ovum size**: 100 μm; **larval habitat**: plankton **larval nutrition**: yolk; **time to maturity**: unknown; **adult nutrition**: omnivore, has been fed *Enteromorpha* sp.; **adult longevity in culture**: 1 year; **number of laboratory life cycles**: three; **location of culture**: California; **use of culture**: life history biology, toxicity testing; **reference**: Richards (1967).

Ophryotrocha diadema: Many species of the genus *Ophrotrocha* have been cultured through several complete life cycles; *O. diadema* is described herein as an example for the genus. This species is known only from southern California where it has been collected from mussel beds collected from boat docks. Since the species measures only 4.5 mm in length, it is obtained in the same manner as described for *D. gyrociliatus* above. Åkesson (1976) described this species from collections transported to Sweden; it has since been distributed to other laboratories. **Maximum adult size:** 4.6 mm: **sexual mode:** hermaphroditic; **spawning method:** fertilized eggs laid in a cocoon; **adult fate:** survives and can reproduce again; **ovum size:** oblong 180×200 μm; **number of ova:** ~28 per capsule; **larval habitat:** within capsule; **larval nutrition:** yolk; **time to maturity:** 4 weeks; **adult food:** fragmented spinach, *Enteromorpha* sp., powdered fish flakes; **longevity in culture:** 7 weeks; **number of laboratory life cycles:** many; **location of culture:** Sweden, other laboratories; **use of culture:** genetics, toxicity; **references:** Åkesson (1976, 1982), Reish (1980).

4.4.6 Nereididae

Neanthes arenaceodentata [=Neanthes caudata, Neanthes acuminata, Nereis crigognatha]: widely distributed species known by various names, but reproductive habits are similar; found in intertidal and subtidal sediments in estuarine waters. **Maximum adult size**: 8 cm; **sexual mode**: separate sexes; **spawning method**: within tube; **adult fate**: female dies, male can reproduce again; **ovum size**: 500–600 μm; **number of ova**: 150–1,000; **larval habitat**: within tube; **larval nutrition**: yolk; **time to maturity**: 3–4 months; **adult food**: omnivore; has been fed *Enteromorpha* and various commercial animal foods; **adult longevity in culture**: female, 4–6 months; male, 1 year; **number of laboratory life cycles**: >100; **location of culture**: California; **use of culture**: behavior, toxicity testing; **references**: Reish (1980), Reish et al. (2005).

Neanthes limnicola: known from the central Oregon coast to San Luis Obispo County, CA, where it has been collected in brackish water and a freshwater lake. **Maximum adult size**: 5 cm; **sexual mode**: hermaphroditic; **spawning method**: self-fertilizing; fertilization occurs within coelom; **adult fate**: female may survive; **ovum size**: 120–210 μm; **number of ova**: few; **larval habitat**: parental coelom, emerge as juveniles; **larval nutrition**: yolk; **time to maturity**: 6–13 months; **adult nutrition**: omnivore, has been fed brine shrimp; **adult longevity in culture**: ~1 year; **number of laboratory life cycles**: two; **location of culture**: California; **use of culture**: reproductive biology; **references**: Smith (1950), Fong and Pearse (1992).

Neanthes succinea: cosmopolitan species found in sediments of estuarine waters. **Maximum adult size**: 5 cm; **sexual mode**: separate sexes; **spawning method**: adults swarm, gametes released in the water column; **adult fate**: both die; **ovum size**: 140–150 μm; **number of ova**: many; **larval habitat**: plankton; **larval nutrition**: yolk; **time to maturity**: unknown; **adult nutrition**: omnivore; has been fed *Enteromorpha* and various commercial animal foods; **adult longevity in culture**: <1 year; **number of laboratory life cycles**: two; **location of culture**: California; **use of culture**: life history biology; **reference**: D.J. Reish, unpublished observations.

Nereis grubei: collected in mats of intertidal seaweed from California to Baja California, Mexico. **Maximum adult size**: 8 cm; **sexual mode**: separate sexes; **spawning method**: adults swarm, gametes released in water column; **adult fate**: both die; **ovum size**: 162–380 μm; **number of ova**: many; **larval habitat**: plankton; **larval nutrition**: yolk; **time to maturity**: unknown; **adult nutrition**: omnivore, has been fed green algae; **adult longevity in culture**: 4–6 months; **number of laboratory life cycles**: three; **location of culture**: California; **use of culture**: toxicity testing; **references**: Reish (1954), Reish et al. (2005), Schroeder (1968).

Platynereis dumerilii: widely distributed in European waters, where it is collected from intertidal seaweeds. This species is now an important lophotrochozoan model system, especially in evolution of development studies. **Maximum adult size**: 6 cm; **sexual mode**: separate sexes; **spawning method**: adults swarm, gametes released in water column; **adult fate**: both die; **ovum size**: 130–160 μm; **number of ova**: many; **larval habitat**: plankton; **larval nutrition**: yolk; **time to maturity**: 5–18 months; **adult nutrition**: omnivore; has been fed spinach, algae and fish food; **adult longevity in culture**: 5–18 months; **number of life cycles in laboratory**: many; **location of culture**: Germany, France; **use of culture**: developmental biology, endocrinology; **references**: Hauenschild (1951, 1970), Hauenschild and Fischer (1969), Hofmann and Schiedges (1984).

Platynereis massiliensis: European species collected from intertidal green algae from the Mediterranean coast of France; similar to *P. dumerilii* in adult morphology, but differs in various aspects of reproduction. **Maximum adult size**: 6 cm; **sexual mode**: hermaphroditic ; **spawning method**: within tube; **adult fate**: may survive; **ovum size**: 340 μm; **number of ova**: <1,000; **larval habitat**: within tube; **larval nutrition**: yolk; **time to maturity**: few months; **adult nutrition**: omnivore; has been fed spinach, algae and fish food; **adult longevity in culture**: unknown; **number of life cycles in laboratory**: several; **location of culture**: Germany; **use of culture**: developmental biology; **references**: Hauenschild (1951), Schneider et al. (1992).

4.4.7 Polynoidae

Halosydna johnsoni: This scale worm occurs among mussels growing on floating docks and in intertidal offshore mussel beds in southern California. A related species, *Halosydna brevisetosa*, is common along the rest of the west coast of the USA in similar habitats. Because of their fairly long, feeding planktonic larval stages, polynoids are often fairly difficult to rear successfully in the laboratory (B. Pernet, pers. obs.). **Maximum adult size**: 6 cm; **sexual mode**: separate sexes; **spawning method**: gametes released into water column; **adult fate**: both live; **ovum size**: 88–94 μm; **number of ova**: 10^5; **larval habitat**: planktonic; **larval nutrition**: plankton, has been fed *Dunaliella* sp.; **time to maturity**: 9 months; **adult food**: carnivore, has been fed frozen brine shrimp; **adult longevity in culture**: unknown; **number of laboratory life cycles**: one; **location of culture**: California; **use of culture**: life history biology, toxicity testing; **reference**: Reish (1980).

4.4.8 Serpulidae

Hydroides elegans: very common fouling species in warmer seas, and often occurs in huge aggregations on floating docks or on the hulls of ships; an important model system for the study of metamorphosis. **Maximum adult size**: 2 cm; **sexual mode**: protandrous hermaphrodite; **spawning method**: eggs released into water column; **adult fate**: survive, repeat spawning; **ovum size**: 45 μm; **number of ova**: unknown; **larval habitat**: planktonic; **larval nutrition**: phytoplankton, has been fed *Isochrysis* sp.; **time to maturity**: 16–28 days; **adult nutrition**: phytoplankton, has been fed *Isochrysis* sp.; **adult longevity in culture**: unknown; **number of laboratory life cycles**: ~10; **location of culture**: Hawaii; **use of culture**: life history evolution, physiological and larval ecology, development; **references**: Qiu and Qian (1998), Miles et al. (2007).

Janua (Dexiospira) brasiliensis: This small polychaete is abundant in fouling communities in southern California. Related spirorbid serpulids occur throughout the world ocean and can likely be cultured successfully using similar techniques. **Maximum adult size**: 3 mm; **sexual mode**: hermaphroditic; **spawning method**: not observed but presumed to be external; fertilized ova are moved to an opercular brood chamber; **adult fate**: survive, repeat spawning; **ovum size**: 130 μm; **number of ova**: 60; **larval habitat**: brooded in opercular chamber, then brief planktonic period; **larval nutrition**: yolk; **time to maturity**: 30 days; **adult nutrition**: suspension feeder, has been fed *Dunaliella* sp.; **adult longevity in culture**: 3 months; **number of laboratory life cycles**: three; **location of culture**: California; **use of culture**: life history, settlement behavior; **references**: Reish (1980), Kirchman et al. (1982).

Pomatoceros triqueter: occurs in European waters where it attaches to rocks, shells or larger seaweeds in the intertidal zone. **Maximum adult size**: 3 cm; **sexual mode**: protandric hermaphrodite; **spawning method**: gametes released into water column; **adult fate**: survive, repeat spawning; **ovum size**: 80 μm; **number of ova**: 83–350; **larval habitat**: planktonic; **larval nutrition**: phytoplankton; **time to maturity**: unknown; **adult nutrition**: suspension feeder; has been fed phytoplankton, e.g. *Dunaliella* sp.; **adult longevity in culture**: up to 4 years; **number of laboratory life cycles**: unknown; **location of culture**: Norway; **use of culture**: inheritance of tentacle color; **reference**: Fjøyn and Gjøen (1950).

Salmacina (or Filograna) spp.: The systematics of these tiny serpulids is not well studied. Adults reproduce asexually by paratomy, forming clonal aggregations of individuals that live in calcareous tubes. Aggregations of adults can be obtained from individual adults raised in the laboratory. Most adults in laboratory-reared aggregations result from asexual reproduction, but some may be derived from sexually produced larvae. **Maximum adult size**: 1 cm; **sexual mode**: hermaphroditic; **spawning method**: eggs retained in parental tube; **adult fate**: survive, repeat spawning; **ovum size**: ~150 μm; **number of ova**: unknown, but <10; **larval habitat**: parental tube, followed by a brief planktonic period; **larval nutrition**: yolk; **time to maturity**: unknown, but <2 months; **adult nutrition**: suspension feeder; has been fed phytoplankton, e.g. *Isochrysis* sp.; **adult longevity in culture**: unknown; **location of culture**: Florida; **use of culture**: tube deposition; **reference**: Pernet (2001).

4.4.9 Spionidae

Boccardia proboscidea: Three reproductive types of females occur, varying in the number of "nurse eggs" they provide developing young as nutritional sources and in the type of larvae that emerge from capsules. **Maximum adult size**: 3 cm; **sexual mode**: hermaphroditic; **spawning method**: within capsules attached to the inner wall of the maternal tube; **adult fate**: survive, repeat spawning; **ovum size**: 92–109 μm; **number of ova**: 1,600–2,300; **larval habitat**: brooded, then planktonic (if no or few nurse eggs) or hatch as juvenile (if many nurse eggs); **larval nutrition**: phytoplankton (if no or few nurse eggs) or yolk (if many nurse eggs); **time to maturity**: 4 months; **adult nutrition**: suspension/deposit feeder; has been fed phytoplankton, commercial fish food and *Artemia nauplii*; **adult longevity in culture**: unknown; **number of laboratory life cycles**: >3; **use of cultures**: life history biology, evolution of development; **references**: Gibson (1997), Gibson and Gibson (2004), F. Oyarzun (pers. comm.).

Malacoceros (Scolelepis) fuliginosus: relatively well studied, broadcast-spawning European spionid. **Maximum adult size**: 2 cm; **sexual mode**: separate sexes; **spawning method**: gametes released into water column; **adult fate**: survive, repeat spawning; **ovum size**: ~150 μm; **larval habitat**: planktonic; **larval nutrition**: phytoplankton; **time to maturity**: 39–157 days; **adult nutrition**: suspension / deposit feeder, has been fed phytoplankton or powdered fish food; **adult longevity in culture**: ~1 year; **number of laboratory life cycles**: unknown; **location of culture**: France; **use of culture**: life history biology; **references**: Guerin (1973, 1987).

Polydora ciliata: This European species has been the focus of many studies of adult and larval biology. **Maximum adult size**: 2 cm; **sexual mode**: separate sexes; **spawning method**: within capsules attached to the inner wall of the maternal tube; **adult fate**: survive, repeat spawning; **ovum size**: ~120 μm; **number of ova**: unknown; **larval habitat**: brooded in parental tube, then planktonic for 16–28 days; **larval nutrition**: phytoplankton; **time to maturity**: 12–14 weeks; **adult food**: suspension / deposit feeder, has been fed phytoplankton; **adult longevity in culture**: 1–2 years; **number of laboratory life cycles**: 11; **location of culture**: Germany; **use of culture**: life history biology; **references**: Thorson (1946), Anger et al. (1986).

Polydora cornuta: Formerly known as *Polydora ligni, P. cornuta* is widespread in both Atlantic and Pacific coastal waters. **Maximum adult size**: 3 cm; **sexual mode**: separate sexes; **spawning method**: within capsules attached to the inner wall of the maternal tube; **adult fate**: survive, repeat spawning; **ovum size**: 88×97 μm; **number of ova**: unknown; **larval habitat**: brooded, then planktonic for 16 or more days; **larval nutrition**: phytoplankton; **adult food**: suspension / deposit feeder; has been fed phytoplankton, powdered *Enteromorpha* and commercial fish food; **adult longevity in culture**: 1–2 years; **number of laboratory life cycles**: 16; **location of culture**: Florida, Germany; **use of culture**: life history biology; **references**: Rice (1975), Reish (1980), Anger et al. (1986).

Pygospio elegans: reproduces both asexually (by fission) and sexually. Substantial evidence exists that variation among populations in the presence of nurse eggs leads to variation in whether or not larvae have a planktonic feeding period. **Maximum adult size**: 1 cm; **sexual mode**: separate sexes; **spawning method**: within capsules attached to the inner wall of the maternal tube; **adult fate**: survive, repeat spawning; **ovum size**: ~100 μm; **number of ova**: unknown; **larval habitat**: brooded, then planktonic for 26–60 days (if no nurse eggs are present) or hatch as juvenile (if nurse eggs are present); **larval nutrition**: phytoplankton (if no nurse eggs) or yolk (if nurse eggs are present); **adult food**: suspension / deposit feeder, has been fed phytoplankton; **adult longevity in culture**: 9–52 months; **number of laboratory life cycles**: nine; **location of culture**: Germany; **use of cultures**: life history biology; **references**: Thorson (1946), Anger et al. (1986), Morgan et al. (1999).

Streblospio benedicti: Individual females produce either small eggs that develop into obligately plankton feeding larvae or large eggs that develop into facultatively feeding larvae. **Maximum adult size**: 2 cm; **sexual mode**: separate sexes; **spawning method**: within dorsal brood chamber, **adult fate**: ≥6 broods per lifetime; **ovum size**: 60–70 μm (small egg form) or 100–200 μm (large egg form); **larval habitat**: brooded, then planktonic for 2–7 weeks (small egg form) or ≤8 days (large egg form); **larval nutrition**: phytoplankton (small egg form), yolk±plankton (large egg form); **time to maturity**: 9–14 weeks; **adult nutrition**: deposit feeder; has been fed phytoplankton and commercial fish food; **adult longevity in culture**: 30–75 weeks; **number of laboratory life cycles**: >10;

location of cultures: California, North Carolina, Florida; use of cultures: life history biology, quantitative genetics; references: Levin et al. (1991), Levin and Bridges (1994), Pernet and McArthur (2006).

4.4.10 Syllidae

Autolytus prolifera and *Autolytus brachycephalus*: These two species are similar in biology; Schiedges (1979)—the source of much of the data below—studied both species as well as their hybrids. **Maximum adult size**: unknown; **sexual mode**: separate sexes; **spawning method**: planktonic stolons mate, with subsequent brood care by female stolon; **adult fate**: atokous adults live; stolons die after spawning; **ovum size**: unknown; **number of ova**: unknown, but presumed to be ~50 per stolon; **larval habitat**: brooded by female stolon, then brief planktonic period; **larval nutrition**: yolk; **time to maturity**: ~80 days; **adult nutrition**: predator; has been fed living hydrozoan, *Eirene viridus*; **adult longevity in culture**: ~1 year; **location of culture**: Germany; **use of culture**: life history biology; **reference**: Schiedges (1979).

Typosyllis pulchra: This species is collected from clumps of mussel beds (*Mytilus californiianus*) occurring on the rocky shores of the Pacific Coast. **Maximum adult size**: unknown; **sexual mode**: separate sexes; **spawning method**: eggs released into water column by swimming stolons; **adult fate**: survive, repeat stolon formation and spawning; **ovum size**: not reported, but ~80 μm; **number of ova**: unknown; **larval habitat**: planktonic; **larval nutrition**: yolk; **time to maturity**: 6–9 months; **adult nutrition**: predatory; has been fed diversity of foods, e.g. *A. nauplii*; **adult longevity in culture**: unknown; **number of laboratory life cycles**: unknown; **location of culture**: Washington; **use of culture**: development; **reference**: Heacox (1980).

4.5 LIFE CYCLE CULTURES OF OLIGOCHAETA

All oligochaetes are hermaphroditic, and transfer sperm via "copulation" (though they have no intromittent organs). All deposit their zygotes in capsules that are usually released into the sediment or attached to vegetation. Embryos undergo direct development and hatch as juveniles. The reproduction of oligochaetes has recently been reviewed by Jamieson and Ferraguti (2006).

4.5.1 Enchytraeidae

Enchytraeus albidus: This species (probably a complex of closely related species since morphs with different spermatotheca are known) is widely distributed in moist soils close to freshwater and saltwater bodies, especially at sites rich in organic matter (e.g. seaweed on the beach, compost, etc.). Other *Enchytraeus* species (e.g. *Enchytraeus crypticus, Enchytraeus luxuriosus*) have also been cultured in the laboratory, using similar methods. Detailed information on the culture of other species of the genus can be found in Bell (1958), Bougouenee (1987) and Römbke and Moser (2002). **Maximum adult size**: 1.5–4.0 cm; **sexual mode**: hermaphroditic; **spawning method**: copulation followed by cocoon deposition; **adult fate**: survive, reproduce again; **ovum size**: cocoon 1.0–1.2 mm; **number of ova**: 7–10 per cocoon; **embryonic nutrition**: yolk; **time to maturity**: 33–74 days depending upon temperature; **adult nutrition**: yeast, white bread soaked with mild, boiled potatoes, corn flour; **adult longevity in culture**: ~1 year; **number of laboratory life cycles**: many; **location of culture**: widespread; **use of culture**: developmental biology, toxicity testing, aquarium food; **references**: Bell (1958), Bougouenee (1987), Römbke and Moser (2002).

4.5.2 Lumbricidae

Eisenia fetida: widespread oligochaete often found in manure piles. **Maximum adult size**: unknown; **sexual mode**: hermaphroditic; **spawning method**: copulation followed by cocoon deposition; **adult fate**: survive, reproduce again; **ovum size**: not stated; **number of ova**: 1–11 eggs deposited per cocoon; **embryonic nutrition**: yolk; **time to maturity**: ~2 months, depending on temperature; **adult nutrition**: manure, shredded paper, plant matter; **adult longevity in culture**: 4–5 years; **number of laboratory life cycles**: many; **location of culture**: widespread; **use of culture**: sludge decomposition, fish bait, toxicity testing, soil enrichment; **references**: Reynolds (1977), Jefferies and Audeley (1988).

4.5.3 Naididae

Paranais litoralis: This small oligochaete occurs in marine soft sediments. Though it can reproduce sexually, it commonly reproduces asexually in cultures by paratomous fission, with development of a new individual preceding actual fission. **Maximum adult size**: 1.5 cm; **sexual mode**: hermaphroditic; **spawning method**: copulation followed by cocoon deposition; **adult fate**: divides clonally many times; **ovum size**: not stated; **number of ova**: not stated; **embryonic nutrition**: yolk; **time to maturity**: unknown for embryos, but products of paratomy can divide again within 4 days; **adult nutrition**: sediment from field sites; **adult longevity in culture**: many years; **number of laboratory life cycles**: many; **location of culture**: New York; **use of culture**: developmental biology, life history biology; **references**: Martinez (1996), Bely (1999).

Pristina leidyi: small, freshwater oligochaete widespread in North America; not known to reproduce sexually in laboratory cultures; rather, it commonly reproduces asexually by paratomous fission, with development of a new individual preceding actual fission. **Maximum adult size**: 1 cm; **sexual mode**: hermaphroditic; **spawning method**: copulation followed by cocoon deposition; **adult fate**: divides clonally many times; **ovum size**: not stated; **number of ova**: not stated; **embryonic nutrition**: yolk; **time to maturity**: unknown for embryos, but products of paratomy can divide again within 1 week; **adult nutrition**: green algae, diatoms; **adult longevity in culture**: many years; **number of laboratory life cycles**: many; **location of culture**: Maryland; **use of culture**: developmental biology; **reference**: Bely and Wray (2001).

4.5.4 Tubificidae

Tubifex tubifex and cryptic species: widely distributed in freshwater ponds, lakes and rivers throughout the world, especially in sediments rich in organic matter. Other *Tubifex* species (e.g. *Tubifex hattai*) have also been cultured in the laboratory using similar methods. Detailed information on the culture of *Tubifex* spp. can be found in Shimizu (1982). **Maximum adult size**: 2–3 cm; **sexual mode**: hermaphroditic; **spawning method**: copulation followed by cocoon deposition; **adult fate**: survive, reproduce again; **ovum size**: not stated; **number of ova**: 2–20 eggs per cocoon; **embryonic nutrition**: yolk; **time to maturity**: ~1.5–3.0 months, depending on temperature; **adult nutrition**: yeast, boiled potatoes, corn flour; **adult longevity in culture**: 2.0–2.5 years; **number of laboratory life cycles**: many; **location of culture**: widespread; **use of culture**: developmental biology, toxicity testing, aquarium food; **references**: Kosiorek (1976). Brinkhurst and Cook (1980), Shimizu (1982).

Lumbriculus variegatus: widely distributed species found in silty and sandy sediments from shallow water to depths of 60 m in reservoirs, rivers, lakes, ponds and marshes throughout the USA and Europe. In addition to reproducing sexually, it reproduces asexually by architomy (i.e. transverse fission preceding any differentiation of clonal offspring). **Maximum adult size**: 9 cm; **sexual mode**: hermaphroditic; **spawning method**: copulation followed by cocoon deposition; **adult fate**: divides clonally many times; **ovum size**: not stated; **number of ova**: not stated; **embryonic nutrition**: yolk; **time to maturity**: unknown for embryos, but products of architomy can divide again within 10–14 days; **adult nutrition**: presoaked paper towel with salmon starter food, maple or poplar leaves; **adult longevity in culture**: many years; **number of laboratory life cycles**: many; **location of culture**: widespread; **use of culture**: toxicity testing, regeneration; **reference**: Phipps et al. (1993).

4.6 LIFE CYCLE CULTURES OF HIRUDINEA (LEECHES)

Like oligochaetes, all hirudineans are hermaphroditic and transfer sperm via "copulation" (though they have no intromittent organs). All deposit their zygotes in capsules that are usually released into the sediment or attached to vegetation, though in glossiphoniids, cocoons are sometimes brooded on the parent's ventral body wall. Embryos undergo direct development and hatch as juveniles. The reproduction of hirudineans has recently been reviewed by Siddall et al. (2006).

4.6.1 Glossiphoniidae

Helobdella triserialis and related *Helobdella* sp.: small leech common in freshwater habitats worldwide; used extensively in studies of annelid development, with recent molecular studies suggesting that numerous species have been confused with *H. triserialis* (Bely and Weisblat 2006). Relatively detailed information on the culture of adults can be found in Fernandez and Olea (1982). **Maximum adult size**: 2.5 cm; **sexual mode**: hermaphroditic; **spawning method**: copulation followed by cocoon deposition; **adult fate**: survive, reproduce again; **ovum size**: ~400 μm; **number of ova**: ~50 eggs (~10 eggs in each of four to five cocoons); **embryonic nutrition**: yolk; **time to maturity**: ~6 weeks, depending on temperature; **adult nutrition**: live freshwater snails; **adult longevity in culture**: ~1 year; **number of laboratory life cycles**: many; **location of culture**: widespread; **use of culture**: developmental biology; **references**: Wedeen et al. (1989), Shankland et al. (1992), Bely and Weisblat (2006).

Theromyzon tessulatum and other *Theromyzon* sp.: cold water leeches sangivorous to waterfowl (primarily duck); found sporatically in ponds throughout northwestern regions of North America (e.g. northern California, Montana, Alaska, Alberta) and Europe. **Maximum adult size**: 6 cm; **sexual mode**: hermaphroditic; **spawning method**: copulation followed by cocoon deposition; **adult fate**: lay one clutch of eggs, then die; **ovum size**:

~800 μm; **number of ova**: ~300 eggs (~50 eggs in each of three to eight cocoons); **embryonic nutrition**: yolk; **time to maturity**: ~1 year; **adult nutrition**: can be fed with live ducks (feeding from nares); **adult longevity in culture**: ~1 year; **number of laboratory life cycles**: three to four; **location of culture**: Calgary, Chile, New Jersey; **use of culture**: developmental biology, cocoon studies; **references**: Torrence and Stuart (1986), Mason et al. (2005).

4.6.2 Hirudinidae

Hirudo medicinalis: The medicinal leech is native to Europe, but was introduced to the USA in the 19th century; widely used medicinally, and also as a model system in neurobiology (Kristan et al. 2005). Leeches identified as *H. medicinalis* have recently been shown to comprise several different species (Siddall et al. 2007). **Maximum adult size**: varies according to feeding state, up to 10 cm long; **sexual mode**: hermaphroditic; **spawning method**: copulation followed by cocoon deposition; **adult fate**: survive, reproduce again; **ovum size**: not stated; **number of ova**: 3–36 per capsule; **embryonic nutrition**: yolk in cocoon fluid; **time to maturity**: 12–18 months; **adult nutrition**: blood from living amphibians and mammals; **adult longevity in culture**: several years; **number of laboratory life cycles**: many; **location of culture**: widespread; **use of culture**: neurobiology, medicinal; **reference**: Sawyer (1980).

ACKNOWLEDGMENTS

We thank Jörg Römbke and David Weisblat for providing information on oligochaetes and leeches.

REFERENCES

ADOUTTE, A., BALAVOINE, G., LARTILLOT, N., LESPINET, O., PRUD'HOMME, B., de ROSA, R. 2001. The new animal phylogeny: reliability and implications. Proc. Nat. Acad. Sci. U.S.A. 97, 4453–4456.

ÅKESSON, B. 1976. Morphology and life cycle of *Ophryotrocha diadema* a new polychaete species from California. Ophelia 15, 23–35.

ÅKESSON, B. 1982. The life table study on three genetic strains of *Ophyrotrocha diadema* (Polychaeta: Dorvillidae). Int. J. Invertebr. Reprod. 5, 59–69.

ANGER, K., ANGER, V., HAGMEIER, E. 1986. Laboratory studies on larval growth of *Polydora ligni, Polydora ciliata,* and *Pygospio elegans.* Helgol. Wiss. Meeresunters. 40, 377–395.

BEESLEY, P.L., ROSS, G.J.B., GLASBY, C.J. 2000. *Polychaetes and allies: the Southern synthesis. Fauna of Australia, Volume 4A. Polychaeta, Myzostomida, Pogonophora, Echiura, Sipuncula.* Melbourne: CSIRO Publishing.

BELL, A.W. 1958. The anatomy of the oligochaete *Enchytraeus albidus,* with a key to the species of the genus *Enchytraeus.* Am. Mus. Novit. No. 1902, 13 p.

BELY, A.E. 1999. Decoupling of fission and regenerative capabilities in an asexual oligochaete. Hydrobiologia 406, 243–251.

BELY, A.E., WEISBLAT, D.A. 2006. Lessons from leeches: a call for DNA barcoding in the lab. Evol. Dev. 8, 491–501.

BELY, A.E., WRAY, G. 2001. Evolution of regeneration and fission in annelids: insights from *engrailed* and *orthodenticle*-class gene expression. Development 128, 2781–2791.

BIDWELL, J.P., SPOTTE, S. 1985. *Artifical Seawaters: Formulas and Methods.* Boston: Bartlett Publications.

BOLKER, J.A. 1995. Model systems in developmental biology. Bioessays 17, 451–455.

BOUGOUENEE, V. 1987. L'elevage en masse d'Enchytraeidae (Annelida, Oligochaeta): Etude, bibliographique, mise aux point experimental et tests dans l'alimentation des poissons. Thesis, University of Toulouse III.

BRINKHURST, R.O., COOK, D.G., eds. 1980. *Aquatic Oligochaete Biology.* New York: Plenum Press.

ECKELBARGER, K.J., GRASSLE, J.P. 1987. Interspecific variation in genital spines, sperm and larval morphology in six sibling species of *Capitella.* Bull. Biol. Soc. Wash. 7, 62–76.

FERNANDEZ, J., OLEA, N. 1982. Embryonic development of glossiphonid leeches. In *Developmental Biology of Freshwater Invertebrates,* edited by F.W. Harrison and R.R. Cowden. New York: Alan Liss, pp. 317–361.

FISCHER, A. 1999. Reproductive and developmental phenomena in annelids: a source of exemplary research problems. In *Reproductive Strategies and Developmental Patterns in Annelids*, edited by A.W.C. Dorresteijn and W. Westheide. Dordrecht: Kluwer Academic Publishers.

FISCHER, A., PFANNENSTIEL, H.-D., eds. 1984. Polychaete reproduction. Fortschritte Zool. 29, 1–351.

FJØYN, B., GJØEN, I. 1950. Studies on the serpulid *Pomatoceros triqueter* L. I. Observations on the life history. Nytt Magasin Zool. 2, 73–84.

FONG, P.P., PEARSE, J.S. 1992. Photoperiodic regulation of parturition in the self-fertilizing viviparous polychaete *Neanthes limnicola* from central California. Mar. Biol. 112, 81–89.

GALTSOFF, P.S., LUTZ, F.E., WELCH, P.S., NEEDHAM, J.G. 1937. *Culture Methods for Invertebrate Animals*. Ithaca, N.Y.: Comstock Publishing.

GEST, H. 1995. *Arabidopsis* to zebrafish: a commentary on "Rosetta Stone" model systems in the biological sciences. Perspect. Biol. Med. 37, 77–85.

GIANGRANDE, A. 1997. Polychaete reproductive patterns, life histories: an overview. Oceanogr. Mar. Biol. Annu. Rev. 35, 323–386.

GIBSON, G.D. 1997. Variable development in the spionid *Boccardia proboscidea* is linked to nurse egg production and larval trophic mode. Invertebr. Biol. 116, 213–226.

GIBSON, G.D., GIBSON, A.J.F. 2004. Heterochrony and the evolution of poecilogony: generating larval diversity. Evolution 58, 2704–2717.

GRASSLE, J.P., GRASSLE, J.F. 1976. Sibling species in the marine pollution indicator *Capitella capitata*. Science 192, 567–569.

GUERIN, J.-P. 1973. Premieres donnees sur la longevite, le rythme de ponte et la fecondite de *Scolelepis* cf. *fuliginosa* en elevage. Mar. Biol. 19, 27–40.

GUERIN J.-P. 1987. Elevage de Spionides en cycle complet. I. Technique d'elevage de l'une des trois formes de *Malacoceros fuliginosus* des Cotes Francaises. Aquaculture 62, 215–257.

HARRISON, R.W., COWDEN, R.C., eds. 1982. *Developmental Biology of Freshwater Invertebrates*. New York: Liss.

HAUENSCHILD, C. 1951. Nachweis der sog.atoken Geschlechtsform des Polychaeten *Platynereis dumerilii* als eigene Art Grund von Zuchtversuchen. Zool. Jb. Physiol. 63, 197–127.

HAUENSCHILD, C. 1970. Die Zucht von niederen marinen Wirbellosen und ihre Anwendung in der experimentallen Zoologie. Helgol. Wiss. Meeresunters. 20, 249–263.

HAUENSCHILD, C., FISCHER, A. 1969. Platynereis dumerilii. Grosses Zool. Prakt. 10b, 1–55.

HEACOX, A.E. 1980. Reproduction and larval development of *Typosyllis pulchra*. Pac. Sci. 34, 245–259.

HOFMANN, K.D., SCHIEDGES, I. 1984. Brain hormone levels and feed-back regulation during gametogenesis, metamorphosis and regeneration in Platynereis dumerilii—an experimental approach. Forschritte Zool. 29, 73–79.

JAMIESON, B.G.M., FERRAGUTI, M. 2006. Non-leech Clitellata. In *Reproductive Biology and Phylogeny of Annelida*, edited by G. Rouse and F. Pleijel. Enfield, New Hampshire: Science Publishers, pp. 235–392.

JEFFERIES, I.R., AUDELEY, E. 1988. A population model for the earthworm *Eisenia fetida*. In *Earthworms in Waste and Environmental Management*, edited by C.A. Edwards and E.F. Neuhauser. The Hague, The Netherlands: SPB Academic Publishing, pp. 119–134.

KIRCHMAN, D., GRAHAM, S., REISH, D.J., MITCHELL, R. 1982. Bacteria induce settlement and metamorphosis of *Janua (Dexiospira) brasiliensis* Grube (Polychaete: Spirorbidae). J. Exp. Mar. Biol. Ecol. 56, 153–163.

KOSIOREK, D. 1976. Development cycle of *Tubifex tubifex* Müller in experimental studies. Polish Arch. Hydrobiol. 21, 411–422.

KRISTAN, W.B.Jr, ., CALABRESE, R.L., FRIESEN, W.O. 2005. Neuronal control of leech behavior. Prog. Neurobiol. 76, 279–327.

KROGH, A. 1929. Progress in physiology. Am. J. Physiol. 90, 243–251.

KUDENOV, J.D. 1974. The reproductive biology of *Eurythoe complanata*. PhD Dissertation, University of Arizona.

LEVIN, L.A., ZHU, J., CREED, E. 1991. The genetic basis of life-history characters in a polychaete exhibiting planktotrophy and lecithotrophy. Evolution 45, 380–397.

LEVIN, L.A., BRIDGES, T.S. 1994. Control and consequences of alternative developmental modes in a poecilogonous polychaete. Am. Zool. 34, 323–332.

MARTIN, P. 2001. On the origin of the Hirudinea and the demise of the Oligochaeta. Proc. R. Soc. Lond. B 268, 1089–1098.

MARTINEZ, D.E. 1996. Rejuvenation of the disposable soma: repeated injury extends lifespan in an asexual annelid. Exp. Gerontol. 31, 699–704.

MASON, T.A., SAYERS, C.W., PAULSON, T.L., COLEMAN, J., SHAIN, D.H. 2005. Cocoon deposition and hatching in the aquatic leech, *Theromyzon tessulatum* (Annelida, Hirudinea, Glossiphoniidae). Am. Midl. Nat. 154, 78–87.

MILES, C.M., HADFIELD, M.G., WAYNE, M.L. 2007. Estimates of heritability for egg size in the serpulid polychaete *Hydroides elegans*. Mar. Ecol. Prog. Ser. 340, 155–162.

MORGAN, T.S., ROGERS, A.D., PATERSON, G.L.J., HAWKINS, L.E., SHEADER, M. 1999. Evidence for poecilogony in *Pygospio elegans*. Mar. Ecol. Prog. Ser. 178, 121–132.

National Research Council Committee on Marine Invertebrates. 1981. *Laboratory Animal Management: Marine Invertebrates*. Washington, D.C.: National Academy Press.

PERNET, B. 2001. Escape hatches for the clonal offspring of serpulid polychaetes. Biol. Bull. 200, 107–117.

PERNET, B., MCARTHUR, L. 2006. Feeding by larvae of two different developmental modes in *Streblospio benedicti*. Mar. Biol. 149, 803–811.

PHIPPS, G.L., ANKLEY, G.T., BENOFF, D.A., MATTSON, V.R. 1993. Use of the aquatic oligochaete *Lumbriculus variegatus* for assessing the toxicity and bioaccumulation of sediment-associated contaminants. Environ. Toxicol. Chem. 12, 269–279.

QIAN, P.-Y., CHIA, F.-S. 1992. Effects of diet type on the demographics of *Capitella* sp.: lecithotrophic development vs. planktotrophic development. J. Exp. Mar. Biol. Ecol. 157, 159–179.

QIU, J.-W., QIAN, P.-Y. 1998. Combined effects of salinity and temperature on juvenile survival, growth and maturation in the polychaete *Hydroides elegans*. Mar. Ecol. Prog. Ser. 168, 127–134.

REISH, D.J. 1954. *The Life History and Ecology of the Polychaetous Annelid Nereis Grubei (Kinberg)*. Occasional Paper No. 14. Allan Hancock Foundation, University of Southern California, 46 p.

REISH, D.J. 1961. The use of the sediment bottle collector for monitoring polluted marine waters. Calif. Fish Game 47, 261–272.

REISH, D.J. 1980. *The Effect of Different Pollutants on Ecologically Important Polychaete Worms*. EPA-6003-80-053. United States Environmental Protection Agency, Research and Development. Narraganset, 138 p.

REISH, D.J., ASATO, S.L., LEMAY, J.A. 1989. The effect of cadmium and DDT on the regeneration in the amphinomid polychaete Eurythoe complanata. VII Simposium Internacional Biologia Marina, La Paz, Baja California Sur, pp. 107–110.

REISH, D.J., CHAPMAN, P.M., INGERSOLL, C.G., MOORE, D.W., PESCH, C.E. 2005. Annelids. In *Standard Methods for Examination of Water and Wastewater*, edited by A.D. Eaton, L.S. Clesceri, E.W. Rice and A.E. Greenberg. Washington, DC: American Public Health Association, American Water Works Association and Water Environment Federation, pp. 8–80.

REYNOLDS, J.W. 1977. *The earthworms (Lumbricidae and Sparganophilidae) of Ontario*. Toronto: Life Sciences Miscellaneous Publications, Royal Ontario Museum.

RICE, S.A. 1975. The life history of Polydora ligni, including a summary of reproduction in the family Spionidae. Master's Thesis, California State University, Long Beach.

RICHARDS, T.L. 1967. Reproduction and development of the polychaete Stauronereis rudolphi, including a summary of development in the superfamily Eunicea. Mar. Biol. 1, 124–133.

RÖMBKE, J., MOSER, T. 2002. Validating the enchytraeid reproduction test: organisation and results of an international ring test. Chemosphere 46, 1117–1140.

ROUSE, G.W., PLEIJEL, F. 2001. *Polychaetes*. Oxford: Oxford University Press.

SAWYER, R.T. 1980. *Leech Biology and Behaviour*, vols. 1 and 2. Oxford: Oxford University Press.

SCHIEDGES, K.-L. 1979. Reproductive biology and ontogenesis in the polychaete genus Autolytus (Annelida: Syllidae): observations on laboratory-cultured individuals. Mar. Biol. 54, 239–250.

SCHNEIDER, S., FISCHER, A., DORRESTEIJN, A.W.C. 1992 A morphometric comparison of dissimilar early develop-ment in sibling species of *Platynereis*. Roux's Arch. Dev. Biol. 201, 243–256.

SCHROEDER, P.C. 1968. On the life history of *Nereis grubei*, a polychaete annelid from California. Pac. Sci. 22, 476–481.

SCHROEDER, P.C., HERMANS, C.O. 1975. Annelida: Polychaete. In *Reproduction of Marine Invertebrates*, vol. 3, edited by A.C. Giese and J.S. Pearse. New York: Academic Press, 1–213.

SEAVER, E., KANESHIGE, L. 2006. Expression of "segmentation" genes during larval and juvenile development in the polychaetes *Capitella* sp. I and *Hydroides elegans*. Dev. Biol. 289, 179–194.

SHANKLAND, M., BISSEN, S.T., WEISBLAT, D.A. 1992. Description of the Californian leech *Helobdella robusta*, sp. nov., and comparison with *Helobdella triserialis* on the basis of morphology, embryology, and experimental breeding. Can. J. Zool. 70, 1258–1263.

SHIMIZU, T. 1982. Development in the freshwater oligochaete *Tubifex*. In *Developmental Biology of Freshwater Invertebrates*, edited by F.W. Harrison and R.R. Cowden. New York: Alan Liss, pp. 283–316.

SIDDALL, M.E., APAKUPAKUL, K., BURRESON, E.M., COATES, K.A., ERSÉUS, C., KÄLLERSJÖ, M., GELDER, S.R., TRAPIDO-ROSENTHAL, H. 2001. Validating Livanow's hypothesis: molecular data agree that leeches, branchiobdellidans and Acanthobdella peledina are a monophyletic group of oligochaetes. Mol. Phylogen. Evol. 21, 346–351.

SIDDALL, M.E., BELY, A.E., BORDA, E. 2006. Hirudinida. In *Reproductive Biology and Phylogeny of Annelida*, edited by G. Rouse and F. Pleijel. Enfield, New Hampshire: Science Publishers, pp. 393–429.

SIDDALL, M.E., TRONTELJ, P., UTEVSKY, S.Y., NKAMANY, M., MACDONALD, K.S., III 2007. Diverse molecular data demonstrate that commercially available medicinal leeches are not *Hirudo medicinalis*. Proc. R. Soc. B Biol. Sci. 274, 1481–1487.

SMITH, R.I. 1950. Embryonic development in the viviparous nereid polychaete *Neanthes lighti* Hartman. J. Morph. 87, 417–455.

SMITH, W.L., CHANLEY, M.H. 1975. *Culture of Marine Invertebrate Animals*. New York: Plenum Press.

STRATHMANN, M.F. 1987. *Reproduction and Development of Marine Invertebrates of the Northern Pacific Coast*. Seattle: University of Washington Press.

TESSMAR-RAIBLE, K., ARENDT, D. 2003. Emerging systems: between vertebrates and arthropods, the Lophotrochozoa. Curr. Opin. Genet. Dev. 13, 331–340.

THORSON, G. 1946. Reproduction and larval development of Danish marine bottom invertebrates, with special reference to the planktonic larvae in the sound (Oresund). Meddr. Danm. Fisk.-og Havunders., ser. Plankton 4, 1–523.

TORRENCE, S.A., STUART, D.K. 1986. Gangliogenesis in leech embryos: migration of neural precursor cells. J. Neurosci. 6, 2736–2746.

Wedeen, C.J., Price, D.J., Weisblat, D.A. 1989. Analysis of the life cycle, genome and homeobox genes of the leech, *Helobdella triserialis*. In *The Cellular and Molecular Biology of Pattern Formation*, edited by D.L. Stocum and T.L. Karr. Oxford: Oxford University Press, pp. 145–167.

Weisblat, D.A., Huang, F.Z. 2001. An overview of glossiphoniid leech development. Can. J. Zool. 79, 218–232.

Wilson, W.H. 1991. Sexual reproductive modes in polychaetes: classification and diversity. Bull. Mar. Sci. 48, 500–516.

Part II

Evolution and Development

Chapter 5

Annelids in Evolutionary Developmental Biology

Dian-Han Kuo

Department of Molecular and Cell Biology, University of California, Berkeley, CA

5.1 INTRODUCTION

Annelid embryos were among the first to be used in embryological studies in the 19th century (Wilson 1898), but gradually slipped out of favor in the following century. Ironically, the classical cell lineage papers from the turn of the 19th century (Wilson 1892; Mead 1897; Child 1900; Treadwell 1901) remain among the most precise, detailed and accurate descriptive literature concerning annelid development. On one hand, this witnesses the timeless quality of these works. On the other hand, it suggests the slow progress in the past 100 years compared to the remarkable advances in "new" model systems such as *Drosophila melanogaster* and *Caenorhabditis elegans*. One may hope, however, that this period of slow progress is now coming to an end. The resurgence of interest in evolutionary developmental biology ("evo-devo") has set the stage for the renaissance of annelid development. Now, a number of research groups around the world are investigating annelid development with new molecular and experimental methods that have not been available until recently (Shankland and Seaver 2000; Seaver 2003; Tessmar-Raible and Arendt 2003).

Here, I will not attempt to produce a detailed review of annelid development, but rather look toward the future, with the goal of pointing out how the annelid community can make connections to the larger community of developmental and evolutionary biologists. For a review of the comparative embryology of annelids, I refer the reader to the classical reviews by Anderson (1966a, b, 1973) and a recent review by Dohle (1999). Even though our view of annelid evolution has changed greatly in the past decade, not much new data on annelid comparative embryology have emerged since these articles. Recent progress in annelid embryology is primarily from clitellate annelids, specifically glossiphoniid leeches such as *Helobdella* spp. (Shankland and Savage 1997; Weisblat and Huang 2001) and tubificid oligochaetes, almost exclusively *Tubifex tubifex* (Shimizu et al. 1998; Shimizu 1999; Shimizu and Nakamoto 2001).

5.2 EVO-DEVO TODAY

Central questions in modern evo-devo are how development evolves and how changes in development relate to morphological or behavioral evolution. Even though development has long been recognized as a player in evolution, these questions could not be addressed at the mechanistic level until tools for probing and manipulating gene expression in embryos became available.

The surge of interest in evo-devo was created by converging results from molecular developmental genetic studies of various model systems over the past two decades. As obvious as they seem today, the greatest surprises from this field were that developmental regulatory genes can be identified as distinct from "housekeeping genes," and that homologous genes in distantly related species can be involved in similar developmental processes. Two classic examples are the presence of *Hox* gene clusters in both vertebrates and arthropods and their similar roles in axial patterning (Carroll 1995), and the expression of *Pax-6* gene homologues in fly and vertebrate eyes (Halder et al. 1995). These results suggest "deep homology" between very different body plans and cell types, and open up the possibility of reconstructing the long extinct common ancestors of metazoans using comparative developmental genetics. These and similar findings attracted many molecular developmental geneticists to consider evo-devo questions.

The underlying tenet of evo-devo is that development is a key reagent in evolutionary processes. To practitioners of this field, the egg clearly came before the chicken, i.e. phenotypic changes are caused by developmental changes. Phenotypic change is viewed as a result of developmental change, and developmental change as a consequence of change in the genetic program. In this regard, by considering the evolution of developmental mechanisms at a molecular level, evo-devo can possibly connect to population genetics and can lead to a synthesis of evolutionary theories.

5.3 EVO-DEVO AS COMPARATIVE BIOLOGY

Evo-devo is obsessed with ancestors. Too often, evo-devo biologists appoint a species as a so-called "basal taxon," as a stand-in for the true ancestor. This approach has several logical flaws that arise from the misinterpretation of phylogenetic trees (Crisp and Cook 2005; Jenner 2006; Krell and Cranston 2006). The misconception of a basal taxon is usually due to ignorance of the tree's temporal dimension. A tree represents a genealogy. Ancestors that gave rise to distinct lines of descendants are represented by nodes from which descendant lineages branch, and the terminal nodes represent living taxa. Strictly speaking, basal taxa are those that represent basal nodes, and thus the term refers to ancestors that no longer exist. Absent a time machine, embryological experimentation can only be conducted in living species. Hence, the species selected for developmental study has to be from a terminal node, not a basal one. Moreover, all terminal nodes (modern species) arising from any given basal node have experienced an equal amount of evolutionary divergence from the last common ancestor. Finally, the observation that only one or a few extant species represent a given branch of the tree does not prove that they are more closely related to the nodal ancestor than those on a more speciose branch, because there may have been many intervening branches that died out (Crisp and Cook 2005). Thus, while it is conceivable that the developmental processes in some species may be relatively unchanged from those in the ancestor, this is a question to be determined and not assumed.

To reconstruct evolutionary histories of developmental characters, one has to characterize homologous developmental characters in different taxa, map these character states by phylogenetics and analyze the data statistically (Harvey and Pagel 1991; Harvey and Purvis 1991; Cunningham et al. 1998; Pagel 1999). Thus, evo-devo issues should be addressed in a two-step process. First, developmental mechanisms in representative species of diverse taxa are studied with both descriptive and experimental methods, then the evolutionary history of developmental change is inferred by interpreting the similarities and differences among these extant representatives in the context of their phylogenetic relationships.

Ideally, the ancestral character state should be deduced from a reasonably large size of species samples. A small dataset is prone to sampling bias and makes the analysis unreliable. Due to the difficulties in studying molecular and cellular mechanisms of development, considerable work is needed to acquire each data point, and thus expecting that a sufficiently large dataset can be generated in a single investigation is unrealistic. Furthermore, the choice of representative species is often dictated by practical considerations such as experimental tractability or the ease of collecting suitable materials. In using data from such representatives as the basis of comparison between higher taxonomical groups, such as that between annelids, arthropods and vertebrates or that between polychaetes and clitellates, one must be aware that generalizing data acquired from any one species to represent an entire taxon is risky.

These problems should not prevent us from performing comparative analyses of development, however, nor do they make a good comparative analysis of annelid development impossible. Rather, bearing in mind the constraints imposed by the limited dataset presently available, we should be appropriately circumspect in interpreting our findings, anticipating that the cumulative efforts by the evo-devo community in studying annelid development and phylogeny will permit more sophisticated analyses and more definitive conclusions.

5.4 WHY ANNELID DEVELOPMENT IS INTERESTING FOR METAZOAN EVO-DEVO BIOLOGISTS

In the 1990s, the emergence of molecular phylogenies dramatically changed our view of metazoan evolution. In premolecular days, metazoan phylogeny was built by grouping taxa based on arbitrarily selected characters, such as number of germ layers, coelom and segmentation. A typical metazoan phylogeny from that era is shown in Fig. 5.1A. A fundamental problem with this approach is that it automatically builds assumptions about the evolution of morphological and thus developmental changes, which is precisely what we're hoping to determine by evo-devo. Molecular phylogenies are useful because they avoid this circular logic, and while they support certain components of the traditional trees, they contradict many others (Fig. 5.1B). For example, protostome and deuterostome groupings survive, but placement of individual taxa among, within and between these two major divisions has been greatly altered.

While the resolution of branching order at the phylum level is not yet ideal, three major clusters can be recognized unambiguously: Deuterostoma, Ecdysozoa and Lophotrochozoa (Aguinaldo et al. 1997). Superphylum Deuterostoma includes the traditionally recognized deuterostomes, such as chordates, echinoderms and hemichordates, but lophophorates have all been moved to Protostomia. The protostome branch, moreover, has undergone a major shake-up of its own, and now comprises two new superphyla,

Ecdysozoa and Lophotrochozoa. Ecdysozoa comprises animals that molt between their life cycle stages (e.g. arthropods and nematodes), while Lophotrochozoa includes annelids, mollusks, flatworms and the lophophorate phyla that were traditionally grouped with other deuterostomes.

How does this altered view of metazoan phylogeny influence the role of annelids in understanding metazoan evolution? In the past, a simple view would depict the two power-ful molecular genetic model species, *D. melanogaster* and *C. elegans*, as representing protostomes and a deeper pseudocoelomate branch of metazoans, and the various verte-brate models (along with echinoderms) as representative deuterostomes. But placing these model species on the new metazoan phylogeny shows an obvious gap in terms of how the three superphyla are represented. *D. melanogaster* and *C. elegans* are now viewed as morphologically divergent ecdysozoans, while Lophotrochozoa is practically an orphan taxon by comparison (Fig. 5.1).

The uneven distribution of model species in the phylogenetic tree as we now under-stand it has two consequences. One is that some major and long-standing questions in evo-devo have not yet been illuminated by insights emerging from the molecular mecha-nisms of development. For example, spiral cleavage only occurs in embryos of a lophotro-chozoan subgroup (Spiralia) and nowhere else. Studies of cleavage mechanisms in *D. melanogaster* or *C. elegans* certainly are not informative regarding the mechanism of spiral cleavage, though they can shed light on the basic mechanisms of cell division that are shared by all metazoan species. The other consideration is that, given the millions of years of independent evolution represented in the Lophotrochozoa, interesting and even useful innovations at cellular and molecular levels may yet be discovered. In the following sec-tions, some unique questions in annelid evo-devo are highlighted.

5.5 CASE STUDY 1: SEGMENTATION

The most prominent feature of the annelid body plan is segmentation. Segmentation refers to multiple, serially iterated anatomical units comprising multiple tissue layers arranged

Figure 5.1 Impacts of molecular phylogeny on the value of model organisms for understanding animal body plan evolution. (A) Generalized examples of classical metazoan phylogeny based on morphological characters. The tree topology reflects general assumptions regarding metazoan body plan evolution. At the base of the tree, the clade "Eumetazoa" signifies the formation of germ layer organization. The clade "Bilateria" signifies the bilaterally symmetric body plan and the appearance of mesoderm. The clades "Acoelomate" and "Pseudocoelomate" signify the gradual evolution of coelome organization. The deuterostome and protostome split reflects a change in blastopore fate. The clade "Articulata" reflects the assumption that segmentation in annelids and arthropods had a common evolutionary origin. Based on this tree, the "big three" of developmental genetics—*Caenorhabditis elegans, Drosophila melanogaster* and *Mus musculus*—were thought to cover a great majority of metazoan evolutionary history by representing Pseudocoelomata, Protostomia and Deuterostoma. (B) Generalized example of modern metazoan phylogeny based on molecular data. Compared to the classical scenario, some presumably primitive groups, such as flatworm (Platyhelminthes) and Nematoda, are in fact more derived. The "simplicity" in the body plan organization of these groups may reflect a secondary loss of complex structures. Another major change is the grouping within protostomes such that annelids and arthropods are no longer sister taxa. This change has forced evo-devo biologists to rethink the origins of segmentation in Metazoa (see text for details). In this new phylogeny, a major clade, Lophotrochozoa, is not represented by any model species in developmental genetics. Note that the phylum Platyhelminthes is polyphyletic with acoel branching off near the root of the Bilateria clade, and the rest as members of Spiralia.

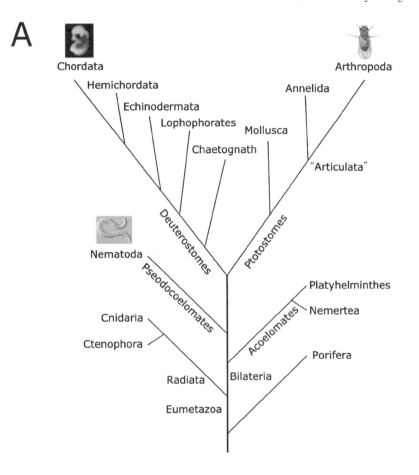

A

Chordata

Hemichordata

Echinodermata

Lophophorates

Chaetognath

Deuterostomes

Nematoda

Pseodocoelomates

Cnidaria

Ctenophora

Radiata

Eumetazoa

Arthropoda

Annelida

Mollusca

"Articulata"

Ptotostomes

Platyhelminthes

Nemertea

Acoelomates

Porifera

Bilateria

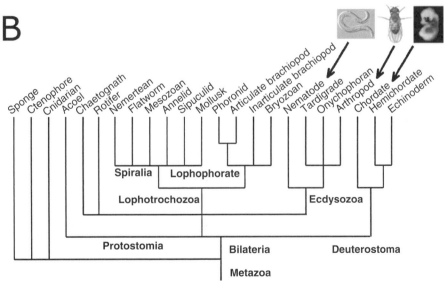

B

Sponge
Ctenophore
Cnidarian
Acoel
Chaetognath
Rotifer
Nemertean
Flatworm
Mesozoan
Annelid
Sipuculid
Mollusk
Phoronid
Articulate brachiopod
Inarticulate brachiopod
Bryozoan
Nematode
Tardigrade
Onychophoran
Arthropod
Chordate
Hemichordate
Echinoderm

Spiralia   Lophophorate

Lophotrochozoa   Ecdysozoa

Protostomia   Bilateria   Deuterostoma

Metazoa

along the elongated anteroposterior (AP) axis. While metamerism (i.e. repetition of single organ or single-tissue-layer structure) is widely distributed among metazoan phyla, segmentation is only observed in annelids, arthropods and vertebrates. Traditionally, annelids and arthropods are grouped as sister clades and their segmentation has been considered homologous, while vertebrate segmentation is believed to have evolved independently. However, in recent molecular studies, annelids and arthropods are placed in Lophotrochozoa and Ecdysozoa, respectively, with the vertebrates in Deuterostoma (Aguinaldo et al. 1997). Thus, each of the three superphyla contains both segmented and unsegmented members.

Under this new phylogenetic context, the homology of segmentation in annelids and arthropods is called into question with three plausible scenarios for the evolutionary origin of Metazoan segmentation (Davis and Patel 1999): Segmentation may have three independent origins, one for each of the three groups; segmentation may have one origin with the unsegmented groups experiencing a secondary loss of segmental body plan, or finally, segmentation may have evolved twice, once in the vertebrates of the deuterostome lineage and once in the protostome ancestor with the unsegmented protostomes representing secondary losses. To begin to differentiate between these hypotheses, one can compare the developmental mechanisms of segmentation in representatives of these three groups. If segmentation mechanisms are profoundly different between two distinct groups, segmentation is more likely to have evolved independently. Conversely, if the segmentation mechanisms are similar, then segmentation is more likely homologous, and derived from a common ancestor.

In general, the cellular process of segmentation has two components, the first of which is formation of the elongated trunk primordium. In many metazoan taxa, the AP axis elongates by posterior growth (i.e. posterior addition to an existing trunk primordium) during embryogenesis. But cases also exist in which an elongated AP axis is already prominent in the egg and does not require posterior growth, such as that of the long germ band insects (e.g. hymenopteran and dipteran insects, including *D. melanogaster*). Posterior growth is generally considered the ancestral mechanism of AP axis elongation in Metazoa (Jacobs et al. 2005). The second component of segmentation is generating morphological periodicity in the form of segmental primordia. These two components are usually coupled temporally and spatially. For example, morphological segmentation progresses from anterior to posterior in those annelids, arthropods and vertebrates that exhibit posterior growth. In long germ band insects where posterior growth does not occur, segmentation takes place simultaneously along the AP axis (Patel 1994; Davis and Patel 2002). Indeed, similarity in the morphological processes of segmentation between annelids, arthropods and vertebrates may hint at a common ancestry, but this may be superficial, due to convergent evolution, possibly built on an ancestral process of posterior growth.

To further test these three hypotheses, we must compare cellular and molecular mechanisms of segmentation, as well as the process(es) of posterior growth of unsegmented animals, in the three superphyla. Although molecular and cellular mechanisms of segmentation are fairly well understood in arthropods (Patel 1994; Lall and Patel 2001; Peel 2004; Peel et al. 2005) and vertebrates (Saga and Takeda 2001; Vasiliauskas and Stern 2001; Dubrulle and Pourquie 2004), our rudimentary understanding of the molecular mechanism of annelid segmentation becomes a roadblock to resolving the evolutionary pattern of Metazoan segmentation. Hence, to the greater evo-devo community, segmentation appears to be among the most important subjects in annelid evo-devo.

5.5.1 Cellular Mechanism of Annelid Segmentation

Among annelids, the cellular basis of segmentation has been studied most extensively in glossiphoniid leeches of the genus *Helobdella*. As in most clitellate annelid embryos, the posterior growth zone of the leech embryo comprises five bilateral pairs of embryonic stem cells, the teloblasts. Each of the five teloblast pairs (i.e. M, N, O, P and Q) contributes to a definitive set of segmentally repeating pattern elements. In some species, the stem cells and their immediate progeny are initially equipotent between the O and P lineages, in which case the stem cells may be designated as O/P teloblasts (Weisblat and Blair 1984; Weisblat and Shankland 1985; Huang and Weisblat 1996; Kuo and Shankland 2004). Among these five teloblast pairs, four contribute exclusively ectodermal tissue, with N teloblasts generating primarily neural tissues of the ventral ectoderm, Q teloblasts the dorsal ectoderm, and O and P teloblasts the lateral ectoderm (Weisblat and Shankland 1985). The remaining pair of M teloblasts generates predominantly mesodermal tissues, along with a small number of neurons (Weisblat and Shankland 1985).

In leech, the teloblasts are positioned in the prospective posterior domain of the embryo and undergo repeated rounds of stem cell-like asymmetric cell divisions, with the smaller daughter cells forming coherent columns of cells oriented toward the prospective anterior. These smaller blast cells are the founder cells of segmental tissue. The column of blast cells produced by each teloblast is called a "bandlet." Within each bandlet, the oldest blast cell clone occupies the anterior end, and the youngest the posterior end. The five bandlets originating from the five ipsilateral teloblasts merge at the dorsoposterior surface of the embryo to form a "germinal band," which elongates along the AP axis by two mechanisms: (1) the posterior addition of new blast cells by the teloblasts, and (2) oriented cell divisions of the primary blast cells and their descendants (Shankland 1999). As the germinal bands elongate, they move laterally across the embryo, from the prospective dorsal domain toward the ventral midline. Movement of the germinal bands and the epibolic spreading of the micromere cap that covers the germinal bands are the most prominent features of gastrulation in the leech embryo (Smith et al. 1996). Eventually, the left and right germinal bands "zip" together from anterior to posterior along the ventral midline to form the "germinal plate," the primordial trunk tissue.

The origin of segmental periodicity can be traced to the production of blast cells (Zackson 1982, 1984; Weisblat and Shankland 1985). In the M, O and P lineages, each of the primary blast cell clones in the same bandlet gives rise to serially homologous pattern elements. Thus, each of the blast cells produced by the iterated asymmetric cell divisions of the M, O and P teloblasts represents one segmental repeat in each of these lineages. In the N and Q lineages, two classes of primary blast cells form in an alternating pattern. Hence, each segmental unit comprises two blast cell clones in the N and Q lineages. Since production of blast cells by teloblasts is also a mechanism of AP axis elongation, AP elongation and segmentation are mechanistically coupled by asymmetric cell divisions of the teloblasts.

Leeches are now recognized as a monophyletic group arising within the oligochaetes (earthworms and their allies), which form a monophyletic group (Clitellata) arising within the polychaetes. The above description of leech segmentation appears comparable in most clitellate species; for example, teloblast lineages of the earthworm *Eisenia fetida* and the sludge worm *T. tubifex* have also been subjected to cell lineage and fate mapping analyses (Storey 1989; Goto et al. 1999; Nakamoto et al. 2000), suggesting that the cell lineage-driven segmentation process observed in *Helobdella* sp. is representative of the clitellate annelids.

Unlike leeches, in which a complete set of 32 body segments is in place by the end of embryogenesis, other oligochaetes can undergo postembryonic segmental addition, which is generally thought to be related to its well-known regenerative capability, and this same patterning mechanism of postembryonic segmentation and regeneration may also be utilized in asexual reproduction by fission in certain oligochaete species (Bely and Wray 2001). Although cellular dynamics of segmentation in embryonic stages is very similar between oligochaetes and leeches, the cellular dynamics of postembryonic segmentation and regeneration, and their relation to embryonic development of segmental tissue remain largely unknown. Certainly, another population of stem cells is present in those oligochaete species capable of regenerating after a portion of the animal is amputated (Bely 2006). One particularly interesting question is the source of stem cells for postembryonic segment addition in these oligochaetes. In the leech *Helobdella robusta*, teloblasts and "supernumerary" blast cells fuse with yolk-rich macromeres to form a syncytial yolk cell after the full complement of segment-forming blast cells has been produced (Liu et al. 1998; Desjeux and Price 1999). The fate of teloblasts that have ceased to give rise to blast cells in oligochaetes is unknown, but they conceivably give rise to postembryonic segmentation stem cells. Under this scenario, the fusion of leech teloblasts with the endoderm would be an evolutionary novelty correlated with the loss of regeneration capacity in this clitellate sublineage. Furthermore, the mechanistic link between postembryonic segmental addition and regeneration can be further elucidated by characterizing the origins of segments arising postembryonically or by regeneration. In classical accounts (Herlant-Meewis 1964), clusters of undifferentiated stem cell reserves distributed throughout the body length are reported in some oligochaete species, and these may be reactivated to produce new tissues in the regenerating animal.

In polychaetes, embryos form trochophore larvae before morphological segmentation takes place. The prospective AP axis of the trochophore elongates by posterior growth, and morphological segmentation occurs progressively in an anterior-to-posterior order as new posterior tissues form. In some polychaete species, the first few anterior segments (larval segments) form simultaneously during the larval stage, while more posterior segments (juvenile segments) form by posterior addition in the juvenile stage (Anderson 1966b, 1973). Similar to oligochaetes, many polychaetes also have the capability to regenerate segments. The mechanism for de novo segmentation by posterior addition is thought to be similar to segment formation in regeneration, and thus segmentation of regenerated tissue is used in lieu of larval segmentation in some molecular studies (Prud'homme et al. 2003; de Rosa et al. 2005). However, the similarity between de novo segmentation and regeneration has yet to be empirically proven.

The cellular dynamics involved in posterior addition and segmentation in polychaete species is not as well described as in clitellates, but the cleavage programs leading to the precursors of segmental tissue appear to be conserved among clitellates and polychaetes (Dohle 1999). In classical works of polychaete embryology (Anderson 1966b, 1973; Wilson 1892), a teloblastic mechanism similar to clitellate segmentation has been reported, but these observations do not appear to be entirely reliable. Compared to the ectodermal lineage, the segmental mesoderm precursor cells begin teloblast-like budding relatively early during embryogenesis in both polychaetes and clitellates, and thus the mesodermal teloblasts can be more reliably documented in both taxa. In contrast, the progenitor of segmental ectoderm goes through a series of nonteloblast divisions before assuming teloblastic fate. In clitellates, the relatively large size of ectodermal teloblasts allows one to follow the lineage leading to ectoteloblasts with a high level of confidence. In polychaetes, however, the cell lineage leading to segmental ectoderm comprises relatively small cells.

By the time segmentation is about to occur, many relatively small cells are already in the trochophore larvae, making it difficult to follow the cellular dynamics of segmentation with satisfactory resolution, thus reducing one's confidence regarding the presence of ectodermal teloblasts in polychaete embryos.

Perhaps the best way to follow cellular dynamics in development is to label a given cell with an intracellular lineage tracer and to follow the progeny of that cell during the course of development. This method has been useful for dissecting cellular mechanisms of segmentation in clitellate annelids (Zackson 1982; Weisblat and Shankland 1985; Goto et al. 1999; Nakamoto et al. 2000). Unfortunately, attempts to use lineage tracers to follow the development of cells 2d and 4d, the progenitors of segmental ectoderm and mesoderm in polychaetes (see below), have not been as successful (Ackermann et al. 2005). As an alternative, BrdU incorporation during the S phase of the cell cycle has been used to label proliferating cells in a given time window. Development of larval segments does not appear to involve a clearly defined posterior growth zone; and while the development of juvenile segments is correlated with a population of proliferating cells residing in the posterior end (Seaver et al. 2005), no evidence suggests that these cells are teloblasts sensu stricto. In vertebrate embryos, proliferating cells in the posterior end are also responsible for AP axis elongation (Nicolas et al. 1996; Mathis and Nicolas 2002), but these cells do not explicitly follow a teloblast-like behavior (i.e. self-renewal asymmetric division and determinate cell fates). Similarly, the posterior growth zone in polychaetes may not be produced by a highly determinate teloblastic mechanism, but rather by a more stochastically regulated pool of stem cells. If so, comparing the mechanism of posterior growth among polychaetes, oligochaetes and leeches may shed light on the evolutionary origin of the spectacularly precise mechanism of teloblast segmentation in clitellate embryogenesis.

5.5.2 Molecular Mechanism of Annelid Segmentation

Another way to investigate annelid segmentation is a candidate gene approach, namely, characterizing the expression, and ideally the function, of annelid homologues known to play roles in *D. melanogaster* or in vertebrate segmentation (Seaver 2003). In the past two decades, homologues of many *D. melanogaster* or vertebrate segmentation genes have been isolated and studied in annelid species (Wedeen and Weisblat 1991; Kostriken and Weisblat 1992; Lans et al. 1993; Savage and Shankland 1996; Iwasa et al. 2000; Seaver et al. 2001; Shimizu and Savage 2002; Song et al. 2002, 2004; Kang et al. 2003; Prud'homme et al. 2003; de Rosa et al. 2005; Rivera et al. 2005; Kerner et al. 2006; Seaver and Kaneshige 2006).

Even though the candidate gene approach is a powerful tool to investigate development in nonmodel organisms (Palopoli and Patel 1996), the conservation of developmental regulatory significance, in comparison to structure and biochemical function, erodes quickly as phylogenetic distance increases (Haag and True 2001). For example, expression patterns of *D. melanogaster* "gap gene" homologues in intermediate germ band insects suggest a conserved function, but functional tests reveal otherwise (e.g. Schroder et al. 2000; Liu and Kaufman 2004; Peel et al. 2005). Considering that the distance between annelids and *D. melanogaster* is much greater than that between long- and intermediate germ band insects, one should be careful not to draw conclusions immediately from gene expression data but should instead carefully explore other options to clarify the uncertainty about gene function.

In some cases, conservation or divergence of gene function can be deduced from expression patterns by falsification. For example, when a putative segmentation gene is not expressed during segment formation, one can deduce that this gene is not involved in segmentation (Kostriken and Weisblat 1992; Seaver et al. 2001; Kang et al. 2003; Seaver and Kaneshige 2006). But the real problem arises when expression patterns look similar to what we expect based on model species. This problem can sometimes be resolved by referencing gene expression patterns to experimental embryology data. The *D. melanogaster engrailed*-class gene in leech (Wedeen and Weisblat 1991; Lans et al. 1993) and some polychaete embryos (Prud'homme et al. 2003), for example, is expressed in a segmentally iterated pattern of stripes. Prior to the revolution of molecular phylogenies, these expression patterns seemed convincing evidence that segmentation is homologous between annelids and arthropods, and indeed this may yet turn out to be the case. But the assignment of annelids and arthropods to distinct superphyla dictates that we should consider other possibilities. More recent studies have shown that in the N lineage of leech, morphological segmentation occurs before *engrailed*-class gene expression (Shain et al. 1998; Shain et al. 2000). Moreover, ablation of *engrailed*-expressing cells in the O and P lineage does not disrupt pattern formation in other sublineages (Seaver and Shankland 2000, 2001). Together, these data suggest that *engrailed* expression does not play a role in generating segmental repeats in leech embryos and that the segmental pattern of *engrailed* expression is a consequence, rather than a cause, of leech segmentation.

In polychaetes, an *engrailed* homologue is expressed in a segmentally iterated pattern during segment formation in some species (Prud'homme et al. 2003) but only after morphological segmentation has taken place in others (Seaver et al. 2001; Seaver and Kaneshige 2006). Even though one might assume that the molecular mechanism of generating segmental periodicity is conserved among polychaetes, and hence that distinct *engrailed* expression patterns among different polychaete species suggest a nonsegmentation role for *engrailed*-class genes, the possibility that *engrailed* has a role in segmentation in some species but not in others cannot be ruled out. Without experimental or functional analyses, the controversy over the roles of *D. melanogaster* segmentation gene homologues in annelid segmentation can hardly be resolved.

5.6   CASE STUDY 2: SPIRAL CLEAVAGE AND AXIS SPECIFICATION

Annelids are members of Spiralia, a traditionally postulated grouping that is now supported by molecular phylogenies, at least to the extent that they are all members of Lophotrochozoa. Spiralian embryos have a unique early cleavage program—spiral cleavage. During cleavage stages, the axis of cell division tilts away from the animal–vegetal axis by ~45°. The axis of the next division makes a 90° turn away from the previous division. This off-axis and alternating pattern of cleavage orientations, which in its unmodified form generates a blastomere arrangement resembling a spiral tower, is considered an ancestral feature of spiralian development and is shared by the majority of annelids, mollusks, polyclad flatworms, echiurans, sipunculids and nemerteans (Costello and Henley 1976). A general common "ground plan" of spiral cleavage is described below, but note that a great deal of variation in spiral cleavage exists; such variation usually arises from differences in blastomere size ratios and cell cycle lengths among different lineages (e.g. Freeman and Lundelius 1992; van den Biggelaar 1993; Boyer and Henry 1998).

5.6.1 Spiral Cleavage and Second Axis Specification

The first two rounds of cell divisions give rise to four blastomeres: A, B, C and D. The zygote first divides into cell AB and cell CD, and then cells A and B arise from cell AB, and cells C and D arise from cell CD. However, A–B and C–D axes are not in the same plane but rather intersect at a small angle. Thus, although the four cells sit roughly at the same equatorial plane, cells A and C are displaced slightly toward, and contact each other, at the animal pole relative to the B and D cells, which contact each other at the vegetal pole. At this stage (i.e. four-cell), each blastomere is called a "quadrant"; in stages that follow, descendants of each of the four blastomeres can be collectively called a quadrant as well.

At the third division, four pairs of unequally sized blastomeres arise from the divisions of A, B, C and D, with the larger macromere on the vegetal-pole side and the smaller micromere on the animal-pole side. The macromeres are denoted 1A, 1B, 1C and 1D, and the micromeres 1a, 1b, 1c and 1d, respectively. Together, the four micromeres (1a–d) are called the "first quartet" micromeres. In the next division, 1A gives rise to micromere 2a and macromere 2A; 1B to 2b and 2B, and so on. Typically, four quartets of primary micromeres (i.e. micromeres arising from the unequal cleavage of macromeres) form by the end of the spiral cleavage phase. However, variations are observed in the number of primary micromeres arising from spiral cleavage in various evolutionary lineages.

An important developmental consequence of micromere quartet formation is the segregation of various developmental fates into different blastomeres. Each blastomere quartet contributes a specific set of cells and/or cell types to the embryo. For example, the first quartet micromeres are major contributors to the anteriormost portion of the central nervous system, and macromeres 4A–D are endodermal precursors (Anderson 1966b, 1973; Render 1991; Nardelli-Haefliger and Shankland 1993; Dictus and Damen 1997; Liu et al. 1998; Huang et al. 2002; Henry et al. 2004; Ackermann et al. 2005). In fact, this pattern of cell fate segregation is a manifestation of asymmetry along the first embryonic axis—the animal–vegetal axis. Recent work in the mud snail *Ilyanassa obsoleta* reveals that transcripts of certain developmental regulatory genes, such as *dpp, eve* and *tld*, show a punctate subcellular distribution, corresponding to their localization at specific centrosomes during mitosis and resulting in their segregation into specific tiers of micromere quartets (Lambert and Nagy 2002). Whether these asymmetrically segregated transcripts are involved in specifying different micromere quartets remains unknown. More recently, works in the polychaete *Platynereis dumerilii* show that asymmetry in nuclear beta-catenin accumulation is involved in the specification of cell fates among the sister blastomeres along the animal–vegetal axis during the cleavage stages (Schneider and Bowerman, 2007). Together, these observations suggest that cell fate specification of micromere tiers may be a complex process that involves multiple molecular pathways.

Formation of a bilaterally symmetric organism from a cylindrically symmetric oocyte requires the establishment of a second embryonic axis at some point in development. In spiralian embryos, this process is referred to as "D quadrant specification," which can occur in several ways (Freeman and Lundelius 1992). In some equal cleavers (e.g. the limpet *Patella vulgate* and the pond snail *Lymnaea stagnalis*), the morphological difference between D and A–C quadrants does not become apparent until the latter part of the spiral cleavage phase. The first two cleavages are equal in these embryos and thus yield four equally sized and presumably equipotent blastomeres, though demonstrated for only a few molluskan species (van den Biggelaar and Guerrier 1979; Arnolds et al. 1983; Martindale et al. 1985). Embryos remain radially symmetric for the next few cell cycles, until an apparently stochastic process results in an inductive interaction between micromeres and one of the vegetal macromeres, which becomes specified as the D quadrant, thus establishing the second embryonic axis (van den Biggelaar 1977; van den Biggelaar and Guerrier 1979; Arnolds et al. 1983; Martindale et al. 1985; Lambert and Nagy 2003).

In unequal cleavers (e.g. polychaetes such as *Nereis* spp., *P. dumerilii* and *Capitella* spp., and clitellate annelids such as *T. tubifex* and *Helobdella* spp.), one blastomere is different from the others (usually the largest and bearing some sort of cytoplasmic specialization) and thus can be designated as the D quadrant at the four-cell stage. This is a consequence of two unequal cleavages: the first gives rise to a smaller AB and a larger CD, and in the second, CD divides unequally to give rise to a smaller C and a larger D while AB divides equally.

A variant of unequal cleavage is the production and reabsorption of a polar lobe by the prospective D quadrant. In this form of unequal cleavage, the cytokinetic furrow proceeds precisely along the animal–vegetal axis, but at the vegetal pole, a protrusion of cytoplasm (i.e. polar lobe) forms as the cleavage furrow progresses, producing a trefoil-shaped embryo with two equally sized blastomeres at the animal-pole side and a polar lobe at the vegetal pole. Near the end of cytokinesis, one of the blastomeres reabsorbs the polar lobe material and thus becomes the larger CD cell. This process is repeated as the CD cell undergoes second cleavage, to produce the larger D cell. Due to the ease of manipulating polar lobes, many classical embryological experiments have been performed on species belonging to this category, including the mud snail *I. obsoleta* and polychaete worms *Chaetopterus* spp. and *Sabellaria cementarium*.

A special pool of cytoplasm is segregated into the D quadrant of embryos that undergo typical unequal cleavage, and this is critical for D quadrant specification in some species (Astrow et al. 1987; Dorresteijn et al. 1987; Dorresteijn and Eich 1991; Nelson and Weisblat 1991, 1992; Pilon and Weisblat 1997). Similarly, material associated with the polar lobe contains cell fate determinants for D quadrant progeny and is required for patterning the second axis in polar lobe-forming species (Clement 1952; Cather and Verdonk 1974; van Dam et al. 1982; Render 1983, 1989; Henry 1986). However, the polar lobe has various sizes and developmental significance in different species. Hence, variations that have escaped detection may yet still exist among these three major categories of D quadrant specification mechanisms (i.e. equal cleavage, unequal cleavage and polar lobe-dependent unequal cleavage).

Recent work on cell signaling in leech embryos reveals a complex signaling network operating in two-cell embryos (Huang et al. 2001; Gonsalves and Weisblat 2007). Leech undergoes unequal cleavage, and based on the classical model, the specification of the D quadrant in unequally cleaving embryos depends on the segregation of cytoplasmic determinants into the prospective D quadrant. Surprisingly, a dynamic signaling system operates in two-cell embryos that otherwise is a perfect fit for the classical description of mosaic development. This brings up the possibility that, at a molecular level, part of the determinant is a localized component of the cell signaling pathways. Hence, the seemingly deterministic early developmental paradigm in unequally cleaving spiralian embryos may in fact be a combined result of stereotypic cell division patterns and localized induction.

In recent years, a similar realization has been made regarding early development in *C. elegans* (Rose and Kemphues 1998).

5.6.2 Evolution of D Quadrant Specification Mechanisms

A major question regarding the evolution of D quadrant specification mechanisms is its phyletic distribution. As judged by morphological criteria, equal cleaving, unequal cleaving and polar lobe producing annelids and mollusks exist. Freeman and Lundelius (1992) proposed that equal cleavage is ancestral for Spiralia, and unequal cleavage and polar lobe mechanism have evolved independently many times. Note, however, that not all equal-cleaving species have been studied by embryological experiments. Possibly, the so-called equal cleavage in some species is only equal in terms of cell size, but the D quadrant has already been specified by the four-cell stage.

An important determination resulting from comparative analyses of development is that evolutionary conservation of morphological features does not guarantee conservation of the underlying mechanisms by which they arise (True and Haag 2001), and annelids have provided some interesting examples. For instance, the leech *H. robusta* and the sludge worm *T. tubifex* use different mechanisms to form teloplasm and to achieve their first unequal cleavages (Shimizu 1982; Astrow et al. 1989; Ishii and Shimizu 1995, 1997; Shimizu et al. 1998; Ren and Weisblat 2006), two features that are conserved among clitellates (Dohle 1999). A similar sort of cryptic evolution has occurred in the cell fate specification mechanism among teloblast lineages of clitellate annelids (Keleher and Stent 1990; Huang and Weisblat 1996; Arai et al. 2001; Kuo and Shankland 2004), whereby variation can be detected experimentally between species belonging to the same genus (Kuo and Shankland 2004). Given these results, D quadrant specification mechanisms may be anticipated to also vary among spiralians, and the extent of these variations may not accurately reflect phylogenetic distance. This is an interesting area for comparative studies, but also dictates that extrapolations between species are fraught with hazard.

5.6.3 Generating Bilateral Symmetry from a Spiral Pattern

In both equal- and unequal-cleaving embryos, the spiral cleavage phase ends with the bilateral division of cells within the D quadrant. The first cell to divide bilaterally is cell 4d, followed by a descendant of cell 2d. In annelids, the 4d lineage will develop into segmental mesoderm, and the 2d sublineage will become segmental ectoderm. As these two particular bilaterally symmetric cell divisions proceed, bilateral symmetry begins to spread from the 2d–4d axis toward other parts of the embryo. This axial patterning activity is induced by the propagation of a series of signaling activities originating from the so-called "D-quadrant organizer," which appears to be blastomere 3D in mollusks (Clement 1962; Labordus and van der Wal 1986; Lambert and Nagy 2001) and 4d in annelids (Treadwell 1901; Lambert and Nagy 2003). However, comparable experimental data for the D quadrant organizer in polychaete embryos are lacking. In clitellates, due to the modification of the cleavage program, bilateral symmetry arises relatively early during embryogenesis (Weisblat 1999). Whether an organizing signal is required for patterning the non-2d/4d ecto- and mesodermal lineages is not clear, but available evidences suggest that cell–cell interactions are required for formation of the endodermal syncytium by fusion of macromeres in *H. robusta* embryos (Isaksen et al. 1999) and bilateral symmetry in the 2d lineage of *T. tubifex* embryos (Nakamoto et al. 2004).

Gastrulation occurs in the concluding phase of spiral cleavage. In polychaetes, depending on the relative sizes of blastomeres and cellular arrangements, the macromeres (4A–D) are internalized, either moving from the vegetal-pole surface into the blastocoel space or becoming covered by the spreading micromere progeny (Anderson 1966b, 1973). In either case, following internalization of the macromeres, the progeny of 4d also moves to the interior of the embryo. By the time gastrulation movement has been completed, cell differentiation is already underway, and the trochophore larva is soon functioning.

5.6.4   Diversity of Annelid Cleavage Programs

Traditionally, three groups are recognized within Phylum Annelida: polychaetes, oligochaetes and leeches. In the traditional morphology-based view, the split of Polychaeta and Clitellata (comprising oligochaetes and leeches) occurs at a basal node of the annelid tree (Fauchald and Rouse 1997). However, results from molecular phylogenetic analysis suggest a different scenario wherein Polychaeta are paraphyletic, and Clitellata and the traditionally recognized phyla Echiura and Pogonophora are derived from polychaetes (McHugh 1997, 2000; Kojima 1998). Moreover, within Clitellata, Oligochaeta is also paraphyletic with the monophyletic group of leech derived from an ancestral oligochaete (Siddall et al. 2001; Erseus and Kallersjo 2004). Compared to polychaetes, clitellates are exclusively direct developers, and clitellate species seem to share a similar embryogenic ground plan that has undergone considerable modification from the ancestral spiral cleavage program (Anderson 1966a, 1973). Hence, the novel developmental pattern of clitellates that appeared during the polychaete-to-clitellate transition can be viewed as a major evolutionary event in annelid development.

Compared to cleavage-stage polychaete embryos, clitellate embryos differ most significantly in the size ratio between micromeres and macromeres, with macromeres retaining most of the yolk. In the D quadrant, the yolk contents along with a special pool of cytoplasm (teloplasm) segregate into the teloblast lineages. Even in unequally cleaving polychaetes, the yolk is distributed more evenly, so the teloblasts are not as visually distinct as in most clitellates. This deviation of cleavage patterns can be understood in terms of the major difference between life history patterns of polychaetes and clitellates. In polychaetes, growth of larval tissues, which arise from non-2d/4d lineages, precedes the onset of feeding and thus is dependent on maternal resources. In contrast, the growth of segmental tissue, which arises from cells 2d and 4d, occurs in postembryonic stages and thus does not rely on maternal resources. In clitellates, which lack a feeding larval stage, growth of the non-2d/4d lineage is minimized during embryogenesis; conversely, development of the 2d and 4d lineages is accelerated and is entirely dependent on maternal resources. Hence, the reallocation of yolk contents and other maternal resources into the 2d and 4d lineages in clitellates can be viewed as a major developmental adaptation to a new life history strategy.

Corresponding to the accelerated development in segmental tissues, modifications in the cellular mechanisms of morphogenesis are evident between different annelid lineages. For example, gastrulation is coupled to axis elongation in clitellate embryos, while it occurs before axis elongation in polychaetes. In polychaete embryos, internalization of 4d progeny occurs after macromeres move into the interior of embryos. In leech, a similar movement by the mesodermal lineage has recently been identified (D.H. Kuo, unpublished observation), but this movement occurs before the internalization of macromeres. All of

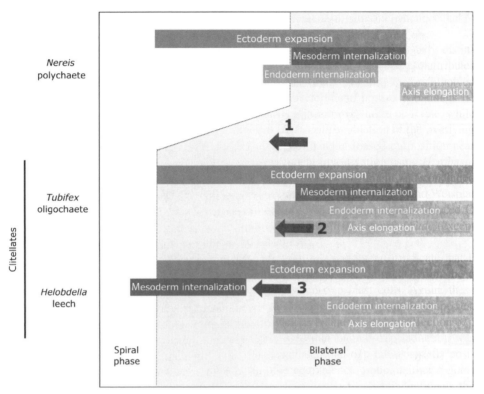

Figure 5.2   Schematic of the heterochronic shift of developmental events in major evolutionary transitions of annelids. The proliferation and expansion of ectodermal lineages, and the internalization of mesoderm and endoderm lineages are mechanisms of gastrulation. The mechanism of axis elongation is the posterior addition of daughter cells by a pool of stem cell populations (see text for details). Gastrulation starts at the later phase of spiral cleavage in polychaetes but in the bilateral cleavage phase in oligochaetes. If the mode of gastrulation in polychaetes is considered ancestral, it would signify a delay in gastrulation in clitellates (arrow 1). In polychaetes, gastrulation precedes axis elongation, but in clitellates, gastrulation is coupled with axis elongation. This change suggests a relative acceleration of segment development in clitellate annelids (arrow 2). Compared with oligochaetes, mesoderm internalization occurs relatively early (in the late spiral cleavage phase) in glossiphoniid leeches (arrow 3). (See color insert.)

these differences suggest that heterochronic shifts of developmental events in 2d/4d lineages and non-2d/4d lineages may have taken place during the polychaete-to-clitellate transition, producing a novel developmental mode in clitellate annelids (Fig. 5.2). Further investigation into the molecular mechanisms of gastrulation cell movements in polychaetes and clitellates should bring interesting insights into the developmental shift that has occurred in the clitellate lineage.

5.7   TOOLS FOR ANALYZING MOLECULAR MECHANISMS OF DEVELOPMENT

The case studies considered above demonstrate the need for annelid evo-devo biologists to investigate the functional roles of developmental regulatory genes. In such analyses, gene function can be revealed by disrupting normal gene expression or activity with genetic or biochemical manipulations. The most powerful methodology for gene function analysis is the so-called classical or forward genetics approach, in which mutant versions of genes associated with specific phenotypes are isolated and characterized. Forward genetics is the key method for the success of model species such as *D. melanogaster* and *C. elegans*. Even though forward genetics is not currently applicable to any annelid species, tools for performing reverse genetics in annelid species have become increasingly available.

In contrast to forward genetics, reverse genetics refers to analyzing gene function by disrupting expression or activity of specific candidate genes. For example, RNA interference (RNAi) and antisense morpholino oligos can be used to knock down gene expression, and microinjection of DNA vectors and *in vitro* synthesized mRNA can be used to overexpress chosen genes. In another approach, with the advent of combinatorial chemistry combined with high-throughput methods for screening chemical libraries, small molecule inhibitors for certain biochemical pathways are available. When carefully and creatively used, these tools can be employed for analyzing gene functions in organisms where forward genetic approaches are not available.

Currently, tools for functional analysis have not yet been widely used in annelid developmental studies. For example, the RNAi technique has been applied successfully to problems of relatively late development in the hirudinid leech *Hirudo medicinalis* (Baker and Macagno 2000). In *Helobdella* species, gene knockdown by antisense morpholinos has been used with some success (Song et al. 2002; Agee et al. 2006) and microinjection of synthetic mRNAs has been used to drive ectopic gene expression (Zhang and Weisblat 2005; Agee et al. 2006). In the smaller polychaete embryos in which microinjection-dependent methods are less applicable, the combinatorial use of small molecule inhibitors and molecular analyses of gene expression has shown its potential as a useful method for investigating developmental processes (Steinmetz et al. 2007). In addition to reverse genetic tools, advances in imaging technology should provide a better description of cellular dynamics during annelid development, which will facilitate our understanding of the phenotypes generated by the methods described above. Despite the ongoing development of reverse genetic tools, reverse genetics can only be used for candidate genes. As mentioned, candidate genes are expected to become less informative as the evolutionary distance between the taxa under comparison increases. This is particularly problematic for students of annelid development because no well-developed genetic model species is available within the annelid clade or, for that matter, the entire lophotrochozoan superphylum. On a brighter side, however, progress in the area of whole genome sequencing has made obtaining the genome sequence of an organism increasingly cheaper and faster. At this writing, whole genome sequence projects have been completed for four lophotrochozoan species, among them two annelid species, the polychaete *Capitella* sp. I and the leech *H. robusta*, along with a molluskan species, the limpet *Lottia gigantea* and a flatworm species, the planarian *Schmidtea mediterranea*. Genome sequencing is also underway for another molluskan species, the sea hare *Aplysia californica*. As part of this undertaking, expressed sequence tags (ESTs) are also being produced from several annelid systems, and EST databases have been generated independently for a larger group of lophotrochozoans. Compared to the candidate gene approach, these approaches are not biased toward *D. melanogaster* or vertebrate development, and thus are more likely to uncover unique aspects of annelid development.

5.8 THE FUTURE OF THE ANNELID MODEL SYSTEMS FOR EVO-DEVO

The emerging new molecular and high-throughput technologies have arguably shown their potential in facilitating better analyses of annelid development. Understanding the actual mechanisms of development and its evolution also requires a detailed understanding of the cellular processes during embryogenesis, a more highly resolved phylogenetic tree for annelids and their allies, and the inclusion of more species in comparative studies. Ideally, many species would be studied simultaneously in comparable detail. However, given the constraint in technological, financial and human resources, this is unlikely to happen. Furthermore, embryos of certain species are more suitable than others for developmental studies. Hence, a compromise must be reached.

Two successful evo-devo systems are those focused on arthropods and nematodes (Sommer 2000; Schierenberg 2001; Simpson 2002; Peel et al. 2005). A defining characteristic in both is the availability of forward genetics and other resources for studying embryonic development of central model species, namely, *D. melanogaster* and *C. elegans*. In each case, knowledge obtained from the detailed developmental genetic analysis of these powerful models provides a reference for the study of satellite species with various evolutionary distances from the central model, and which are often more difficult to study (analogous to "hub-and-spoke" networks in the airlines industry). Although the utility of this approach declines as the phylogenetic distance from the central model system increases, it has generally proven useful as long as the satellite species and the model species share the same body plan.

We have already seen that a difficulty in annelid evo-devo is the absence of a genetic model species (i.e. all the annelid species are currently being studied as satellite species with *D. melanogaster, C. elegans* or vertebrates as the reference model). This is not to say that no one is interested in building an annelid-centered system for evo-devo studies, and progress of annelid whole genome sequencing projects testifies to the success of such ongoing efforts. However, among the annelid species that have been routinely used in developmental studies, including the polychaetes *P. dumerilii, Capitella* sp. I, *Chaetopterus* sp. and *Hydroides elegans*, the oligochaete *T. tubifex* and leeches *Helobdella* spp. and *Theromyzon tessulatum*, none shows much promise of becoming a suitable system for the implementation of forward genetics due to various technical and logistic considerations. Moreover, the emergence of another large network of researchers using a single species comparable to *D. melanogaster* and *C. elegans* communities seems unlikely, given the restraints on growth of the basic science infrastructure.

Thus, instead of attempting to follow the example of nematode and arthropod evo-devo by adopting the satellite species system, the annelid evo-devo community should consider a more decentralized approach, in which homologous developmental genes or developmental processes are analyzed at comparable levels in several key species. By accumulating data simultaneously in a number of species, a valid consensus regarding the outlines of annelid evo-devo may emerge via a more efficient use of limited resources (if the arthropod and nematode evo-devo systems are analogous to "hub-and-spoke" networks, the system proposed for annelid evo-devo would be analogous to a "point-to-point" network that can also be efficient if properly implemented).

Another foreseeable benefit of adopting a decentralized system is to open up new study subjects in annelid evo-devo. When using the satellite species approach, the only available reference model species are *D. melanogaster, C. elegans* and vertebrates. Consequently, the only viable study subject for annelid satellite species is metazoan body plan evolution, which, although exciting, is not the only topic of interest in evo-devo. By viewing annelid experimental systems as an independent collective entity, rather than the poor relatives of genetic model species, we can free annelid evo-devo from the confines of metazoan body plan evolution as the exclusive or primary topic of interest. Here, I would like to offer a couple of examples for such new areas of study.

First, a recent study of molecular phylogeny of *Helobdella* spp. laboratory cultures (Bely and Weisblat 2006) establishes the potential use of *Helobdella* species complexes to study the microevolution of developmental mechanisms. In the past decade, differences in cell fates and cell fate specification mechanisms have been uncovered among laboratory colonies of *Helobdella* species with various geographic origins (Huang et al. 2002; Kuo and Shankland 2004). One particularly important potential for evo-devo is to reconcile macro- and microevolution by identifying the molecular changes that underlie phenotypic evolution across taxa with various evolutionary distances. These types of studies are most informative when conducted between closely related species (Simpson 2002). Given these conditions, the developmental differences between *Helobdella* species would be a suitable subject of study for the purpose of understanding how the evolution of DNA sequences leads to developmental changes.

Another potentially new area of annelid evo-devo is that of behavioral neurobiology. For some decades, the medicinal leech *H. medicinalis* has been a powerful model for studying neural circuits that generate behavior (Kristan et al. 2005). The neural circuits underlying behaviors such as crawling and swimming have been dissected to the level of individual neurons. Notably, different behavioral patterns are exhibited by different leech species. For example, while *H. medicinalis* is perfectly capable of swimming, *H. robusta* does not swim at all. Yet the neuronal arrangement pattern in ventral ganglia appears to be conserved among leech species, and many exhibit a rhythmic ventilating behavior very similar to swimming. Interesting questions, including how homologous neurons change function with behavior and what molecular, cellular and developmental changes are required for behavioral evolution, are largely unexplored. Evo-devo biologists have focused their attention on the evolution of morphology, while the behavioral dimension of phenotypic evolution has been relatively ignored. Note that behavioral evolution is an important component of adaptation, and must be integrated with morphological evolution to achieve a well-concerted evolutionary adaptation. Thus, the growing body of knowledge in the neural basis of leech behavior and leech development may join to form the basis of a new area of neurobehavioral evo-devo.

The two examples provided have a clear "bdellocentric" bias because the author works on leech development and thus is most familiar with leech. Nonetheless, the goal here is to promote the use of a decentralized system for studying annelid evo-devo in order to expand the scope of annelid evo-devo. The most pressing issue today is to deepen and broaden our understanding of annelid development by improving our ability to study the details of developmental mechanisms and the functions of developmental genes in more annelid species. This goal can only be achieved with efforts of the entire community, and I hope that this chapter will serve as a primer for building a more diverse and collaborative annelid evo-devo community.

ACKNOWLEDGMENTS

I thank Deirdre Lyons and David Weisblat for their critical reading of earlier versions of this manuscript. Professor Weisblat's ingenious airline analogy was extremely helpful for delivering the message. I also thank Dan Shain and Sonia Krutzke for their comments and suggestions on later versions of the manuscript.

REFERENCES

ACKERMANN, C., DORRESTEIJN, A., FISCHER, A. 2005. Clonal domains in postlarval *Platynereis dumerilii* (Annelida: Polychaeta). J. Morph. 266, 258–280.

AGEE, S.J., LYONS, D.C., WEISBLAT, D.A. 2006. Maternal expression of a NANOS homolog is required for early development of the leech *Helobdella robusta*. Dev. Biol. 298, 1–11.

AGUINALDO, A.M.A., TURBEVILLE, J.M., LINFORD, L.S., RIVERA, M.C., GARET, J.R., RAFF, R.A, LAKE, J.A. 1997. Evidence for a clade of nematodes, arthropods and other moulting animals. Nature 387, 489–493.

ANDERSON, D.T. 1966a. The comparative early embryology of the Oligochaeta, Hirudinea and Onychophora. Proc. Linn. Soc. N. S. W. 91, 10–43.

ANDERSON, D.T. 1966b. The comparative embryology of the Polychaeta. Acta Zool. 47, 1–42.

ANDERSON, D.T. 1973. *Embryology and Phylogeny in Annelids and Arthropods*. Oxford: Pergamon.

ARAI, A., NAKAMOTO, A., SHIMIZU, T. 2001. Specification of ectodermal teloblast lineages in embryos of the oligochaete annelid *Tubifex*: involvement of novel cell-cell interactions. Development 128, 1211–1219.

ARNOLDS, W.J.A., van den BIGGELAAR, J.A.M., VERDONK, N.H. 1983. Spatial aspects of cell interactions involved in the determination of dorsoventral polarity in equally cleaving gastropods and regulative abilities of their embryos, as studied by micromere deletions in *Lymnaea* and *Patella*. Roux's Arch. Dev. Biol. 192, 75–85.

ASTROW, S.H., HOLTON, B., WEISBLAT, D.A. 1987. Centrifugation redistributes factors determining cleavage patterns in leech embryos. Dev. Biol. 120, 270–283.

ASTROW, S.H., HOLTON, B., WEISBLAT, D.A. 1989. Teloplasm formation in a leech, *Helobdella triserialis*, is a microtubule-dependent process. Dev. Biol. 135, 306–319.

BAKER, M.W., MACAGNO, E.R. 2000. RNAi of the receptor tyrosine phosphatase *Hm*LAR2 in a single cell of an intact leech embryo leads to growth-cone collapse. Curr. Biol. 7, 1071–1074.

BELY, A.E. 2006. Distribution of segment regeneration ability in the Annelida. Integr. Comp. Biol. 46, 508–518.

BELY, A.E., WEISBLAT, D.A. 2006. Lessons from leeches: a call for DNA barcoding in the lab. Evol. Dev. 8, 491–501.

BELY, A.E., WRAY, G.A. 2001. Evolution of regeneration and fission in annelids: insights from *engrailed*- and *orthodenticle*- class gene expression. Development 128, 2781–2791.

van den BIGGELAAR, J.A.M. 1977. Development of dorsoventral polarity and mesentoblast determination in *Patella vulgata*. J. Morph. 154, 157–186.

van den BIGGELAAR, J.A.M. 1993. Cleavage pattern in embryos of *Haliotis tuberculata* (Archaeogastropoda) and gastropod phylogeny. J. Morph. 216, 121–139.

van den BIGGELAAR, J.A.M., GUERRIER, P. 1979. Dorsoventral polarity and mesentoblast determination as concomitant results of cellular interactions in the mollusk *Patella vulgata*. Dev. Biol. 68, 462–471.

BOYER, B.C., HENRY, J.Q. 1998. Evolutionary modifications of the spiralian developmental program. Am. Zool. 38, 621–633.

CARROLL, S.B. 1995. Homeotic genes and the evolution of arthropods and chordates. Nature 376, 479–485.

CATHER, J.N., VERDONK, N.H. 1974. The development of *Bithynia tentaculata* (Prosobranchia, Gastropoda) after removal of the polar lobe. J. Embryol. Exp. Morph. 31, 415–422.

CHILD, C.M. 1900. The early development of *Arenicola* and *Sternaspis*. Wilhelm Roux Arch. Entwicklugsmech. Org. 9, 587–722.

CLEMENT, A.C. 1952. Experimental studies on germinal localization in *Ilyanassa*. I. the role of the polar lobe in determination of the cleavage pattern and its influence in later development. J. Exp. Zool. 121, 593–626.

CLEMENT, A.C. 1962. Development of *Ilyanassa* following removal of the D macromere at successive cleavage stages. J. Exp. Zool. 149, 193–216.

COSTELLO, D.P., HENLEY, C. 1976. Spiralian development: a perspective. Am. Zool. 16, 277–291.

CRISP, M.D., COOK, L.G. 2005. Do early branching lineages signify ancestral traits? Trends Ecol. Evol. 20, 122–128.

CUNNINGHAM, C.W., OMLAND, K.E., OAKLEY, T.H. 1998. Reconstructing ancestral character states: a critical reappraisal. Trends Ecol. Evol. 13, 361–366.

van DAM, W.I., DOHMEN, M.R., VERDONK, N.H. 1982. Localization of morphogenetic determinants in a special cytoplasm present in the polar lobe of *Bithynia tentaculata* (Gastropoda). Roux's Arch. Dev. Biol. 191, 371–377.

DAVIS, G.K., PATEL, N.H. 1999. The origin and evolution of segmentation. Trends Genet. 15, M68–M72.

DAVIS, G.K., PATEL, N.H. 2002. Short, long, and beyond: molecular and embryological approaches to insect segmentation. Ann. Rev. Entomol. 47, 669–699.

DESJEUX, I., PRICE, D.J. 1999. The production and elimination of supernumerary blast cells in the leech embryo. Dev. Genes Evol. 209, 284–293.

DICTUS, W.J.A.G., DAMEN, P. 1997. Cell-lineage and clonal-contribution map of the trochophore larva of *Patella vulgata* (Mollusca). Mech. Dev. 62, 213–226.

DOHLE, W. 1999. The ancestral cleavage pattern of the clitellates and its phylogenetic deviations. Hydrobiologia 402, 267–283.

DORRESTEIJN, A.W.C., EICH, P. 1991. Experimental change of cytoplasmic composition can convert determination of blastomeres in *Platynereis dumerilii* (Annelida, Polychaeta). Roux's Arch. Dev. Biol. 200, 342–351.

DORRESTEIJN, A.W.C., BORNEWASSER, H., FISCHER, A. 1987. A correlative study of experimentally changed first cleavage and Janus development in the trunk of *Platynereis dumerilii* (Annelida, Polychaeta). Roux's Arch. Dev. Biol. 196, 51–58.

DUBRULLE, J., POURQUIE, O. 2004. Coupling segmentation to axis formation. Development 131, 5783–5793.

ERSEUS, C., KALLERSJO, M. 2004. 18S rDNA phylogeny of Clitellata (Annelida). Zool. Scr. 33, 187–196.

FAUCHALD, K., ROUSE, G. 1997. Polychaete systematics: past and present. Zool. Scr. 26, 71–138.

FREEMAN, G., LUNDELIUS, J.W. 1992. Evolutionary implications of the mode of D quadrant specification in coelomates with spiral cleavage. J. Evol. Biol. 5, 205–247.

GONSALVES, F.C., WEISBLAT, D.A. 2007. MAPK regulation of maternal and zygotic Notch transcript stability in early development. Proc. Natl. Acad. Sci. U.S.A. 104, 531–536.

GOTO, A., KITAMURA, K., SHIMIZU, T. 1999. Cell lineage analysis of pattern formation in the *Tubifex* embryo. I. Segmentation in the mesoderm. Int. J. Dev. Biol. 43, 317–327.

HAAG, E.S., TRUE, J.R. 2001. From mutants to mechanisms? Assessing the candidate gene paradigm in evolutionary biology. Evolution 55, 1077–1084.

HALDER, G., CALLAERTS, P., GEHRING, W.J. 1995. New perspectives on eye evolution. Curr. Opin. Genet. Dev. 5, 602–609.

HARVEY, P.H., PAGEL, M.D. 1991. *The Comparative Method in Evolutionary Biology*. Oxford: Oxford University Press.

HARVEY, P.H., PURVIS, A. 1991. Comparative methods for explaining adaptations. Nature 351, 619–624.

HENRY, J.J. 1986. The role of unequal cleavage and the polar lobe in the segregation of developmental potential during first cleavage in the embryo of *Chætopterus variopedatus*. Roux's Arch. Dev. Biol. 195, 103–116.

HENRY, J.Q., OKUSU, A., MARTINDALE, M.Q. 2004. The cell lineage of the polyplacophoran, *Chaetopleura apiculata*: variation in the spiralian program and implications for molluscan evolution. Dev. Biol. 272, 145–160.

HERLANT-MEEWIS, H. 1964. Regeneration in annelids. Adv. Morphog. 4, 155–215.

HUANG, F.Z., WEISBLAT, D.A. 1996. Cell fate determination in an annelid equivalence group. Development 122, 1839–1847.

HUANG, F.Z., BELY, A.E., WEISBLAT, D.A. 2001. Stochastic WNT signaling between nonequivalent cells regulates adhesion but not fate in the two-cell leech embryo. Curr. Biol. 11, 1–7.

HUANG, F.Z., KANG, D., RAMIREZ-WEBER, F.-A., BISSEN, S.T., WEISBLAT, D.A. 2002. Micromere lineage in the glossiphoniid leech *Helobdella*. Development 129, 719–732.

ISAKSEN, D.E., LIU, N-J.L., WEISBLAT, D.A. 1999. Inductive regulation of cell fusion in leech. Development 126, 3381–3390.

ISHII, R., SHIMIZU, T. 1995. Unequal first cleavage in the *Tubifex* egg: involvement of a monastral mitotic apparatus. Dev. Growth Differ. 37, 687–701.

ISHII, R., SHIMIZU, T. 1997. Equalization of unequal first cleavage in the *Tubifex* egg by introduction of an additional centrosome: implications for the absence of cortical mechanisms for mitotic spindle asymmetry. Dev. Biol. 189, 49–56.

IWASA, J.H., SUVER, D.W., SAVAGE, R.M. 2000. The leech *hunchback* protein is expressed in the epithelium and CNS but not in the segmental precursor lineages. Dev. Genes Evol. 210, 277–288.

JACOBS, D.K., HUGHES, N.C., FITZ-GIBBON, S.T., WINCHELL, C.J. 2005. Terminal addition, the Cambrian radiation and the Phanerozoic evolution of bilaterian form. Evol. Dev. 7, 498–514.

JENNER, R.A. 2006. Unburdening evo-devo: ancestral attractions, model organisms, and basal baloney. Dev. Genes Evol. 216, 385–394.

KANG, D., HUANG, F., LI, D., SHANKLAND, M., GAFFIELD, W., WEISBLAT, D.A. 2003. A *hedgehog* homolog regulates gut formation in leech (*Helobdella*). Development 130, 1645–1657.

KELEHER, G.P., STENT, G.S. 1990. Cell position and developmental fate in leech embryogenesis. Proc. Natl. Acad. Sci. U.S.A. 87, 8457–8461.

KERNER, P., ZELADA GONZALEZ, F., Le GOUAR, M., LEDENT, V., ARENDT, D., VERVOORT, M. 2006. The expression of a hunchback ortholog in the polychaete annelid *Platynereis dumerilii* suggests an ancestral role in mesoderm development and neurogenesis. Dev. Genes Evol. 216, 821–828.

KOJIMA, S., 1998. Paraphyletic status of Polychaeta suggested by phylogenetic analysis based on the amino acid sequences of elongation factor-1a. Mol. Phylogenet. Evol. 9, 255–261.

KOSTRIKEN, R., WEISBLAT, D.A. 1992. Expression of a *Wnt* gene in embryonic epithelium of the leech. Dev. Biol. 151, 225–241.

KRELL, F.T., CRANSTON, P.S. 2006. Which side of the tree is more basal? Syst. Entomol. 29, 279–281.

KRISTAN, W.B.Jr, ., CALABRESE, R.L., FRIESEN, W.O. 2005. Neuronal control of leech behavior. Prog. Neurobiol. 76, 279–327.

KUO, D-H., SHANKLAND, M. 2004. Evolutionary diversification of specification mechanisms within the O/P equivalence group of the leech genus *Helobdella*. Development 131, 5859–5869.

LABORDUS, V., van der WAL, U.P. 1986. The determination of the shell field cells during the first hour in the sixth cleavage cycle of eggs of *Ilyanassa obsoleta*. J. Exp. Zool. 239, 65–76.

LALL, S., PATEL, N.H. 2001. Conservation and divergence in molecular mechanisms of axis formation. Ann. Rev. Genet. 35, 407–437.

LAMBERT, J.D., NAGY, L.M. 2001. MAPK signaling by the D quadrant embryonic organizer of the mollusc *Ilyanassa obsoleta*. Development 128, 45–56.

LAMBERT, J.D., NAGY, L.M. 2002. Asymmetric inheritance of centrosomally localized mRNAs during embryonic cleavages. Nature 420, 682–686.

LAMBERT, J.D., NAGY, L.M. 2003. The MAPK cascade in equally cleaving spiralian embryos. Dev. Biol. 263, 231–241.

LANS, D., WEDEEN, C.J., WEISBLAT, D.A. 1993. Cell lineage analysis of the expression of an *engrailed* homolog in leech embryos. Development 117, 857–871.

LIU, N-J.L., ISAKSEN, D.E., SMITH, C.M., WEISBLAT, D.A. 1998. Movements and stepwise fusion of endodermal precursor cells in leech. Dev. Genes Evol. 208, 117–127.

LIU, P.Z., KAUFMAN, T.C. 2004. *hunchback* is required for suppression of abdominal identity, and for proper germband growth and segmentation in the intermediate germband insect *Oncopeltus fasciatus*. Development 131, 1515–1527.

MCHUGH, D. 1997. Molecular evidence that echiurans and pogonophorans are derived annelids. Proc. Natl. Acad. Sci. U.S.A. 94, 8006–8009.

MCHUGH, D. 2000. Molecular phylogeny of the Annelida. Can. J. Zool. 78, 1873–1884.

MARTINDALE, M.Q., DOE, C.Q., MORRILL, J.B. 1985. The role of animal-vegetal interaction with respect to the determination of dorsoventral polarity in the equal-cleaving spiralian, *Lymnaea palustris*. Roux's Arch. Dev. Biol. 194, 281–295.

MATHIS, L, NICOLAS, J.-F. 2002. Cellular patterning of the vertebrate embryo. Trends Genet. 18, 627–635.

MEAD, A.D. 1897. The early development of marine annelids. J. Morph. 13, 227–327.

NAKAMOTO, A., ARAI, A., SHIMIZU, T. 2000. Cell lineage analysis of pattern formation in the *Tubifex* embryo. II. Segmentation in the ectoderm. Int. J. Dev. Biol. 44, 797–805.

NAKAMOTO, A., ARAI, A., SHIMIZU, T. 2004. Specification of polarity of teloblastogenesis in the oligochaete annelid *Tubifex*: cellular basis for bilateral symmetry in the ectoderm. Dev. Biol. 272, 248–261.

NARDELLI-HAEFLIGER, D., SHANKLAND, M. 1993. *Lox10*, a member of the *NK-2* homeobox gene class, is expressed in a segmental pattern in the endoderm and in the cephalic nervous system of the leech *Helobdella*. Development 118, 877–892.

NELSON, B.H., WEISBLAT, D.A. 1991. Conversion of ectoderm to mesoderm by cytoplasmic extrusion in leech embryos. Science 253, 435–438.

NELSON, B.H., WEISBLAT, D.A. 1992. Cytoplasmic and cortical determinants interact to specify ectoderm and mesoderm in the leech embryo. Development 115, 103–115.

NICOLAS, J.F., MATHIS, L., BONNEROT, C. 1996. Evidence in the mouse for self-renewing stem cells in the formation of a segmented longitudinal structure, the myotome. Development 122, 2933–2946.

PAGEL, M. 1999. Inferring the historical patterns of biological evolution. Nature 401, 877–884.

PALOPOLI, M.F., PATEL, N.H. 1996. Neo-Darwinian developmental evolution: can we bridge the gap between pattern and process? Curr. Opin. Genet. Dev. 6, 502–508.

PATEL, N.H. 1994. Developmental evolution: insights from studies of insect segmentation. Science 266, 581–590.

PEEL, A. 2004. The evolution of arthropod segmentation mechanisms. Bioessays 26, 1108–1116.

PEEL, A.D., CHIPMAN, A.D., AKAM, M. 2005. Arthropod segmentation: beyond the *Drosophila* paradigm. Nat. Rev. Genet. 6, 905–916.

PILON, M., WEISBLAT, D.A. 1997. A *nanos* homolog in leech. Development 124, 1771–1780.

PRUD'HOMME, B., de ROSA, R., ARENDT, D., JULIEN, J.-F., PAJAZITI, R., DORRESTEIJN, A.W.C., ADOUTTE, A., WITTBRODT, J., BALAVOINE, G. 2003. Arthropod-like expression patterns of *engrailed* and *wingless* in the annelid *Platynereis dumerilii* suggest a role in segment formation. Curr. Biol. 13, 1876–1881.

REN, X., WEISBLAT, D.A. 2006. Asymmetrization of first cleavage by transient disassembly of one spindle pole aster in the leech Helobdella robusta. Dev. Biol. 292, 103–115.

RENDER, J. 1991. Fate map of the first quartet micromeres in the gastropod *Ilyanassa obsoleta*. Development 113, 495–501.

RENDER, J.A. 1983. The second polar lobe of *Sabellaria cementarium* plays an inhibitory role in apical tuft formation. Roux's Arch. Dev. Biol. 192, 120–129.

RENDER, J.A. 1989. Development of *Ilyanassa obsoleta* embryos after equal distribution of polar lobe material at first cleavage. Dev. Biol. 132, 241–250.

RIVERA, A.S., GONSALVES, F.C., SONG, M.H., NORRIS, B.J., WEISBLAT, D.A. 2005. Characterization of Notch-class gene expression in segmentation stem cells and segment founder cells in *Helobdella robusta* (Lophotrochozoa; Annelida; Clitellata; Hirudinida; Glossiphoniidae). Evol. Dev. 7, 588–599.

de ROSA, R., PRUD'HOMME, B., BALAVOINE, G. 2005. *caudal* and *even-skipped* in the annelid *Platynereis dumerilii* and the ancestry of posterior growth. Evol. Dev. 7, 574–587.

ROSE, L.S., KEMPHUES, K.J. 1998. Early patterning of the *C. elegans* embryo. Annu. Rev. Genet. 32, 521–545.

SAGA, Y., TAKEDA, H. 2001. The making of somite: molecular events in vertebrate segmentation. Nat. Rev. Genet. 2, 835–845.

SAVAGE, R.M., SHANKLAND, M. 1996. Identification and characterization of a *hunchback* orthologue, *Lzf2*, and its expression during leech embryogenesis. Dev. Biol. 175, 205–217.

SCHIERENBERG, E. 2001. The sons of fortune: early embryogenesis, evolution and ecology of nematode. Bioessays 23, 841–847.

SCHNEIDER, S.Q., BOWERMAN, B. 2007. b-catenin asymmetries after all animal/vegetal-oriented cell divisions in *Platynereis dumerilii* embryos mediate binary cell-fate specification. Dev. Cell 13, 73–86.

SCHRODER, R., ECKERT, C., WOLFF, C., TAUTZ, D. 2000. Conserved and divergent aspects of terminal patterning in the beetle *Tribolium castaneum*. Proc. Natl. Acad. Sci. U.S.A. 97, 6591–6596.

SEAVER E.C. 2003. Segmentation: mono- or polyphyletic? Int. J. Dev. Biol. 47, 583–595.

SEAVER, E.C., KANESHIGE, L.M. 2006. Expression of 'segmentation' genes during larval and juvenile development in the polychaetes *Capitella* sp. I and *H. elegans*. Dev. Biol. 289, 179–194.

SEAVER, E.C., SHANKLAND, M. 2000. Leech segmental repeats develop normally in the absence of signals from either anterior or posterior segments. Dev. Biol. 224, 339–353.

SEAVER, E.C., SHANKLAND, M. 2001. Establishment of segment polarity in the ectoderm of the leech *Helobdella*. Development 128, 1629–1641.

SEAVER, E.C., PAULSON, P.A., IRVINE, S.Q., MARTINDALE, M.Q. 2001. The spatial and temporal expression of *Ch-en*, the *engrailed* gene in the polychaete *Chaetopterus*, does not sopport a role in body axial segmentation. Dev. Biol. 236, 195–209.

SEAVER, E.C., THAMM, K., HILL, S.D. 2005. Growth patterns during segmentation in the two polychaete annelids, *Capitella sp.* I and *Hydroides elegans*: comparisons at distinct life history stages. Evol. Dev. 7, 312–326.

SHAIN, D.H., RAMIREZ-WEBER, F.-A., HSU, J., WEISBLAT, D.A. 1998. Gangliogenesis in leech: morphogenetic processes leading to segmentation in the central nervous system. Dev. Genes Evol. 208, 28–36.

SHAIN, D.H., STUART, D.K., HUANG, F.Z., WEISBLAT, D.A. 2000. Segmentation of the central nervous system in leech. Development 127, 735–744.

SHANKLAND, M. 1999. Anteroposterior pattern formation in the leech embryo. In *Cell Lineage and Fate Determination*, edited by S.A. Moody. San Diego, CA: Academic Press, pp. 207–224.

SHANKLAND, M., SAVAGE, R.M. 1997. Annelids, the segmented worms. In *Embryology: Constructing the Organism*, edited by S.F. Gilbert and A.M. Raunio. Sunderland, MA: Sinauer, pp. 219–235.

SHANKLAND, M., SEAVER, E.C. 2000. Evolution of the bilaterian body plan: what have learned from annelids. Proc. Natl. Acad. Sci. U.S.A. 97, 4434–4437.

SHIMIZU, T. 1982. Ooplasmic segregation in the *Tubifex* egg: mode of pole plasm accumulation and possible involvement of microfilaments. Roux's Arch. Dev. Biol. 191, 246–256.

SHIMIZU, T. 1999. Cytoskeletal mechanisms of ooplasmic segregation in annelid eggs. Int. J. Dev. Biol. 43, 11–18.

SHIMIZU, T., NAKAMOTO, A. 2001. Segmentation in annelids: cellular and molecular basis for metameric body plan. Zool. Sci. 18, 285–298.

SHIMIZU, T., SAVAGE, R.M. 2002. Expression of *hunchback* protein in a subset of ectodermal teloblasts of the oligochaete annelid *Tubifex*. Dev. Genes Evol. 212, 520–525.

SHIMIZU, T., ISHII, R., TAKAHASHI, H. 1998. Unequal cleavage in the early *Tubifex* embryo. Dev. Growth Differ. 40, 257–266.

SIDDALL, M.E., APAKUPAKUL, K., BURRESON, E.M., COATES, K.A., ERSEUS, C., GELDER, S.R., KALLERSJO, M., TRAPIDO-ROSENTHAL, H. 2001. Validating Livanow: molecular data agree that leeches, branchiobdellidans, and *Acanthobdella peledina* form a monophyletic group of oligochaetes. Mol. Phylogenet. Evol. 21, 346–351.

SIMPSON, P. 2002. Evolution of development in closely related species of flies and worms. Nat. Rev. Genet. 3, 907–917.

SMITH, C.M., LANS, D., WEISBLAT, D.A. 1996. Cellular mechanisms of epiboly in leech embryos. Development 122, 1885–1894.

SOMMER, R.J. 2000. Evolution of nematode development. Curr. Opin. Genet. Dev. 10, 443–448.

SONG, M.H., HUANG, F.Z., CHANG, G.Y., WEISBLAT, D.A. 2002. Expression and function of an *even-skipped* homolog in the leech *Helobdella robusta*. Development 129, 3681–3692.

SONG, M.H., HUANG, F.Z., GONSALVES, F.C., WEISBLAT, D.A. 2004. Cell cycle-dependent expression of a *hairy* and *Enhancer of split (hes)* homolog during cleavage and segmentation in leech embryos. Dev. Biol. 269, 183–195.

STEINMETZ, P., ZELADA-GONZÁLES, F., BURGTORF, C., WITTBRODT, J., ARENDT, D. 2007. Polychaete trunk neuroectoderm converges and extends by mediolateral cell intercalation. Proc. Natl. Acad. Sci. U.S.A. 104, 2727–2732.

STOREY, K.G. 1989. Cell lineage and pattern formation in the earthworm embryo. Development 107, 519–531.

TESSMAR-RAIBLE, K., ARENDT, D. 2003. Emerging systems: between vertebrates and arthropods, the Lophotrochozoa. Curr. Opin. Genet. Dev. 13, 331–340.

TREADWELL, A.L. 1901. Cytogeny of *Pordake obscura* Verrill. J. Morph. 17, 399–487.

TRUE, J.R., HAAG, E.S. 2001. Developmental system drift and flexibility in evolutionary trajectories. Evol. Dev. 3, 109–119.

VASILIAUSKAS, D., STERN, C.D. 2001. Patterning the embryonic axis: FGF signaling and how vertebrate embryos measure time. Cell 106, 133–136.

WEDEEN, C.J., WEISBLAT, D.A. 1991. Segmental expression of an *engrailed*-class gene during early development and neurogenesis in an annelid. Development 113, 805–814.

WEISBLAT, D.A. 1999. Cellular origins of bilateral symmetry in glossiphoniid leech embryos. Hydrobiologia 402, 285–290.

WEISBLAT, D.A., BLAIR, S.S. 1984. Developmental interderterminacy in embryos of the leech *Helobdella triserialis*. Dev. Biol. 101, 326–335.

WEISBLAT, D.A., HUANG, F.Z. 2001. An overview of glossiphoniid leech development. Can. J. Zool. 79, 218–232.

WEISBLAT, D.A., SHANKLAND, M. 1985. Cell lineage and segmentation in the leech. Philos. Trans. R. Soc. Lond. B 312, 39–56.

WILSON, E.B. 1892. A cell-lineage of *Nereis*. A contribution to the cytogeny of the annelid body. J. Morph. 6, 361–481.

WILSON, E.B. 1898. Considerations on cell-lineage and ancestral reminiscence. Annals N. Y. Acad. Sci. 11, 1–27.

ZACKSON, S.L. 1982. Cell clones and segmentation in leech development. Cell 31, 761–770.

ZACKSON, S.L. 1984. Cell lineage, cell-cell interaction, and segment formation in the ectoderm of a glossiphoniid leech embryo. Dev. Biol. 104, 143–160.

ZHANG, S.O., WEISBLAT, D.A. 2005. Applications of mRNA injections for analyzing cell lineage and asymmetric cell divisions during segmentation in the leech *Helobdella robusta*. Development 132, 2103–2113.

Chapter 6

Evolution, Development and Ecology of *Capitella* sp. I: A Waxing Model for Polychaete Studies

Susan D. Hill* and Robert M. Savage†

*Department of Zoology, Michigan State University, East Lansing MI, and MBL, Woods Hole, MA
†Department of Biology, Williams College, Williamstown, MA, and MBL, Woods Hole, MA

6.1 INTRODUCTION

The recognition of *Capitella capitata* (Grassle and Grassle 1976) as a complex of sibling species rather than as a single cosmopolitan species has opened the door for the use of *Capitella* sp. I as a model organism in studies of evolution and development. The evolution of any group of organisms must be tied with its ecological ability to survive as a reproducing population and to be more fit in certain circumstances than other competing populations or species. As we hope to show, capitellids are the poster children of polychaetes for evo-devo studies.

Since the recognition of the lophotrochozoan clade (reviewed by Halanych 2004), polychaetes have assumed a central role in shaping our understanding of bilaterian evolution. Segmentation, one of the most obvious annelid characteristics, is apparent in each of the major protostome branches and in the deuterostomes—i.e. in the lophotrochozoans, the ecdysozoans and the chordates. Did segmentation arise once, twice or three times? The presence of a trochophore-like larva, the feature for which the trochozoans have been named, is widespread. Is it plesiomorphic or a matter of convergence? Even more far-reaching, the mesodermal layer, by definition, present in all triploblastic animals, appears to develop quite differently in protostomes and deuterostomes, yet involves conserved genes present in both Ecdysozoa and Deuterostomia. What role do these genes play in the Lophotrochozoa? And what about coelom formation? Are the differences as great as we have thought, or are they variations upon a theme, involving conserved genes and familiar regulatory sequences? For these, as well as for many other reasons, the evolution of the polychaetes, probably the most basal annelids, deserves particular attention. The capitellid

Annelids in Modern Biology, Edited by Daniel H. Shain
Copyright © 2009 John Wiley & Sons, Inc.

complex provides a system in which to investigate both the evolution of large phylogenetic differences and the microevolution of characteristics, especially life history traits that have probably led to speciation. In this chapter, we will attempt to show the potential of new developmental studies to cast light on old questions regarding the evolution of Bilateria and lophotrochozoans, and also speciation within the *Capitella* complex.

6.2 SPECIATION STUDIES

Capitella sp. I has advantages that make it a useful model for studies of evolution and development. The most opportunistic of the *Capitella* complex, it can be reared with relative ease in the laboratory. Eggs are deposited in brood tubes allowing siblings (usually 50–200 per brood) of the same age to be collected. Larvae emerge from the brood tube as nonfeeding metatrochophores that can be readily induced to settle. Metamorphosis is rapid and juveniles immediately commence feeding on the reduced mud substrate utilized by adults. Generation time is relatively short (4–6 weeks at 20 °C). Sexes are distinguishable so specific matings can be designed if desired. An important development resulting from the independent work of E. Seaver and R. Savage is the sequencing of the *Capitella* species I genome by the US Department of Energy Joint Genome Institute.[1]

Capitellids have long been of considerable ecological and economic interest. As early and rapid colonizers of disturbed areas of high organic content, they are often used as indicator species of anthropogenic influence. They may reach densities of up to 250,000/m^2 within 1 month (Grassle and Grassle 1974); at the same time, they are able to find and colonize areas as widely dispersed as polluted harbors and sewer outfalls (a major source of the Woods Hole cultures), fish farms (Tsutsumi 1990; Yokoyama 2002; Tomassetti and Porrello 2005), oil spills (Sanders et al. 1980), deep-sea habitats (Grassle and Grassle 1978) and glacier-fed estuaries (Wlodarska-Kowalczuk et al. 2007), and even whale falls (Fugiwara et al. 2007). Until the 1970s, most capitellids were thought to belong to a single, ubiquitous species with worldwide distribution.

6.2.1 Allozyme Patterns and Species Designations

J. Grassle's pioneering work on capitellids in the 1970s and 1980s foreshadowed evo-devo work in the 21st century. Many questions she raised concerning the events that led to speciation within the complex can now be addressed in terms of gene expression and regulation. Grassle and Grassle (1976) described *C. capitata* as a complex of often co-occurring sibling species, many of which were morphologically so similar that field identification at the species level was very difficult or impossible. In this way, they resemble the tubificids and other cryptic clitellates discussed by Erséus and Gustafsson (see Chapter 3). The six species included in their seminal paper (Grassle and Grassle 1976), which ascribes species status to populations within the *Capitella* complex, were collected from five populations in the Woods Hole vicinity and two from Gloucester, MA. As many as five species, characterized on the basis of allozyme patterns and life history traits, were found in a single sample, although their frequencies varied greatly within the annual cycle. Attempts to cross animals of different allelic compositions were unsuccessful. When electrophoretic migration rates of seven allozymes were determined, different alleles were found for all enzymes in all species except two—phosphoglucomutase in species I and Ia

---

[1] US Department of Energy Joint Genome Institute (http://www.jgi.doe.gov/).

and malate dehydrogenase-1 in species II and IIa. No common phosphohexose isomerase (PHI) alleles were detected among any of the species. The species designations (*Capitella* sp. II, *Capitella* sp. III, etc.) that are still used today are based on the electrophoretic mobility of their PHI alleles relative to the most common PHI allele of *Capitella* sp. I. Capitellids not included in the allozyme comparison are referenced by their geographic designation (e.g. *Capitella* ORL is from Orleans, MA; *Capitella* sp. B is from Barcelona, Spain).

In an extension of this work, Grassle and Grassle (1978) compared a wider range of polymorphic loci in a slightly different set of six species (I, Ia, II, III, IIIa and a deep-sea species collected from sunken wood panels left on the sea floor for 9 months). Evolutionary ecology models predict that species with greater physiological/behavioral adaptations and greater mobility should show less allelic variation, while species in stable, less temporally variable habitats should show higher variation. To some extent, this prediction was borne out. The authors used the length of time larvae spent in the water column as a measure of dispersal ability. When the percentage of loci polymorphic for the allozymes under investigation and dispersal ability were considered, the least genetic diversity was found in species Ia and III, which have the longest observed larval life of the six species examined. The highest genetic diversity was found in the deep-sea species, which had a broad but patchy distribution. However, the species generally considered the most opportunistic, *Capitella* sp. I, and one of the more restricted species, *Capitella* sp. IIIa, a direct developer with larvae that spend no time in the water column, unexpectedly showed the same intermediate percentage of polymorphic loci. As the authors point out, allelic differences can only reflect variation at the loci under investigation and may bear no relationship to variation within the genome as a whole (Grassle and Grassle 1978). With the sequencing of the *Capitella* sp. I genome, these issues can now be investigated in more detail as capitellid material again becomes available.

6.2.2 Chromosome Numbers

Karyotypes (Grassle et al. 1987) have revealed that different capitellid species have very different chromosome numbers, not easily explained by polyploidy. For example, *Capitella* sp. I, Ia and ORL have a diploid number of 20; sp. II and IIIa, 26 and sp. III, 14. These differences do not seem to correlate with larval lifestyle (lecithotrophy/planktotrophy) or reproductive mode (male, female, hermaphrodite) (Grassle et al. 1987). Clearly, such differences would explain a lack of interbreeding, and perhaps explain percentage differences in polymorphic loci, but what genetic material has been lost or gained in species appearing so similar?

6.2.3 Life History Traits

Of the dozen or more capitellid species now recognized, many show divergent life history traits such as differences in reproductive mode, egg size, breeding season and larval dispersal characteristics. Some of these are the result of developmental differences that have a profound influence on the distribution and success of populations, and are probably significant in speciation. Differences in life history traits are well documented, although in some instances, the details remain unclear (Grassle and Grassle 1976, 1978). For example, larvae of sp. III have not been induced to settle in the laboratory. These small, planktonic larvae spend more than 2 weeks in the water column in culture without settling, and efforts to induce metamorphosis have been unsuccessful (Grassle and Grassle 1976). The dispersal capability of larvae of the deep-sea species is unknown.

Sexuality of adults seems genetically determined in most capitellid species but, in some, can be environmentally influenced. Samples of most species collected in the wild consist primarily of males and females in roughly 1:1 ratios. In culture, males of sp. I, sp. II and possibly IIa, if maintained at low density, become simultaneous (at least for a while) hermaphrodites (Grassle and Grassle 1976; Holbrook and Grassle 1984; Petraitis 1985, 1988, 1991). Interestingly, hermaphrodites frequently spawn a greater number of viable oocytes than females. Offspring of a hermaphroditic egg bearer develop first as males, which in turn can become hermaphrodites. The above sex change appears not to occur in all species. Species III consists only of obligate outcrossing hermaphrodites, and species IIIa is a true gonochoristic species (Grassle and Grassle 1976).

The opportunistic, co-occurring species *Capitella* sp. I and II both produce lecithotrophic eggs, which they deposit in brood tubes. The maternal parent remains with the developing oocytes, presumably aerating the tube. Larvae of both species are nonfeeding and are competent to metamorphose almost immediately upon emergence from the tube. Both have a short generation time. These characteristics have allowed *Capitella* sp. I and sp. II to be cultured through many generations; unfortunately, we are unaware of any species II currently maintained in culture.

As pointed out by Grassle et al. (1987), it is difficult to distinguish between characters that led to speciation from those that are the consequence of post-speciation evolution. The capitellid species are particularly interesting because they show differences at almost all levels examined except their external adult morphology.

6.3 *CAPITELLA* SP. 1 MORPHOLOGY

Capitella sp. I (and sp. II) adults appear quite homonomous, with an anterior region consisting of a prostomium, an asetigerous segment containing the mouth, and nine thoracic setigers, followed by approximately 18–24 ovigerous abdominal setigers in the female. Non-egg-bearing abdominal segments terminate in a pygidium. Segments are probably added throughout life by a growth zone immediately anterior to the pygidium. With the onset of oogenesis, sex determination can be made. Males bear two sets of highly modified genital spines in segments 8 and 9; developing oocytes beginning in segment 10 can be seen through the ventral abdominal wall of females. In hermaphrodites, both spines and oocytes are present. A complete description of *Capitella* sp. I accompanied by an annotated bibliography is in preparation (Blake et al., in press).

Early development involves unequal spiral cleavage with polar lobe production and epibolic gastrulation by day 3 (Eisig 1899; Werbrock et al. 2001). A classic ciliated, unsegmented trochophore never forms; rather, a segmented metatrochophore emerges from the brood tube about 9 days after spawning. In keeping with the terminology of Seaver et al. (2005), we will consider embryonic development to cease and the onset of larval stages to begin with the appearance of trochal bands (Fig. 6.1). At day 4, anterior segments begin to appear. Using BrdU labeling and other techniques, Seaver et al. (2005) found that initial segments are produced by a midbody ventrolateral growth zone. A posterior growth zone generates later larval segments and persists in juveniles and adults (Fig. 6.2).

The larval stage of *Capitella* sp. I (and sp. II) is a nonfeeding, ~13-segmented, ciliated metatrochophore. It has red eyes (the development of which has been described; Rhode

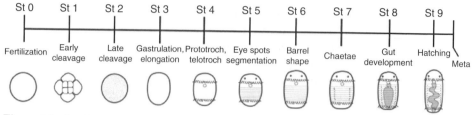

Figure 6.1 Developmental stages in *Capitella* sp. I. St 1—early cleavage; St 2—late cleavage; St 3—gastrulation (stomodeum apparent, ventrolateral nuclear condensation); St 4—appearance of trochal bands, segmentation begins at anterior; St 5—eyespots, segmentation apparent; St 6—segmentation is clearly visible, larva more elongate than in previous stage; St 7—midbody completely segmented, chaetae present; St 8—midgut clearly visible and greenish; St 9—segmented metatrochophore ready to emerge from brood tube (Seaver et al. 2005). Reprinted with permission from Wiley-Blackwell Publishers.

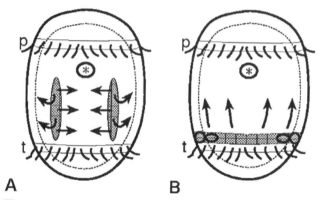

Figure 6.2 Growth patterns in *Capitella*. sp. I. (A) Early larval segments arise from a ventrolateral band of cells extending from stomodeum to telotroch. (B) Later larval growth from posterior growth zone. Asterisks mark the position of the developing stomodeum. Arrows indicate direction of segmentation. Abbreviations: p = prototroch; t = telotroch (modified from Seaver et al. 2005 with permission from Wiley-Blackwell Publishers).

1993), is strongly phototropic, bears setae and is equipped with a well-developed prototroch, telotroch and neurotroch. An arc of cilia with rays extending outward surrounds the anus laterally and ventrally. [Ultrastuctural differences in ciliation patterns of species within the complex have been described (Eckelbarger and Grassle 1987a)]. Metatrochophores of both species I and II are competent to settle and metamorphose almost immediately upon emergence from the brood tube but seem capable of dispersing widely if a suitable cue is not provided. The sensory apparatus used to detect the cue is not known, but Biggers and Laufer (1999) suggest that cells in the prototroch and developing prostomium are involved.

6.4 REPLACEMENT OF LOST SEGMENTS AND REPRODUCTIVE TRADE-OFFS

Anterior regeneration has not been evoked experimentally in either species I or II, nor has it been observed in the field; however, posterior segments, if amputated, regenerate well. Individuals of both species undergoing posterior regeneration are often collected in the field and setae have been retrieved from the guts of flounder, crabs and plaice, so segment loss in nature is a common occurrence.

In the Grassle laboratory at the Marine Biological Laboratory in Woods Hole, where both *Capitella* sp. I and sp. II were available, we carried out intensive experiments investigating the reproductive cost of replacing lost segments as a contributing factor to the relative success and co-occurrence of the two species (Hill et al. 1982, 1987, 1988). Since both regeneration and reproduction are energetically costly, we compared reproductive trade-offs in maximally and minimally regenerating and reproducing females of species I and II. Following amputation of posterior segments, the blastema that forms gives rise to a new pygidium and growth zone that generates new successive segments. In summary, after undergoing repeated simultaneous amputations and reproductive cycles, *Capitella* sp. I females maintained their high reproductive output and reduced the number of segments they regenerated in keeping with their extremely opportunistic lifestyle. *Capitella* sp. II banked a little more on the future; after repeated amputations, they produced fewer eggs per clutch but replaced posterior segments at almost as high a level as after the first amputation.

6.5 METATROCHOPHORES, CILIARY BANDS AND MUSCULATURE

A planktotrophic trochophore larva with downstream feeding was long considered plesiomorphic for the trochozoan group (Neilson and Nørrevang 1985; Peterson et al. 1997). Rouse (1999, 2000a), however, has called this into question, proposing that lecithotrophy preceded planktotrophy, and opposed band feeding evolved numerous times (reviewed by Henry et al. 2007). Clearly, this is an area in need of further study. *Capitella* sp. 1, with its metatrochophore (and no true trochophore stage), and other sibling *Capitella* species may provide ideal materials to examine this question. Indeed, Rouse (2000a, b) includes Capitellidae with downstream feeding clades (although only circumstantial evidence suggests that any of the complex feed as larvae) and specifically calls for further research on this group.

We (Hill and Boyer) are currently investigating dispersal characteristics of *Capitella* sp. I larvae. Since the metatrochophores are nonfeeding and represent the dispersal phase, ciliary bands are agents of locomotion, rather than food collection. Metamorphosis in *Capitella* sp. I is not dramatic but includes loss of cilia as well as other morphological and behavioral changes. If cued with appropriate, reduced mud (Grassle and Grassle 1976) or triggered with methyl farnesoate or other juvenile hormones (Biggers and Laufer 1992, 1999), larvae are competent to settle and metamorphose upon emergence. In our hands, larvae can remain viable in Millipore-filtered seawater for up to 2 weeks. Work by other investigators has shown that they retain their ability to detect suitable habitat (Butman et al. 1988) and to produce undiminished broods of offspring at maturity (Cohen and Pechenik 1999) after remaining for at least 5 or 6 days as swimming larvae.

Embryos begin to develop musculature that will persist through metamorphosis and into the juvenile stage (Hill and Boyer 2001; Seaver et al. 2005) about 4 days after deposition in the brood tube during stage 4 (staging based on Seaver et al. 2005; see Fig. 6.1). As muscle bands are developing, larvae are also forming the ciliary bands necessary for larval dispersal (Fig. 6.3A, B; Hill and Boyer 2005). Cilia of the prototroch are the first to appear followed by the telotroch, by an arc of cilia beneath the forming stomodeum and by the anal cilia. Bands of cilia thicken and the neurotroch develops. At least some

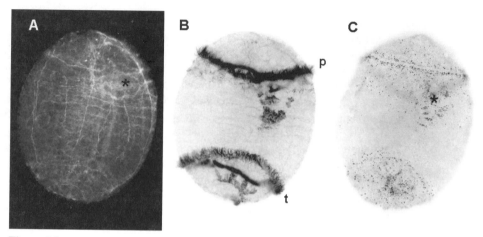

Figure 6.3   *Capitella*. sp. I late stage 4 larvae. (A) Early muscle development, fibers stained with rhodamine-phalloidin. (B) Same embryo labeled with anti-acetylated alpha tubulin. Trochal bands and muscles develop almost simultaneously. (C) Metatrochophore larva labeled with anti-HNK-1. Labeling corresponds to developing ciliary bands. Ventral view. Asterisks mark the position of the developing stomodeum. Abbreviations: p = prototroch; t = telotroch (Hill and Boyer 2001, 2003, 2005).

cells in each trochal band and the anal cilia label with an antibody against HNK-1 (Fig. 6.3C; Hill and Boyer 2003), an antigen expressed in many cell adhesion and neural recognition molecules. HNK-1 is used to trace migrating neural crest cells in vertebrates. Its appearance in *Capitella* sp. I follows formation of the cilia, perhaps reflecting the precursors of early innervation sites of trochal bands.

Arenas-Mena et al. (2007) have found that ciliary band markers in *Hydroides elegans* are expressed very early in development. In this species, the prototroch is the first of the bands to develop and trochoblast precursors are the first blastomeres to move out of cell proliferation and into differentiation mode. *In situ* experiments demonstrate that genes similar to *caveolin, alpha-* and *beta-tubulin* and *tektin* are expressed in the eight primary trochoblast precursors, in $1q^{221}$ and $1q^{212}$ (q designating quadrant unknown in this equally cleaving egg). Similar studies have not yet been conducted in *Capitella* sp. I but interesting comparisons could be made. *H. elegans* is an indirect developer that feeds in the larval stage, and becomes ciliated relatively earlier in its development than *Capitella* sp. I. Cilia appear in the blastula and the prototroch is clearly visible during gastrulation (Seaver et al. 2005, Arenas-Mena et al. 2007); in *Capitella* sp. I, the prototroch and telotroch develop almost simultaneously, after many more cell divisions than in *H. elegans*, with the neurotroch lagging slightly behind (Hill and Boyer 2005; Seaver et al. 2005). An antibody against acetylated alpha tubulin was used by all of the above investigators to label cilia. Arenas-Mena et al. (2007) report that the alpha-tubulin homologue is expressed transiently in apical tuft cells, adoral ciliary cells and the metatroch as well as in the trochal blastomeres that will form the prototroch. Further investigation including cell lineage studies (Henry et al. 2007) on other polychaetes could do much to resolve the ambiguity that surrounds trochal bands.

In *Capitella* sp. I, less than 24 h after the first appearance of cilia, precursors of the first longitudinal muscles are detected by phalloidin binding. Two longitudinal ventral muscles and four longitudinal dorsal fibers, a ring around the stomodeum and an arc that extends into the episphere appear almost simultaneously in a loosely controlled sequence

(Hill and Boyer 2001). Fibers first appear in the prototrochal region, then elongate posteriorly. Additional longitudinal fibers form and circular segmental musculature appears. Development of segmental muscles proceeds from anterior to posterior and from lateral to ventral, finally completing the dorsal quadrant in each segment (Hill and Boyer 2001; Seaver et al. 2005). By emergence from the brood tube, the metatrochophore has well-developed longitudinal, circular and oblique muscles, as well as short, intersegmental and setal sac muscles; these become the musculature of the juvenile worm at metamorphosis.

Early longitudinal and segmental muscles develop in a manner similar to that seen in *Capitella* sp. I in the oligochaetes *Enchytraeus coronatus* and *Limnodrilus* sp. (Bergter and Paululat 2007; Bergter et al. 2007), and in the polychaete *Ophryotrocha diadema* (Bergter et al. 2008). The described pattern of muscle development differs from that seen in leeches, leading Bergter et al. (2008) to suggest that the basic pattern described above is plesiomorphic in polychaetes.

Hill and Boyer (2005) suggest that a heterochronic shift has occurred in at least *Capitella* sp. I, and perhaps in others of the *Capitella* complex, so that the trochophore stage is bypassed. Rather, the two locomotory systems—the ciliary bands of the swimming larva and the musculature of the crawling/burrowing juvenile—develop simultaneously quite early in development, thus poising the larva either to settle immediately upon emergence from the brood tube or to remain in the water column until it encounters suitable substrate.

6.6 GENE EXPRESSION DURING THE SPECIFICATION AND DIFFERENTIATION OF GERM LAYERS

Triploblasty and the presence of a coelom cannot be overemphasized when considering bilaterian evolution. The developmental mechanisms by which these form seem very different in deuterostomes and protostomes, and indeed contribute to our recognition of the distinction between the groups (reviewed by Reiger and Purschke 2005). Molecular genetic techniques now allow comparisons of gene functions in controlling tissue differentiation. A growing body of literature suggests that while many genes have a conserved function in ecdysozoans and deuterostomes, an even greater similarity exists between lophotrochozoans and the latter (Tessmar-Raible and Arendt 2003). The availability of *Capitella* sp. I embryos/larvae, as well as other characteristics described above, makes capitellids an ideal lophotrochozoan model for investigating these and related questions. Even within polychaetes, variability seems the rule.

6.6.1 Mesoderm and Nervous System

The origin of mesoderm in *C. capitata* was traced by Eisig in 1899 as forming from three micromeres—3c, 3d and 4d. Asymmetric divisions of two large teloblasts located at the posterior of the developing embryo generate two anteriorly extending bands of cells that apparently give rise to segmental mesoderm. Using BrdU labeling, Seaver et al. (2005) found two regions of cell proliferation: bilateral growth zones at stages 4–6 that contribute to the embryonic midbody ectoderm and mesoderm, and a later posterior growth zone in stages 6–9 that gives rise to the larval and juvenile posterior ectoderm and mesoderm (Fig. 6.2). The relationships between micromeres, teloblasts and growth zones are unclear. Developing *Capitella* sp. I oocytes are protected by a close-fitting membrane that makes cell deletion/labeling experiments difficult and, to date, fate maps have not been completed.

Experiments to investigate the role of genes known to be influential in mesoderm determination and differentiation are being undertaken by the Seaver laboratory. The transcription factors twist and snail regulate mesoderm formation in many bilaterians and are widely conserved (Leptin 1991; Technau and Scholz 2003). Two homologues of each have been isolated in *Capitella* sp. I, and their *in situ* expression patterns in embryos and larvae have been localized (Dill et al. 2007).

The role of *twist* in mesoderm differentiation, especially in mesoderm patterning and muscle formation, has been well documented in both ecdysozoans (insects—Handel et al. 2005; Sokol and Ambros 2005; nematodes—Harfe et al. 1998) and chordates (*Xenopus laevis*—Hopwood et al. 1989; amphioxus—Yasui et al. 1998). In both *Drosophila melanogaster*, a long germ band insect (Thisse et al. 1988) and the short germ band beetle *Tribolium castaneum* (Handel et al. 2005), *twist* also plays a role in the early specification of the mesoderm itself, although this role has not been seen in other groups.

In *Capitella* sp. I, *twist* expression seems to correlate with areas where the mesoderm is undergoing differentiation but not with its initial specification. *CapI-twt1* is expressed slightly earlier than *CapI-twt2* (stage 3 versus 4) and, although areas of expression overlap, clear differences are apparent (Dill et al. 2007). *CapI-twt2* is expressed most prominently in, but is not restricted to, regions of *twt1* expression. Expression patterns suggest that both are involved in the development of coelomic mesoderm and/or late-forming larval musculature and in the formation of the prostomial muscle(s). Since neither is expressed during cleavage stages, Dill et al. (2007) conclude that, in *Capitella* sp. I, *twist* homologues do not specify mesodermal cell lineages; rather, they suggest that the plesiomorphic function of *twist* is its role in mesoderm differentiation and that its role in mesoderm specification may be unique to insects.

Snail homologues encode zinc finger transcriptional repressors that regulate cell movement, adhesion, division and survival (Barrallo-Gimeno and Nieto 2005). Snail proteins are known to regulate cell movements of individual cells by repressing E-cadherins, but they also play a role in tissue movements where cells maintain contact, as in mesoderm migration during formation of the ventral furrow in *D. melanogaster* (Leptin 1991) and gastrulation in several vertebrates where they seem to act by inducing the movement of presumptive mesoderm rather than by specifying its patterning (Barrallo-Gimeno and Nieto 2005). Among lophotrochozoans, in the limpet *Patella vulgata*, expression of *snail* is restricted almost entirely to the ectoderm (Lespinet et al. 2002); in the leech *Helobdella robusta* (Goldstein et al. 2001), snail proteins were detected in precursors of different tissue layers prior to gastrulation. In *Capitella* sp. I, *CapI-sna1* is expressed in the first four blastomeres and in precursors of all germ layers throughout cleavage and early gastrulation (Dill et al. 2007). By the end of gastrulation, expression is seen only around the closing blastopore. At later stages, *CapI-sna1* is expressed in the developing nervous system, in the lateral mesoderm where it partially overlaps with *CapI-twt1* and *CapI-twt2* expression, in the growth zone in both ectoderm and mesoderm and in the foregut. *CapI-sna2* expression is much more restricted both in time and location. Expression is detected in the developing nervous system and foregut, and by stage 9, only in the posterior ventral nerve cord. Dill et al. (2007) conclude that *snail* genes seem to regulate cell movement during gastrulation and suggest that, in annelids, *snail* may play a role in mesoderm differentiation, although this has not been reported in ecdysozoans or deuterostomes. Furthermore, in *Capitella* sp. I, *snail* genes seem to play a role in neurogenesis, perhaps related to the internalization of these cells from epithelial ectoderm (Dill et al. 2007).

In *Capitella* sp. I, expression of a member of the GATA1/2/3 family identified as *CapIGATAA1* (Boyle and Seaver 2008a) is expressed in the developing brain and ventral nerve cord as it is in chordates. At stage 6, they report a ladder-like expression pattern extending the length of the ventral nerve cord. Similarly, Gillis et al. (2007) isolated two GATA factors from *Platynereis dumerilii* and report expression of *Pd GATA1/2/3* in ectodermal derivatives including the nervous system. Using BLAST searches of available genomic data, they identified one GATA factor gene in the cnidarian *Hydra magnipapillata*, agreeing with the work of Martindale et al. (2004) on another cnidarian *Nematostella vectensis*. Interestingly, in the sponge *Reniera* sp., Gillis et al. found no GATA factor gene sequences.

Several of the arthropod segmentation genes discussed later are expressed in the *Capitella* sp. I nervous system: *hunchback* (*Cc-hb*—a gap gene) during development and in adults (Werbrock et al. 2001), *engrailed* and *hedgehog* (segment polarity genes) in the developing ventral nerve cord (*hedgehog* also in the developing cerebral commissures) (Seaver and Kaneshige 2006), and *CapI-hes2* and *CapI-hes3* (part of notch signaling pathway) during early brain development (Thamm and Seaver 2008).

Fröbius and Seaver (2006a) have isolated a novel paired-like homeobox gene that they named homeobrain-like (*CapI-hbnl*) because of its similarity to *D. melanogaster* homeobrain, noting that it is a useful neural marker of the annelid nervous system. *CapI-hbnl* is expressed first in the developing larval brain at stage 5 during the later phases of cerebral ganglia formation and in developing eyes by stage 6 (eyes are visible at stage 5). Gene orthology analysis reveals that *CapI-hbnl* clusters with *D. melanogaster hbn* in a node that contains paired-like homeobrain genes from mosquito, sea urchin, hemichordate and cnidarian. Fröbius and Seaver (2006a) suggest that these genes form part of a new family of paired-like homeodomain genes. *D. melanogaster hbn* has not been reported in eye development and, in the cnidarians *Hydra vulgaris* and *H. magnipapillata*, the related gene *HyAlx* was expressed only in the tentacle zone, not in the nerve net.

6.6.2 Gut Formation

The foregut of the metatrochophore comprises a buccal cavity, a pharynx that, with the buccal cavity, will form the eversible proboscis of the adult, and a ciliated esophagus (Boyle and Seaver 2008a). During development, the foregut roof thickens and develops a cuticular fold that surrounds a ciliated pad, gland-like structures and proboscis-retracting muscles. This "dorsal pharynx" is characteristic of oligochaetes, whose feeding pattern *Capitella* sp. I share and has not been reported previously in a marine polychaete (Boyle and Seaver 2008b). The extensive, folded midgut is followed by a short hindgut that extends through the pygidium and terminates in the anus. Foregut and hindgut are traditionally believed to be of ectodermal derivation, while the midgut is lined with endoderm. Table 6.1 summarizes gene expression patterns identified to date during the development of the *Capitella* sp. I gut.

Four *GATA* genes and one *FoxA*, gene families known to be involved in patterning ectodermal and endodermal components of the gut as well as other tissues in many bilaterians, were isolated in *Capitella* sp. I and their expression patterns mapped (Boyle and Seaver 2008a). In vertebrates, GATA1/2/3 factors have patterning functions in erythroid and ectodermal lineages, while GATA4/5/6 factors are expressed in endomesodermal lineages; their role in protostome development is less clear. Using available genome sequences, Gillis et al. (2008) identified the complete *GATA* complement of 53 genes

Table 6.1   Gene expression during gut development in *Capitella* sp. I

Developmental Stages of *Capitella* sp. I	Blastomeres	Foregut	Midgut	Hindgut
Stage 1: early cleavage	*foxA—A-C quadrants, gataB1, hb, sna1, vasa, nanos*			
Stage 2: late cleavage	*foxA, hb, vasa nanos*			
Stage 3: gastrulation, elongation		*blastopore: foxA, gataB1, sna1* *stomodeum: hb, hh, vasa, nanos*		
Stage 4: proto- and telotroch (larvae)		*foxA, cdx, hb, hh, sna1, twt2 (epithelium), hes3*	*gataB1* *hb*	*foxA* *wnt1* *cdx* *hb*
Stage 5: visible segmentation, eyespots		*foxA, gataA1, cdx, hb, hh sna1* *twt1 (muscle)* *twt2 (epithelium)* *delta* *hes1, hes2, hes3*	*gataB1* *gataB2* *hb*	*fox A* *wnt1* *cdx* *hh*
Stage 6: barrel shape		*foxA, gataA1, cdx, delta, sna1, sna2, twt1 (muscle)* *twt2 (epithelium) notch, hes2, hes 3*	*gataB1* *gataB2* *hb*	*foxA* *wnt1* *hb* *cdx* *vasa, nanos*
Stage 7: setae appear		*foxA, gataA1, cdx, sna2* *twt1 (muscle)* *twt2 (epithelium)* *notch, hes2, hes3*	*xlox* *gataB1* *gataB2* *hb*	*foxA, wnt1, cdx, hb, hh, twt1*
Stage 8: foregut visible; midgut expansive, green in transmitted light		*foxA, gataA1, cdx, hb*	*gataB1* *gataB2* *xlox, hb*	*foxA, wnt1, hb, hh, cdx*
Stage 9: ready-to-emerge metatrochophores		*foxA, gataA1, cdx, hb, hh*	*Xlox* *hb*	*foxA* *cdx, hb, hh*

The list will lengthen as more genes are investigated. Gene expression during different embryonic/larval stages is shown, but sequential gene interactions are not implied.

in nine protostomes (five phyla) including *Capitella* sp. I and established the orthology of the two *GATA* classes (*GATA1/2/3* or *GATA4/5/6*) in protostomes. They conclude that in protostomes, while *GATA1/2/3* remained unduplicated, *GATA4/5/6* homologues underwent protostome-specific duplications. Boyle and Seaver (2008a) suggest that *GATAb* genes underwent independent duplication in annelids or perhaps in the capitellid lineage.

In *Capitella* sp. I, Boyle and Seaver (2008a) isolated four *GATA* genes, three of which (*CapI-gataB1, CapI-gataB2, CapI-gataB3*) group within the GATA4/5/6 subfamily and one that groups with GATA1/2/3 (*CapI-gataA1*). Both subfamilies are expressed during gut development. *CapI-gataB1* is expressed in the developing endoderm of the midgut; *CapI-gataB2* expression overlaps with *CapI-gataB1* but is also expressed in visceral mesoderm; *CapI-gataB3* is coexpressed with *CapI-gataB2* only in the visceral mesoderm. The only GATA1/2/3 orthologue, *CapI-gataA1*, is expressed primarily in ectodermal tissues of the foregut and the developing nervous system. While the expression patterns are clearly delineated, it cannot be concluded that they follow strict germline boundaries, since fate mapping in *Capitella* sp. I is incomplete. However, the pattern is similar to that seen in *P. dumerilii* in which two GATA factors have been found, and in a wide range of protostomes and deuterostomes (Gillis et al. 2007).

Fox, a member of the fork-head box family of transcription factors, is expressed during gut formation in deuterostomes, from echinoderms and hemichordates to mice (summarized by Boyle and Seaver 2008a). Among edysozoans, *Fox* expression has been reported during the formation of parts of the digestive system in *D. melanogaster* and in *Caenorhabditis elegans*. Boyle and Seaver (2008a) report that a *foxA* orthologue (i.e. *CapI-foxA*) is expressed in the ectodermal portions of the gut, namely, the stomodeum and its derivatives, the pharynx and esophagus, and in the hindgut derived from the proctodeum (Fig. 6.4A). Its expression pattern suggests that *CapI-foxA* may be involved in establishing the foregut/midgut and midgut/hindgut transition at germ-layer boundaries. A similar expression pattern of two *foxA* genes in *H. elegans* (Arenas-Mena 2006) leads Boyle and Seaver (2008a) to propose that a single *foxA* gene in *Capitella* performs the function of the two genes in *H. elegans*. Since a *foxA* orthologue is expressed in the ectodermal pharynx of the cnidarian *N. vectensis* (Martindale et al. 2004), as well as a *GATA* orthologue in the endoderm, it seems to be a pattern that was present before the evolution of "Urbilateria," and whose function has been roughly conserved.

Genes known to be involved in insect segmentation are also expressed in the developing gut in *Capitella* sp. I. For example, the segment polarity gene *wingless* (*CapI–wnt1*) initially is expressed in *Capitella* sp. I at stage 4 at the extreme posterior of the embryo, marking the position of the future anus. As the gut develops and the lumen forms, expression is seen at the ectodermal–endodermal junction. When larvae are ready to emerge from the brood tube, competent to metamorphose and commence feeding, no further expression

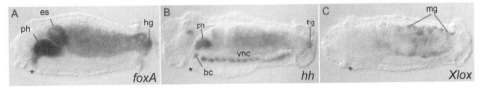

Figure 6.4 Expression of *Capitella* sp. I molecular markers in restricted domains along the length of the gut as analyzed by whole-mount *in situ* hybridization. All animals are stage 8, lateral views (anterior is to the left, ventral is down). Gene names are marked in bottom right corner. (A) *FoxA* is expressed in ectodermal portions of the gut including the pharynx, esophagus and hindgut (Boyle and Seaver 2008a). (B) *hh* expression in the foregut is restricted to the dorsal face of the pharynx, a portion of the buccal cavity, and the hindgut. Note that *hh* is also expressed in the ventral nerve cord (Seaver and Kaneshige 2006). (C) Expression of the ParaHox gene *Xlox* is restricted to the endodermal midgut (Fröbius and Seaver 2006b). Asterisks mark the position of the mouth. Abbreviations: bc = buccal cavity; es = esophagus; hg = hindgut; mg = midgut; ph = pharynx; vnc = ventral nerve cord. (See color insert.)

of *CapI-wnt1* is seen, although new segments continue to be added by the growth zone (Fig. 6.4B; Seaver and Kaneshige 2006). Another segment polarity gene *hedgehog* (*CapI-hh*) is also expressed in a ring of hindgut tissue and in anterior gut tissue (Seaver and Kaneshige 2006). *CapI-twt1* and *twt2* are expressed in the developing fore- and hindgut (Dill et al. 2007) and *CapI-sna1* is expressed in surface cells as the blastopore closes, and in the developing foregut. Members of the Notch signaling pathway are expressed in the developing foregut correlating to an area of high cell proliferation (Thamm and Seaver 2008).

Two of three *ParaHox* genes (*CapI-Cdx* and *CapI-Xlox*) are also expressed in the developing gut (Fröbius and Seaver 2006b) and will be discussed separately.

6.7   SEX AMONG THE VERMES

Sex in *Capitella* sp. I seems to be genetically controlled. The sexes are separate but males at low density have the ability to produce eggs (i.e. to become hermaphroditic). Expression of *vasa* and *nanos* genes, genes involved in germline development in many phyla, is now being investigated in the Seaver lab. Originally considered to be a segmentation gene, *nanos* appears to be associated with the primordial germ line and perhaps with early embryonic patterning. Both *CapI-vasa* and *CapI-nanos* are expressed in adults in developing gametes. Both are also expressed in developing somatic tissues—during cleavage, in most blastomeres of all four quadrants, and later in the presumptive brain, mesodermal bands and foregut. *CapI-nanos* expression at the fifth to sixth cleavage is enriched in 4d descendants. Perhaps the most interesting observation is its expression in a group of cells in larval segment 4 and in adults in thoracic segments 4 and 5. Since sperm ducts are in segments 7 and 8 and ovaries start in segment 10 (Eckelbarger and Grassle 1983, 1987b), expression of *vasos* and *nanos* in segments 4 and 5 raises interesting questions. Dill and Seaver (2008) propose that these are primordial germ cells that migrate to their definitive site. A new structure in a similar location called the primary gonad was described by Rebscher et al. (2007) in *P. dumerilii*. In neither case were these "organs" recognized prior to *in situ* experimentation.

In *P. dumerilii*, maternal vasa protein was found in developing oocytes and in early cleavage stages, where it became segregated into the micromeres, including 4d (Rebscher et al. 2007). Labeling of the mesodermal presumptive growth zone (MPGZ) by tracers injected into 4d indicates that this micromere is the founder cell of the mesodermal bands in this species. Expression of *Pdu-vasa* overlapped with BrdU labeling confirming identification of the MPGZ. At 4 days postfertilization, four cells expressing *Pdu-vasa* migrated anteriorly from the posteriorly located MPGZ to a region in segment 3 just behind the pharynx. In early juveniles, these cells divided to form a newly discovered cluster named the primary gonad (Rebscher et al. 2007). This structure was pushed posteriorly by the developing jaws and pharynx to segment 5 and gave rise to numerous gonial clusters. Is this structure homologous to the capitellid cluster of primordial germ cells noted above? In *P. dumerilii*, expression of *Pdu-vasa* became gradually restricted during larval development to the MPGZ and the four migratory cells. In juveniles, expression continues only in the developing gonads.

Rebscher et al. (2007) propose a two-step process of germ cell specification and differentiation in *P. dumerilii*. First, a population of undifferentiated cells is established in the mesodermal posterior growth zone that is probably descendent of 4d resulting from maternal cytoplasmic determinants. Cells in the MPGZ are potentially able to differentiate either into mesoderm or germ cells. The second step is the migration of the four cells that will give rise to the primary germ cells. Rebscher et al. (2007) suggest that the differentiation of primordial germ cells from undifferentiated, multipotent mesodermal cells as seen in *P. dumerilii* is the ancestral mechanism of germ cell specification.

The *vasa* orthologue is expressed in developing sperm as well as in oocytes in *Capitella* sp. I but not in *P. dumerilii*. Using an anti-vasa antibody, Rebscher et al. (2007) detected vasa protein in whole body extracts at all developmental stages except in males as they approached maturity and spawned females, indicating a germ cell specificity in this species. Perhaps the difference in *vasa* expression in *Capitella* sp. I and *P. dumerilii* is a reflection of differences in life history traits. *P. dumerilii* is semelparous—males and females are separate, and both die soon after spawning. Oogenesis in *Capitella* sp. I is more plastic; both sexes are multiparous and if females are well fed, the next generation of oocytes is mature by the time the tended brood of metatrochophores emerges. Additionally, males may undergo oogenesis. Perhaps the continued presence of *vasa* and *nanos* helps to maintain tissues "at the ready" in this species.

Zygotically expressed *nanos* is also associated with both primordial germ cell differentiation and with normal development in the leech *H. robusta*. The use of morpholino knockdown strategy and subsequent rescue (Agee et al. 2006) confirmed that maternal *Hro-nos* is required for epiboly and normal development.

6.8 ANNELIDS AND THE SEGMENTATION DEBATE

The ease of identifying and characterizing developmental regulatory genes in little-studied phyla is contributing answers to long-standing questions in metazoan evolution. One such question focuses on the molecular and cellular origin(s) of segmentation. Developmental and evolutionary biologists have debated the origin of three segmented phyla (chordates, annelids and arthropods) for the past century, and this debate continues to the present. The three phyla are not considered closely related; therefore, segmentation has either been lost or gained multiple times. The wealth and breadth of new comparative data from versatile functional knockdown techniques and genomic analyses will contribute worthy insights into the developmental mechanisms that underlie the evolutionary origins of both segmented and unsegmented animal body plans.

Much of our understanding of segmental pattern formation comes from studies in the fruit fly *D. melanogaster*. In dipteran flies, a hierarchy of zygotically active segmentation genes is responsible for generating the metameric units along the anteroposterior axis. The segmentation gene products differentially regulate *Hox* gene expression in a segment-specific manner, thereby coordinating segment generation with segmental identity. Large numbers of transcription factors, signaling factors and their pathway components have been identified and, for most proteins, their function in fly development determined. Information from this model is an important point of reference as developmental regulators are identified in other animals using cloning by homology approaches. As a result of such comparative studies, a core set of animal body plan genes appears to have been present in the urbilaterian. These genes comprise what is now called the genetic toolbox for development shared by all animals (Carroll et al. 2005).

The upside of the toolbox concept is that we have identified a subset of gene products within a given genome that are responsible for generating the animal body plan and all its parts. The challenge is to distinguish shared components of the developmental regulatory network operating in each animal from divergent components and to determine how these changes mediate the diversity of form observed in the natural world today. The variation of developmental programs in arthropods exemplifies this challenge in the following question: to what extent is the fly segmentation gene pathway conserved in other insects? Segmentation gene orthologues have been identified in several insect orders and are expressed in patterns similar to that of the fruit fly (Lui and Kaufman 2005; Choe et al. 2006; He et al. 2006) with notable exceptions (Damen 2007). The conservation of the segmentation gene pathway is noteworthy given the variation in developmental programs across insect orders. Although short germ band insects (e.g. grasshoppers) pattern the embryo in a cellular context in contrast to long germ band insects (e.g. flies), which pattern the embryo in an acellular or syncytial context, they both appear to utilize the segmentation gene hierarchy first described in the long germ band insect *D. melanogaster*. One can infer that the variations observed in the segmentation gene circuits presumably underlie the differences between short and long germ insect development.

Although much of the information available in arthropods comes from insects, data suggest that arthropods share key components of the segmentation machinery and therefore a common evolutionary history. In flies, the segmentation hierarchy comprises maternal genes and three tiers of zygotically active genes called gap, pair rule and segment polarity that are responsible for generating the metameric units along the anterior–posterior axis. A segment's axial identity is determined by Hox proteins. The segment polarity proteins engrailed (en), wingless (wg) and hedgehog (hh) are the most highly conserved tier of the segmentation hierarchy based on expression data obtained from a wide range of arthropods. The three proteins act as a circuit whose interactions define segmental boundaries at the segmented germ band stage also viewed as the phylotypic stage of Arthropoda (von Dassow et al. 2000; Damen et al. 2005; Gabriel and Goldstein 2007). The upstream regulators of the segment polarity genes are the pair-rule proteins whose role is to translate broad areas of expression into periodic patterns. The expression patterns of the pair-rule proteins, in a one-to-one comparison of insects to other arthropods, exhibit significant variation but appear to be expressed in either single or double segmental periodicity before the expression of the segment polarity genes and any morphological appearance of segments (Chipman et al. 2004; Davis et al. 2005; Damen 2007). These data suggest that the segment polarity and the more variable pair-rule classes of genes are likely to play conserved roles in arthropod segmentation. Because of lack of data, it is not clear how the orthologues to the segmentation genes from other tiers (e.g. gap class) function in noninsect arthropods.

6.8.1 Annelid Orthologues to Fly Segmentation Genes

6.8.1.1 Gap Genes

The orthologue of the insect gap gene *hunchback* was characterized in capitellids by the Savage laboratory (Werbrock et al. 2001). In *D. melanogaster*, *hunchback* participates in anterior pattern formation by regulating the expression of other segmentation and *Hox* genes in a concentration-dependent manner. In clitellate annelids, *hunchback* is expressed in micromeres and their derivatives but not in the segment region of the trunk or in any other tissue as an AP gradient (Savage and Shankland 1996; Shimizu and Savage 2002). In the polychaete *Capitella* sp. *I*, *hb* orthologue (*Cc-hb*) expression patterns are similar to those in clitellates and do not follow the insect pattern. *Cc-hb* is expressed in the oocyte germinal vesicle and in every macromere and micromere throughout early cleavage. At gastrulation, expression becomes restricted but is apparent in cells undergoing epiboly at the leading edge of the blastopore. In developing larva, *Cc-hb* is expressed in the presumptive trochal bands, foregut, developing brain and ventral nerve cord, and in the prostomium and pygidium. Importantly, it is not expressed in segmental precursor cells (Werbrock et al. 2001; Pinnell et al. 2006).

In contrast, maternal *hb* expression is not seen in *P. dumerilii*. Kerner et al. (2006) report that, in this species, *hunchback* is not expressed in unfertilized eggs or early developmental stages; however, after its first expression at 13 h postfertilization, by which time the prototroch has formed, *Pdu-hb* is expressed in all larval stages and in juvenile worms. *Pdu-hb* is first expressed in one cell in the episphere that probably contributes to the brain, and in a broad region of the hyposphere. By 15 h, the dorsal ectoderm and the mesodermal bands in the trunk (hyposphere) show expression. As development continues, *Pdu-hb* is expressed in the developing nervous system, the stomodeum, chaetal sacs, anal region and posterior growth zone. The differences in annelid *hb* expression observed between *P. dumerilii* and the other annelids remain unexplained.

6.8.1.2   *Pair-Rule Genes*

Expression patterns of the second tier of segmentation genes recognized in *D. melanogaster*, the pair-rule genes, are being investigated in *Capitella* sp. I. Three genes—*pax3/7*, *runt* and *even-skipped*—are expressed during segmentation but in a pattern that differs from that observed in insects (Seaver, pers. comm.). Expression of *pax3/7* is initiated prior to segment formation and is observed as two longitudinal bands in the ventrolateral ectoderm rather than as segmental stripes. *Runt* expression appears in the ventral nerve cord then in other tissues including the gut. Expression in the trunk is initiated after segmentation has occurred and again is not seen as segmental stripes. The investigators suggest that *even-skipped* is expressed in the mesodermal stem cells that probably give rise to trunk mesoderm. As mesodermal bandlets form and later expand around the circumference of the larva, mesodermal *eve* expression persists. In *P. dumerilii*, the pair-rule gene *even-skipped* is expressed in the posterior growth zone (de Rosa et al. 2005). Because of the grouping of genes studied by different investigators, *even-skipped* in *Platynereis* will be discussed further with *caudal* and *brachyury*.

6.8.1.3   *Segment Polarity Genes*

Seaver and Kaneshige (2006) isolated orthologues of three segment polarity genes, *engrailed, hedgehog* and *wingless*, in *Capitella* sp. I. Expression patterns of *CapI-en*, *CapI-hh* and *CapI-wnt1* in embryonic, larval and juvenile stages of *Capitella* sp. I and of *Hel-en* in *H. elegans* were followed. Because most of the segments of the adult polychaetes are generated after metamorphosis, expression patterns in *Capitella* sp. I juvenile worms as well as in developing larvae were localized. All three genes are expressed during development in *Capitella* sp. I. *CapI-wnt1* marks the future anus. *CapI-hh* is expressed in the hindgut, just anterior to *CapI-wnt1* epression, and in the ventral nerve cord, anterior gut tissue and mesoderm. In *Capitella* sp. I and *H. elegans*, as well as in *Chaetopterus* (Seaver et al. 2001), only *en* expression is initiated prior to the overt appearance of segments and so might be a candidate "segmentation" gene; however, in none of the three species does it show the characteristic "striped" pattern before or after segmentation. A reiterated segmental expression in ectodermal patches in pre- and post-segmented tissue never encircles the body and so is not considered a "stripe" (Seaver and Kaneshige 2006).

Experiments using *P. dumerilii* are in direct contrast. The first expression of *Pdu-wnt1* and *Pdu-en* preceded segmentation and continued in well-defined segments (Prud'homme et al. 2003). As in *Capitella* sp. I, initial segmentation in developing larvae and segment addition from the posterior growth zone in juveniles were investigated. Because post-metamorphic growth is slow in *P. dumerilii* (one segment every five days), animals undergoing regeneration following caudal amputation were used as a rapid source of somitogenesis, after confirmation that expression patterns were the same during regeneration and juvenile growth. Both *Pdu-en* and *Pdu-wnt1* were expressed in presomitic ectoderm in a pattern of alternating stripes corresponding to localization of *Pdu-en* at the anterior segment boundary and *Pdu-wnt* at the posterior boundary. We applaud the use of regenerating *P. dumerilii* for the study of gene expression; this should be an area of fruitful experimentation in the future. We caution, however, that until clearly shown, one cannot assume that the sequence of gene expression during regeneration recapitulates that seen in normal development.

The contrasting comparative data underscore the need for a highly resolved annelid phylogeny. Species of *Capitella* and *Playtnereis* may or may not be basal polychaetes. Both are considered homonomous because they generate segments that seem quite similar from anterior to posterior; however, just as snakes and caecilians are not assumed to be basal vertebrates with hagfish and lampreys based on the homonomous appearance of their visible segmentation, neither can we make the same assumption with polychaetes. In capitellids, differences are evident between thoracic and abdominal segments, and between gametogenic and other segments. In *P. dumerilii*, the production of a specialized gamete-bearing epitoke at maturity is probably derived. In recent annelid phylogenies based on large datasets, Rousset et al. (2007) do not find support for the monophyly of either Annelida or Polychaeta while Struck et al. (2007) find that the monophyly of the Annelida, including the Echiura and Sipuncula, is supported. Further studies will resolve these and other issues, but, in the meantime, the basal polychaete remains unknown.

6.8.2   Interpreting Conflicting Comparative Data from Segmentation Gene Homologues in Annelids, Arthropods and Chordates

6.8.2.1   *Engrailed*, *Hedgehog* and *Wingless* Circuits

A cautionary tale concerns the role that segment polarity gene homologues may play in annelids. In arthropods, the *hh*, *en* and *wg* regulatory circuit is expressed in periodic segmental stripes along the length of the AP axis prior to observed morphological segmentation. Does the regulatory circuit of segment polarity genes characterized in arthropods operate in annelids and, if so, does it define segmental boundaries? In leeches (*H. robusta*, *Theromyzon tessulatum*), oligochaetes (*Pristina leidyi*) and polychaetes (*Chaetopterus*, *Capitella* sp. I, *H. elegans*), the spatiotemporal expression patterns of the *engrailed*, *wingless* and *hedgehog* orthologues are not consistent with a role in the delineation of annelid segments (Shain et al. 1998; Wedeen and Weisblat 1991; Bely and Wray 2001; Seaver et al. 2001; Kang et al. 2003; Seaver and Kaneshige 2006). However, in the polychaete *P. dumerilii*, *wg* and *en* gene expression is observed in continuous ectodermal stripes on either side of segment boundaries prior to morphological segmentation. These data support the view that they are playing a role in segment formation, similar to that observed in arthropods (Prud'homme et al. 2003). At least two possible interpretations may help to explain these conflicting data. First, the *en/wg* circuit may have been independently recruited in arthropods and *P. dumerilii* and therefore represents an example of convergent evolution. Second, the *en/wg* circuit was lost in most annelids examined to date except *P. dumerilii* and the current differences in the annelid segmentation gene cascade mirror what has happened in arthropods. The way forward in this debate is to examine spatiotemporal distributions of protein in addition to the mRNA accumulation patterns generated by *in situ* hybridization, to undertake functional studies and to examine a broader range of taxa.

6.8.2.2 *Notch Signaling*

The Notch signaling pathway, *Delta, Notch* and *hairy*, plays a role in vertebrate segmentation and in some noninsect arthropods, suggesting a shared role (Damen 2007). Although these genes are expressed in insects, their role in this group seems related to mesoderm differentiation rather than segmentation. Consequently, they are not part of the segmentation pathway originally defined in *D. melanogaster*. Thamm and Seaver (2008) isolated, characterized and traced the expression patterns of several genes in the Notch signaling pathway in *Capitella* sp. I. Sequences for *Delta, Notch*, three *hairy and enhancer of split* genes (*CapI-hes1, CapI-hes2, CapI-hes3*) and two *hairy and enhancer of split* related genes (*CapI-hesr1, CapI-hesr2*) were determined. Because most adult *Capitella* sp. I segments are generated after metamorphosis, they studied expression patterns in juvenile worms as well as in developing larvae. In general, *CapI-Delta, CapI-Notch, CapI-hes2* and *CapI-hes3* were expressed in regions of segment formation, although their expression was quite broad. In most cases, expression began in presumptive tissues slightly before visible segmentation at stage 5, but was strongest after segments were apparent, arguing against a role in segmentation. This was true for the first embryonic segments that differentiate in the anterior midbody, for larval segments added from the posterior growth zone and for segments added after metamorphosis at the persistent posterior growth zone. Furthermore, expression was uniform and did not reveal the iterated, striped expression pattern seen in vertebrates and in the spider *Cupiennius salei* (Damen et al. 2005, Damen 2007).

The one exception was *CapI-hes1*, a pair-rule gene, which is expressed prior to segmentation. It appears at stage 4 in the first two presumptive segments and is expressed ahead of embryonic segmentation during stage 5. Similarly, in both developing larvae and juveniles, *CapI-hes1* expression is restricted to the presegmental region of the posterior growth zone. There, the anterior edge of *CapI-hes1* expression and the posterior boundary of *CapI-Delta* and *CapI-Notch* correspond closely to the posterior boundary of the nascent segment. Thamm and Seaver (2008) note that both *hairy* in *D. melanogaster* and *hes* genes in vertebrates are activated in a *Notch*-independent manner. They conclude that *CapI-hes1* is the most likely candidate of the genes investigated thus far to have a role in segmentation in *Capitella* sp. I.

Since fewer segmentation genes have been characterized in vertebrates and annelids than in arthropods, the possibility remains that elements of the segmentation gene pathway operating in arthropods may be conserved in the other two groups. The involvement of *Notch* signaling in basal arthropods and vertebrate segmentation supports this idea (Patel 2003). Like other cell–cell signaling pathways, the Notch signaling pathway is expressed and functions in numerous developmental events throughout embryogenesis in animals (Shi and Stanley 2006). Experiments using RNA-mediated interference (Schoppmeier and Damen 2005) indicate that in basal arthropods such as chelicerates (spiders) and myriapods (centipedes), the Notch/Delta signal transduction cascade is required to form segments.

Since Notch signaling is not involved in *D. melanogaster* segmentation, one cannot determine whether Notch signaling observed in basal arthropods and vertebrate segmentation (Dequeant and Pourquie 2008) is a result of common origin or of convergent evolution. In annelids, the role of Notch signaling remains unclear. Although the leech *notch* class and *hairy/enhancer of split* genes (*Hro-hes*) are expressed in a cell cycle-dependent manner (Song et al. 2004), the transcripts are distributed uniformly in the segmental precursor cells along the entire AP axis and do not appear to delineate segments. If Notch signaling is involved in leech and *Capitella* sp. I segmentation, it is functioning in a different manner than in vertebrates and basal arthropods. Recent *Notch* data from representatives of segmented phyla remind us to be cautious when generating hypotheses concerning origins of animal patterning mechanisms. What seems clear is that the best way to resolve questions regarding homology in genetic regulatory networks is to characterize more gene products in more taxa.

To what extent is the fly segmentation gene pathway conserved in annelids and vertebrate chordates? The current available data presented in this chapter suggest that the expression of the majority of segmentation gene orthologues identified in nonarthropod but segmented animals appear to function in different ways to that of their counterparts in flies. Therefore, the comparative data suggest independent origins of segmentation during the course of metazoan evolution. If this indeed is the case, then vertebrates, arthropods and annelids may have independently generated a segmentation gene cascade that served two tightly coordinated functions: first, the segmentation gene products created the metameric units along the body axis, and second, they functioned as upstream regulators of *Hox* gene expression responsible for specifying segmental identity. Clearly, genes canonically associated with segmentation are expressed differently in species of *Capitella* and *Platynereis*. Currently, we feel that the body of evidence does not support a single origin of segmentation for the Bilateria. Further work on polychaetes as well as other lophotrochozoans, a further resolution of the polychaete clade and an identification of factors that *do* regulate segmentation in capitellids are called for.

6.9 A-P POLARITY—*Hox* AND *ParaHox* GENES

6.9.1 *Hox* Expression

The role of the *Hox* cluster, believed to have arisen by duplication from a "ProtoHox" cluster, in establishing anterior–posterior patterning has been well documented especially in insects and vertebrates (reviewed by Garcia-Fernández 2005; Butts et al. 2008). The cluster is noted for the linearity of *Hox* genes on the chromosome, which is reflected in both their spatial and temporal expression along the AP axis during development. This colinearity seems to be an ancestral trait seen in bilaterians and, at least to some extent, in the diploblastic cnidarian *N. vectensis* (Martindale et al. 2004). In *Capitella* sp. I, 11 *Hox* genes seem to represent the ancestral lophotrochozoan complement (Fröbius et al. in press). Expression patterns of these genes reflect both temporal and spatial colinearity with the exception of *CapI-Post1*. Before the appearance of larval segments, expression is initiated in three anterior genes. Initiation of *Hox* gene expression located more posteriorly either coincides with segmentation or closely follows it. Several genes show an anterior or posterior expression pattern that corresponds to the thoracic-abdominal boundary. At metamorphosis, expression patterns alter to reflect new posterior boundaries.

In *P. dumerilii and Nereis virens, Hox* genes are expressed in a pattern that, to some extent, reflects spatial and temporal colinearity. Kulakova et al. (2007) studied expression patterns of 11 genes previously reported as belonging to the *Hox* cluster in *N. virens* and cloned nine *Hox* genes from *P. dumerilii*. Expression was seen mostly in 2d micomere descendants (i.e. integument, ventral ganglia, growth zone and pygidium). Both temporal and spatial expression patterns differed somewhat from the canonical pattern (Kulakova et al. 2007). In another polychaete, *Chaetopterus*, which has a body plan highly modified for tubiculous life, Irvine and Martindale (2001) found a basic conservation of colinearity of anterior expression boundaries. Expression of *Hox1-Hox5* genes showed a clear temporal linearity, with the exception of *CH-Hox2*, beginning before overt segmentation and continuing through the larval stage. This expression in very different polychaetes supports the hypothesis that the broad function of *Hox* genes in development is conserved, although their precise role in determining morphological regionalization remains an interesting question.

6.9.2 *ParaHox* Expression

Homeobox genes that lie outside the *Hox* cluster but code for homeodomain transcription factors also play a role in polarity determination during development. Brooke et al. (1998), studying the cephalochordate amphioxus, found three genes (*Gsx, Xlox* and *Cdx*) that are tightly clustered and display the spatial and temporal colinearity of *Hox* genes (i.e. the spatial order of genes match expression sites along the body axis). They proposed the name "*ParaHox*" for the cluster and suggested that *Hox* and *ParaHox* genes had arisen from an ancestral "*ProtoHox*" gene cluster by duplication events, making the two evolutionary sisters. *ParaHox* genes are noted for their role in patterning the gut and, to some extent, the nervous system (reviewed by Furlong et al. 2007).

Fröbius and Seaver (2006b) mapped expression patterns of *CapI-Gsx, CapI-Xlox* and *CapI-Cdx* during development in *Capitella* sp. I. Their identity as *ParaHox* orthologues of three distinct classes was confirmed by several phylogentic analyses; however, expression patterns differed from the spatial/temporal colinearity of chordates. Instead of the expected *Gsx, Xlox, Cdx* temporal sequence, the posterior gene *CapI-Cdx* was expressed first, appearing shortly after blastopore closure and persisting until emergence. Of the three, it was the most broadly expressed, appearing in all three germ layers, especially in the forming nervous system, belly plates and gut. The anterior gene, *CapI-Gsx*, was expressed briefly during stage 4 in association with brain development. The central gene *CapI-Xlox* was next expressed in the developing midgut (Fig. 6.4C), again out of sequence with the amphioxus model. Spatially, two of the genes showed the anticipated linearity, but the broad expression of *CapI-Cdx* does not match the chordate model. Expression of all three has been reported in *N. virens* (Kulakova et al. 2008). As in *Capitella* sp. I, *Gsx* (*Nvi-Gsh*) is expressed only during brain development, but expression patterns of the other two differ during gut and nervous system formation and growth. Interestingly, *Gsx* is expressed in the cnidarian *N. vectensis* in some ectoderm cells in the oral region perhaps indicating an ancient neural function (Finnerty et al. 2003).

Expression of a *Cdx* orthologue was reported in the polychaete *P. dumerilii* in the growth zone and in regenerating posterior segments (de Rosa et al. 2005) and in the oligochaete *Tubifex tubifex*, while *Xlox* expression has been described in the endoderm in the leeches *Hirudo medicinalis* and *Helobdella triserialis* (cited in Ferrier and Holland 2001). All three genes were identified in the sipunculids *Phascolion strombus* and

Nephasoma minuta (Ferrier and Holland 2001). In the ecdysozoans *D. melanogaster* and *C. elegans*, only subsets of *ParaHox* genes have been identified. The *ParaHox* cluster thus seems ancestral, but questions remain. Understanding the evolution and functional role of *ParaHox* genes in development will clearly require further investigation.

Three genes that play a role in posterior development in other organisms, namely, *caudal* (a ParaHox gene), *brachyury* (a T-box gene) and *even-skipped* (a pair-rule gene) all seem to play an important role in the growth of *P. dumerilii* (Arendt et al. 2001; de Rosa et al. 2005). From the time of blastopore closure, all three are expressed in a U shape around the proctodeal portion of the developing gut. Arendt and coworkers propose that the blastopore in this species is amphistomous, giving rise to both mouth and anus. During gastrulation, it first becomes elongated and slit-like, then assumes the configuration of a figure 8. Growth of ventral ectoderm then leads to formation of the stomodeum and proctodeum as the two extremities are pushed further apart (Arendt et al. 2001). de Rosa et al. (2005) propose that all three orthologues are involved in establishing the anteroposterior axis around the blastopore and in patterning the hindgut-pygidial posterior region. *Caudal* and *even-skipped* are expressed during posterior segment addition in both larvae and regenerating animals. They suggest that the expression of these widely distributed genes supports the hypothesis that growth by addition of posterior segments is an ancestral trait.

6.10 ANNELID GENOMICS: DRAFT GENOME SEQUENCE

Investigations into a sampling of annelid taxa that include potential basal groups of polychaetes such as *Capitella* sp. I and *P. dumerilii*, and clitellates including leeches and oligochaetes, offer important comparative advantages based on their phylogenetic history and distinctive patterning programs. The characterization of segmentation mechanisms in basal polychaetes will provide critical background information from which broader comparisons can be made within the annelid phylum and between phyla. The leech *H. robusta* and polychaetes *Capitella* sp. I and *P. dumerilii* are ideal annelid systems because they breed prolifically throughout the year and have relatively large (*H. robusta*, 400 μm; *Capitella* sp. I, 250 μm; *P. dumerilii*, 180 μm), experimentally accessible embryos that facilitate both molecular and cellular studies. Additionally, each of these annelids has a distinctive body plan: the presumptive basal polychaetes have remarkably diverse and adaptable body plans reflecting their habitats, while the specialized or derived leeches have a fixed number of segments and lack the ability to regenerate segments. Using representative annelid taxa in comparative studies will allow us to distinguish general developmental mechanisms of cell determination from phylum-specific ones, and will provide insights into how morphological diversity is generated in annelids.

In the past decade, remarkable progress in DNA sequencing technology and its associated tools, combined with more sophisticated and reliable gene finding and annotation algorithms, has accelerated the pace of genomic sequencing. The recent release of *Capitella* sp. I and *H. robusta* draft genomic sequences by the Joint Genome Institute (Department of Energy, Walnut Creek, CA, USA) and a large-scale expressed sequence tags (EST) project on *P. dumerilii* led by the Arendt lab (EMBL, Heidelberg, Germany) will continue to provide benefits to the scientific community because of their broad impact on many disciplines of biological research. Most importantly, the annelid data begin to fill a major gap in genomic resources in the understudied superphylum known as the Lophotrochozoa, and will thus serve as a critical reference point for comparative genomics. Currently, little is known about the cellular and molecular basis of segment pattern formation in annelids.

The characterization of developmental regulatory genes in species outside traditional model systems addresses fundamental questions regarding their role in divergent body plans and will help to clarify questions regarding orthology. The combination of accessible genomic data and the use of functional knockdown techniques in understudied animal systems that are not conducive to traditional genetic approaches now make it possible to address how regulatory genes and their networks have changed in the course of annelid developmental evolution.

6.11 THE FUTURE—WHERE IS THIS GOING?

6.11.1 A Few Practical Applications

The ability of *Capitella* sp. I to colonize polluted areas has long made them a group of interest not only as an indicator species but also as potential agents of remediation. An extensive literature, beyond the scope of this review, can be found in ecological/environmental journals. As mentioned, *Capitella* sp. I is an early colonizer of fish farming areas and now is being introduced as an agent of remediation in some Japanese farms (Kinoshita et al. 2008). Forbes et al. (2001) found that *Capitella* sp. I is able to metabolize polycyclic aromatic hydrocarbons, widespread, extremely toxic, environmental pollutants. They identified two novel cytochrome P450 (CYP) genes in *Capitella* sp. I, one belonging to a new family CYP331 and the other to a new subfamily CYP4AT, that seem to play a role in detoxification of polycyclic aromatic hydrocarbons (Li et al. 2004). The authors point out that understanding both the evolution of the CYP gene superfamily and the molecular basis of the biotransformations by capitellids requires further work; obviously, interest in the role of capitellids in bioremediation will increase as we understand more of their genetic tools.

Several of the capitellid species are virtually impossible to identify at the species level using standard techniques. As outlined above, differences occur in allozyme patterns, chromosome numbers and life history traits. Cryptic morphological differences have been found among many capitellid species. For example, male genital spines of species I, Ia, II, IIIa and ORL differ slightly, but the differences are subtle and, because of intraspecific variation, are of little taxonomic help (Eckelbarger and Grassle 1987a). Differences in sperm and in egg and egg envelope morphology (Eckelbarger and Grassle 1983, 1987a, b) can be seen at an ultrastructural level. Differences in the distribution of larval cilia are apparent: prototroch, telotroch, neurotroch and pygidial cilia in the larvae of species I, Ia, II and ORL show characteristic differences in their cilia density and pattern. Surprisingly, the small planktonic larvae of species III that persist longest in the water column have a weakly developed prototroch and neurotroch and lack pygidial cilia (Eckelbarger and Grassle 1987a). While these are important differences, they, like allozyme differences, karyotypes and life history traits, are of limited value in species identification in the field or even in the laboratory.

Recently, Du et al. 2007 characterized microsatellite loci from the *C. capitata* complex. Using genomic DNA sequences deposited in GenBank before April 20, 2006 and animals collected from three sampling sites in Qingdao, Shandong province of China, they found 10 polymorphic loci judged suitable for studies of population genetics. They considered all animals to be *C. capitata* (although they recognize the existence of different species); consequently, it is unknown whether they were dealing with different populations or different species. Hebert et al. (2003) outline the advantages of utilizing DNA sequences as genetic "bar codes" for taxon diagnosis. Bely and Weisblat (2006) make a plea that all wild-collected specimens are bar coded and voucher specimens preserved. They sequenced part of the mitochondrial gene *cytochrome oxidase 1* and found that cryptic species occur in leech, as in capitellids. *H. robusta* was found to represent three species, two of which coexist. The *Capitella* complex is surely a group where the development of microarrays or bar-coding tools to aid in species identification would be extremely welcome.

Interest in the field of regeneration has been stimulated by recent advances in stem cell research. Following formation of the initial blastema, regeneration often mimics the growth processes seen during early development. At some level, genes used during early development must be reactivated during regeneration, but how and in what sequence? And what signals indicate that a part is missing? Prud'homme et al. (2003) conclude that the expression pattern of *en* and *wnt1* is the same during the addition of larval segments and regeneration in *P. dumerilii* and uses the two interchangeably. Now that appropriate tools are available, further work comparing not just posterior regeneration but the seemingly more complex anterior regeneration observed in some polychaetes could have far-reaching medical applications.

6.11.2 Understanding Phylogeny

Perhaps the most important role that further work on capitellids will have is in our understanding of phylogeny. The *Capitella* complex, with its co-occurring sibling species, is an ideal system for speciation studies. Unfortunately, only *Capitella* sp. I is currently maintained in culture; however, molecular tools will allow productive investigation as other species again become available. The current expansion of annelid studies is long overdue. This is a group with diverse body plans and habits. Clearly, more work and broader sampling are required to resolve the polychaete hierarchy, the position of lophotrochozoans and the nature of the Urbilateria (Dunn et al. 2008). Struck et al. (2007) and Dunn et al. (2008) place capitellids as the sister group of the controversial echiurans. Rouse (2000a) calls for more work on *Capitella* to resolve the controversy surrounding the trochophore and includes capitellids in a group that engages in downstream feeding. While this may well be true of some species (perhaps for *Capitella* sp. III?), it is not the case for *Capitella* sp. I or for many others of the complex.

Probably, the most important role that a model organism can play, however, is not as a "problem solver"; rather, the data collected from the "model" lay the groundwork with which other organisms can be compared. The differences, as well as the similarities, between *Capitella* sp. I and *P. dumerilii* call for further investigation, not just on these species but on other polychaetes and nonpolychaete lophotrochozoans. Are either capitellids or nereids "basal" polychaetes? How would we recognize a basal polychaete? As Jenner (2006) points out, species that display characters that we consider derived are just as closely related to a common ancestor as are "sister" species that lack these traits. The traits of a species may change with time and selection, but its history does not. Everyone recognizes that chaetopterids and sabellids, for example, have specialized tagmata whose evolution has allowed them to establish their successful, but differing, tubiculous lifestyles. Homonomous body plans also can result from adaptive evolution leading to convergence. What role has either convergence or character loss played in polychaete evolution? Our current thought that segmentation, as an adult character, has been lost at least twice among polychaetes is supported by the inclusion of echiuroids and sipunculids in the clade (Struck et al. 2007; Dunn et al. 2008). As Chipman (2008) points out, the segmentation question in annelids, the "third segmented phylum," is interesting, not just because it may illuminate the "basal" ancestor, but in its own right. All of the questions raised above are fascinating for the same reason—they encourage us to explore how familiar genes with mechanistically similar pathways lead to the great diversity of life.

ACKNOWLEDGMENTS

The authors wish to thank Elaine Seaver for contributing Figure 6.4 and for sharing unpublished results, Barbara Boyer for proofreading the manuscript and Judy Grassle for introducing us to *Capitella* sp. I. S.D.H. thanks all of the above for their insightful collaboration over the years.

REFERENCES

AGEE, S.J., LYONS, D.C., WEISBLAT, D.A. 2006. Maternal expression of a NANOS homolog is required for early development of the leech *Helobdella robusta*. Dev. Biol. 298, 1–11.

ARENAS-MENA, C. 2006. Embryonic expression *of HeFoxA1* and *HeFoxA2* in an indirectly developing polychaete. Dev. Genes Evol. 216, 727–736.

ARENAS-MENA, C., SUK-YING WONG, K., ARANDI-FOROSANI, N. 2007. Ciliary band gene expression patterns in the embryo and trochophore larva of an indirectly developing polychaete. Gene Expr. Patterns 7, 544–549.

ARENDT, D., TECHNAU, U., WITTBRODT, J. 2001. Evolution of the bilaterian foregut. Nature 409, 81–85.

BARRALLO-GIMENO, A., NIETO, M.A. 2005. The Snail genes as inducers of cell movement and survival: implications in development and cancer. Development 132, 3151–3161.

BELY, A.E., WEISBLAT, D.A. 2006. Lessons from leeches: a call for DNA barcoding in the lab. Evol. Dev. 8, 491–501.

BELY, A.E., WRAY, G.A. 2001. Evolution of regeneration and fission in annelids: insights from engrailed- and orthodenticle-class gene expression. Development 128, 2781–2791.

BERGTER, A., PAULULAT, A. 2007. Pattern of body wall muscle differentiation during embryonic development of *Enchytraeus coronatus* (Annelida: Oligochaeta; Enchytraeidae). J. Morph. 268, 537–549.

BERGTER, A., HUNNEKUHL, V.S., SCHNIEDERJANS, M., PAULULAT, A. 2007. Evolutionary aspects of pattern formation during clitellate muscle development. Evol. Dev. 9, 602–617.

BERGTER, A., BRUBACHER, J.L., PAULULAT, A. 2008. Muscle formation during embryogenesis of the polychaete *Ophryotrocha diadema* (Dorvalidae)—new insights into annelid muscle patterns. Front. Zool. 5. doi:10.1186/174209994-5-1.

BIGGERS, W.J., LAUFER, H. 1992. Chemical induction of settlement and metamorphosis of *Capitella capitata* Sp. I

(Polychaeta) larvae by juvenile hormone-active compounds. Invertebr. Reprod. Dev. 22, 39–46.

BIGGERS, W.J., LAUFER, H. 1999. Settlement and metamorphosis of *Capitella* larvae induced by juvenile hormone-active compounds is mediated by protein kinase C and ion channels. Biol. Bull. 196, 187–198.

BLAKE, J.A., GRASSLE, J.P., ECKELBARGER, K.J. In press. A new species designation for the experimental capitellid, Capitella sp. I, with a review of the literature for confirmed records. In: Maciolek, N.J., J.A. Blake (Eds). Proceedings of the Ninth International Polychaete Conference. *Zoosymposia* 2. Auckland, New Zealand: Magnolia Press.

BOYLE, M.J., SEAVER, E.C. 2008a. Developmental expression of *Fox A* and *GATA* genes during gut formation in the polychaete annelid, *Capitella* sp. I. Evol. Dev. 10, 89–105.

BOYLE, M.J., SEAVER, E.C. 2008b. Evidence of a dorsal pharynx in the marine polychaete, Capitella sp. I (Polychaeta: Capitellidae). Zootaxa (in press).

BROOKE, N.M., GARCIA-FERNANDEZ, J., HOLLAND, P.W.H. 1998. The ParaHox gene cluster is an evolutionary sister of the Hox gene cluster. Nature 392, 920–922.

BUTMAN, C.A., GRASSLE, J.P., WEBB, C.M. 1988. Substrate choices made by marine larvae settling in still water and in a flume flow. Nature 333, 771–773.

BUTTS, T., HOLLAND, P.W.H., FERRIER, D.E.K. 2008. The urbilaterian Super-Hox cluster. Trends Genet. 24, 259–262.

CARROLL, S.B., GRENIER, J.K., WEATHERBEE, S.D. 2005. *From DNA to Diversity Molecular Genetics and the Evolution of Animal Design*, 2nd edn. Malden, MA: Blackwell Publishing.

CHIPMAN, A.D. 2008. Annelids step forward. Evol. Dev. 10, 141–142.

CHIPMAN, A.D., ARTHUR, W., AKAM, M. 2004. A double segment periodicity underlies segment generation in centipede development. Curr. Biol. 14, 1250–1255.

CHOE, C.P., MILLER, S.C., BROWN, S.J. 2006. A pair-rule gene circuit defines sequentially in the short-germ insect

Tribolium castaneum. Proc. Natl. Acad. Sci. U.S.A. 103, 6560–6564.

COHEN, R.A., PECHENIK, J.A. 1999. Relationship between sediment organic content, metamorphosis, and postlarval performance in the deposit-feeding polychaete *Capitella* sp. I. J. Exp. Mar. Biol. Ecol. 240, 1–18.

DAMEN, W.G. 2007. Evolutionary conservation and divergence of segmentation process in arthropods. Dev. Dyn. 236, 1379–1391.

DAMEN, W.G., JANSSEN, R., PRPIC, N.M. 2005. Pair rule gene orthologs in spider segmentation. Evol. Dev. 7, 618–628.

von DASSOW, G., MEIR, E., MUNRO, E., ODELL, G. 2000. The segment polarity network is a robust developmental module. Nature 406,188–192.

DAVIS, G.K., D'ALESSIO, J.A., PATEL, N.H. 2005. Pax3/7 genes reveal conservation and divergence in the arthropod segmentation hierarchy. Dev. Biol. 285, 169–184.

DEQUEANT, M.L., POURQUIE, O. 2008. Segmental patterning of the vertebrate embryonic axis. Nat. Rev. Genet. 9, 370–382.

DILL, K.K., SEAVER, E.C. 2008. *Vasa* and *nanos* are co-expressed in somatic and germ line tissue from early embryonic cleavage stages through adulthood in the polychaete *Capitella* sp. I. Dev. Genes Evol. 218, 453–463.

DILL, K.K., THAMM, K., SEAVER, E.C. 2007. Characterization of *twist* and *snail* gene expression during mesoderm and nervous system development in the polychaete annelid *Capitella* sp. I. Dev. Genes Evol. 217, 435–447.

DU, H., HAN, J., LIN, K., QU, X., WANG, W. 2007. Characterization of 11 microsatellite loci derived from genomic sequences of polychaete *Capitella capitata* complex. Mol. Ecol. Notes 7, 1144–1146.

DUNN, C.W., HEJNOL, A., MATUS, D.Q., PANG, K., BROWNE, W.E., SMITH, S.A., SEAVER, E., ROUSE, G.W., OBST, M., EDGECOMBE, G.D., SØRENSEN, M.V., HADDOCK, S.H.D., SCHMIDT-RHAESA, A., OKUSU, A., KRISTENSEN, R.M., WHEELER, W.C., MARTINDALE, M.Q., GIRIBET, G. 2008. Broad phylogenetic sampling improves resolution of the animal tree of life. Nature 452, 745–749.

ECKELBARGER, K.J., GRASSLE, J.P. 1983. Ultrastructural differences in the eggs and ovarian follicle cells of *Capitella* (Polychaeta) sibling species. Biol. Bull. 165, 379–393.

ECKELBARGER, K.J., GRASSLE, J.P. 1987a. Interspecific variation in genital spine, sperm and larval morphology in six sibling species of *Capitella*. Bull. Biol. Soc. Wash. 7, 62–76.

ECKELBARGER, K.J., GRASSLE, J.P. 1987b. Spermatogenesis, sperm storage and comparative sperm morphology in nine species of *Capitella, Capitomastus* and *Capitellides* (Polychaeta: Capitellidae). Mar. Biol. 95, 415–429.

EISIG, H. 1899. Zur Entwicklungsgeschichte der Capitelliden. Mitt. Zool. Station Neapel 13, 1–292.

FERRIER, D.E.K., HOLLAND, P.W.H. 2001. Sipunculan ParaHox genes. Evol. Dev. 3, 263–270.

FINNERTY, J.R., PAULSON, D., BURTON, P., PANG, K., MARTINDALE, M.Q. 2003. Early evolution of a homeobox gene: the parahox gene Gsx in the Cnidaria and Bilateria. Evol. Dev. 5, 331–345.

FORBES, V.E., ANDREASSEN, M.S.H., CHRISTENSEN, L. 2001. Metabolism of the polycyclic aromatic hydrocarbon fluoranthene by the polychaete *Capitella capitata* species I. Environ. Toxicol. Chem. 20, 738–747.

FRÖBIUS, A.C., SEAVER, E.C. 2006a. *Capitella* sp.I homeobrain-like, the first lophotrochozoan member of a novel paired-like homeobox gene family. Gene Expr. Patterns 6, 985–991.

FRÖBIUS, A.C., SEAVER, E.C. 2006b. *ParaHox* gene expression in the polychaete annelid *Capitella* sp. I. Dev. Genes Evol. 216, 81–88.

FRÖBIUS, A.C., MATUS, D.Q., SEAVER, E.C. In press. Genomic organization and expression demonstrate spatial and temporal Hox gene colinearity in the lophotrochozoan *Capitella* sp. I. PloS ONE.

FUGIWARA, Y., KAWATO, M., YAMAMOTO, T., YAMANAKA, T., SATO-OKOSHI, W., NODA, C., TSUCHIDA, S., KOMAI, T., CUBELIO, S.S., SASAKI, T., JACOBSEN, K., KUBOKAWA, K., FUJIKURA, K., MARUYAMA, T., FURUSHIMA, Y., OKOSHI, K., MIYAKE, H., MIYAZKI, M., NOGI, Y., YATABE, A., OKUTANI, T. 2007. Three-year investigations into sperm whale-fall ecosystems in Japan. Mar. Ecol. 38, 1–14.

FURLONG, R.F., YOUNGER, R., KASAHARA, M., REINHARDT, R., THORNDYKE, M., HOLLAND, P.W.H. 2007. A degenerate Parahox gene cluster in a degenerate vertebrate. Mol. Biol. Evol. 24, 2681–2686.

GABRIEL, W.N., GOLDSTEIN, B. 2007. Segmental expression of Pax3.7 and engrailed homologs in tardigrade development. Dev. Genes Evol. 217, 421–33.

GARCIA-FERNÁNDEZ, J. 2005. The genesis and evolution of homeobox gene clusters. Nat. Rev. Genet. 6, 881–92.

GILLIS, W.J., BOWERMAN, B., SCHNEIDER, S.Q. 2007. Ectoderm- and endoderm-specific GATA transcription factors in the marine annelid *Platynereis dumerilii*. Evol. Dev. 9, 39–50.

GILLIS, W.Q., BOWERMAN, B.A., SCHNEIDER, S.Q. 2008. The evolution of protostome GATA factors: Molecular phylogenetics, synteny, and intron/exon structure reveal orthologous relationships. BMC Evol. Biol. 8, 112.

GOLDSTEIN, B., LEVITEN, M.W., WEISBLAT, D.A. 2001. *Dorsal* and *snail* homologs in leech development. Dev. Genes Evol. 211, 329–337.

GRASSLE, J.F., GRASSLE, J.P. 1974. Opportunistic life histories and genetic systems in marine benthic polychaetes. Jour. Mar. Res. 32, 253–284.

GRASSLE, J.F., GRASSLE, J.P. 1978. Life histories and genetic variation in marine invertebrates. In *The Genetics, Ecology, and Evolution of Marine Organisms*, edited by J.A. Beardmore and B. Battaglia. New York: Plenum Press, pp. 347–364.

GRASSLE, J.P., GRASSLE, J.F. 1976. Sibling species in the marine pollution indicator *Capitella* (Polychaeta). Science 192, 567–569.

GRASSLE, J.P., GELFMAN, C.E., MILLS, S.W. 1987. Karyotypes of *Capitella* sibling species, and of several species in the related genera *Capitellides* and *Capitomastus* (C. Polychaeta). Bull. Biol. Soc. Wash. 7, 77–88.

HALANYCH, K.M. 2004. A new view of animal phylogeny. Ann. Rev. Ecol. Syst. 35, 229–256.

HANDEL, K., BASAL, A., FAN, X., ROTH, S. 2005. *Tribolium castaneum* twist: gastrulation and mesoderm formation in a short germ-band beetle. Dev. Genes Evol. 215, 13–31.

HARFE, B.D., VAZ GOMES, A., KENYON, C., LIU, J., KRAUSE, M., FIRE, A. 1998. Analysis of *Caenorhabditis elegans Twist* homologue identifies conserved and divergent patterning. Genes Dev. 12, 2623–2635.

HE, Z.B., CAO, Y.Q., WANG, Z.K., CHEN, B., PENG, G.X., XIA, Y.X. 2006. Role of hunchback in segment patterning of *Locusta migratoria manilensis* revealed by parental RNAi. Dev. Growth Differ. 48, 439–45.

HEBERT, P.D.N., CYWINSKA, A., BALL, S.L., deWAARD, J.R. 2003. Biological identifications through DNA barcodes. Proc. R. Soc. Lond. B 270, 313–321.

HENRY, J.Q., HEJNOL, A., PERRY, K.J., MARTINDALE, M.Q. 2007. Homology of ciliary bands in spiralian trochophores. Integr. Comp. Biol. 47, 865–871.

HILL, S.D., BOYER, B.C. 2001. Phalloidin labeling of developing muscle in embryos of the polychaete *Capitella* sp. I. Biol. Bull. 201, 257–258.

HILL, S.D., BOYER, B.C. 2003. HNK-1/N-CAM immunoreactivity correlates with ciliary patterns during development of the polychaete *Capitella* sp. I. Biol. Bull. 205, 182–184.

HILL, S.D., BOYER, B.C. 2005. The role of heterochrony in the evolution of the life cycle of the opportunistic polychaete *Capitella* sp. I. *Developmental Basis of Evolutionary Change Symposium*, Chicago, October 2005 (Abstract).

HILL, S.D., GRASSLE, J.P., MILLS, S.W. 1982. Regeneration and maturation in two sympatric *Capitella* (Polychaeta) sibling species. Biol. Bull. 163, 366.

HILL, S.D., GRASSLE, J.P., FERKOWICZ, M.J. 1987. Effect of segment loss on reproductive output in *Capitella* species I. Biol. Bull. 173, 430.

HILL, S.D., FERKOWICZ, M.J., GRASSLE, J.P. 1988. Effect of tail regeneration on early fecundity in *Capitella* sp. I and II. Biol. Bull. 175, 311.

HOLBROOK, M.J.L., GRASSLE, J.P. 1984. The effect of low density on the development of simultaneous hermaphroditism in male *Capitella* sp. I (Polychaeta). Biol. Bull. 166, 103–109.

HOPWOOD, N.D., PLUCK, A., GURDON, J.B. 1989. A *Xenopus* mRNA related to *Drosophila twist* is expressed in response to induction in the mesoderm and the neural crest. Cell 59, 893–903.

IRVINE, S.Q., MARTINDALE, M.Q. 2001. Comparative analysis of Hox gene expression in the polychaete *Chaetopterus*: implications for the evolution of body plan regionalization. Am. Zool. 41, 640–651.

JENNER, R.A. 2006. Unburdening evo-devo: ancestral attractions, model organisms, and basal baloney. Dev. Genes Evol. 216, 385–394.

KANG, D., HUANG, F., LI, D., SHANKLAND, M., GAFFIELD, W., WEISBLAT, D.A. 2003. A hedgehog homolog regulates gut formation in leech (*Helobdella*). Development 130, 1645–1657.

KERNER, P., ZELADA GONZALES, F., Le GOUAR, M., LEDENT, V., ARENDT, D., VERVOORT, M. 2006. The expression of a hunchback ortholog in the polychaete annelid *Platynereis dumerilii* suggests an ancestral role in mesoderm development and neurogenesis. Dev. Genes Evol. 216, 821–828.

KINOSHITA, K., TAMAKI, S., YOSHIOKA, M., SRITHONGUTHAI, S., KUNIHIRO, T., HAMA, D., OHWADA, K., TSUTSUMI, H. 2008. Bioremediation of organically enriched sediment deposited below fish farms with artificially mass-cultured colonies of a deposit-feeding polychaete *Capitella* sp. I. Fish. Sci. 74, 77–87.

KULAKOVA, M., BAKALENKO, N., NOVIKOVA, E., COOK, C.E., ELISEEVA, E., STEINMETZ, P.R.H., KOSTYUCHENKO, R.P., DONDUA, A., ARENDT, D., AKAM, M., ANDREEVA, T. 2007. Hox gene expression in larval development of the polychaetes *Nereis virens* and *Platynereis dumerilii* (Annelida, Lophotrochozoa). Dev. Genes Evol. 217, 39–54.

KULAKOVA, M.A., COOK, C.E., ANDREEVA, T.F. 2008. *ParaHox* gene expression in larval and postlarval development of the polychaete *Nereis virens* (Annelida, Lophotrochozoa). BMC Dev. Biol. 8, Article 61.

LEPTIN, M. 1991. Twist and snail as positive and negative regulators during *Drosophila* mesoderm development. Genes Dev. 5, 1568–1576.

LESPINET, O., NEDERBRAGT, A.J., CASSAN, M., DICTUS, W.J., van LOON, A.E., ADOUTTE, A. 2002. Characterization of two *snail* genes in the gastropod mollusk *Patella vulgata*. Implications for understanding the ancestral function of *snail* related genes in Bilateria. Dev. Genes Evol. 212, 186–195.

LI, B., BISGAARD, H.C., FORBES, V.E. 2004. Identification and expression of two novel cytochome P450 genes, belonging to CYP4 and a new CYP331 family, in the polychaete *Capitella capitata* sp. I. Biochem. Biophys. Res. Com. 325, 510–517.

LUI, P.Z., KAUFMAN, T.C. 2005. *Even-skipped* is not a pair-rule gene but has segmental and gap-like functions in *Oncopeltus fasciatus*, an intermediate germband insect. Development 1323, 2081–2092.

MARTINDALE, M.Q., PANG, K., FINNERTY, J.R. 2004. Investigating the origins of triploblasty: "mesodermal" gene expression in a diploblastic animal, the sea anemone *Nematostella vectensis* (phylum, Cnidaria; class, Anthozoa). Development 131, 2463–2474.

NEILSON, C., NØRREVANG, A. 1985. The Trochaea theory: an example life cycle phylogeny. In *The Origins and Relationships of Lower Invertebrates*, edited by S. Conway Morris, D. George, R. Gibson, H.M. Platt. Oxford: Clarendon Press, pp. 28–41.

PATEL, N.H. 2003. The ancestry of segmentation. Dev. Cell. 5, 2–4.

PETERSON, K.J., CAMERON, R.A., DAVIDSON, E.H. 1997. Set-aside cells in maximal indirect development: Evolutionary and developmental significance. Bioessays 19, 623–631.

PETRAITIS, P.S. 1985. Digametic sex determination in the marine polychaete, *Capitella capitata* (species type I). Heredity 55, 151–156.

PETRAITIS, P.S. 1988. Occurrence and reproductive success of feminized males in the polychaete *Capitella capitata* (species type I). Mar. Biol. 97, 403–412.

PETRAITIS, P.S. 1991. The effects of sex ratio and density on the expression of gender in the polychaete *Capitella capitata*. Evol. Ecol. 5, 393–404.

PINNELL, J., LINDEMAN, P.S., COLAVITO, S., LOWE, C., SAVAGE, R.M. 2006. The divergent roles of the segmentation gene *hunchback*. Integr. Comp. Biol. 46, 519–532.

PRUD'HOMME, B., de ROSA, R., ARENDT, D., JULIEN, J.-F., PAJAZITI, R., DORRESTEIJN, A.W.C., ADOUTTE, A., WITTBRODT, J., BALAVOINE, G. 2003. Arthropod-like expression patterns of engrailed and wingless in the annelid *Platynereis dumerilii* suggest a role in segment formation. Curr. Biol. 13, 1876–1881.

REBSCHER, N., ZELADA-GONZÁLES, F., BANISCH, T.U., RAIBLE, F., ARENDT, D. 2007. Vasa unveils a common origin of germ cells and of somatic stem cells from the posterior growth zone in the polychaete *Platynereis dumerilii*. Dev. Biol. 306, 599–611.

REIGER, R.M., PURSCHKE, G. 2005. The coelom and origin of the annelid body plan. Hydrobiologia 535/536, 127–138.

RHODE, B. 1993. Laval and adult eyes in *Capitella* spec. I (Annelida, Polychaeta). J. Morph. 217, 327–335.

de ROSA, R., PRUD'HOMME, B., BALAVOINE, G. 2005. *caudal* and *even*-skipped in the annelid *Platyneries dumerilii* and the ancestry of posterior growth. Evol. Dev. 7, 574–587.

ROUSE, G.W. 1999. Trochophore concepts: ciliary bands and the evolution of larvae in spiralian metazoans. Biol. J. Linn. Soc. 66, 411–464.

ROUSE, G.W. 2000a. Bias? What bias? Gain and loss of downstream larval-feeding in animals. Zool. Scr. 29, 213–236.

ROUSE, G.W. 2000b. The epitome of hand waving? Larval feeding and the hypotheses of metazoan phylogeny. Evol. Dev. 2, 222–223.

ROUSSET, V., PLEIJEL, F., ROUSE, G.W., ERSÉUS, C., SIDDALL, M.E. 2007. A molecular phylogeny of annelids. Cladistics 23, 41–63.

SANDERS, H.L., GRASSLE, J.F., HAMPSON, G.R., MORSE, L.S., GARNER-PRICE, S., JONES, C.C. 1980. Anatomy of an oil spill: Long term effects from the grounding of the barge *Florida* off West Falmouth, Massachusetts. J. Mar. Res. 38: 265–380.

SAVAGE, R.M., SHANKLAND, M. 1996. Identification and characterization of a hunchback orthologue, Lzf2, and its expression during leech embryogenesis. Dev. Biol. 175, 205–217.

SCHOPPMEIER, M., DAMEN, W.G. 2005. *Suppressor of Hairless* and *Presenlin* phenotypes imply involvement of canonical *Notch*-signaling in segmentation of the spider *Cupiennius salie*. Dev. Biol. 280, 211–224.

SEAVER, E.C., KANESHIGE, L.M. 2006. Expression of "segmentation" genes during larval and juvenile development in the polychaetes *Capitella* sp. I and *H. elegans*. Dev. Biol. 289, 179–194.

SEAVER, E.C., PAULSON, D.A., IRVINE, S.Q., MARTINDALE, M.Q. 2001. The spatial and temporal expression of *Ch-en*, the engrailed gene in the polychaete *Chaetopterus*, does not support a role in body axis segmentation. Dev. Biol. 236, 195–209.

SEAVER, E.C., THAMM, K., HILL, S.D. 2005. Growth patterns during segmentation in the two polychaete annelids, *Capitella* sp. I and *Hydroides elegans*: comparisons at distinct life history stages. Evol. Dev. 7, 312–326.

SHAIN, D.H., RAMIREZ-WEBE, F.A., HSU, J., WEISBLAT, D.A. 1998. Gangliogenesis in leech: morphogenetic processes leading to segmentation in the central nervous system. Dev. Genes Evol. 208, 28–36.

SHI, S., STANLEY, P. 2006. Evolutionary origins of Notch signaling in early development. Cell Cycle 5, 274–278.

SHIMIZU, T., SAVAGE, R.M. 2002. Expression of hunchback protein in a subset of ectodermal teloblasts of the oligochaete annelid *Tubifex*. Dev. Genes Evol. 212, 520–525.

SOKOL, N.S., AMBROS, V. 2005. Mesodermally expressed *Drosophila microRNA-1* is regulated by Twist and is required in muscles during larval growth. Genes Dev. 19, 2343–2354.

SONG, M.H., HUANG, F.Z., GONSALVES, F.C., WEISBLAT, D.A. 2004. Cell cycle-dependent expression of a *hairy* and *Enhancer of split (hes)* homolog during cleavage and segmentation in leech embryos. Dev. Biol. 269, 183–195.

STRUCK, T.H., SCHULT, N., KUSEN, T., HICKMAN, E., BLEIDORN, C., McHUGH, D., HALANYCH, K.M. 2007. Annelid phylogeny and the status of Sipuncula and Echiura. BMC Evol. Biol. 7, Article 57.

TECHNAU, U., SCHOLZ, C.B. 2003. Origin and evolution of endoderm and mesoderm. Int. J. Dev. Biol. 47, 531–539.

TESSMAR-RAIBLE, K., ARENDT, D. 2003. Emerging systems: between vertebrates and arthropods, the Lophotrochozoa. Curr. Opin. Genet. Dev. 13, 331–340.

THAMM, K., SEAVER, E.C. 2008. Notch signaling during larval and juvenile development in the polychaete annelid *Capitella* sp. I. Dev. Biol. 320, 304–318.

THISSE, B., STOETZEL, C., GOROSTIZA-THISSE, C., PERRIN-SCHMITT, F. 1988. Sequence of the *twist* gene and nuclear localization of its protein in endomesodermal cells of early *Drosophila* embryos. EMBO J. 7, 2175–2183.

TOMASSETTI, P., PORRELLO, S. 2005. Polychaetes as indicators of marine fish farm organic enrichment. Aquac. Int. 13, 109–128.

TSUTSUMI, H. 1990. Population persistence of *Capitella* sp. (Polychaeta, Capitellidae) on a mud flat subject to environmental disturbance by organic enrichment. Mar. Ecol. Prog. Ser. 63, 147–156.

WEDEEN, C.J., WEISBLAT, D.A. 1991. Segmental expression of an engrailed-class gene during early development and neurogenesis in an annelid. Development 113, 805–814.

WERBROCK, A.H., MEIKLEJOHN, D.A., SAINZ, A., IWASA, J.H., SAVAGE, R.M. 2001. A polychaete *hunchback* ortholog. Dev. Biol. 235, 476–488.

WLODARSKA-KOWALCZUK, M., SZYMELFENIG, M., ZAJAC-ZKOWSKI, M. 2007. Dynamic sedentary environments of an Arctic glacier-fed river estuary (Adventfjorden, Sval-bard). II: Meio- and macrobenthis fauna. Estuar. Coast. Shelf Sci. 74, 274–284.

YASUI, K., ZHANG, S.C., UEMURA, M., AIZAWA, S., UEKI, T. 1998. Expression of a twist-related gene, *Bbtwist*, during the development of a lancelet species and its relation to cephalochordate anterior structures. Dev. Biol. 195, 49–50.

YOKOYAMA, H. 2002. Impact of fish and pearl farming on the benthic environments in Gokasho Bay: Evaluation from seasonal fluctuations of the macrobenthos. Fish. Sci. 68, 258–268.

Chapter 7

Stem Cell Genesis and Differentiation in Leech

Shirley A. Lang and Daniel H. Shain

Biology Department, Rutgers, The State University of New Jersey, Camden, NJ

7.1 INTRODUCTION

Although the field of stem cell research is relatively young, its roots can be traced to the mid-19th century. Advancements in microscopy enabled scientists to examine samples in far greater detail than previously possible, culminating in development of the cell theory. The first two tenets, proffered by Schleiden and Schwann in the 1830s, stated that (1) all organisms consist of one or more cells, and (2) the cell is the basic unit of structure for all organisms. The third tenet, added by Rudolph Virchow in 1855, describes a fundamental property of stem cells, namely, that all cells arise from preexisting cells. The implication of this property was obvious; how does a single cell (e.g. a fertilized egg) displaying a limited set of phenotypic characters generate the 200+ cell types found in most animal species? Thus was born the concept of cells from which all others "stem."

Indeed, the mid-1800s were a dynamic and exciting era in biological research. In addition to improved microscopy and the establishment of cell theory, Darwin's *On the Origin of Species*, published in 1859, opened the door to new avenues of exploration. At Darwin's behest, many biologists delved into comparative embryology in an attempt to understand the morphological evolution of extant organisms (Bowler 1996; Hall 1999). Armed with little more than a light microscope and a great deal of patience, investigators described in exquisite detail the developmental phenomena of various taxa, and then compared notes (the birth of evolutionary development or evo-devo). Organisms chosen for these endeavors tended to have embryos that developed rapidly and had relatively few (and therefore large) cells to facilitate viewing—prominently represented were several species of leech.

Leeches have been recognized as valuable models for the investigation of developmental processes for well over a century. Charles O. Whitman's (1878) pioneering cell lineage study, *The Embryology of Clepsine*, wherein he describes the highly stereotyped and strongly determinate pattern of leech development, inspired intense interest in this simple invertebrate system toward the end of the 19th century. Despite successful studies in both embryology and neuroanatomy, the use of leech as a model for biological investi-

Annelids in Modern Biology, Edited by Daniel H. Shain
Copyright © 2009 John Wiley & Sons, Inc.

gation was largely abandoned within the first decade of the 20th century. Contributing to the decline was a shift in focus toward genetics, prompted by the rediscovery of Mendel's observations and the shortcomings of comparative embryology as a primarily observational and historical (thus untestable) discipline (Allen 2007).

Interest in leech neurobiology was rekindled in the 1960s when neurophysiologists Alan Hodgkin and Andrew Huxley (1952) won the Nobel Prize for their quantitative description of ion flux in squid giant axons. Researchers looking for other simple invertebrate species amenable to this application turned to *Hirudo medicinalis* and *Haemopis* spp., leeches that had proven so fruitful for turn-of-the-century neurobiologists Leydig (1862), Retzius (1891) and Ramon y Cajal (1904). By the 1970s, the introduction of microinjectable intracellular lineage tracers (Weisblat et al. 1978) sparked renewed interest in the role of cell lineage in development. Since the albumenotrophic embryos of *H. medicinalis* are small and cannot develop outside of their cocoon, they are not well suited to this technology. Consequently, the glossiphoniid leeches championed by Whitman (1878), with their large, yolky eggs, were restored to research prominence.

Over the past several decades, species of glossiphoniid leeches, particularly those of *Helobdella* and *Theromyzon*, have provided model experimental systems for investigating various developmental processes (e.g. polarity, epiboly, gastrulation). Their relatively large eggs (~400 μm in *Helobdella* spp., ~800 μm in *Theromyzon* spp.), rapid development times (about 1 h/cell cycle), independent development (i.e. outside their egg capsule), hardiness and experimental accessibility has made them particularly suitable for modern cellular and molecular techniques. In this chapter, we highlight aspects of leech embryonic development that contribute to the formation and specification of embryonic stem cells or teloblasts; other developmental processes that have been investigated in leech are reviewed elsewhere (e.g. Pilon and Weisblat 1997; Shankland and Bruce 1998; Weisblat and Huang 2001; Weisblat 2007).

7.2 STEM CELL GENESIS AND DEVELOPMENT

The developmental pattern known as spiral cleavage is widespread throughout protostomes (e.g. annelids, mollusks, rotifers) and is considered to be evidence of a common evolutionary ancestry (Wilson 1898). The stereotyped pattern is characterized by two cleavage divisions parallel to the animal–vegetal (A-V) axis, resulting in four primary blastomeres designated A, B, C and D. Shortly thereafter, each quadrant cleaves asymmetrically to produce a smaller daughter cell (micromere) at the animal pole, each of which is offset by 45° from its parent blastomere. Subsequent rounds of cleavage generate additional quartets of micromeres with offset orientations alternating between clockwise and counterclockwise. As a result of this predictability, individual cells from a given species can be reliably identified and their fates compared to corresponding cells in other spiralians (e.g. Table 7.1).

The classic spiralian cleavage mode outlined above is generally regarded as the ancestral pattern; however, numerous phylogenetic alterations of this pattern have occurred over evolutionary time. Within Hirudinea, cleavage deviations have accompanied changes in egg size and content, as well as the feeding habits of developing embryos. Comparisons to more basal annelids, such as *Tubifex* (Oligochaeta), *Acanthobdella* (Acanthobdellida) and certain aspects of polychaete development, indicate that only glossiphoniid leeches have retained the fundamental characteristics of the ancestral spiral cleavage pattern among leech taxa (Sandig and Dohle 1988; Dohle 1999). Additionally, cell pedigree

Table 7.1 Leech blastomere designations with corresponding spiralian nomenclature (in birth order)

Bissen and Weisblat (1989)	Sandig and Dohle (1988)[a]
d′	1d
c′	1c
a′	1a
b′	1b
c″	2c
DNOPQ	2d
DM	2D
dnopq′	$2d^1$
DNOPQ′	$2d^2$
dm′	3d
DM′	3D
a″	2a
b″	2b
dnopq″	$2d^{21}$
DNOPQ″	$2d^{22}$
dm″	4d
DM″	4D
c‴	3c
a‴	3a
b‴	3b
dnopq‴	$2d^{221}$
DNOPQ‴	$2d^{222}$
NOPQ(L and R)	T(L and R)
nopq′	t^I
NOPQ′	T^I
nopq″	t^{II}
NOPQ″	T^{II}
opq′	opq^I
OPQ′	OPQ^I

[a]Micromere designation dm″ switched to 4D, and macromere DM″ to 4d.

analyses of the most commonly employed glossiphoniidae genera have found no major developmental differences between them (Sandig and Dohle 1988; Bissen and Weisblat 1989). Therefore, the following description of leech embryogenesis is specifically applicable to both *Theromyzon* and *Helobdella* spp., and likely is representative of glossiphoniids in general (Whitman 1878); any known differences will be mentioned.

Leeches may be either protandrous or cosexual hermaphrodites (Davies and Singhal 1988). Some species are capable of self-fertilization (*Glossiphonia complanata*, Whitman 1878; *Helobdella* spp., Wedeen et al. 1989), though cross-fertilization is more typical. Meiosis is initiated upon fertilization (which occurs internally), but is arrested at metaphase I until oviposition (Fernandez and Olea 1982). Since egg deposition can occur over the course of several hours, and eggs are quiescent until such time, a single clutch contains embryos that are developmentally asynchronous. Meiosis resumes once the egg has been deposited within the cocoon and development proceeds in a highly stereotyped manner (note that cell cycle length is both temperature- and species-specific).

An overview of leech embryogenesis is presented schematically in Fig. 7.1. The staging system and blastomere designations, which differ from standard spiralian nomen-

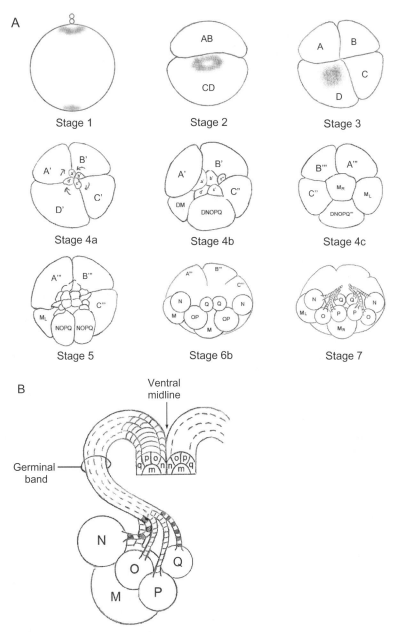

Figure 7.1 Schematic of leech development. (A) Selected embryonic stages showing blastomere divisions, teloplasm location (darkened areas) and teloblast positions. All views are from the animal pole except stage 1 (lateral view with polar bodies at animal pole) and stage 4c (vegetal pole view). Micromeres in stages 6b and 7 are omitted for clarity. Based on images from Sandig and Dohle (1988). (B) Stage 8 schematic depicting teloblast arrangement, germinal band formation and coalescence (based on *Theromyzon tessulatum* embryos). Both N and Q teloblasts produce two classes of blast cells (shown as alternating light and dark bands), while M, O and P each generate only one blast cell type. The rostral end of the germinal bands shows ectodermal bandlets around the subjacent mesodermal bandlet, and their juncture defines the ventral midline of the growing germinal plate. Based on sketches in Weisblat et al. (1984) and Keleher and Stent (1990).

clature, were introduced by Fernandez (1980) and were modified by Bissen and Weisblat (1989; Table 7.1). At the time of deposition, eggs already possess a polarity based on the eccentric location of the meiotic spindle, which defines the animal pole and marks the location of polar body extrusion (Fig. 7.1A, stage 1; Whitman 1878). A well-characterized series of contractions and deformations involving both microtubules and actin filaments results in expulsion of the two polar bodies and segregation of yolk-free cytoplasm (teloplasm) to the poles (Fernandez et al. 1987, 1990). Additionally, during the course of these cytoplasmic rearrangements, the male and female pronuclei move toward the center of the zygote and, shortly after completion of teloplasm formation, they fuse (Fernandez et al. 1990).

First cleavage is positioned meridionally but offset from the poles, generating a smaller AB blastomere and a larger CD blastomere; CD inherits most of the polar teloplasm (Fig. 7.1A, stage 2). Second cleavage is meridional as well, with blastomere CD dividing unequally to produce a larger D macromere (which receives the teloplasm) and a smaller C macromere. The ensuing equal cleavage of blastomere AB results in a four-cell embryo with three equal-sized macromeres: A, B and C (which later fuse to form the midgut endoderm), and one larger macromere D, which gives rise to all 10 embryonic stem cells/teloblasts (Fig. 7.1A, stage 3). At third cleavage, which is equatorial and highly unequal, each macromere buds off a much smaller micromere toward the animal pole. This primary micromere quartet (designated a′, b′, c′ and d′) is the first of three quartets produced by the four macromeres, each of which is designated with a "prime" for each division (e.g. at fourth cleavage, A′ produces macromere A″ and micromere a″). Subsequent cleavages from the D′ lineage add 13 more micromeres (total of 25), most of which congregate at the animal pole forming a "micromere cap" and later play a role in gastrulation.

Although in most respects the classic spiralian cleavage pattern is conserved, glossiphoniid leeches display an apparently derived anomaly beginning with the third cleavage. Whereas quadrants A, C and D follow the standard pattern, alternating between dexiotropically and laeotropically oriented cleavages, the B quadrant spindle is always positioned contrary to that of the other three (Fig. 7.1A, stage 4a). Consequently, leech embryos exhibit bilateral symmetry very early in development since the A and B quadrants produce mirror-image progeny with the cleavage plane between them defining the midline. This deviation in the leech B macromere cleavage orientation has not been described in any other annelid taxa (Sandig and Dohle 1988).

Macromeres A‴, B‴ and C‴ cease to divide further after contributing three micromeres each to the micromere cap, but undergo karyokineses to become multinucleate before fusing in a two-step process to form the syncytial yolk cell (Liu et al. 1998; Isaksen et al. 1999). Macromere D′, which gives rise to the segmented mesoderm and ectoderm, divides almost equally at the fourth cleavage to produce cell DM, located vegetally, and cell DNOPQ, situated near the animal pole (Fig. 7.1A, stage 4b). DM undergoes two unequal cleavages, resulting in two additional micromeres plus the mesodermal protoblast DM″ before dividing equally into right and left mesodermal stem cells (M_R and M_L, respectively). DNOPQ produces three micromeres and macromere DNOPQ‴ prior to an equal division into right and left ectodermal proteloblasts $NOPQ_R$ and $NOPQ_L$ (Whitman 1878; Sandig and Dohle 1988).

The bilaterally paired NOPQ proteloblasts add two micromeres each to the animal cap before dividing laterally into teloblast N and proteloblast OPQ; right and left OPQ lie adjacent to each other at the midline with N_R and N_L positioned distally. Two additional micromeres are budded off of each OPQ before division into teloblasts Q_R and Q_L (which now occupy the midline position nearer the animal pole), and OP_R and OP_L (obliquely vegetal to Q) (Sandig and Dohle 1988). Unlike the M, N and Q teloblasts that have fixed developmental potential from birth, O and P constitute an equivalence group capable of assuming either O or P fate. Therefore, even after each OP divides equally, they are referred to as O/P teloblasts (four per embryo). Once their progeny adopt either the O or P pattern, they are singularly distinguished as O or P (Weisblat and Blair 1984; Zackson 1984).

By the end of stage 6, cleavage is complete and all five bilaterally paired stem cells, M, N, O/P, O/P and Q, have been born. Each teloblast undergoes a series of highly unequal divisions at consistent intervals (approximately one per hour at 23 °C) resulting in a column of primary blast cells called a bandlet. Lineage tracing analyses have revealed that each bandlet can be definitively attributed to a specific teloblast based on the characteristic spatiotemporal pattern of its cell divisions (Zackson 1982). Seven classes of primary blast cells are present, one each for M, O and P and two each, in alternation, for N and Q. Since blast cells arise in a segmentally repeated manner, each hemisegment approximately comprises the progeny of a single M, O and P, and two N and Q primary blast cells.

Bandlets are arranged such that the mesodermal (M) bandlets lie subjacent to the four ectodermal bandlets (N, O, P and Q), which are stereotypically positioned on the surface of the embryo beneath a provisional, micromere-derived epithelial integument. Ipsilateral bandlets join to form right and left germinal bands and, as new blast cells are added from the posterior growth zone, move vegetally beneath the expanding edge of the epithelium. Eventually, the distal ends of each germinal band fuse (underlying M_L and M_R bandlets are connected from their birth) in the future cephalic region and the two bands conjoin along the ventral midline in a rostrocaudal progression to form the germinal plate (Fig. 7.1B). Within the germinal plate, the blast cells proliferate laterally, displacing the provisional integument, until joining at the dorsal midline to enclose the syncytial yolk cell, teloblast remnants and supernumerary blast cells.

7.3 FACTORS AFFECTING STEM CELL GENESIS

7.3.1 Teloplasm

The importance of teloplasm in stem cell genesis was first recognized by Whitman (1878). After observing that teloplasm was always segregated to the D (called "x" by Whitman) macromere at second cleavage, he concluded that "the vital energies of the egg seem concentrated" within it. Given the unique status of the D macromere as the sole progenitor of all 10 teloblasts, and thus all segmental tissue, determining the role of teloplasm in specifying D-cell fate has been a top priority. Investigations in this area have focused on three aspects: (1) cytoskeletal mechanisms involved in teloplasm formation and localization, (2) primary components of teloplasm and 3) the intrinsic capacity of the teloplasm to stipulate D-cell fate.

Cytoplasmic rearrangement resulting in heterogeneous distributions of cellular material is a routine occurrence in embryonic cells and a hallmark of stem cells. Segregation of morphogenetic determinants into domains, eccentrically situated with regard to the mitotic spindle (as in the bipolar teloplasm pools), ensures divergent fates for resulting daughter cells. Investigations into the biomechanical processes underlying teloplasm formation and localization in annelids have implicated three cytoskeletal elements: actin microfilaments, microtubules and astral microtubules. In most cases, one element is the primary operator for teloplasm formation, followed by a different element for segregation. Teloplasmic domains develop in nearly all annelid eggs, but considerable diversity is observed in the size, shape and location of the pools, as well as in the cytoskeletal mechanisms involved in their formation (Shimizu 1999). Even among species of *Helobdella* and *Theromyzon*, different mechanistic elements have been noted, though the end result is the same. Therefore, *Theromyzon* spp., the better characterized in this regard, will serve as the leech model.

In leech, teloplasm formation commences immediately after extrusion of the second polar body (Whitman 1878). Centrifugal movement of organelle-rich cytoplasm from deep within the egg forms a subcortical layer of yolk-deficient ectoplasm (Fernandez et al. 1987). This process is microtubule-dependent and may be driven by a single, centrally located microtubule organizing center (cytaster). Contractile rings form around the poles, followed by meridional furrows extending from each polar ring toward the equator. As the furrows deepen, cytoskeletal elements, organelles (particularly mitochondria) and maternally derived polyadenylated RNAs and proteins accumulate within them. As the contraction rings advance poleward, the meridional furrows retract from the equator, drawing the organelle-rich ectoplasm along and concentrating it at the poles. Teloplasm at the animal pole forms a conspicuous ring, while vegetal pole teloplasm appears as a more modest circular spot. In *Theromyzon* spp., actin microfilaments are largely responsible for the poleward progression, whereas in *Helobdella robusta*, it seems to be a microtubule-dependent process (Astrow et al. 1989; Fernandez et al. 1998).

In an effort to establish a causal link between teloplasm inheritance and D-cell fate, Astrow et al. (1987) used mild centrifugation to demonstrate that C macromeres are competent to assume D-cell fate if the teloplasm, normally segregated to D at second cleavage, is instead partitioned in part to C. Additionally, they found a direct correlation between developmental fate and the amount of teloplasm inherited by the blastomere; the greater the volume of teloplasm, the more likely it was to adopt D fate. The observation that both C and D follow the D cleavage pattern if each receives an equal partition of teloplasm indicates that the role of teloplasm in specifying D-cell fate may not be due to an inherent D-fate determinant, but rather to the sheer quantity of cellular materials (i.e. mitochondria, ribosomes, mRNA) capable of carrying out the D-fate program (Pilon and Weisblat 1997).

Further investigations have focused on the role of teloplasm inheritance in the specification of ectodermal versus mesodermal fate. Initial hypotheses have assumed that the obliquely equatorial division of D' into mesodermal proteloblast DM and ectodermal proteloblast DNOPQ resulted in DM receiving primarily vegetal teloplasm, while DNOPQ received animal teloplasm. Consequently, investigators have sought to identify specific mesodermal and/or ectodermal determinants within the respective teloplasmic domains. The fallacy of this assumption was illustrated by Holton et al. (1989) with the observation that, prior to fourth cleavage, vegetal teloplasm migrates to the animal pole and becomes confluent with the animal teloplasm. Therefore, when D' cleaves, both daughters (DM and DNOPQ) inherit a similar mixture of teloplasm, thus precluding the presence of individually pooled, fate-determining factors. Shimizu (1999) has speculated that within the Annelida, the development of bipolar teloplasmic domains is merely an evolutionary adaptation to facilitate efficient ooplasmic segregation necessitated by a dramatic increase in egg size. Polychaete eggs, generally half as large as their clitellate counterparts, only form unipolar teloplasmic domains. This theory lends support to the inherent equipotency of the two teloplasm pools in that their bipolar formation is related to time and energy, rather than fate.

Following up on the findings of Astrow and colleagues, Nelson and Weisblat (1991) demonstrated that teloplasmic extrusion from the DNOPQ proteloblast converts it from ectodermal to mesodermal fate. Furthermore, replacing the extruded animal teloplasm with vegetal teloplasm from the DM proteloblast rescues the ectodermal fate of DNOPQ (DM, however, is seemingly unaffected by teloplasm extrusion). These results confirm the equipotency of the bipolar teloplasm pools and indicate that the position of the teloplasm with regard to the A-V axis is a factor in specifying ectodermal fate. Subsequently, these same investigators showed that normal DNOPQ fate is dependent upon a particular spatial arrangement brought about by rigidly specified cleavage planes, which permit ectodermal determinants localized to the animal hemisphere to interact with teloplasmic components (Nelson and Weisblat 1992).

7.3.2 Asymmetric Cell Divisions

Asymmetric cell division resulting in divergent fates of sister cells is fundamental to early development and is achieved through a variety of mechanisms. Asymmetries may be obvious (size, teloplasm pool) or visually undetectable (differentially partitioned determinants), and may be brought about by either intrinsic or extrinsic factors, or by a combination of both (such as described above for the ectodermal specification of DNOPQ). Investigations in this area have implicated transcriptional activity (both maternal and zygotic), mitotic spindle placement, cell–cell contact/interaction and positional cues as playing roles in ensuring fate divergence.

Leech embryos, like those of most metazoans, are endowed with numerous maternally derived messenger RNAs that are largely responsible for directing early development. The role of maternal mRNAs as fate determinants has been established in several model systems (e.g. Sardet et al. 2003). Holton et al. (1994) showed that most homogeneously distributed mRNAs present in stage 1 leech embryos colocalize with the teloplasm prior to first cleavage and remain associated with it throughout cleavage. Bissen and Weisblat (1991) demonstrated that stage 1 embryos injected with the transcriptional inhibitor α-amanitin develop normally through stage 4, indicating that the asymmetric cleavages resulting in the designation of the D macromere as the segmental progenitor are maternally regulated. However, they also determined that normal stem cell genesis cannot occur without zygotic mRNA synthesis.

Early onset of zygotic transcription occurs in a variety of animal phyla, though most species do not require zygotic mRNAs prior to gastrulation. In leech, unlike most taxa (with the prominent exception of mice), zygotic mRNAs are necessary for normal development before completing cleavage (i.e. cell divisions before growth). The D' lineage of transcriptionally inhibited *H. robusta* embryos divides abnormally, beginning at stage 5, producing numerous medium-sized, yolky cells rather than the normal complement of micromeres and teloblasts (Bissen and Weisblat 1991). Even though mRNA synthesis is necessary for stem cell birth, once born, blast cell production ensues even in the presence of α-amanitin. Subsequently, Bissen and Smith (1996) determined that zygotic gene products are necessary for positioning the mitotic spindle in the large D'-derived cells so that asymmetric cleavages resulting in micromeres can occur; in the absence of such gene products, cleavage rates are unchanged, but every division is symmetric.

Recently, Ren and Weisblat (2006) described the mechanism employed by *H. robusta* in generating asymmetry during first cleavage. They found that gamma-tubulin is transiently downregulated at the prospective AB spindle pole causing it to condense, thus permitting the prospective CD spindle pole to enlarge. As a result, the initially symmetrically positioned mitotic spindle assembly shifts toward the prospective AB blastomere and the first cleavage plane is eccentrically located. This mechanism does not depend on a polarized cortex (as in *Caenorhabditis elegans*), nor on an intrinsically asymmetric spindle apparatus (as in *Tubifex*), yet the end result, asymmetric cleavage, is the same.

In addition to transcriptional activity, physical contact between cells as well as positional cues play important roles in ensuring asymmetric divisions. Symes and Weisblat (1992) found that the physical constraint imposed upon blastomere CD by contact with AB is necessary for the subsequent asymmetric division of CD into a smaller C and larger D macromere at second cleavage; thus, isolated (unconstrained) CD blastomeres cleave equally. Apparently, the effect of AB on CD is largely mechanical rather than inductive, since a dextran bead in place of the AB macromere also causes CD to cleave unequally. Conversely, a similar experiment using the oligochaete *Tubifex tubifex* found that while direct cell contact is necessary to achieve asymmetric division of CD, dextran bead-imposed constraint does not rescue asymmetric division; rather, alteration of the right spindle of the CD blastomere appears to be induced by asymmetrically distributed cortical factors (Takahashi and Shimizu 1997).

Inductive signaling also plays a role in specifying the distinct fates of NOPQ proteloblast daughter cells in *T. tubifex*. In an elegant series of experiments, Nakamoto et al. (2004) transplanted $NOPQ_L$ to various positions and assessed their subsequent division patterns for evidence of cell polarity. Their results indicate that NOPQ is not initially polarized; early $NOPQ_L$ transplanted to the position of $NOPQ_R$ develops normally (i.e. N is distal and OPQ is proximal to the midline). However, polarity develops gradually such that transplantation of late $NOPQ_L$ (shortly before the birth of N) to the right side generates an N teloblast at the midline and distal OPQ. M teloblasts, anteriolateral micromeres and sister NOPQ cells were identified as positional factors at least partially responsible for inducing polarization.

7.4 STEM CELL DIFFERENTIATION

M_L and M_R are the first-born teloblasts and sole progeny of their parent mesodermal proteloblast, DM″. Ectodermal proteloblast DNOPQ‴, on the other hand, generates four teloblast pairs (N, O, P and Q) with each pair arising consecutively. Studies have demonstrated that, with the exception of the O/P equivalence group, the fate of each teloblast is set from birth and no regulatory or compensatory process can replace a given teloblast once it is ablated. As they arise, teloblast pairs generate primary blast cells at a constant rate (approximately one cell per hour; Whitman 1878), producing lineage-specific bilateral columns of cells (bandlets). Ipsilateral bandlets contact each other in parallel to form right and left germinal bands that abut at their distal ends. First-born blast cells in each bandlet are farthest from the parent teloblast and contribute progeny to the anteriormost segments; later-born blast cells contribute to progressively more posterior segments. Due to the asynchronous genesis of teloblast pairs and the temporal regularity of blast cell production, the bandlets of earlier-born teloblast pairs each contain different numbers of blast cells by the time O/P divides into O and P. Consequently, distalmost blast cells in shorter bandlets must move past cells in longer bandlets so that segmental founder cells are in register.

As described previously, the five types of teloblasts generate seven classes of blast cells, each of which undergoes unique, highly stereotyped division patterns before terminal differentiation into definitive segmentally iterated structures. To date, efforts to identify lineage-specific factors leading to distinctly fated stem cell descendants have been largely unsuccessful. Numerous homologues of developmental regulatory genes (e.g. *Hox* family, *engrailed, even-skipped, wnt*) have been isolated and characterized in leech. Most, however, are broadly expressed throughout all five teloblast lineages, and thus are unlikely to play a role in specifying stem cell fate.

7.4.1 M Lineage

When mesodermal proteloblast DM″ divides, daughter cells M_L and M_R remain in close contact and the initial blast cells of their respective nascent bandlets are joined from birth (Fernandez and Stent 1980). As blast cell production continues, the conjoined bandlets migrate toward the surface and provide a stable substratum for developing ectodermal bandlets. Shortly after germinal band union, the M bandlets display an iterative pattern of discrete cell clusters indicative of protosegmental organization (Zackson 1982). The progeny of the first two cells in each M bandlet contribute to cephalic muscle fibers; most of the remaining blast cells give rise to segmental muscle, nephridia, some ganglionic neurons and probably germline cells. Since mesodermal teloblasts arise relatively early in development, they generate more blast cells than are needed to provide 32 segment's worth of progeny; these supernumerary blast cells later fuse with the syncytial yolk cell (reviewed in Weisblat and Huang 2001).

In *Drosophila melanogaster*, the gene *nanos* is required for normal abdomen formation, germ cell migration and germline stem cell maintenance; however, outside of insects, *nanos* function is largely limited to germline development. In leech, Kang et al. (2002) found that zygotic expression of *Hro-nos*, the *nanos*-class homologue in *H. robusta*, is broadly distributed throughout the germinal plate in stage 8 embryos, but is eventually restricted to putative primordial germ cells (PGCs) of mesodermal origin during stages 9 and 10. Agee et al. (2006) showed that embryos in which HRO-NOS was knocked down displayed abnormal micromere distribution, asymmetrically positioned germinal bands and epiboly failure leading to death. Their findings demonstrate the importance of *Hro-nos* expression in early leech development but fail to link it to fate specification of any particular cell lineage. Conversely, in the spiralian mollusk *Ilyanassa obsoleta, nanos* mRNAs are found to be specifically localized to 4d- (DM″ in leech) derived cells, and *nanos* expression is required for normal cleavage and proliferation of mesodermal blast cells (Rabinowitz et al. 2008). This study is the first to connect *nanos* expression with somatic cell patterning in a spiral-cleaving organism, and identifies a putative difference between the genetic programs of two spiralians.

To date, the only gene positively linked to mesodermal differentiation in leech is *Hau-Pax3/7A*, a member of the widely conserved paired-box (*Pax*) family of transcription factors. *Pax* family genes are divided into four distinct groups (I–IV) based on sequence similarity; all play critical roles in early development throughout metazoan phyla. The *Pax III* gene group, to which *Hau-Pax3/7A* belongs, includes genes involved in segmentation, neurogenesis and myogenesis. Woodruff et al. (2007) isolated *Hau-Pax3/7A* in *Helobdella* sp. (Austin) and investigated its expression pattern and function. They found that maternal expression is localized to the teloplasm, broadly expressed in both mesodermal and ectodermal progenitors, and then disappears prior to blast cell production; zygotic expression, however, is confined to m blast cells and their progeny. Although the investigators were unable to determine the function of maternal transcripts, zygotic *Hau-Pax3/7A* expression was shown to be required for normal segmental organization, nephridial development and dorsal body cavity formation. Due to its restriction to the M lineage, *Hau-Pax3/7A* appears to be an early marker of leech mesoderm.

7.4.2 N and Q Lineages

The N and Q teloblasts each produce two classes of blast cells in alteration, which are designated by the subscripts f (früh, meaning early) or s (spät, meaning late) based on the timing of their first mitoses. Within the N bandlet, n_f divides at a distance of approximately 22 blast cells from the N teloblast, while n_s divides approximately 28 cells away (Zackson 1984). Both sublineages contribute progeny primarily to the ventral nerve cord; within each segment, however, n_s is the founder cell for anterior hemiganglion, while n_f contributes to posterior hemiganglion as well as non-CNS (central nervous system) tissue (Shain et al. 1998). In the Q lineage, q_f and q_s undergo their first mitoses ~28 and ~33 cells from teloblast Q, respectively, contributing primarily to dorsal ectoderm (Zackson 1984). Again, however, the two sublineages (q_f and q_s) make specific segmental contributions; one generates only ectodermal tissue, while the other contributes to both ectodermal and neuronal structures.

As a consequence of producing two different classes of blast cells, both of which are required per segmental repeat, N and Q teloblasts must generate twice as many blast cells as M, O and P. Also, the distalmost cells in the N and Q bandlets must continually migrate past more posteriorly fated cells in the M, O and P bandlets in order to maintain their segmental register. Currently, nothing is known about regulatory mechanisms controlling this movement, nor have factors specifically associated with either N or Q differentiation been described.

7.4.3 O and P Lineages

Since first being described as an equivalence group (Shankland and Weisblat 1984; Weisblat and Blair 1984), the O and P lineages have been the subject of numerous studies. Early investigations revealed that in the absence of the P lineage, the O bandlet adopts P-pattern fate. Since M, N and Q lineages acquire their fates independently from extrinsic factors, the pluripotency of o/p blast cells is intriguing. For some time after their birth, even after undergoing first mitoses typical of either an o-type or a p-type cleavage pattern, o/p primary blast cells retain the ability to transfate in response to changes in their external environment (Ho and Weisblat 1987). Efforts to elucidate mechanisms involved in o/p fate specification have yielded a variety of often conflicting results. Recent studies have implicated developmental system drift (DSD) (Kuo and Shankland 2004) and misidentified leech specimens (Bely and Weisblat 2006) as the primary reasons for these inconsistencies.

Pursuant to stage 6 cleavage resulting in teloblast Q, the remaining OP proteloblast pair produces several o/p primary blast cells before dividing equally into O/P sister teloblasts. In *Helobdella triserialis, H. robusta* (Austin) and the oligochaete *Tubifex hattai*, four o/p blast cells are produced before O/P teloblast birth (Bissen and Weisblat 1989; Goto et al. 1999; Kuo and Shankland 2004), while five cells are produced in *Theromyzon tessulatum* (Sandig and Dohle 1988). Due to these initial o/p cells, ipsilateral O/P teloblasts generate individual bandlets that are joined at their anterior ends (Fig. 7.1B). The distalmost o/p cells contribute definitive progeny to the rostral segments, while succeeding o/p cells contribute differentiated descendants to midbody and caudal segments (Shankland 1987). The segmentally iterated structural elements (e.g. neurons, epidermal cell florets, nephridial tubule cells) derived from a single o primary blast cell are distinct from those descended from a single p primary blast cell (Weisblat and Blair 1984). In rostral segments, each o/p blast cell generates a complement of pattern elements equal to one o and one p blast cell (Shankland 1987).

In addition to giving rise to discrete progeny, o and p blast cells are easily distinguished by their characteristic mitotic cleavage patterns (Zackson 1984). These attributes have enabled the use of lineage tracers and cell ablations for investigating factors involved in the specification of o/p bandlet fate. In normal development, both o/p bandlets lie between bandlets n and q; the more dorsal of the two (adjacent to q) takes on P fate, while the more ventral (adjacent to n) takes on O fate. In *H. triserialis*, ablation of the overlying micromere-derived epithelium causes the presumptive o bandlet (based on position near n) to assume P fate (Ho and Weisblat 1986). A subsequent investigation using *H. robusta* (Sacramento) confirmed the role of the provisional epithelial integument and concluded that it specified an O fate default state in the absence of local signals (Huang and Weisblat 1996). However, Kuo and Shankland (2004) found no evidence of epithelial induced O fate in *H. robusta* (Austin), nor did Keleher and Stent (1990) in *Theromyzon rude* (except when loss of compression permitted positional changes).

Findings by Huang and Weisblat (1996) in *H. robusta* (Sacramento) indicate that P fate induction is due to short-range signals (which need only be transient) from the q bandlet, and that no inductive interactions occur between o/p blast cells. Again, these results are both confirmed and contradicted in studies using other glossiphoniid species. Zackson (1984) found that Q teloblast ablation (as well as N and M) had no effect on o/p fate specification in *Helobdella stagnalis* and proposed that mutual o/p induction was responsible. In *T. rude*, Keleher and Stent (1990) concluded that o/p fate is strictly positional, and although induced by signal(s) outside the equivalence group, neither q nor n influence o/p fate. Kuo and Shankland (2004) concur that q bandlet induction is the primary factor in deciding o/p fate in both *Helobdella* sp. (Galt) and *H. robusta* (Austin). However, they report that *H. robusta* (Austin) has a redundant o/p fate specification pathway that functions in the absence of the Q lineage and is dependent on both the M lineage and o/p–o/p interactions. To complicate matters further, in the oligochaete *T. tubifex*, O and P do not represent an equivalence group. P fate is primary and O fate is induced by the p bandlet; thus, the P lineage, like M, N and Q, has its fate set from birth (Arai et al. 2001).

Efforts to reconcile conflicting results such as these have shed light on a previously unrecognized evolutionary process, known as DSD. We generally assume that phenotypic similarities between closely related species are governed by homologous developmental pathways, but changes in underlying developmental mechanisms without concomitant changes in morphology appear ubiquitous (True and Haag 2001). The results above demonstrate how prevalent this phenomenon may be, particularly among *Helobdella* species.

Despite the appeal of DSD as the answer to the confounding results preventing our full understanding of O/P fate mechanisms, all of the problems cannot be laid on its doorstep. In the decades since leech has become an established experimental system, several glossiphoniid species (most from *Helobdella* or *Theromyzon* genera) have been employed in various studies. Although long-term, laboratory-bred colonies exist, many species are difficult to maintain and their health is easily compromised. Consequently, nearly all colonies have relied on field-gathered supplemental specimens. Until quite recently, species identification was based solely on morphological features; since visual distinctions between closely related species can be notoriously difficult, misidentified species were common (e.g. Siddall et al. 2007)

Recently, Bely and Weisblat (2006) undertook molecular phylogenetic analyses of cytochrome oxidase I (COI) genes from numerous *Helobdella* laboratory colonies, as well as field-gathered specimens from their original collection sites. Although they definitively identified only one isolate [*Helobdella* sp. (Galt) as *Helobdella europaea*], their results indicate that the numerous isolates examined likely represent five different species. Additionally, specimens identified in several published papers as "*H. robusta*" are in fact three distinct species, two of which coexist at the same locale (Sacramento). Furthermore, the species referred to as "*H. triserialis*," and extensively employed throughout the 1980s, does not group with the known *H. triserialis* COI sequence from a Bolivian specimen. Thus, for several published studies, the O/P fate mechanisms described cannot be definitively assigned to a particular species; therefore, some apparently contradictory results may not actually exist.

7.5   GENE EXPRESSION

A great triumph of modern biology has been the realization that fundamental aspects of development (e.g. genes and genetic pathways) are often conserved across metazoan phyla. Products of the *Hox* gene cluster, for example, establish the anteroposterior axis in disparate animals ranging from arthropods to chordates (Ferrier and Holland 2001), and *distal-less* homologues appear to promote morphological outgrowth throughout the Animalia (Zerucha and Ekker 2000). On the other hand, some developmental genes have retained strong sequence identity, while their functions have diverged in various animal lineages. Segmentation, for example, requires *engrailed* gene expression in arthropods (Kornberg 1981), but *engrailed*-class gene expression does not seem required in at least some annelids (Shain et al. 1998, 2000; Seaver and Shankland 2000; Seaver et al. 2001; see Chapters 5 and 6).

In the context of identifying genes associated with stem cell genesis and differentiation, the popular methodology of cloning putative homologues identified in other animal systems (i.e. the "candidate gene approach") seems to fall short for a few reasons. First, collective data from comparisons of developmental processes—even within genera (e.g. *Helobdella*)—reveal a striking divergence in the mechanisms by which animals arrive at similar end points (e.g. asymmetric cleavages, O/P fate determination, segmentation), and secondly, the search for animal stem cell-specific genes has been somewhat elusive (e.g. Evsikov and Solter 2003; Vogel 2003), even though several genes are generally linked with stem cell-specific properties (e.g. self-renewal; *esg1, nanog, oct4, piwi*). Note, however, that these genes are also expressed in founder cell populations and/or are broadly expressed during development (Bierbaum et al. 1994; Pesce and Scholer 2001; Tanaka et al. 2002; Chambers et al. 2003). Various transcriptional profiling efforts have identified candidate genes involved with stem cell self-renewal and pluripotency (Terskikh et al. 2001; Ivanova et al. 2002; Ramalho-Santos et al. 2002), but little overlap occurs between gene datasets, which has called these analyses into question (Evsikov and Solter 2003; Fortunel et al. 2003; Vogel 2003). More recently, a unique combination of transcription factors has been identified that appears sufficient to promote mammalian ES (embryonic stem) cell fate when ectopically expressed in some nonstem cell types (Takahashi and Yamanaka 2006; Park et al. 2008), but these are unlikely to represent the endogenous mechanism by which ES cells acquire their fate.

Thus, to understand biological processes at mechanistic and evolutionary angles, one should perhaps independently unravel the process in distinct animal taxa before making too many comparisons. Indeed, we have pursued this tact in our quest to identify stem cell-specific genes by employing a nonbiased, differential display-PCR (DD-PCR) approach in the glossiphoniid leech, *T. tessulatum* (Hohenstein and Shain 2004). Embryos of *T. tessulatum* are particularly suited for this purpose due to their extraordinary size

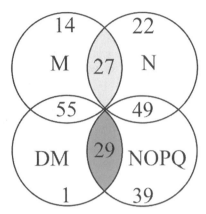

Figure 7.2   Categories of differentially displayed cDNAs identified in developing *Theromyzon tessulatum* embryos. Proteloblast-specific cDNAs were expressed in both DM and NOPQ, but not in M or N teloblasts (29; dark gray); teloblast-specific cDNAs were in both M and N cells, but not in DM or NOPQ (27, light gray). Note that DM, which gives rise to the bilateral M teloblasts, contained only one differentially expressed cDNA, while NOPQ, which gives rise to teloblasts N, O, P and Q, contained 39; these latter differentially expressed cDNAs are possibly a mixture of O, P and Q determinants. Reprinted from Hohenstein and Shain (2004) with permission from AlphaMed Press.

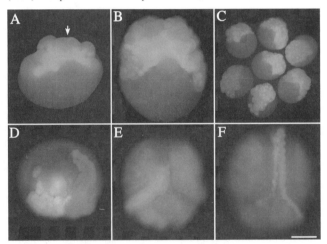

Figure 7.3   Phenotypic effects of microinjecting K110 oligonucleotides into *Theromyzon tessulatum* embryos. (A) Normal embryo after injecting fluorescein-dextran amine (FDA) (green) or K110 sense oligos into the D quadrant macromere, +20 h further development (arrow defines a plane of symmetry among ectodermal teloblasts). (B) Representative embryo after injecting any of four independent K110 antisense oligos into the D macromere, +24 h. (C) Embryos from the same clutch injected with K110 antisense oligo into D, +24 h. (D) Normal bandlets following injection of K110 sense oligo (coinjected with tetramethylrhodamine-dextrin amine [RDA], red) into NOPQ (right); NOPQ (left) injected with K110 antisense oligo (coinjected with FDA, +48 h). (E) M (right) teloblast injected with K110 sense (+50 h, red, normal); M (left) teloblast injected with K110 antisense (+48 h, green, apparently normal). (F) N (right) teloblast injected with K110 sense (+72 h, red, normal); N (left) teloblast injected with K110 antisense (+70 h, green, apparent cell deficiencies). Scale bar = 100 μm. (See color insert.)

(~800 μm diameter) and developmental asymmetries (see Fig. 7.1), which permit relatively simple dissections of stem cell precursors (proteloblasts) and embryonic stem cells (teloblasts) alike.

As described by Hohenstein and Shain (2004), the comparison of proteloblast and telobast gene expression profiles identified relatively small gene sets specific to each cell type, as well as identified numerous candidate lineage-specific genes in mesodermal and neuroectordermal lineages (Fig. 7.2). The characterization of proteloblast- and teloblast-specific cDNAs has revealed an assortment of genes that appear enriched in translation initiation factors and transcription factors, respectively. The functional knockdown of one teloblast-specific mRNA, K110, inhibits normal teloblast genesis by causing the abnormal proliferation of proteloblasts, while the same manipulation does not seem to affect development once a teloblast is born (Fig. 7.3). Knockdown of another telobast-specific mRNA, K46, permitted teloblast birth, but bandlets were truncated after a few teloblastic divisions. The remaining differentially expressed genes at the transition between proteloblasts and teloblasts remain uncharacterized, but we hope that the availability of these gene pools will ultimately facilitate a mechanistic understanding of stem cell genesis and differentiation in leech.

7.6 CONCLUSION

The prospects and potential of stem cells in biology/medicine are not questioned, yet many fundamental questions remain about their basic biology, development and evolution. The features and experimental accessibility of glossiphoniid leech embryos make them especially suitable research subjects for exploring these questions. And just as other simple model systems have made valuable contributions to science (e.g. *D. melanogaster, C. elegans* to development, genetics and human disease), so too could leeches make landmark contributions to stem cell biology.

REFERENCES

AGEE, S.J., LYONS, D.C., WEISBLAT, D.A. 2006. Maternal expression of a NANOS homolog is required for early development of the leech *Helobdella robusta*. Dev. Biol. 298, 1–11.

ALLEN, G.E. 2007. A century of evo-devo: the dialectics of analysis and synthesis in twentieth-century life science. In *From Embryology to Evo-Devo: a History of Developmental Evolution*, edited by M.D. Laubichler and J. Maienschein. Cambridge: MIT Press, pp. 123–167.

ARAI, A., NAKAMOTO, A., SHIMIZU, T. 2001. Specification of ectodermal teloblast lineages in embryos of the oligochaete annelid *Tubifex*: involvement of novel cell-cell interactions. Development 128, 1211–1219.

ASTROW, S.H., HOLTON, B., WEISBLAT, D.A. 1987. Centrifugation redistributes factors determining cleavage patterns in leech embryos. Dev. Biol. 120, 270–283.

ASTROW, S.H., HOLTON, B., WEISBLAT, D.A. 1989. Teloplasm formation in a leech, *Helobdella triserialis*, is a microtubule-dependent process. Dev. Biol. 135, 306–319.

BELY, A.E., WEISBLAT, D.A. 2006. Lessons from leeches: a call for DNA barcoding the lab. Evol. Dev. 8, 491–501.

BIERBAUM, P., MACLEAN-HUNTER, S., EHLERT, F., MOROY, T., MULLER, R. 1994. Cloning of embryonal stem cell-specific genes: characterization of the transcriptionally controlled gene *esg-1*. Cell Growth Differ. 5, 37–46.

BISSEN, S.T., SMITH, C.M. 1996. Unequal cleavage in leech embryos: zygotic transcription is required for correct spindle orientation in subset of early blastomeres. Development 122, 599–606.

BISSEN, S.T., WEISBLAT, D.A. 1989. The durations and compositions of cell cycles in embryos of the leech, *Helobdella triserialis*. Development 105, 105–118.

BISSEN, S.T., WEISBLAT, D.A. 1991. Transcription in leech: mRNA synthesis is required for early cleavages in *Helobdella* embryos. Dev. Biol. 146, 12–23.

BOWLER, P.J. 1996. *Life's Splendid Drama: Evolutionary Biology and the Reconstruction of Life's Ancestry 1860-1940*. Chicago: University of Chicago Press.

CHAMBERS, I., COLBY, D., ROBERTSON, M., NICHOLS, J., TWEEDIE, S., SMITH, A. 2003. Functional expression cloning of *Nanog*, a pluripotency sustaining factor in embryonic stem cells. Cell 113, 643–655.

DARWIN, C.D. 1859. *On the Origin of Species by Means of Natural Selection*. London: John Murray.

DAVIES, R.W., SINGHAL, R.N. 1988. Cosexuality in the leech *Nephelopsis obscura* (Erpodellidae). Int. J. Invertebr. Reprod. Dev. 13, 55–64.

DOHLE, W. 1999. The ancestral cleavage pattern of the clitellates and its phylogenetic deviations. Hydrobiologia 402, 267–283.

EVSIKOV, A.V., SOLTER, D. 2003. Comment on "'Stemness': transcriptional profiling of embryonic and adult stem cells" and "a stem cell molecular signature." Science 302, 393.

FERNANDEZ, J. 1980. Embryonic development of the glossiphoniid leech *Theromyzon rude*: characterization of developmental stages. Dev. Biol. 76, 245–262.

FERNANDEZ, J., OLEA, N. 1982. Embryonic development of glossiphoniid leeches. In *Developmental Biology of Freshwater Invertebrates*, edited by F.W. Harrison and R.R. Cowden. New York: A. R. Liss, pp. 317–366.

FERNANDEZ, J., STENT, G.S. 1980. Embryonic development of the glossiphoniid leech *Theromyzon rude*: structure and development of the germinal bands. Dev. Biol. 78, 407–434.

FERNANDEZ, J., OLEA, N., MATTE, C. 1987. Structure and development of the egg of the glossiphoniid leech *Theromyzon rude*: Characterization of developmental stages and structure of the early uncleaved egg. Development 100, 211–225.

FERNANDEZ, J., OLEA, N., TELLEZ, V., MATTE, C. 1990. Structure and development of the egg of the glossiphoniid leech *Theromyzon rude*: reorganization of the fertilized egg during completion of the first meiotic division. Dev. Biol. 137, 142–154.

FERNANDEZ, J., OLEA, N., UBILLA, A., CANTILLANA, V. 1998. Formation of polar cytoplasmic domains (teloplasms) in the leech egg is a three-step segregation process. Int. J. Dev. Biol. 42, 149–162.

FERRIER, D.E., HOLLAND, P.W. 2001. Ancient origin of the *Hox* gene cluster. Nat. Rev. Genet. 2, 33–38.

FORTUNEL, N.O., OTU, H.H., NG, H., CHEN, J. 2003. Comment on: "'Stemness': transcriptional profiling of embryonic and adult stem cells" and "A stem cell molecular signature" (I). Science 302, 393.

GOTO, A., KITAMURA, K., ARAI, A., SHIMIZU, T. 1999. Cell fate analysis of teloblasts in the *Tubifex* embryo by intercellular injection of HRP. Dev. Growth Diff. 41, 703–713.

HALL, B.K. 1999. *Evolutionary Developmental Biology*, 2nd edn. Dordrecht, The Netherlands: Kluwer Academic Publishers.

HO, R.K., WEISBLAT, D.A. 1987. A provisional epithelium in leech embryo: cellular origins and influence on a developmental equivalence group. Dev. Biol. 120, 520–534.

HODGKIN, A.L., HUXLEY, A.F. 1952. A quantitative description of membrane current and its application to conduction and excitation in nerve. J. Physiol. 117, 500–544.

HOHENSTEIN, K.A., SHAIN, D.H. 2004. Changes in gene expression at the precursor→stem cell transition in leech. Stem Cells 22, 514–521.

HOLTON, B., ASTROW, S.H., WEISBLAT, D.A. 1989. Animal and vegetal teloplasms mix in the early embryo of the leech *Helobdella triserialis*. Dev. Biol. 131, 182–188.

HOLTON, B., WEDEEN, C.J., ASTROW, S.H., WEISBLAT, D.A. 1994. Localization of polyadenylated RNAs during teloplasm formation and cleavage in leech embryos. Roux's Arch. Dev. Biol. 204, 46–53.

HUANG, F.Z., WEISBLAT, D.A. 1996. Cell fate determination in an annelid equivalence group. Development 122, 1839–1847.

ISAKSEN, D.E., LIU, N.L., WEISBLAT, D.A. 1999. Inductive regulation of cell fusion in leech. Development 126, 3381–3390.

IVANOVA, N.B., DIMOS, J.T., SCHANIEL, C., HACKNEY, J.A., MOORE, K.A., LEMISCHKA, I.R. 2002. A stem cell molecular signature. Science 298, 601–604.

KANG, D., PILON, M., WEISBLAT, D.A. 2002. Maternal and zygotic expression of a *nanos*-class gene in the leech *Helobdella robusta*: primordial germ cells arise from segmental mesoderm. Dev. Biol. 245, 28–41.

KELEHER, G.P., STENT, G.S. 1990. Cell position and developmental fate in leech embryogenesis. Proc. Natl. Acad. Sci. U.S.A. 87, 8457–8461.

KORNBERG, T. 1981. *engrailed*: a gene controlling compartment and segment formation in *Drosophila*. Proc. Natl. Acad. Sci. U.S.A. 78, 1095–1099.

KUO, D-H., SHANKLAND, M. 2004. Evolutionary diversification of specification mechanisms within the O/P equivalence group of the leech genus *Helobdella*. Development 131, 5859–5869.

LEYDIG, F. 1862. Über das Nervensystem der Anneliden. Arch. Anat. Physiol. Lpz. 1862, 90–124.

LIU, N.L., ISAKSEN, D.E., SMITH, C.M., WEISBLAT, D.A. 1998. Movements and stepwise fusion of endodermal precursor cells in leech. Dev. Genes Evol. 208, 117–127.

NAKAMOTO, A., ARAI, A., SHIMIZU, T. 2004. Specification of polarity of teloblastogenesis in the oligochaete annelid *Tubifex*: cellular basis for bilateral symmetry in the ectoderm. Dev. Biol. 272, 248–261.

NELSON, B.H., WEISBLAT, D.A. 1991. Conversion of ectoderm to mesoderm by cytoplasmic extrusion in leech embryos. Science 253, 435–438.

NELSON, B.H., WEISBLAT, D.A. 1992. Cytoplasmic and cortical determinants interact to specify ectoderm and mesoderm in the leech embryo. Development 115, 103–115.

PARK, I., ZHAO, R., WEST, J.A., YABUUCHI, A., HUO, H., INCE, T.A., LEROU, P.H., LENSCH, M.W., DALEY, G.Q. 2008. Reprogramming of human somatic cells to pluripotency with defined factors. Nature 451, 135–136.

PESCE, M., SCHOLER, H.R. 2001. Oct-4: gatekeeper in the beginnings of mammalian development. Stem Cells 19, 271–278.

PILON, M., WEISBLAT, D.A. 1997. Early events leading to fate decisions during leech embryogenesis. Semin. Cell Dev. Biol. 8, 351–358.

RABINOWITZ, J.S., CHAN, X.Y., KINGSLEY, E.P., DUAN, Y., LAMBERT, J.D. 2008. Nanos is required in somatic blast cell lineages in the posterior of a mollusk embryo. Curr. Biol. 18, 331–336.

RAMALHO-SANTOS, M., YOON, S., MATSUZAKI, Y., MULLLIGAN, R.C., MELTON, D.A. 2002. "Stemness": Transcriptional profiling of embryonic and adult stem cells. Science 298, 597–600.

RAMON Y CAJAL, S. 1904. Variaciones morfólogicas de reticulo nervioso de invertebrados y vertebrados sometidos a la acción de condiciones naturals (nota preventiva). Trabajos Lab. Invest. Biol. 3, 287–297.

REN, X., WEISBLAT, D.A. 2006. Asymmetrization of first cleavage by transient disassembly of one spindle pole aster in the leech *Helobdella robusta*. Dev. Biol. 292, 103–115.

RETZIUS, G. 1891. Zur Kenntniss des centralen Nervensystem der Würmer. Biol. Unters. (N.F.) 2, 1–28.

SANDIG, M., DOHLE, W. 1988. The cleavage pattern in the leech *Theromyzon tessulatum* (Hirudinea Glossiphoniidae). J. Morphol. 196, 217–252.

SARDET, C., NISHIDA, H., PRODON, F., SAWADA, K. 2003. Maternal mRNAs of *PEM* and *macho 1*, the ascidian muscle determinant, associate and move with a rough endoplasmic reticulum network in the egg cortex. Development 130, 5839–5849.

SEAVER, E.C., SHANKLAND, M. 2000. Leech segmental repeats develop normally in the absence of signals from either anterior or posterior segments. Dev. Biol. 224, 339–353.

SEAVER, E.C., PAULSON, D.A., IRVINE, S.Q., MARTINDALE, M.Q. 2001. The spatial and temporal expression of *Ch-en*, the *engailed* gene in the pollychaete *Chaetopterus*, does not support a role in body axis segmentation. Dev. Biol, 236, 195–209.

SHAIN, D.H., RAMIREZ, F.A., HSU, J., WEISBLAT, D.A. 1998. Gangliogenesis in leech: morphogenetic processes leading to segmentation in the central nervous system. Dev. Genes Evol. 208, 28–36.

SHAIN, D.H., STUART, D., HUANG, F.Z., WEISBLAT, D.A. 2000. Segmentation of the central nervous system in leech. Development 127, 735–744.

SHANKLAND, M. 1987. Cell lineage in leech embryogenesis. Trends Genet. 3, 314–319.

SHANKLAND, M., BRUCE, A.E. 1998. Axial patterning in the leech: developmental mechanisms and evolutionary implications. Biol. Bull. 195, 370–372.

SHANKLAND, M., WEISBLAT, D.A. 1984. Stepwise commitment of blast cell fates during the positional specification of the O and P cell lines in the leech embryo. Dev Biol. 106, 326–342.

SHIMIZU, T. 1999. Cytoskeletal mechanisms of ooplasmic segregation in annelid eggs. Intl. J. Dev. Biol. 43, 11–18.

SIDDALL, M.E., TRONTELI, P., UTEVSKY, S.Y., NKAMANY, M., MACDONALD, K.S. 2007. Diverse molecular data demonstrate that commercially available medicinal leeches are not *Hirudo medicinalis*. Proc. Biol. Sci. 274, 1481–1487.

SYMES, K., WEISBLAT, D.A. 1992. An investigation of the specification of unequal cleavages in leech embryos. Dev. Biol. 150, 203–218.

TAKAHASHI, H., SHIMIZU, T. 1997. Role of intercellular contacts in generating an asymmetric mitotic apparatus in the *Tubifex* embryo. Dev. Growth. Differ. 39, 351–362.

TAKAHASHI, K., YAMANAKA, S. 2006. Induction of pluripotent stem cells from mouse embryonic and adult fibroblast cultures by defined factors. Cell 126, 663–676.

TANAKA, T.S., KUNATH, T., KIMBER, W.L., JARADAT, S.A., STAGG, C.A., USUDA, M., YOKOTA, T., NIWA, H., ROSSANT, J., KO, M.S. 2002. Gene expression profiling of embryo-derived stem cells reveals candidate genes associated with pluripotency and lineage specificity. Genome Res. 12, 1921–1928.

TERSKIKH, A.V., EASTERDAY, M.C., LI, L., HOOD, L., KORNBLUM, H.I., GESCHWIND, D.H., WEISSMAN, I.L. 2001. From hematopoiesis to neuropoiesis: evidence of overlapping genetic programs. Proc. Natl. Acad. Sci. U.S.A. 98, 7934–7939.

TRUE, J.R., HAAG, E.S. 2001. Developmental system drift and flexibility in evolutionary trajectories. Evol. Dev. 3, 109–119.

VOGEL, G. 2003. 'Stemness' genes still elusive. Science 302, 371.

WEDEEN, C.J., PRICE, D.J., WEISBLAT, D.A. 1989. Analysis of the life cycle, genome and homeo box genes of the leech, *Helobdella triserialis*. In *The Cellular and Molecular Biology of Pattern Formation*, edited by D.L. Stocum and T.L. Karr. New York: Oxford University Press, pp. 145–167.

WEISBLAT, D.A. 2007. Asymmetric cell division in the early embryo of the leech *Helobdella robusta*. Prog. Mol. Subcell. Biol. 45, 79–95.

WEISBLAT, D.A., BLAIR, S.S. 1984. Developmental indeterminacy in embryos of the leech *Helobdella triserialis*. Dev. Biol. 101,326–335.

WEISBLAT, D.A., HUANG, F.Z. 2001. An overview of glossiphoniid leech development. Can. J. Zool. 79, 218–232.

WEISBLAT, D.A., SAWYER, R.T., STENT, G.S. 1978. Cell lineage analysis by intracellular injection of a tracer enzyme. Science 202, 1295–1298.

WEISBLAT, D.A., KIM, S.Y., STENT, G.S. 1984. Embryonic origins of cells in the leech *Helobdella triserialis*. Dev. Biol. 104, 65–85.

WHITMAN, C.O. 1878. The embryology of *Clepsine*. Q. J. Microsc. Sci. 18, 215–315.

WILSON, E.B. 1898. Considerations on cell-lineage and ancestral reminiscence, based on a re-examination of some points in the early development of annelids and polyclades. Ann. N. Y. Acad. Sci. 11, 1–27.

WOODRUFF, J.B., MITCHELL, B.J., SHANKLAND, M. 2007. *Hau-Pax3/7A* is an early marker of leech mesoderm involved in segmental morphogenesis, nephridial development, and body cavity formation. Dev. Biol. 306, 824–837.

ZACKSON, S.L. 1982. Cell clones and segmentation in leech development. Cell 31, 761–770.

ZACKSON, S.L. 1984. Cell lineage, cell-cell interaction, and segment formation in the ectoderm of a glossiphoniid leech embryo. Dev. Biol. 104, 143–160.

ZERUCHA, T., EKKER, M. 2000. Distal-less-related homeobox genes of vertebrates: evolution, function, and regulation. Biochem Cell Biol. 78, 593–601.

Part III

Neurobiology and
Regeneration

Chapter 8

Cellular and Behavioral Properties of Learning in Leech and Other Annelids

Kevin M. Crisp* and Brian D. Burrell†

*Department of Biology and Neuroscience Program, St. Olaf College, Northfield, MN
†Neuroscience Group, Division of Basic Biomedical Sciences, Sanford School of Medicine, University of South Dakota, Vermillion, SD

[W]orms, although standing low in the scale of organization, possess some degree of intelligence. This will strike everyone as very improbable; but it may be doubted whether we know enough about the nervous system of the lower animals to justify our natural distrust of such a conclusion.

—*Charles Darwin* (1881)[1]

8.1 INTRODUCTION

Annelids represent a powerful model system for studying the cellular and molecular processes mediating learning and memory. As noted 40 years ago, "annelids offer some of the most promising material for study of learning mechanisms in simple systems...their availability, simplicity, phylogenetic position, learning capacity, and tolerance [for manipulation] suggest that they are well worth new attention" (Bullock and Quarton 1966). Studies on classical conditioning of defensive reflexes in annelids flourished in the 1950s and 1960s. The study of learning in earthworms continued until the mid-1970s, when earthworm pheromones were demonstrated to bias performance in certain types of behavioral paradigms (Rosenkoetter and Boyce 1975). More recently, attention has focused on the cellular basis of learning and memory in leeches, largely because of their suitability for modern neurophysiological recording techniques.

The medicinal leech *Hirudo medicinalis* has a relatively simple nervous system in which many neurons have been uniquely identified with respect to their morphology,

---

[1] Quote from Charles Darwin's *The Formation of Vegetable Mould through the Action of Worms with Observations on Their Habits*, published in 1966 as *Darwin on Humus and the Earthworm* by Faber and Faber (London).

---

connectivity, function and location in the nervous system. The presence of identifiable neurons in leech has enabled behaviorally relevant neural circuits to be characterized in detail. Large, easily visualized neurons can be individually penetrated with microelectrodes for electrophysiological study both in isolated ganglia and in partially dissected (semi-intact) preparations that permit simultaneous monitoring of behaviors and their neural correlates (Fig. 8.1). The accessibility of this system to experimental manipulation

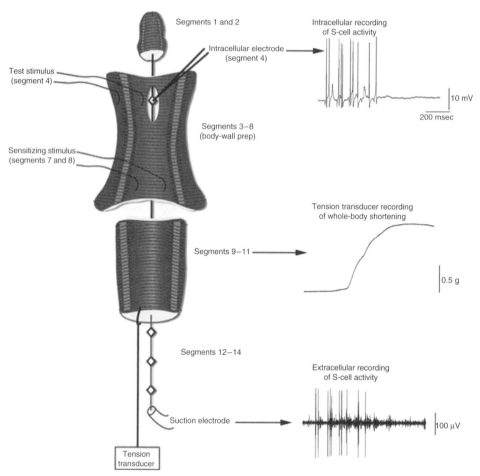

Figure 8.1 Semi-intact preparation used to study learning in the whole-body shortening reflex. A portion of the preparation is dissected to form a flat sheet of skin that is pinned to the bottom of the recording dish. Implanted silver wire electrodes deliver both shortening-eliciting stimuli and sensitizing stimuli to the skin. A single ganglion in segment 4 is exposed to permit intracellular recording from individual neurons. The middle portion of the preparation is left intact and connected to a tension transducer so that the intensity of the shortening response can be measured and recorded. The nerve cord and ganglia from the posterior-most portion of the preparation are completely exposed so that extracellular recordings from the connective nerve can be made using a suction electrode. Traces on the right are representative recordings during an elicited whole-body shortening response representing an intracellular recording from an S cell (top), a tension transducer recording of the shortening response (middle) and an extracellular recording of connective nerve activity (bottom). The largest units in the connective nerve recording are S-cell action potentials (modified from Burrell and Sahley 2005, with permission from the American Physiological Society). (See color insert.)

permits studies of the mechanisms underlying whole animal behavior at levels of complexity ranging from ion channels and individual neurons to neural circuits. Over the past four decades, a wealth of information about the neural basis of leech behavior has accrued, with investigators exploiting the unique advantages of this system in diverse areas, such as sensory processing, behavioral choice and the generation of rhythmical behaviors (see Kristan et al. 2005). Additionally, the cellular and molecular properties of leech neurons exhibit the same degree of complexity as their vertebrate counterparts and utilize biochemical signaling processes that have been highly conserved across millions of years of evolution (see Burrell and Sahley 2001). In this chapter, we discuss how annelids have been used to study the physiological basis of learning and memory, while demonstrating the unique advantages of the annelid nervous system for studying neurophysiological processes at cellular and molecular levels.

8.2 LEARNING IN THE LEECH WHOLE-BODY SHORTENING REFLEX AND ROLE OF THE S INTERNEURON

Whole-body shortening in leech is a defensive withdrawal reflex that protects the anterior of the animal, including the feeding apparatus (anterior sucker, jaws and pharynx) from damage. Throughout this chapter, we will use the term "shortening" to refer to whole-body shortening. However, a localized withdrawal response (local shortening) has also been described by Wittenberg and Kristan (1992a, b) that is elicited by sensory stimulation of one body segment, producing a localized shortening or a direction-specific bend away from the stimulus by shortening a subset of the longitudinal muscle fibers of the same segment ipsilateral to the touch. Further details about the local shortening reflex and its behavioral plasticity are discussed later.

Mechanosensory or photic stimuli trigger the shortening reflex, which involves the nearly simultaneous contraction of all the body segments. Mechanosensory-elicited whole-body shortening requires activation of several pressure (P) and touch (T) sensory neurons; the nociceptive (N) cells sometimes contribute, too (Shaw and Kristan 1995). Whole-body shortening is most efficiently produced by stimuli applied to the anterior third of the animal, while more posterior stimuli tend to elicit swimming.

Similar to other annelids, the leech has a hydrostatic skeleton (Skierczynski et al. 1996) and two muscle fiber systems that control its length. The circular muscles form rings perpendicular to the animal's anterior–posterior axis, and their contraction causes elongation. The longitudinal muscles run parallel to the animal's anterior–posterior axis, and contraction of this muscle system causes shortening. The segmentally iterated motor neurons in ventral nerve cord ganglia, including the L motor neurons and a population of ventral and dorsal excitatory motor neurons, control the tension of the longitudinal muscle fiber system (Stuart 1969, 1970; Zoccolan et al. 2002). These motor neurons receive monosynaptic and polysynaptic excitation from T, P and N cells (Nicholls and Purves 1970; Lockery and Kristan 1990; Wittenberg and Kristan 1992a), but the mechanosensory neurons can only stimulate motor neurons in the same ganglion or adjacent ganglia. Whole-body shortening requires the activation of motor neurons along the entire length of the animal; therefore, the motor neurons must be stimulated by an intersegmental input that extends throughout the leech central nervous system (CNS). In the leech, the S cell (discussed in more detail later) is an intersegmental neuron that receives input from mechanosensory cells, has output to L motor neurons throughout the CNS and is active during shortening. The S cell, however, is not necessary for shortening (Shaw and Kristan

1999), although it appears to be critical for learning in this reflex. This implies the presence of an additional intersegmental element (or elements) that is necessary for initiating the shortening response (Fig. 8.2). Unfortunately, the neuron or neurons that comprise this pathway have not been identified.

The shortening reflex exhibits both nonassociative learning (e.g. habituation, sensitization and dishabituation) and associative learning (e.g. classical conditioning). Repeated presentations of appropriate mechanosensory stimuli elicit habituation, which can be reversed or dishabituated by delivering a stronger, more sustained mechanosensory stimulus at a different location on the leech's body. The same intense or noxious mechanosensory stimulus when delivered to a nonhabituated leech elicits sensitization of the shortening reflex. Both dishabituation and sensitization depend, at least in part, on serotonin (5-HT) since depletion of 5-HT from the CNS with the toxin 5,7-dihydroxytryptamine (5,7-DHT) eliminates sensitization and partially reduces dishabituation of shortening (Ehrlich et al. 1992). The effects of 5-HT depletion are specific to learning and are not due to a decrease

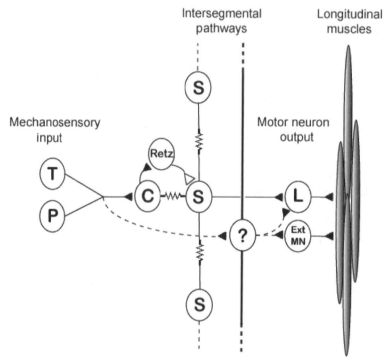

Figure 8.2 Whole-body shortening neural circuit. Mechanosensory input from T and P cells activates the S cell and, presumably, the unidentified interneuron or interneurons ("?" in figure) responsible for initiating the shortening reflex. The S cell is known to have synaptic input onto the L motor neurons, and the unidentified shortening interneuron is presumed to activate L motor neurons and other excitatory motor neurons that innervate the longitudinal muscles (listed jointly as Ext MN). The resistor symbols between the S cells and between the S and C cells represent electrical synapses that link these interneurons. Dashed lines represent presumptive inputs to and outputs from the shortening interneuron. Although not directly part of the shortening circuit, the Retzius cells (Retz) have been included to show the S-cell/Retzius cell feedback circuit. Filled triangles represent fast excitatory synaptic input, while open triangles represent modulatory excitatory input. Synaptic connections between mechanosensory cells and motor neurons are not shown. Abbreviations: T = touch; P = pressure; C = coupling neurons; L = longitudinal.

in the leech's ability to shorten (Ehrlich et al. 1992; Burrell and Sahley 1999). The difference in the effect of 5-HT depletion on sensitization and dishabituation indicates that while both forms of nonassociative learning have 5-HT-dependent components, they are mediated by distinct physiological processes, an observation that has also been made in *Aplysia*, a marine mollusk (Rankin and Carew 1988).

Classical conditioning of the leech shortening reflex can be produced by pairing a weak stimulus that elicits shortening (the conditioned stimulus or CS) with a stronger stimulus similar to that used to elicit sensitization (the unconditioned stimulus or US). Classical conditioning of shortening in leech has received less attention than nonassociative learning of this reflex, but exhibits the same properties of associative learning in other animals, e.g. a dependence on the proper CS–US order, CS pre-exposure effect and extinction (Sahley et al. 1994a). Depletion of 5-HT disrupts, but does not completely eliminate classical conditioning of the shortening reflex (Sahley 1994), indicating that associative learning in this behavior involves 5-HT-dependent and -independent forms of neuromodulation.

8.3 ROLE OF THE S INTERNEURON: MODULATION OF EXCITABILITY

Although learning in the shortening reflex is likely to involve changes at multiple sites in the leech CNS, considerable evidence suggests a critical role for the S interneuron (S cell). The "S" refers to the large action potential or spike produced by this interneuron (Frank et al. 1975). S cells form a fast-conducting network that is analogous to the giant fiber systems of other annelids (Bagnoli et al. 1972). These fibers have large diameters for rapid conduction velocity and for reducing current load imposed by (typically) numerous electrical synapses (Dorsett 1980). In several species of aquatic and terrestrial oligochaetes, but not leech, these fibers are surrounded by a loose myelin-like sheath with dorsal nodes (functionally related to the nodes of Ranvier) that allow saltatory conduction and even faster impulse propagation (Gunther 1973, 1976; Zoran and Drewes 1987). Habituation and sensitization of withdrawal reflexes in polychaetes and oligochaetes have been attributed to fatigue of mechanoreceptor-to-giant fiber and giant fiber-to-motor neuron synapses [with the exception of sabellids in which branches of the giant fiber directly innervate longitudinal muscle fibers (Horridge 1959; Roberts 1962a, b, 1966; Krasne 1965)]. The unpaired, median giant nerve fibers of oligochaetes and some other annelids mediate shortening of the anterior segments in response to noxious stimulation. Single impulses in these fibers elicit an all-or-nothing anterior withdrawal response in aquatic species, including the glossiphoniid leech *Haementeria ghilianii* (Kramer 1981), but not in terrestrial species of oligochaetes that burrow (Zoran and Drewes 1987) or in the medicinal leech *H. medicinalis* (Shaw and Kristan 1999).

In *H. medicinalis*, each segmental ganglion contains a single S cell that projects a primary neurite that bifurcates a short distance from the soma (Fig. 8.3), sending one axon anteriorly and the other posteriorly into the medial fascicle of the connective nerve (Faivre's nerve; Frank et al. 1975). The S-cell axon terminates at the midpoint of the connective nerve fiber bundle where it forms an electrical synapse with the axon of its counterpart from the neighboring ganglion. Thus, S cells form a linear network of electrically coupled interneurons that extends throughout the leech CNS. The degree of electrical coupling is so strong that action potentials initiated in any one S cell reliably propagate to all other S cells throughout the network.

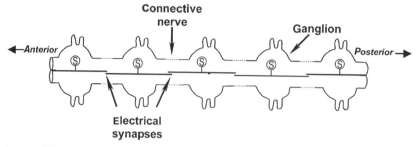

Figure 8.3 The S-cell network. Each ganglion contains a single S interneuron that projects an axon into the anterior and posterior connective nerves. At the midpoint of each connective, these axons form an S-to-S electrical synapse, resulting in a chain or network that extends throughout the leech CNS.

Because the S cell is strongly activated during shortening, it was once thought to play a role in initiating the shortening reflex (Bagnoli et al. 1972; Magni and Pellegrino 1975, 1978a, b). The S cell receives synaptic input from all the major mechanosensory neurons, namely, the T cells (Muller and Scott 1981) and the P and N cells (Laverack 1969; Bagnoli et al. 1975; Baccus et al. 2000; B.D. Burrell, pers. obs.). The S cell itself has electrical (Magni and Pellegrino 1978) and chemical (B.D. Burrell, pers. obs.) synaptic input onto the L motor neurons. However, activation of the S-cell network is insufficient to elicit shortening, and lesions of the S-cell network do not affect the leech's capacity to shorten (Shaw and Kristan 1999). This distinguishes the S cell in *H. medicinalis* from the unpaired median giant fibers of oligochaetes, which mediate a similar shortening of the anterior segments in response to noxious stimulation.

Rather, the S cell appears to play a role in the plasticity of shortening. When the S cell is lesioned, the capacity for sensitization of reflexive shortening is lost, whereas the capacity for dishabituation is only partially disrupted. This critical role for the S cell in learning has been demonstrated by ablating a single S-cell soma, mechanical cutting of Faivre's nerve (resulting in S-cell axotomy) or selective axotomy of the S cell using laser photoablation (Sahley et al. 1994b; Modney et al. 1997; Burrell et al. 2003). Importantly, these lesions preserve the shortening reflex and only eliminate the plasticity inherent to the reflex.

Like other annelids, the leech CNS has an impressive capacity for regeneration (Duan et al. 2005), making it possible to examine whether regeneration of the lesioned S-cell network is accompanied by functional recovery of sensitization. S cells regenerate their cut axon, which accurately reconnects with the axon of its homologue from the adjacent ganglion within 2–3 weeks, reforming the electrical synapse midway between ganglia that couple the two neighboring S cells (Carbonetto and Muller 1977; Muller and Carbonetto 1979; Mason and Muller 1996). In situations where an axotomized S cell regenerates, the capacity for sensitization of the shortening reflex is completely restored (Modney et al. 1997; Burrell et al. 2003). These lesion studies illustrate one of the experimental strengths of the leech in that the effects of neural injury *and* recovery can be examined at both anatomical and functional/behavioral levels. This represents one of the few examples in which lesioning of a single axon eliminates a form of learning that is then restored by regenerating that axon and reconnecting to its appropriate postsynaptic target (operant conditioning in the mollusk *Lymnaea* is also disrupted by the ablation of a single neuron; Scheibenstock et al. 2002). Other annelids, including some oligochaetes (e.g. *Lumbriculus variegatus*), have even more impressive regeneration capabilities and are able to replace lost cells and even whole segments, something that leeches cannot do.

Note that these lesion studies demonstrate the role of S cells in the *initiation* of sensitization since all lesions were made at sites between the test stimulus (used to elicit shortening) and the sensitizing stimuli. That is, the lesions were in a position to prevent S-cell activity elicited by the sensitizing stimuli from reaching all parts of the leech CNS, but the portion of the preparation that actually shortened had a largely intact S-cell network. Thus, it can be inferred that one role of the S cell is to signal to the rest of the leech CNS the advent of a noxious or sensitizing stimulus and to initiate sensitization. Sensitizing stimuli elicit two to three times more activity in the S cell compared with the test stimuli used to elicit shortening (Burrell and Sahley 2005). Since this activity propagates throughout the S-cell network, it may lead to changes in the leech CNS that result in sensitization of the shortening reflex. How might the S cell initiate changes in the CNS that lead to sensitization or dishabituation? S-cell axotomy and 5-HT depletion produce the same effect on nonassociative learning, namely, the complete disruption of sensitization and a partial disruption of dishabituation. The S cell is not serotonergic, but these results suggest a connection between 5-HT and the S cell. In fact, the S cell has synaptic input onto Retzius cells (Crisp and Muller 2006), the largest serotonergic neurons in the leech CNS. Bursts of S-cell activity produced by a sensitizing (and presumably dishabituating) stimuli likely contribute to the activation of Retzius cells throughout the leech CNS resulting in 5-HT release, leading to sensitization and full dishabituation. P cells have been shown to have polysynaptic input to Retzius neurons throughout the nerve cord (Szczupak and Kristan 1995), and the S cell may be one of the neurons that mediates this P-to-Retzius connection. Additionally, the S cell itself contains at least one neuromodulatory transmitter, a neuropeptide called myomodulin, which has an excitatory effect on Retzius cells (Keating and Sahley 1996; Wang et al. 1999). Bursts of impulses like those produced by sensitizing stimuli are well suited for the release of a peptide cotransmitter from the S cell.

The S cell may also have a role in the expression or maintenance of sensitization by undergoing changes that contribute directly to the production of the sensitized response. Following the delivery of the sensitizing stimuli, S-cell activity during shortening increases, and a correlation between S-cell activity and the intensity of the shortening response is observed, whereas no such correlation is detected in the nonsensitized response (Sahley et al. 1994b; Burrell et al. 2001). This suggests that the S cell contributes to producing the enhanced or sensitized shortening response as a result of its increased activity. Enhanced S-cell activity in the sensitized leech is due, at least in part, to an increase in S-cell excitability (Burrell et al. 2001; Burrell and Sahley 2005). This increase is mediated by 5-HT based on (1) similar increases in excitability that are observed following application of 5-HT or stimulation of the serotonergic Retzius cells and (2) drugs that block 5-HT-induced increases in excitability also prevent sensitization-induced potentiation of excitability (Burrell et al. 2001, 2005). 5-HT-dependent enhancement of S-cell excitability involves activation of metabotropic, type 7 5-HT receptors ($5\text{-}HT_7$ receptor), which subsequently stimulate a cAMP/PKA second messenger pathway (Crisp and Muller 2006). Interestingly, habituation of the shortening reflex is accompanied by a decrease in S-cell excitability (Burrell et al. 2001), but no lesion studies have yet established the importance of S cells to this form of learning. Although low concentrations of 5-HT reduce S-cell excitability (Burrell et al. 2001), whether or not 5-HT is responsible for the habituation-induced decrease in excitability remains unknown.

S-to-Retzius (R) cell communication is mediated through a disynaptic pathway, in which the first synapse is an electrical junction with the coupling (C) interneurons (Figs. 8.2 and 8.4). The S → C synapse is of such high fidelity that impulses originating in the

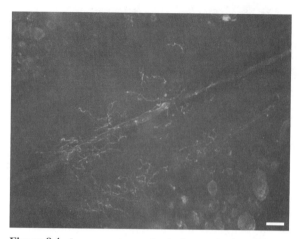

Figure 8.4 Laser scanning confocal micrograph of possible synaptic contact sites between the Retzius cell and the S-cell axon. A Retzius cell and the S cell in the same ganglion were filled with Lucifer yellow dye (green) and a high-molecular-weight dextran conjugated to Texas Red dye (red), respectively. The dextran did not pass through gap junctions, so that the coupling interneurons were not labeled, as they would be with a lower molecular weight dye. A Z series projection of three consecutive optical sections, acquired at 0.3 micron intervals, reveals varicosities of the Retzius cell neurites apparently associated with the S-cell axon. No S-cell varicosities or secondary neurites were observed in these sections. Scale bar = 25 μm. (See color insert.)

S cell invariably initiate impulses in the C interneurons, except under high divalent cation conditions (high extracellular Ca^{2+} and Mg^{2+} concentrations) that reduce the electrical excitability of the postsynaptic cell (Muller and Scott 1981; Crisp and Muller 2006). The C → Retzius synapse is a glutamatergic, CNQX-sensitive (6-cyano-7-nitroquinoxaline-2,3-dione, a competitive antagonist of AMPA/kainate glutamate receptors) synapse, so the disynaptic connection from the S cell to the Retzius cell produces a CNQX-sensitive excitatory postsynaptic potential (EPSP) at a short latency in the Retzius cell resembling a monosynaptic connection. Previously, vesicle-filled presynaptic terminals and associated synapses were imaged in horseradish peroxidase-injected C interneurons using electron microscopy (Muller and Scott 1981), but no chemical synaptic targets of the C interneurons had been found until the discovery of the C → Retzius connection. Putative synaptic contacts between fluorescent dye-filled C- and Retzius cells have been identified using confocal microscopy (Crisp and Muller 2006). Some experiments in saline containing high concentrations of divalent cations also suggest the presence of a weak monosynaptic connection between the S and Retzius cells, as well as a longer-latency (observable after ~25 ms) polysynaptic connection (Crisp and Muller 2006; K.M. Crisp, unpublished observations).

As mentioned earlier, Retzius cell activity leads to enhanced S-cell excitability through the release of 5-HT. The particular mode of 5-HT release from the Retzius cells that modulates the S cell remains unclear. In culture, Retzius cells have been shown to secrete 5-HT extrasynaptically from vesicles fusing with the somatic membrane where no active zones are found, as well as from active zones localized to presynaptic terminals (Trueta et al. 2003, 2004; De Miguel and Trueta 2005). When an isolated leech ganglion is contained in a saline volume of ~50 μL, stimulation of a single Retzius cell at 3–6 Hz for 10 min results in a 5-HT concentration of ~37 nM in the bathing saline (Willard 1981). However, 5-HT circulating as a neurohormone is not necessarily the only means by which Retzius cells signal to S cells. When the S cell and the Retzius cell are filled with fluorescent tracers, potential presynaptic contact sites from the Retzius cell onto the S cell primary and secondary axons can be observed using confocal microscopy (Fig. 8.4). Many of these putative contacts are within a 0.3-μm-thick plane that includes the S-cell axon. Furthermore, vesicles in the S-cell axon are never observed in electron micrograph sections in which the S cell has been injected with horseradish peroxidase (Scott and Muller 1980), while vesicles and presynaptic release sites can be observed in similarly prepared Retzius cell sections (Liu and Nicholls 1989). Although it is impossible to be certain that apparent contact sites observed in confocal micrographs using the above methods are functional presynaptic sites, the data suggest that direct synaptic release of 5-HT from the Retzius cell onto the S cell plays a role in the regulation of S-cell excitability.

Interestingly, sensitization of reflexive shortening may depend on increased neuronal excitability given that modulation of excitability is important to learning in a variety of vertebrate and invertebrate species (see reviews by Daoudal and Debanne 2003; Zhang and Linden 2003; Frick and Johnston 2005). The biophysical basis for increased S-cell excitability has not been thoroughly investigated, but sensitization- and 5-HT-induced increases in S-cell excitability involve both an increase in the number of action potentials elicited by a stimulus (increased firing rate) and a decrease in the amount of stimulus current required to elicit a single action potential (Belardetti et al. 1982; Burrell et al. 2001, 2005; Crisp and Muller 2006). Enhanced firing is due, in part, to a decrease in S-cell afterhyperpolarization (AHP) that is largely mediated by Ca^{2+}-dependent K^+ channels (Burrell and Crisp 2008). However, 5-HT-induced enhancement of S-cell excitability is likely to involve changes in a variety of ion channels besides those that mediate the AHP.

If it is true that the S cell plays a role in the expression of sensitization (i.e. generation of the enhanced or sensitized response), then why does this interneuron make no contribution to shortening in the nonsensitized state? One possible explanation is that during sensitization, the S cell becomes "recruited" into the shortening neural circuit by virtue of its increased firing rate. The S cell does have synaptic input onto the L motor neurons and contributes much of this motor neuron's activity early in an elicited shortening response (Shaw and Kristan 1995, 1999). Possibly in the nonsensitized state, the rate of S-cell activity during reflexive shortening is too low to have any measurable influence on L motor neuron activity during shortening. However, increased S-cell activity during shortening, as occurs during sensitization, may allow the S cell to contribute to L motor neuron activity in a way that it could not prior to sensitization.

An alternative, albeit not mutually exclusive, mechanism is that the S cell may also act by maintaining an increased arousal state during sensitization. One way that the S cell may be able to regulate arousal is through its signaling onto the serotonergic Retzius neurons (discussed earlier). Strong activation of the S-to-Retzius connection could elicit 5-HT release throughout the leech CNS, leading to a heightened arousal state that would contribute to the sensitized response. S-cell excitability (and therefore activity) is directly enhanced by 5-HT from the Retzius cell, creating a positive feedback loop by the S-cell/R-cell circuit (Crisp and Muller 2006), which may act to maintain elevated 5-HT levels in the CNS. Similarly, several serotonergic neurons in *Aplysia* are excited by bath-applied 5-HT, suggesting that positive feedback plays a role in setting and maintaining a state of arousal in response to a sensitizing stimulus (Marinesco et al. 2004). In this situation, the serotonergic CC3/CB1 cells excite downstream serotonergic cells, and the resultantly high 5-HT levels in hemolymph excite CC3/CB1 cells.

This use of the S-cell/Retzius circuit to maintain elevated levels of 5-HT may contribute to sensitization of other behaviors in leech. For example, sensitization of swimming is 5-HT dependent (Zaccardi et al. 2004), suggesting that the S cell contributes to the arousal state for this behavior during nonassociative learning (Weeks 1982; Debski and Friesen 1986). Although the S cell is not part of the swimming circuit, it does have input to swim-gating neuron 205 (Weeks 1982). Also, the possibility that the S-cell/Retzius feedback circuit is bidirectional cannot be excluded. Low concentrations of 5-HT can reduce S-cell excitability (Burrell et al. 2001), and the resulting lower levels of S-cell activity will lead to reduced 5-HT release that would maintain reduced S-cell excitability. 5-HT also has an inhibitory effect on the Retzius cells themselves, providing an additional level of complexity to the S-cell/Retzius feedback circuit (Kerkut and Walker 1967; Walker and Smith 1973; Acosta-Urquidi et al. 1989).

Indeed, there may be additional ways in which the S cell contributes to leech arousal through the S-cell/Retzius circuit. The S cell is known to contain myomodulin, which depolarizes Retzius cell resting potential, thereby increasing its activity and presumably the amount of 5-HT release (Keating and Sahley 1996; Wang et al. 1999). Myomodulin may affect other components in the leech CNS that modulate arousal. For example, myomodulin depolarizes the S-cell resting potential (B.D. Burrell, unpublished observation), although it is not known if the S cell is automodulatory.

8.4 LEARNING IN THE LEECH SWIM CIRCUIT

Tactile stimulation of leech skin, especially on the more posterior portions of the animal, promotes swimming that is controlled by a well-characterized circuit (for review, see Kristan et al. 2005). Strong cutaneous stimulation activates T, P and N cells, which in turn excite trigger neurons (e.g. cell Tr1) whose somata reside in the head ganglion and whose axons descend through the entire nerve cord (Brodfuehrer and Friesen 1986). Brief intracellular stimulation of these trigger neurons elicits swimming that outlasts the duration of the stimulus. The trigger neurons provide part of the excitatory input to gating neuron 204, an unpaired interneuron found in midbody ganglia 9-16 (Weeks 1981). Intracellular stimulation of cell 204 drives the swim central pattern generator (CPG), but swimming only lasts as long as the excitation of cell 204 (except in the presence of 5-HT; Angstadt and Friesen 1993). Tr1 also inactivates inhibitory inputs to the oscillator neurons of the swim CPG, though the identity of these inputs is unknown (Brodfuehrer and Friesen 1986).

Plasticity intrinsic to this neural circuit may play an important role in determining which behavior (i.e. swimming, shortening or crawling) is expressed in response to tactile stimulation. The expression of swimming in response to tactile stimulation gradually habituates, and eventually touch is insufficient to evoke swimming (Debski and Friesen 1985). That habituation is gradual can be seen by examining the decline in swim burst frequency during successive swim episodes (Fig. 8.5). Swim length also declines (defined as the number of swim cycles evoked by a single stimulus) by ~50% and then swimming ceases completely. A progressive increase in the delay from the stimulus to swim onset is also observed during habituation training (Debski and Friesen 1985; Zaccardi et al. 2001). This habituation shows seven of the nine characteristics outlined by Thompson and Spencer (1966) as being characteristic of habituation across species. Swimming in response to touch recovers spontaneously (if the animal remains undisturbed) 40 min after habituation, but can be restored within 1 min if the animal is dishabituated by a novel, more intense stimulus, such as pinching the skin (Debski and Friesen 1985). The "subzero effect," whereby continued stroking after the animal has stopped responding delays spontaneous recovery, can also be demonstrated. In contrast to most characterized examples of habitu-

A

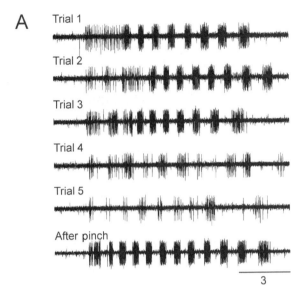

Trial 1

Trial 2

Trial 3

Trial 4

Trial 5

After pinch

3

B

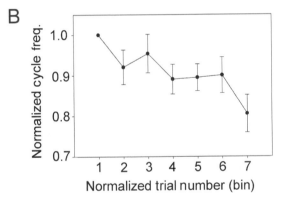

Figure 8.5 Habituation of the touch-based swim response in leech. (A) Tactile stimulation of a skin flap with a fine wire loop induced fictive swimming activity, defined as high-frequency bursts of motor neuron DE-3 impulses recorded from the dorsal posterior nerve with a cycle period of 0.5–1.2 s for at least three consecutive bursts (Kristan and Calabrese 1976). When stimulation was repeated at 30-s intervals, the response habituated to criterion, defined as two consecutive stimulations that failed to produce swimming (trials 4 and 5; Debski and Friesen 1985). Swim responsiveness for this preparation was quantified as 5 (trials to reach criterion). The swim response was easily dishabituated by pinching the skin flap with a pair of coarse forceps (bottom trace). (B) Average decrease in swim burst frequency over successive trials ($n = 5$). The first swim episode evoked (trial 1) has the highest frequency. During later trials, stimulation of the skin resulted in a slower fictive swimming rhythm. Frequency (in Hz) was normalized to the average frequency recorded during the first trial (1.0); a subsequent frequency of 0.8 represents a 20% decrease in swim cycle frequency averaged over the swim episode evoked during the relevant trial. Because of differences in the number of trials in which swimming could be evoked between preparations, mean frequencies for each trial were distributed into seven trial bins. The mean burst frequency obtained during the first trial was placed into bin no. 1, and the frequency obtained during the last evoked swim episode was placed into bin no. 7. Frequencies of intermediate evoked-swim trials were distributed in order among the remaining five bins.

ation in other organisms (Thompson and Spencer 1966), habituation to stroking of the skin is segment specific and does not generalize to other segments.

Although a neural pathway leading from tactile stimulation to the expression of swimming is well-known, the loci responsible for the habituation and dishabituation of swimming remain obscure. Light stroking of the skin used to habituate the swim response selectively activates the T mechanoreceptors, but habituation does not coincide with a decrease in T-cell impulse activity, suggesting that habituation must involve loci downstream in the swim circuit (Debski and Friesen 1987). Further support for this conclusion comes from the observation that swimming habituates after repeated trials in which intracellular stimulation of the T cell (resulting in a fixed number of impulses across trials) is used as the stimulus.

The response of cell 204 to tactile stimulation declines gradually over the course of habituation, corresponding perhaps to the decline in swim length during successive trials (Debski and Friesen 1986). However, no change in the response of cell 204 corresponds with the cessation of swimming when the animal is habituated. Furthermore, although the S cell responds both to light brushing of the skin (activating T cells only) and to noxious pinches of the skin, S-cell activity during skin stimulation does not correlate with habituation or dishabituation of swimming activity (Debski and Friesen 1987; K.M. Crisp, pers. obs.). Changes associated with habituation, therefore, must be localized in parallel pathways conveying T-cell input to the swim CPG.

5-HT may play a role in the dishabituation and sensitization of swimming (Zaccardi et al. 2004). Stimulation of single or pairs of Retzius cells in leech ganglia in a relatively large bath volume do not elicit swimming (a 5-HT activated behavior), but stimulation of Retzius cells in a small bath of only tens of microliters does activate swimming; this finding has been used to suggest that the Retzius cells activate swimming by elevating 5-HT levels in the blood (as a circulating neurohormone) rather than by synaptic mechanisms (Willard 1981; Nusbaum and Kristan 1986; Crisp and Mesce 2006). In intact leeches, injections of high concentrations of 5-HT (200 μm) mimics the effects of dishabituation, while injections of methysergide, a 5-HT receptor blocker, impairs both dishabituation and sensitization (Catarsi et al. 1990; Zaccardi et al. 2004). Interestingly, injections of membrane-permeable analogues of cAMP also mimic the effects of sensitization and dishabituation, while injections of adenylyl cyclase inhibitors prevent sensitization and dishabituation in these experiments (Zaccardi et al. 2004). When bath-applied to the isolated CNS, 100 μM 5-HT delays habituation; this action is blocked by inhibitors of adenylyl cyclase or protein kinase A (Alkatout and Crisp 2007). Together, these studies suggest that 5-HT may counteract habituation through a PKA-signaling pathway, similar to 5-HT's actions during sensitization (and partial dishabituation) of the shortening reflex. Such modulation would be well suited to maintain the leech's responsiveness to stimulation during periods of hunger.

8.5 USING THE LEECH TO STUDY INTRINSIC FORMS OF SENSITIZATION

Nonassociative learning is usually straightforward: repeated stimulation produces habituation, and delivery of a strong, salient or noxious stimulus produces sensitization or dishabituation. However, instances occur in both vertebrates and invertebrates where repeated presentation of the test stimulus (the stimulus used to elicit the behavior of interest) produces sensitization instead of the expected habituation (Davis and Wagner 1969; Groves et al. 1970; Bashinski et al. 1985; Schanbacker et al. 1996; Prescott and Chase 1999). This has been referred to as "wind-up," "warm-up" or "intrinsic" sensitization (the latter term coined by Davis and File 1984) and is distinguished from the more traditional or "extrinsic" form of sensitization by the fact that intrinsic (or wind-up) sensitization is due to repeated delivery of the test stimulus, while the extrinsic form requires the delivery of an additional stimulus (the sensitizing stimulus) that is distinct from the test stimulus (Hinde 1970; Davis and File 1984).

One example of intrinsic sensitization in leech involves the local shortening response. Local shortening reflex is not simply a reduced version of the whole-body shortening reflex, but actually involves a distinct neural circuit. Interestingly, many neurons that mediate local shortening are also responsible for producing local bending (Wittenberg and Kristan 1992a, b). Light-evoked local shortening (measured as the probability of light eliciting a response) readily habituates with repeated stimulation and dishabituates following delivery of a noxious mechanosensory stimulus (Lockery et al. 1985). Local shortening also undergoes extrinsic sensitization following delivery of a noxious mechanosensory stimulus. Although short-term habituation ($\approx$13 min) of local shortening is readily produced, repeated habituation training sessions do not produce long-term habituation. Instead, the probability of local shortening significantly increases with each day of habituation training (Lockery et al. 1985). Therefore, the repeated stimulation protocol elicits both habituating processes that dominate in the short term and sensitizing processes that take longer (~20 days) to develop.

The local bending reflex in leech also exhibits both intrinsic and extrinsic forms of sensitization. Local bending is elicited by stimulation of P cells, and repeated activation leads to a significant increase in the intensity of the reflex that is likely due to intrinsic sensitization (Lockery and Kristan 1991). Extrinsic sensitization of the bending reflex could also be produced by stimulation of the nociceptive mechanosensory cells or by stimulation of serotonergic neurons 21 and 61, but not Retzius cells (Lockery and Kristan 1991). These two techniques for eliciting extrinsic sensitization may represent two parts of the same sensitizing pathway; that is, stimulation of the N cells activates neurons 21 and 61, with the subsequent 5-HT release mediating extrinsic sensitization of the bending reflex.

Intrinsic sensitization is also observed in the whole-body shortening reflex. Repeated delivery of shortening-inducing stimuli (test stimuli) normally induces habituation. However, habituation does not occur if shortening is elicited by a separate but nearby stimulus (novel stimulus) prior to repetitive delivery of the test stimulus (Burrell and Sahley 1998). The ability of the novel stimulus to elicit intrinsic sensitization depends on its proximity to the subsequent test stimulus; the closer the two stimuli are to each other, the more likely that sensitization will be induced (Burrell and Sahley 1998). Intrinsic sensitization does not require 5-HT, indicating that it is distinct from extrinsic sensitization of this behavior at a physiological level (Ehrlich et al. 1992; Burrell and Sahley 1999). Intrinsic sensitization of whole-body shortening, however, requires the presence of an intact S-cell network (Burrell et al. 2000).

8.6 SYNAPTIC PLASTICITY IN LEECH CNS

Changes in synaptic transmission are thought to play a critical role in mediating both vertebrate and invertebrate learning and memory (Malenka and Bear 2004; Glanzman 2006; Hawkins et al. 2006). As mentioned previously, the synapse between sensory neurons and annelid giant fibers appears to be an important locus for synaptic plasticity. The sabellid worm *Branchiomma vesiculosum* (a sedentary polychaete), for example, withdraws its branchial crown into its tube in response to tactile stimulation. The withdrawal wanes and postwithdrawal reemergence quickens with repeated trials until extinction of the response, which spontaneously recovers to about 50% of its initial magnitude within 8 h. The giant fibers respond to each successive touch with fewer impulses until they fail to respond at all. The labile site appears to be the connection between the sensory neurons and the giant fibers in *B. vesiculosum* because branchial crown withdrawal does not fatigue when giant fibers are repeatedly stimulated directly (Krasne 1965).

Although no direct connection between synaptic plasticity and learning has been demonstrated in the leech, synaptic connections made by mechanosensory neurons onto the S cell exhibit a variety of forms of activity-dependent plasticity that bear striking resemblance to synaptic changes observed in vertebrates. The leech S cell is an attractive candidate for examining such synaptic plasticity given its role in sensitization of the shortening reflex and other behaviors. Additionally, because the S-cell network extends throughout the leech CNS and receives afferent input from all regions of the leech body, it has the potential to act as a coincidence detector or "association network" that recognizes the coincidence of conditioned and unconditioned stimuli.

Most studies of activity-dependent synaptic plasticity have focused on long-term potentiation (LTP) and long-term depression (LTD) that is mediated by activation of postsynaptic NMDA receptors (NMDA-R). NMDA-Rs are ionotropic glutamate receptors that require both glutamate binding (which indicates presynaptic activity) *and* membrane depolarization (which indicates postsynaptic activity) to open, features that make NMDA-Rs particularly efficient in detecting the coincidence of pre- and postsynaptic activity (Malenka and Bear 2004). This capacity for coincidence detection makes NMDA-R-dependent synaptic plasticity an attractive cellular mechanism for mediating associative learning. At the glutamatergic $P \rightarrow S$ synapse (Baccus et al. 2000), tetanic stimulation elicits LTP that requires NMDA-R activation, is restricted to the tetanized synapse (synapse specific) and appears to involve postsynaptic changes (Burrell and Sahley 2004). NMDA-R-dependent potentiation is also observed at the $P \rightarrow AP$ synapse (AP = anterior pagoda neuron) following the application of forskolin (Grey and Burrell 2008). This chemical LTP (cLTP) is dependent on postsynaptic increases in intracellular Ca^{2+}, requires activation of PKA and CamKII, and involves trafficking of glutamate receptors. All of these features of LTP at the $P \rightarrow S$ and $P \rightarrow AP$ synapses are shared with the NMDA-R-dependent LTP observed in a number of vertebrate and invertebrate synapses (Malenka and Bear 2004; Glanzman 2006). Additionally, the leech version of the gene encoding the NR1 subunit of the leech NMDA-R has been partially sequenced and found to be expressed in a number of identified neurons within the leech CNS (K.B. Grey, B.L. Moss and B.D. Burrell, unpublished data).

Although NMDA-R mediated synaptic plasticity has dominated studies of LTP and LTD, it is now clear that some forms of LTP and LTD are NMDA-R independent and utilize other molecules for coincidence detection, e.g. metabotropic glutamate receptors and/or voltage-gated Ca^{2+} channels (Malenka and Bear 2004; Raymond 2007). At the $T \rightarrow S$ synapse, which is also glutamatergic (Li and Burrell 2006), tetanic stimulation elicits LTP that is NMDA-R independent (Burrell and Sahley 2004; Burrell and Li 2008). Instead, $T \rightarrow S$ LTP is mediated by the activation of metabotropic glutamate receptors, voltage-gated Ca^{2+} channels and protein kinase C (Burrell and Li 2008). Interestingly, LTP at the tetanized synapse is accompanied by LTD at nontetanized $T \rightarrow S$ connections (same postsynaptic S cell, different presynaptic T cells), and this heterosynaptic LTD is NMDA-R dependent (Burrell and Sahley 2004). This simultaneous use of homosynaptic LTP and heterosynaptic LTD has been observed in a variety synapses in the vertebrate brain (see review by Bi and Poo 2001) and likely acts to enhance the contrast between the active (tetanized) and inactive (nontetanized) synapses, enhancing transmission at one connection while reducing transmission in parallel inputs onto the same postsynaptic target.

Another form of activity-dependent synaptic plasticity that may be relevant to learning and memory involves changes in action potential propagation within sensory cell axonal branch points. This form of plasticity is bidirectional in that it involves both increases and decreases in synaptic transmission. Decreased synaptic transmission occurs as a result of conduction block, the failure of an action potential(s) to propagate past a branch point and therefore not reach a significant portion of a neuron's presynaptic terminals. Conduction block occurs following repeated or sustained activation of the neuron (this accounts for the activity dependence) that leads to hyperpolarization of the branch point, most likely as a result of increased activity by the Na^+/K^+ pump and/or the Ca^{2+}-gated K^+ channels (Baylor and Nicholls 1969; Jansen and Nicholls 1973; Van Essen 1973). Branch point hyperpolarization makes it more difficult for action potentials traveling from smaller diameter fibers to bring the membrane at the relatively larger diameter branch point to threshold (Yau 1976).

Increased synaptic transmission occurs as a result of branch point reflection, as first described by Baccus (1998). Like conduction block, reflection occurs as a result of branch point hyperpolarization and likely involves the same biophysical mechanism. However, the level of branch point hyperpolarization during reflection is less than that required to produce conduction block. Therefore, reflection represents an intermediate state—hyperpolarized relative to the full conducting state of the branch point, but not as hyperpolarized as in the blocked state. During reflection, action potential propagation at the branch point is delayed, but not blocked. The additional time required for the action potential to bring the branch point to threshold exceeds the absolute refractory period of the axon. This allows initiation of a second action potential in the prebranch point axon, activating a portion of the neuron's presynaptic terminals twice. Changes in synaptic transmission due to conduction block and/or reflection have been observed in synapses made by the T and P cells onto a variety of postsynaptic targets including L motor neurons, the S cell, anterior pagoda cells and even other T cells (Muller and Scott 1981; Macagno et al. 1987; Gu et al. 1989, 1991; Baccus 1998; Baccus et al. 2000; Scuri et al. 2002, 2007; Cataldo et al. 2005).

Conduction block is inhibited by 5-HT (Catarsi and Brunelli 1991; Mar and Drapeau 1996) due, at least in part, to a cAMP-dependent process that reduces the activity of the Na^+/K^+ pump (Catarsi et al. 1993). Conduction block can also be inhibited by octopamine, although this effect is not direct and instead appears to be mediated by the activation of serotonergic neurons (Catarsi et al. 1995). In T cells, repetitive activity potentiates the hyperpolarization that is thought to produce conduction block (Scuri et al. 2002, 2005, 2007; Cataldo et al. 2005). This increase in activity-dependent hyperpolarization is mediated by an increase in intracellular Ca^{2+} (likely due to influx through voltage-dependent Ca^{2+} channels and release from intracellular stores) and the subsequent production of arachidonic acid that results in inhibition of the Na^+/K^+ pump (Scuri et al. 2005). It has been suggested that conduction block contributes to learning in the swimming response (Brunelli et al. 1997; Zaccardi et al. 2004), but as with LTP and LTD, this has not been directly observed during learning in behaviorally intact preparations. Conduction block and reflection have been observed in a number of other vertebrate and invertebrate neurons, indicating widespread use of this form of neuroplasticity (Debanne et al. 1999; Antic et al. 2000; Amir and Devor 2003; Evans et al. 2003).

8.7   CONCLUSIONS

Annelids such as the leech are particularly useful for studies of learning and memory; they provide the advantages of an invertebrate nervous system in which it is possible to ascertain functional properties of individual neurons while exhibiting many of the same forms of neuroplasticity thought to contribute to learning in vertebrate brains, e.g. LTP/LTD, conduction block and modulation of neuronal excitability. As demonstrated with experiments utilizing bending and whole-body shortening reflexes, the activity or changes in activity of individual neurons can be linked with learning-related behavioral plasticity.

One deficiency in the study of learning and memory in annelids has been the lack of a greater range, and for that matter, more modern research methods. Methodologies such as the use of voltage- and calcium-sensitive dyes to detect neural activity (e.g. Dierkes et al. 2004 or Briggman et al. 2005), dyes such as FM1-43 that can detect synaptic release (Trueta et al. 2004), electrophysiological methods that isolate membrane currents (e.g. Tobin and Calabrese 2005; Barsanti et al. 2006) and computational modeling (e.g. Moss et al. 2005; Zheng et al. 2007) have already been used in leech and other annelids, and simply need to be applied to questions of learning and memory. Voltage-sensitive dyes have been used with great success in identifying neurons whose activities play important roles in the generation and regulation of a variety of leech behaviors (Taylor et al. 2003; Briggman et al. 2005; Briggman and Kristan 2006). In principle, this same technique could be used to identify neurons that contribute learning-related behavioral changes. Additionally, molecular-based techniques such as RNA interference (Baker and Macagno 2000; Shefi et al. 2006) and expression-tag sequencing of the *H. medicinalis* and *Lumbricus rubellus* genomes are being developed and will provide new tools for investigating the molecular and cellular basis of learning in annelids. Clearly, attempts should be made to develop annelid models to study the cellular and molecular mechanisms of human cognitive and behavioral disorders. Other invertebrates, such as *Caenorhabditis elegans* and *Drosophila melanogaster*, have been used successfully to study both human neurological disorders (e.g. Alzheimer's, Parkinson's and Huntington's diseases; Thompson and Marsh 2003; Bilen and Bonini 2005) as well as behavioral disorders (e.g. addiction; Wolf and Heberlein 2003), and the leech is already used for studies involving neural injury and repair (Duan et al. 2005). Relatively few laboratories are currently conducting annelid-based research on learning and memory, and many attractive avenues for investigation are available; thus, the time is ripe for a new generation of investigators to recognize the benefits of annelids in this arena and to make new discoveries on the cellular basis of learning and memory.

ACKNOWLEDGMENTS

The authors are indebted to the helpful comments of Drs. K.J. Muller and D. Shain during the preparation of this chapter. A portion of the work described in this chapter was supported by grants from the National Science Foundation (NSF) (IBN-0432683, B.D.B.), by a subproject of the National Institutes of Health grant (P20 RR015567, B.D.B.), which is designated as a Center of Biomedical Research Excellence (COBRE), and by a grant from the Support of Mentors and their Students (SOMAS, NSF and Davidson College) in the Neurosciences Program (SOMAS-1356, K.M.C.).

REFERENCES

ACOSTA-URQUIDI, J., SAHLEY, C.L., KLEINHAUS, A.L. 1989. Serotonin differentially modulates two K+ currents in the Retzius cell of the leech. J. Exp. Biol. 145, 403–417.

ALKATOUT, B.A., CRISP, K.M. 2007. Serotonin delays habituation of leech swim response to touch. Behav. Brain Res. 182, 145–149.

AMIR, R., DEVOR, M. 2003. Extra spike formation in sensory neurons and the disruption of afferent spike patterning. Biophys. J. 84, 2700–2708.

ANGSTADT, J.D., FRIESEN, W.O. 1993. Modulation of swimming behavior in the medicinal leech. I. Effects of serotonin on the electrical properties of swim-gating cell 204. J. Comp. Physiol. [A] 172, 223–234.

ANTIC, S., WUSKELL, J.P., LOEW, L., ZECEVIC, D. 2000. Functional profile of the giant metacerebral neuron of Helix aspersa: temporal and spatial dynamics of electrical activity in situ. J. Physiol. 527, 55–69.

BACCUS, S.A. 1998. Synaptic facilitation by reflected action potentials: enhancement of transmission when nerve impulses reverse direction at axon branch points. Proc. Natl. Acad. Sci. U.S.A. 95, 8345–8350.

BACCUS, S.A., BURRELL, B.D., SAHLEY, C.L., MULLER, K.J. 2000. Action potential reflection and failure at axon branch points cause stepwise changes in EPSPs in a neuron essential for learning. J. Neurophysiol. 83, 1693–1700.

BAGNOLI, P., BRUNELLI, M., MAGNI, F. 1972. A fast conducting pathway in the central nervous system of the leech Hirudo medicinalis. Arch. Ital. Biol. 110, 35–51.

BAGNOLI, P., BRUNELLI, M., MAGNI, F., PELLEGRINO, M. 1975. The neuron of the fast conducting system in hirudo medicinalis: identification and synaptic connections with primary afferent neurons. Arch. Ital. Biol. 113, 21–43.

BAKER, M.W., MACAGNO, E.R. 2000. RNAi of the receptor tyrosine phosphatase HmLAR2 in a single cell of an intact leech embryo leads to growth-cone collapse. Curr. Biol. 10, 1071–1074.

BARSANTI, C., PELLEGRINI, M., PELLEGRINO, M. 2006. Regulation of the mechanosensitive cation channels by ATP and cAMP in leech neurons. Biochim. Biophys. Acta 1758, 666–672.

BASHINSKI, H.S., WERNER, J.S., RUDY, J.W. 1985. Determinants of infant visual fixation: evidence for a two-process theory. J. Exp. Child. Psychol. 39, 580–598.

BAYLOR, D.A., NICHOLLS, J.G. 1969. After-effects of nerve impulses on signalling in the central nervous system of the leech. J. Physiol. 203, 571–589.

BELARDETTI, F., BIONDI, C., COLOMBAIONI, L., BRUNELLI, M., TREVISANI, A. 1982. Role of serotonin and cyclic AMP on facilitation of the fast conducting system activity in the leech Hirudo medicinalis. Brain Res. 246, 89–103.

BI, G., POO, M. 2001. Synaptic modification by correlated activity: Hebb's postulate revisited. Annu. Rev. Neurosci. 24, 139–166.

BILEN, J., BONINI, N.M. 2005. Drosophila as a model for human neurodegenerative disease. Annu. Rev. Genet. 39, 153–171.

BRIGGMAN, K.L., KRISTAN, W.B., Jr. 2006. Imaging dedicated and multifunctional neural circuits generating distinct behaviors. J. Neurosci. 26, 10925–10933.

BRIGGMAN, K.L., ABARBANEL, H.D., KRISTAN, W.B., Jr. 2005. Optical imaging of neuronal populations during decision-making. Science 307, 896–901.

BRODFUEHRER, P.D., FRIESEN, W.O. 1986. Initiation of swimming activity by trigger neurons in the leech subesophageal ganglion. I. Output connections of Tr1 and Tr2. J. Comp. Physiol. [A] 159, 489–502.

BRUNELLI, M., GARCIA-GIL, M., MOZZACHIODI, R., SCURI, R., ZACCARDI, M.L. 1997. Neurobiological principles of learning and memory. Arch. Ital. Biol. 135, 15–36.

BULLOCK, T.H., QUARTON, G.C. 1966. Simple systems for the study of learning mechanisms. Neurosci. Res. Program Bull. 4, 105–233.

BURRELL, B.D., CRISP, K.M. 2008. Serotonergic modulation of afterhyperpolarization in a neuron that contributes to learning in the leech. J. Neurophysiol. 99, 605–616.

BURRELL, B.D., LI, Q. 2008. Co-induction of long-term potentiation and long-term depression at a central synapse in the leech. Neurobiol. Learn. Mem. 90, 275–279.

BURRELL, B.D., SAHLEY, C.L. 1998. Generalization of habituation and intrinsic sensitization in the leech. Learn. Mem. 5, 405–419.

BURRELL, B.D., SAHLEY, C.L. 1999. Serotonin depletion does not prevent intrinsic sensitization in the leech. Learn. Mem. 6, 509–520.

BURRELL, B.D., SAHLEY, C.L. 2001. Learning in simple systems. Curr. Opin. Neurobiol. 11, 757–764.

BURRELL, B.D., SAHLEY, C.L. 2004. Multiple forms of long-term potentiation and long-term depression converge on a single interneuron in the leech CNS. J. Neurosci. 24, 4011–4019.

BURRELL, B.D., SAHLEY, C.L. 2005. Serotonin mediates learning-induced potentiation of excitability. J. Neurophysiol. 94, 4002–4010.

BURRELL, B.D., SAHLEY, C.L., MULLER, K.J. 2000. Laser axotomy of a single interneuron disrupts intrinsic sensitization in the leech. Abstr. Soc. Neurosci.

BURRELL, B.D., SAHLEY, C.L., MULLER, K.J. 2001. Non-associative learning and serotonin induce similar bidirectional changes in excitability of a neuron critical for learning in the medicinal leech. J. Neurosci. 21, 1401–1412.

BURRELL, B.D., SAHLEY, C.L., MULLER, K.J. 2003. Progressive recovery of learning during regeneration of a single synapse in the medicinal leech. J. Comp. Neurol. 457, 67–74.

CARBONETTO, S., MULLER, K.J. 1977. A regenerating neurone in the leech can form an electrical synapse on its severed axon segment. Nature 267, 450–452.

CATALDO, E., BRUNELLI, M., BYRNE, J.H., AV-RON, E., CAI, Y., BAXTER, D.A. 2005. Computational model of touch sensory cells (T Cells) of the leech: role of the afterhyperpolarization (AHP) in activity-dependent conduction failure. J. Comput. Neurosci. 18, 5–24.

CATARSI, S., BRUNELLI, M. 1991. Serotonin depresses the after-hyperpolarization through the inhibition of the Na+/K+ electrogenic pump in T sensory neurones of the leech. J. Exp. Biol. 155, 261–273.

CATARSI, S., GARCIA-GIL, M., TRAINA, G., BRUNELLI, M. 1990. Seasonal variation of serotonin content and nonassociative learning of swim induction in the leech *Hirudo medicinalis*. J. Comp. Physiol. [A] 167, 469–474.

CATARSI, S., SCURI, R., BRUNELLI, M. 1993. Cyclic AMP mediates inhibition of the Na(+)-K+ electrogenic pump by serotonin in tactile sensory neurones of the leech. J. Physiol. 462, 229–242.

CATARSI, S., SCURI, R., BRUNELLI, M. 1995. Octopamine and Leydig cell stimulation depress the afterhyperpolarization in touch sensory neurons of the leech. Neuroscience 66, 751–759.

CRISP, K.M., MESCE, K.A. 2006. Beyond the central pattern generator: amine modulation of decision-making neural pathways descending from the brain of the medicinal leech. J. Exp. Biol. 209, 1746–1756.

CRISP, K.M., MULLER, K.J. 2006. A 3-synapse positive feedback loop regulates the excitability of an interneuron critical for sensitization in the leech. J. Neurosci. 26, 3524–3531.

DAOUDAL, G., DEBANNE, D. 2003. Long-term plasticity of intrinsic excitability: learning rules and mechanisms. Learn. Mem. 10, 456–465.

DARWIN, C.D. 1966. *Darwin on Humus and Earthworms.* London: Faber and Faber, p. 58.

DAVIS, M., FILE, S.E. 1984. Intrinsic and extrinsic mechanisms of habituation and sensitization: implications for the design and analysis of experiments. In *Habituation, Sensitization and Behavior*, edited by H.V.S. Peeke and L.F. Petrinovich. New York: Academic Press, pp. 287–323.

DAVIS, M., WAGNER, M. 1969. Habituation of the startle response under an incremental sequence of stimulus intensities. J. Comp. Physiol. Psychol. 67, 486–492.

DEBANNE, D., KOPYSOVA, I.L., BRAS, H., FERRAND, N. 1999. Gating of action potential propagation by an axonal A-like potassium conductance in the hippocampus: a new type of non-synaptic plasticity. J. Physiol. Paris 93, 285–296.

DEBSKI, E.A., FRIESEN, W.O. 1985. Habituation of swimming activity in the medicinal leech. J. Exp. Biol. 116, 169–188.

DEBSKI, E.A., FRIESEN, W.O. 1986. Role of central interneurons in habituation of swimming activity in the medicinal leech. J. Neurophysiol. 55, 977–994.

DEBSKI, E.A., FRIESEN, W.O. 1987. Intracellular stimulation of sensory cells elicits swimming activity in the medicinal leech. J. Comp. Physiol. [A] 160, 447–457.

DE MIGUEL, F.F., TRUETA, C. 2005. Synaptic and extra-synaptic secretion of serotonin. Cell Mol. Neurobiol. 25, 297–312.

DIERKES, P.W., WENDE, V., HOCHSTRATE, P., SCHLUE, W. R. 2004. L-type Ca2+ channel antagonists block voltage-dependent Ca2+ channels in identified leech neurons. Brain Res. 1013, 159–167.

DORSETT, D.A. 1980. Design and function of giant fibre systems. Trends Neurosci. 3, 205–208.

DUAN, Y., PANOFF, J., BURRELL, B.D., SAHLEY, C.L., MULLER, K.J. 2005. Repair and regeneration of functional synaptic connections: cellular and molecular interactions in the leech. Cell Mol. Neurobiol. 25, 441–450.

EHRLICH, J.S., BOULIS, N.M., KARRER, T., SAHLEY, C.L. 1992. Differential effects of serotonin depletion on sensitization and dishabituation in the leech, *Hirudo medicinalis*. J. Neurobiol. 23, 270–279.

EVANS, C.G., JING, J., ROSEN, S.C., CROPPER, E.C. 2003. Regulation of spike initiation and propagation in an Aplysia sensory neuron: gating-in via central polarization. J. Neurosci. 23, 2920–2931.

FRANK, E., JANSEN, J.K., RINVIK, E. 1975. A multisomatic axon in the central nervous system of the leech. J. Comp. Neurol. 159, 1–13.

FRICK, A., JOHNSTON, D. 2005. Plasticity of dendritic excitability. J. Neurobiol. 64, 100–115.

GLANZMAN, D.L. 2006. The cellular mechanisms of learning in Aplysia: of blind men and elephants. Biol. Bull. 210, 271–279.

GREY, K.B., BURRELL, B.D. 2008. Forskolin induces NMDA receptor-dependent potentiation at a central synapse in the leech. J. Neurophysiol. 99, 2719–2724.

GROVES, P.M., LEE, D., THOMPSON, R.F. 1970. Effects of stimulus frequency and intensity on habituation and sensitization in acute spinal cat. Physiol. Behav. 4, 383–388.

GU, X.N., MACAGNO, E.R., MULLER, K.J. 1989. Laser microbeam axotomy and conduction block show that electrical transmission at a central synapse is distributed at multiple contacts. J. Neurobiol. 20, 422–434.

GU, X.N., MULLER, K.J., YOUNG, S.R. 1991. Synaptic integration at a sensory-motor reflex in the leech. J. Physiol. 441, 733–754.

GUNTHER, J. 1973. A new type of "node" in the myelin sheath of an invertebrate nerve fiber. Experientia 29, 1263–1265.

GUNTHER, J. 1976. Impulse conduction in the myelinated giant fibers of the earthworm. Structure and function of the dorsal nodes in the median giant fiber. J. Comp. Neurol. 168, 505–531.

HAWKINS, R.D., KANDEL, E.R., BAILEY, C.H. 2006. Molecular mechanisms of memory storage in Aplysia. Biol. Bull. 210, 174–191.

HINDE, R.A. 1970. Behavioral habituation. In *Short-Term Changes in Neural Activity*, edited by G. Horn and R.A. Hinde. London: Cambridge University Press, pp. 3–40.

HORRIDGE, G.A. 1959. Analysis of the rapid responses of *Nereis* and *Harmothoe* (Annelida). Proc. R. Soc. Lond. B Biol. Sci. 150, 245–262.

JANSEN, J.K., NICHOLLS, J.G. 1973. Conductance changes, an electrogenic pump and the hyperpolarization of leech neurones following impulses. J. Physiol. 229, 635–655.

KEATING, H.H., SAHLEY, C.L. 1996. Localization of the myomodulin-like immunoreactivity in the leech CNS. J. Neurobiol. 30, 374–384.

KERKUT, G.A., WALKER, R.J. 1967. The action of acetylcholine, dopamine and 5-hydroxytryptamine on the spontaneous activity of the cells of Retzius in the leech *Hirudo medicinalis*. Br. J. Pharmacol. Chemother. 30, 644–654.

KRAMER, A.P. 1981. The nervous system of the glossiphoniid leech *Haementeria ghilianii*. II. Synaptic pathways controlling body wall shortening. J. Comp. Physiol. [A] 144, 449–457.

KRASNE, F.B. 1965. Escape from recurring tactile stimulation in *Branchiomma vesiculosum*. J. Exp. Biol. 42, 307–322.

KRISTAN, W.B., Jr., CALABRESE, R.L. 1976. Rhythmic swimming activity in neurones of the isolated nerve cord of the leech. J. Exp. Biol. 65, 643–668.

KRISTAN, W.B., Jr., CALABRESE, R.L., FRIESEN, W.O. 2005. Neuronal control of leech behavior. Prog. Neurobiol. 76, 279–327.

LAVERACK, M.S. 1969. Mechanoreceptors, photoreceptors and rapid conduction pathways in the leech, *Hirudo medicinalis*. J. Exp. Biol. 50, 129–140.

LI, Q., BURRELL, B.D. 2006. CNQX inhibits transmission at rectifying but not non-rectifying electrical synapses. Soc. Neurosci. (Abstr.132.4).

LIU, Y., NICHOLLS, J.G. 1989. Steps in the development of chemical and electrical synapses by pairs of identified leech neurons in culture. Proc. R. Soc. Lond. B Biol. Sci. 236, 253–268.

LOCKERY, S.R., RAWLINS, J.N., GRAY, J.A. 1985. Habituation of the shortening reflex in the medicinal leech. Behav. Neurosci. 99, 333–341.

LOCKERY, S.R., KRISTAN, W.B., Jr. 1990. Distributed processing of sensory information in the leech. I. Input-output relations of the local bending reflex. J. Neurosci. 10, 1811–1815.

LOCKERY, S.R., KRISTAN, W.B., Jr. 1991. Two forms of sensitization of the local bending reflex of the medicinal leech. J. Comp. Physiol. [A] 168, 165–177.

MACAGNO, E.R., MULLER, K.J., PITMAN, R.M. 1987. Conduction block silences parts of a chemical synapse in the leech central nervous system. J. Physiol. 387, 649–664.

MAGNI, F., PELLEGRINO, M. 1975. Nerve cord shortening induced by activation of the fast conducting system in the leech. Brain Res. 90, 169–174.

MAGNI, F., PELLEGRINO, M. 1978a. Neural mechanisms underlying the segmental and genaralized cord shortening reflexes in the leech. J. Comp. Physiol. [A] 124, 339–351.

MAGNI, F., PELLEGRINO, M. 1978b. Patterns of activity and the effects of activation of the fast conducting system on the behaviour of unrestrained leeches. J. Exp. Biol. 76, 123–135.

MALENKA, R.C., BEAR, M.F. 2004. LTP and LTD: an embarrassment of riches. Neuron 44, 5–21.

MAR, A., DRAPEAU, P. 1996. Modulation of conduction block in leech mechanosensory neurons. J. Neurosci. 16, 4335–4343.

MARINESCO, S., WICKREMASINGHE, N., KOLKMAN, K.E., CAREW, T.J. 2004. Serotonergic modulation in aplysia. II. Cellular and behavioral consequences of increased serotonergic tone. J. Neurophysiol. 92, 2487–2496.

MASON, A., MULLER, K.J. 1996. Accurate synapse regeneration despite ablation of the distal axon segment. Eur. J. Neurosci. 8, 11–20.

MODNEY, B.K., SAHLEY, C.L., MULLER, K.J. 1997. Regeneration of a central synapse restores nonassociative learning. J. Neurosci. 17, 6478–6482.

MOSS, B.L., FULLER, A.D., SAHLEY, C.L., BURRELL, B.D. 2005. Serotonin modulates axo-axonal coupling between neurons critical for learning in the leech. J. Neurophysiol. 94, 2575–2589.

MULLER, K.J., CARBONETTO, S. 1979. The morphological and physiological properties of a regenerating synapse in the CNS of the leech. J. Comp. Neurol. 185, 485–516.

MULLER, K.J., SCOTT, S.A. 1981. Transmission at a "direct" electrical connexion mediated by an interneurone in the leech. J. Physiol. 311, 565–583.

NICHOLLS, J.G., PURVES, D. 1970. Monosynaptic chemical and electrical connexions between sensory and motor cells in the central nervous system of the leech. J. Physiol. 209, 647–667.

NUSBAUM, M.P., KRISTAN, W.B., Jr. 1986. Swim initiation in the leech by serotonin-containing interneurones, cells 21 and 61. J. Exp. Biol. 122, 277–302.

PRESCOTT, S.A., CHASE, R. 1999. Sites of plasticity in the neural circuit mediating tentacle withdrawal in the snail *Helix aspersa*: implications for behavioral change and learning kinetics. Learn. Mem. 6, 363–380.

RANKIN, C.H., CAREW, T.J. 1988. Dishabituation and sensitization emerge as separate processes during development in Aplysia. J. Neurosci. 8, 197–211.

RAYMOND, C.R. 2007. LTP forms 1, 2 and 3: different mechanisms for the "long" in long-term potentiation. TINS 30, 167–175.

ROBERTS, M.B.V. 1962a. The rapid response of *Myxicola infundibulum* (Grubed). J. Mar. Biol. Assoc. U.K. 42, 527–539.

ROBERTS, M.B.V. 1962b. The giant fibre reflex of the earthworm, *Lumbricus terrestris* L. I. The rapid response. J. Exp. Biol. 39, 219–227.

ROBERTS, M.B.V. 1966. Facilitation in the rapid response of the earthworm, *Lumbricus terrestris* L. J. Exp. Biol. 45, 141–150.

ROSENKOETTER, J.S., BOYCE, R. 1975. Earthworm pheremones and T-maze performance. J. Comp. Physiol. Psychol. 88, 904–910.

SAHLEY, C.L. 1994. Serotonin depletion impairs but does not eliminate classical conditioning in the leech *Hirudo medicinalis*. Behav. Neurosci. 108, 1043–1052.

SAHLEY, C.L., BOULIS, N.M., SCHURMAN, B. 1994a. Associative learning modifies the shortening reflex in the semi-intact leech *Hirudo medicinalis*: effects of pairing, predictability, and CS pre-exposure. Behav. Neurosci. 108, 340–346.

SAHLEY, C.L., MODNEY, B.K., BOULIS, N.M., MULLER, K.J. 1994b. The S cell: an interneuron essential for sensitization and full dishabituation of leech shortening. J. Neurosci. 14, 6715–6721.

SCHANBACKER, A., KOCH, M., PILZ, P.K.D., SCHNITZLER, H.U. 1996. Lesions of the amygdala do not affect the enhancement of the acoustic startle response by background noise. Physiol. Behav. 60, 1341–1346.

SCHEIBENSTOCK, A., KRYGIER, D., HAQUE, Z., SYED, N., LUKOWIAK, K. 2002. The Soma of RPeD1 must be present for long-term memory formation of associative learning in Lymnaea. J. Neurophysiol. 88, 1584–1591.

SCOTT, S.A., MULLER, K.J. 1980. Synapse regeneration and signals for directed axonal growth in the central nervous system of the leech. Dev. Biol. 80, 345–363.

SCURI, R., MOZZACHIODI, R., BRUNELLI, M. 2002. Activity-dependent increase of the AHP amplitude in T sensory neurons of the leech. J. Neurophysiol. 88, 2490–2500.

SCURI, R., MOZZACHIODI, R., BRUNELLI, M. 2005. Role for calcium signaling and arachidonic acid metabolites in the activity-dependent increase of AHP amplitude in leech T sensory neurons. J. Neurophysiol. 94, 1066–1073.

SCURI, R., LOMBARDO, P., CATALDO, E., RISTORI, C., BRUNELLI, M. 2007. Inhibition of Na+/K+ ATPase potentiates synaptic transmission in tactile sensory neurons of the leech. Eur. J. Neurosci. 25, 159–167.

SHAW, B.K., KRISTAN, W.B., Jr. 1995. The whole-body shortening reflex of the medicinal leech: motor pattern, sensory basis, and interneuronal pathways. J. Comp. Physiol. [A] 177, 667–681.

SHAW, B.K., KRISTAN, W.B., Jr. 1999. Relative roles of the S cell network and parallel interneuronal pathways in the whole-body shortening reflex of the medicinal leech. J. Neurophysiol. 82, 1114–1123.

SHEFI, O., SIMONNET, C., BAKER, M.W., GLASS, J.R., MACAGNO, E.R., GROISMAN, A. 2006. Microtargeted gene silencing and ectopic expression in live embryos using biolistic delivery with a pneumatic capillary gun. J. Neurosci. 26, 6119–6123.

SKIERCZYNSKI, B.A., WILSON, R.J., KRISTAN, W.B., Jr., SKALAK, R. 1996. A model of the hydrostatic skeleton of the leech. J. Theor. Biol. 181, 329–342.

STUART, A.E. 1969. Excitatory and inhibitory motoneurons in the central nervous system of the leech. Science 165, 817–819.

STUART, A.E. 1970. Physiological and morphological properties of motoneurones in the central nervous system of the leech. J. Physiol. 209, 627–646.

SZCZUPAK, L., KRISTAN, W.B., Jr. 1995. Widespread mechanosensory activation of the serotonergic system of the medicinal leech. J. Neurophysiol. 74, 2614–2624.

TAYLOR, A.L., COTTRELL, G.W., KLEINFELD, D., KRISTAN, W.B., Jr. 2003. Imaging reveals synaptic targets of a swim-terminating neuron in the leech CNS. J. Neurosci. 23, 11402–11410.

THOMPSON, L.M., MARSH, J.L. 2003. Invertebrate models of neurologic disease: insights into pathogenesis and therapy. Curr. Neurol. Neurosci. Rep. 3, 442–448.

THOMPSON, R.F., SPENCER, W.A. 1966. Habituation: a model phenomenon for the study of the neuronal substrates of behavior. Psychol. Rev. 73, 16–43.

TOBIN, A.E., CALABRESE, R.L. 2005. Myomodulin increases Ih and inhibits the NA/K pump to modulate bursting in leech heart interneurons. J. Neurophysiol. 94, 3938–3950.

TRUETA, C., MENDEZ, B., DE MIGUEL, F.F. 2003. Somatic exocytosis of serotonin mediated by L-type calcium channels in cultured leech neurones. J. Physiol. 547, 405–416.

TRUETA, C., SANCHEZ-ARMASS, S., MORALES, M.A., De MIGUEL, F.F. 2004. Calcium-induced calcium release contributes to somatic secretion of serotonin in leech Retzius neurons. J. Neurobiol. 61, 309–316.

VAN ESSEN, D.C. 1973. The contribution of membrane hyperpolarization to adaptation and conduction block in sensory neurones of the leech. J. Physiol. 230, 509–534.

WALKER, R.J., SMITH, P.A. 1973. The ionic mechanism for 5-hydroxytryptamine inhibition on Retzius cells of the leech *Hirudo medicinalis*. Comp. Biochem. Physiol. 45A, 979–993.

WANG, Y., STRONG, J.A., SAHLEY, C.L. 1999. Modulatory effects of myomodulin on the excitability and membrane currents in Retzius cells of the leech. J. Neurophysiol. 82, 216–225.

WEEKS, J.C. 1981. Neuronal basis of leech swimming: separation of swim initiation, pattern generation, and intersegmental coordination by selective lesions. J. Neurophysiol. 45, 698–723.

WEEKS, J.C. 1982. Segmental specialization of a leech swim-initiating interneuron, cell 205. J. Neurosci. 2, 972–985.

WILLARD, A.L. 1981. Effects of serotonin on the generation of the motor program for swimming by the medicinal leech. J. Neurosci. 1, 936–944.

WITTENBERG, G., KRISTAN, W.B., Jr. 1992a. Analysis and modeling of the multisegmental coordination of shortening behavior in the medicinal leech. I. Motor output pattern. J. Neurophysiol. 68, 1683–1692.

WITTENBERG, G., KRISTAN, W.B., Jr. 1992b. Analysis and modeling of the multisegmental coordination of shortening behavior in the medicinal leech. II. Role of identified interneurons. J. Neurophysiol. 68, 1693–1707.

WOLF, F.W., HEBERLEIN, U. 2003. Invertebrate models of drug abuse. J. Neurobiol. 54, 161–178.

YAU, K.W. 1976. Physiological properties and receptive fields of mechanosensory neurones in the head ganglion of the leech: comparison with homologous cells in segmental ganglia. J. Physiol. 263, 489–512.

ZACCARDI, M.L., TRAINA, G., CATALDO, E., BRUNELLI, M. 2001. Nonassociative learning in the leech *Hirudo medicinalis*. Behav. Brain Res. 126, 91–92.

ZACCARDI, M.L., TRAINA, G., CATALDO, E., BRUNELLI, M. 2004. Sensitization and dishabituation of swim induction in the leech *Hirudo medicinalis*: role of serotonin and cyclic AMP. Behav. Brain Res. 153, 317–326.

ZHANG, W., LINDEN, D.J. 2003. The other side of the engram: experience-driven changes in neuronal intrinsic excitability. Nat. Rev. Neurosci. 4, 885–900.

ZHENG, M., FRIESEN, W.O., IWASAKI, T. 2007. Systems-level modeling of neuronal circuits for leech swimming. J. Comput. Neurosci. 22, 21–38.

ZOCCOLAN, D., PINATO, G., TORRE, V. 2002. Highly variable spike trains underlie reproducible sensorimotor responses in the medicinal leech. J. Neurosci. 22, 10790–10800.

ZORAN, M.J., DREWES, C. 1987. Rapid escape reflexes in aquatic oligochaetes: variations in design and function of evolutionarily conserved giant fiber systems. J. Comp. Physiol. [A] 161, 729–738.

Chapter 9

Development, Regeneration and Immune Responses of the Leech Nervous System

Michel Salzet* and Eduardo Macagno†

*Laboratoire de Neuroimmunologie des Annélides, Université des Sciences et Technologies de Lille, Villeneuve d'Ascq, France
†Division of Biological Sciences, University of California, San Diego, La Jolla, CA

9.1 INTRODUCTION

During embryogenesis, various developmental programs set up the structures, cell properties and systemic features of the adult nervous system yielding, as the end result, a functional organ that not only detects and analyzes sensory information and produces behavioral responses, but that is also critically involved in regulating physiology and maintaining homeostasis. Normally, the adult nervous system is in a stable functional state and readily responds to perturbations of critical parameters within ranges of variation that are relatively small. Major perturbations, however, including mechanical trauma, which can trigger regeneration, or septic shock, which provokes an innate immune response, activate mechanisms that include an enhanced or reduced expression of specific sets of genes essential for tissue repair and restoration of the normal, homeostatic state. Despite being induced by quite different stresses, response mechanisms appear to involve some of the same genes and molecular pathways, as will be discussed later in this review. Moreover, these stress response mechanisms likely recapitulate aspects of developmental programs that initially have established the normal state, indicating that development, regeneration and neuroimmune responses may share some common elements and themes. This raises many interesting questions that are now accessible experimentally in the leech. For example, to what extent are molecular mechanisms and pathways shared under various conditions, and how are they modified to take into account the differences between embryonic and adult contexts?

In this chapter, we will discuss recent studies of the leech nervous system, focusing on the molecular mechanisms underlying development, regeneration and the immune response. We will consider whether and how current observations provide evidence in favor or against the hypothesis that these processes involve some common themes and perhaps share underlying mechanistic aspects. We will review work published between ~2000 and the present, and refer the reader to past reviews (e.g. Baker and Macagno 2001; also French and Kristan 1994; Blackshaw and Nicholls 1995; Jellies and Johansen 1995; Wong et al. 1995; Kristan et al. 2000, 2005; Salzet 2001a, 2002b; Blackshaw et al. 2004; Duan et al. 2005; Salzet et al. 2006) for earlier work and interesting areas not included here.

9.2 BACKGROUND

The leech central nervous system (CNS) has a fixed number of 32 bilateral neuromeres. The four anteriormost neuromeres fuse to form the subesophageal ganglion (SubEG) and the seven posteriormost fuse to form the tail ganglion; single bilateral neuromeres comprise individual ganglia found in each of the corresponding midbody segments. A supraesophageal ring of nonsegmental origin, together with the SubEG, comprises the head ganglion. Central ganglia are connected by a bilateral pair of nerves (the lateral "connectives") and a single small medial nerve (Faivre's), and to the periphery by two (or three in glossiphoniid leeches) bilateral pairs of nerves ("roots") that branch in stereotypic patterns allowing the identification of fourth-order, and in some cases, fifth-order branches (Fig. 9.1). Additionally, many sensory neurons in the body wall and other internal organs comprise the peripheral nervous system (PNS), providing a variety of sensory information to the CNS.

The leech CNS arises from the segmentally iterated progeny of five embryonic teloblast lineages, M, N, O, P and Q, among which the N lineage makes the most significant contribution. In hirudinid leeches, each segmental ganglionic primordium gives rise to ~400 neurons (Macagno 1980). Most of these occur as bilateral pairs (~180–190 pairs), but perhaps 5–8% are unpaired, with at least some becoming unpaired through cell death (Macagno 1980; Sawyer 1986). Thus, understanding how a leech segmental ganglion functions requires, in principle, detailed knowledge of the function and connectivity of only ~200–220 individual neurons. Moreover, since each segmental ganglion is a variation on a theme (with the exception of the "sex" ganglia of body segments 5 and 6, which have additional complements of neurosecretory cells; Baptista et al. 1990), the leech has one of the most accessible nervous systems from a systems analysis point of view.

Current knowledge of which neurons contribute to the activity of neuronal circuits responsible for generating specific behavioral responses (see review by Kristan et al. 2005) is becoming much more complete as a result of the recent and successful application of multineuronal functional imaging to leech ganglia (Briggman and Kristan 2006). Since gene expression and function can now be assayed and modulated in individual leech neurons or groups of neurons, we propose that a systems-level approach, focused on relating the neural expression of genetic programs to physiological programs, would be timely and perhaps uniquely feasible in leech.

9.3 RECENT WORK ON THE DEVELOPMENT OF THE NERVOUS SYSTEM

Several recent contributions have extended our knowledge and understanding of two general areas related to the formation of the leech nervous system. One area is the formation of peripheral nerves and dendritic arbors. Shain et al. (2004), for example, explored

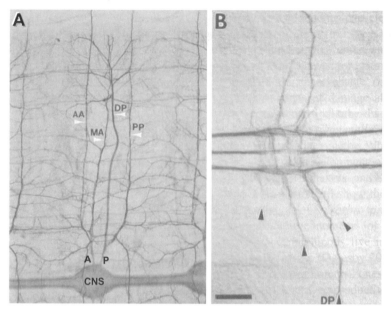

Figure 9.1 Segmental nerves at two stages of development in *Hirudo medicinalis*, labeled with an antibody against ACT. (A) Micrograph of axon tracts that form segmentally iterated peripheral nerves in one hemisegment of an El4 embryo. The ACT antibody labels an epitope expressed by all centrally located neurons, a subset of peripheral neurons and a repeated pattern of longitudinal muscles. The four major nerve roots AA, MA, DP and PP, arise as bifurcations from two nerve roots (A = anterior; P = posterior) and project many finer nerves and branches in the developing germinal plate. Three of the major nerves innervate the ventral, lateral and dorsal body walls, whereas one of them (DP) rises out of the plane of focus and extensively innervates the dorsal body wall. The DP nerve projects along a muscle that inserts near the ventral and dorsal midline and later projects through the lumen. (B) The ACT antibody labels the initial axonal projections within and from the CNS, as shown in a segment from an early-stage embryo (E9). Four projections extend from each side of the ganglion (arrowheads), forming each of the four major nerve pathways independently (the DP, which extends first and more quickly, is labeled). Later they will condense into two roots, as seen in panel A. Anterior is left. Scale bar = 100 μm.

Source: Reproduced from Jellies et al. (1996), with permission from John Wiley & Sons, Inc.

how differences in the patterns of nerves exiting from segmental ganglia arise in species belonging to different leech families (hirudinids and glossiphoniids). Within this theme, Venkitaramani et al. (2004), while exploring the nature of antigens detected by specific monoclonal antibodies (mAbs) raised against leech tissue, also examined mechanisms that give rise to the dorsal-posterior (DP) nerve, a peripheral branch of the posterior nerve root. Complementary to nerve patterning in the body wall are patterns of innervation of peripheral structures, such as sensory tiling of the skin by the peripheral arbors of T and P cells. One current project explores the possible role of the peripheral secretion of netrin as a signal that distinguishes ventral from dorsal territory, a question currently addressed by Shefi et al. (2006) for P-cell arbors using a novel ballistic approach to RNAi and ectopic gene expression. Conversely, time-lapse imaging of afferent projections has been used to examine how central terminal arbors of peripheral sensillar sensory cells are formed in the ganglionic neuropils (Baker et al. 2003). Another study has examined how the disruption of peripheral target contact influences the development of central dendritic branches of a leech motor neuron (Johnson et al. 2000).

An area of leech neuronal development that has received a lot of attention is the role of gap junctions (GJs) in the development of neuronal circuits underlying specific behaviors (Kristan et al. 2000; Marin-Burgin et al. 2005, 2006). The recent cloning of many different leech innexins (Dykes et al. 2004; Dykes and Macagno 2006), which compose the invertebrate GJ, should allow for rapid advancements in this area. In the sections that follow, we discuss these published findings in more detail.

9.3.1 Early Cell Interactions and Genesis of Peripheral Nerves

Central ganglia of the leech nervous system connect to each other and to peripheral targets via a relatively simple network of longitudinal connective nerves and transverse segmental nerves. The segmental nerves have been the subject of many studies in several leech species. In *Hirudo medicinalis*, they initially consist of four nerves, anterior-anterior (AA), medial-anterior (MA), posterior-posterior (PP) and DP, with DP forming first, followed by MA and later AA and PP. These nerves eventually condense into two nerve roots (AA and MA form the anterior root, PP and DP the posterior root) that bifurcate outside of the ganglion (Jellies et al. 1996; see Fig. 9.1). This sequence is different in other leeches; for example, the glossiphoniid leech *Theromyzon tessulatum* has five major segmental nerves (AA, MA, PP, DP and UP), and their order of appearance is AA and MA first, followed by PP, with DP and UP branching later from the PP nerve beyond the ganglionic perimeter (Shain et al. 2004). Therefore, three pairs of nerve roots are in each ganglion, and condensation is not observed (Shain et al. 2004). The genesis of the segmental nerve pattern in *T. tessulatum* and its dependence on certain cell–cell interactions was assayed by immunostaining embryonic preparations with antibody to acetylated tubulin (ACT) (stains axons in the leech preferentially) to track process outgrowth, and by injecting fluorescent tracers into individual teloblasts of the mesodermal (M) and ectodermal (N, O, P and Q) lineages to mark their progeny (see Shain et al. 2004, for references to earlier related work). Dependence on specific lineages was tested by ablating subsets of cells from each of these lineages in developing embryos. Not surprisingly, ablation of mesodermal derivatives caused the most severe abnormalities, yielding a disorganized array of processes and the absence of all transverse and many longitudinal nerves. This effect is probably compounded by the accompanying disruption of the distribution and localization of ectodermal derivatives, but it may be largely due to the absence of the normal orthogonal pattern of muscle fibers, with which nerves associate closely (Shain et al. 2004, and earlier references therein).

Ectodermal ablations, by contrast, yielded a range of more specific phenotypes, such as the coalescence of two segmental nerves (N deletion), truncation or displacement of a particular segmental nerve (O deletion), or relatively minor effects (P and Q deletions). A specific case examined in greater detail was the formation of the posterior segmental nerve. This nerve, which was abnormal in embryos with an N deletion, has been shown in previous studies to be pioneered by an O lineage-derived neuron, the dorsal pressure sensory cell (P_D). Following unilateral deletion of O-lineage progeny, including this neuron, the PP nerve formed late but followed an abnormal trajectory in the periphery. When N lineage-derived cells were deleted unilaterally, the P_D neuron failed to migrate to its normal position in the posterior compartment of the hemiganglion and projected its axon into an ipsilateral anterior nerve, the MA nerve. Followers of the P_D axon also traveled along the MA nerve, but the group then split off and assumed a more posterior trajectory, soon after exiting the ganglion. These observations, and others not discussed here, led Shain and coauthors to propose that nerve formation relies on an orchestrated set of interactions among particular cells arising from different lineages (Shain et al. 2004).

9.3.2 A Role for a Specific Peripheral Muscle Bundle in DP Nerve Formation

Among the four major segmental nerves mentioned above, three (AA, MA and PP) are pioneered by peripheral neurons growing into the CNS, while the fourth (DP) is pioneered by the central P_D neuron projecting an axon to the body wall (Kuwada 1985; Jellies et al. 1996). Genesis of the DP nerve was studied by Venkitaramani et al. (2004) using mAbs that label all muscles in leech (Lan3-14 and Laz10-1; Zipser and McKay 1981; Thorey and Zipser 1991). Both mAbs recognize the same ~400 kDa antigen, a leech member of the filamin family of actin binding proteins (Venkitaramani et al. 2004). One set of muscles, the medial flattener muscles, which extend from ventral to dorsal on either side of the ventral nerve cord (Venkitaramani et al. 2004), was of particular interest. Using one of the mAbs to visualize these muscles during embryogenesis and anti-ACT to visualize axonal projections, they determined that the P_D axon pioneering the DP nerve grows only along the middle of three discrete bundles of myofibers comprising the medial flatteners in each segment. This suggests that the middle bundle bears some molecular cues recognized by the central projections, and indeed, only the middle muscle bundle is immunoreactive to leech tractin, an L1-family cell adhesion molecule (Huang et al. 1998). Moreover, the immunoreactivity is not only transient, but it also occurs only during the developmental period when the DP nerve is being formed. Attempts to affect DP nerve formation with antibodies to leech tractin were not successful, but that could be due to their being ineffective blockers of tractin function, though a possible alternative explanation is the presence of redundant cues unique to the middle muscle bundle. Regardless, it is safe to conclude that a unique flattener muscle bundle "provides a substrate for early axonal outgrowth and nerve formation and that this function may be associated with differential expression of distinct cell adhesion molecules" (Venkitaramani et al. 2004). Interestingly, while the P_D neuron pioneers the DP nerve, this nerve can form normally without the P_D neuron, as shown by deletion of the P_D cell (Gan et al. 1999). This is not surprising in that the marked peripheral path along the flattener muscle is not affected and can be utilized by later-growing axons, either efferent or afferent. However, removal of the P_D cell can profoundly affect the peripheral arbors of cells that follow this pioneer to their target areas (Gan et al. 1999).

9.3.3 Ectopic Ganglia Promote the Formation of Novel Peripheral Nerve Pathways

Several earlier studies (Johansen et al. 1992; Jellies et al. 1996; Huang et al. 1998) implicated interactions between outgoing projections of central neurons and ingrowing afferents from peripheral neurons in the formation of the segmental nerves in *H. medicinalis* embryos. In one study (Jellies and Johansen 1995), the embryonic CNS was removed and as a consequence, the trajectories of peripheral neuron projections were disordered and abnormally directed. They concluded that CNS-derived guidance cues are necessary for appropriate navigation of afferent projections into the CNS. More recently, Jellies et al. (2000) performed a complementary test by transplanting embryonic central ganglia into ectopic sites in the body wall, some lateral to the ventral midline, in segments from which the resident ganglia had been excised. These abnormally located ganglia not only survived, but extended numerous fasciculated projections in multiple directions, many abnormal, including some that crossed segmental boundaries or projected to the opposite side of the embryo. Interestingly, when these "central" projections met peripheral sensory axons, the latter grew along the former toward and into the ectopic ganglia, regardless of pathway orientation or the crossing of normally respected boundaries (e.g. segmental, midline). The behavior of peripheral sensory axons appeared to require physical contact with the CNS projections, failing to turn toward the ganglia before sometimes making sharp turns in order to fasciculate with afferent bundles. Taken together, the experimental results presented in these two studies (Jellies and Johansen 1995; Jellies et al. 2000) demonstrate that CNS projections are both necessary and sufficient to steer peripheral sensory afferents into the CNS, and that diffusible signals are probably not important. The cues that afferents recognize to select a path into the CNS are not known.

9.3.4 Establishment of the Boundaries of P-Neuron Peripheral Arbors

As mentioned above, the P_D cell pioneers the DP nerve, and extends without branching into dorsal territory, where it establishes a characteristic terminal arbor and, as it matures, a sensory domain in the skin called a "sensory tile." Each segmental ganglion contains two P_D cells, as well as two P_V cells that innervate ventral domains (i.e. four P cells per ganglion), therefore establishing four sensory tiles per segment (Fig. 9.2).

Recent genetic analyses in the fruit fly (Hughes et al. 2007; Matthews et al. 2007; Soba et al. 2007) have determined that an unusually large family of DSCAM isoforms play a key role in tiling, but these molecules appear unique to insects (mammals seem to have only a single isoform, and large families have not been reported in other phyla). Fly mechanosensory cells of certain classes reside in the larval epidermis and establish sensory tiles with sharp boundaries; each expresses a different DSCAM isoform that is used to recognize itself and thereby to prevent an overlap of its own branches within their sensory arbor, as well as creating sharp boundaries with their homologues. The mechanisms that translate these unique properties of fly DSCAM into cellular responses are not presently known.

In leech, a similar role in self-avoidance and arbor tiling is played by another member of the Ig superfamily, the receptor protein tyrosine phosphatase (RPTP) *Hm*LAR2, in comb cells (CCs), a nonneuronal but possibly glial-like cell type (Baker and Macagno 2000a, b, 2006, 2007; Baker et al. 2000; Biswas et al. 2002). *Hm*LAR2 is highly expressed throughout CCs, including growth cones and filopodia, where it appears to associate with integrins and to form puncta that probably represent focal adhesion sites (Baker et al. 2008). Homophilic binding of *Hm*LAR2, which occurs when tips of growth cone filopodia contact another part of the same cell or an adjacent CC, causes filopodia to retract, thereby blocking the possibility of a cell process growing over another part of the same cell or an adjacent CC. Mechanistically, this appears to occur through cleavage of the ectodomain, which binds to the ectodomain of an apposed *Hm*LAR2 molecule and is then internalized and thus removed from the cell surface (Baker et al. submitted). Thus, *Hm*LAR2 mediates self-avoidance as well as cell-type avoidance, leading to tiling of the body wall by CCs.

Although *Hm*LAR2, as well as a related RPTP, *Hm*LAR1, are expressed in some leech central neurons (Gershon et al. 1998), their functions in the development of the CNS have

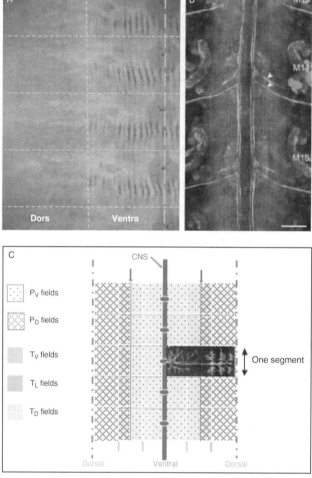

Figure 9.2   Possible role for netrin in the tiling of P mechanosensory neuron sensory fields. (A) Several segments in the left side of a dissected E9 embryo stained for the presence of netrin mRNA (modified from Gan et al. 1999, with permission from John Wiley & Sons, Inc.). A segmentally iterated pattern of longitudinal muscle cells can be observed in the ventral region, but not in the dorsal body wall. Areas of innervation by the dorsal and ventral P cells are delimited by dashed lines. (B) Bilateral distribution of netrin protein (red) secreted by these longitudinal muscle cells (scale bar = 100 μm; modified from Aisemberg et al. 2001, with permission from Springer). The protein is seen only in ventral regions, perhaps loosely bound to laminin in the extracellular matrix, which keeps it from diffusing to dorsal regions. Since dorsal P cells express the leech orthologue of the unc-5 netrin receptor, netrin binding likely inhibits the projections of this neuron from branching in ventral territory. Arrowheads indicate nerve roots. (C) Diagram of body wall tiling by sensory fields of P and T mechanosensory cells for several segments. The epidermis of each hemisegment is subdivided into two tiles by P-cell arbors and into three tiles by T-cell arbors. The developing (E10) sensory fields of local dorsal and ventral P cells are displayed in the right middle segment.
Source: Modified from Gan and Macagno 1995a, with permission from the Society for Neuroscience. (See color insert.)

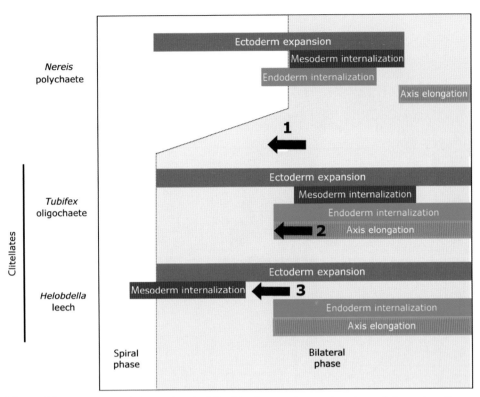

Figure 5.2 Schematic of the heterochronic shift of developmental events in major evolutionary transitions of annelids. (See text for full caption.)

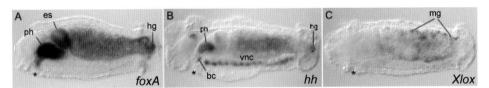

Figure 6.4 Expression of *Capitella* sp. I molecular markers in restricted domains along the length of the gut as analyzed by whole-mount *in situ* hybridization. (See text for full caption.)

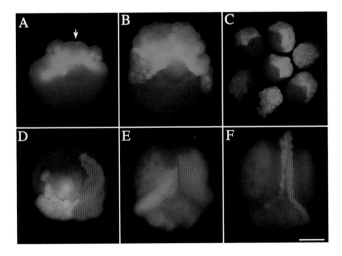

Figure 7.3 Phenotypic effects of microinjecting K110 oligonucleotides into *Theromyzon tessulatum* embryos. (See text for full caption.)

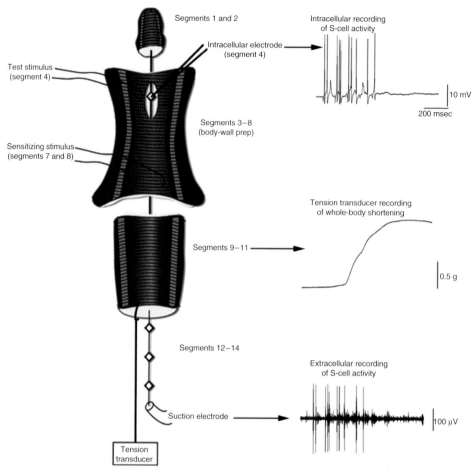

Segments 1 and 2

Intracellular recording
of S-cell activity

Intracellular electrode
(segment 4)

Test stimulus
(segment 4)

10 mV

200 msec

Segments 3–8
(body-wall prep)

Sensitizing stimulus
(segments 7 and 8)

Tension transducer recording
of whole-body shortening

Segments 9–11

0.5 g

Segments 12–14

Extracellular recording
of S-cell activity

Suction electrode

100 μV

Tension
transducer

Figure 8.1 Semi-intact preparation used to study learning in the whole-body shortening reflex. (See text for full caption.)

Figure 8.4 Laser scanning confocal micrograph of possible synaptic contact sites between the Retzius cell and the S-cell axon. (See text for full caption.)

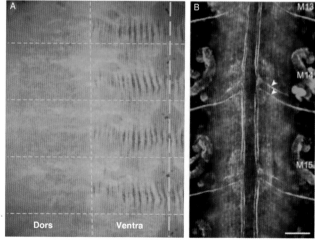

Dors   Ventra

Figure 9.2 Possible role for netrin in the tiling of P mechanosensory neuron sensory fields. (See text for full caption.)

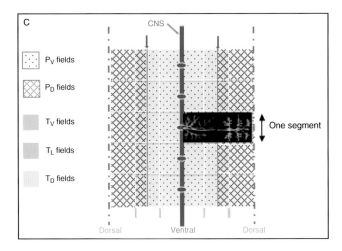

Figure 9.2 *Continued*

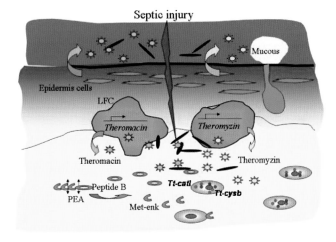

Figure 9.7 Schematic of putative innate immune response of leeches involving antibacterial peptides. (See text for full caption.)

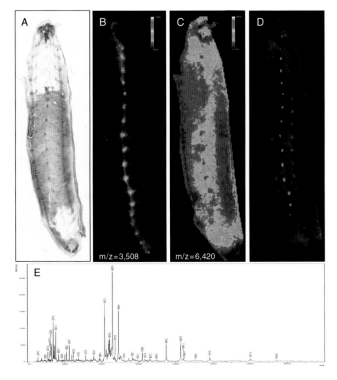

Figure 9.8 MALDI-TOF imaging of an opened E12 leech embryo in whole mount. (See text for full caption.)

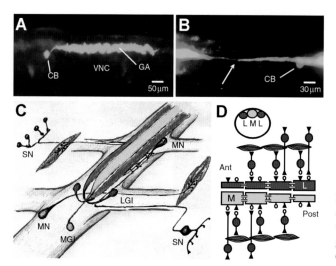

Figure 10.1 Oligochaete escape reflex anatomy. Rapid escape in oligochaete worms is mediated by segmentally arranged giant axons that form through-conducting GF systems. (See text for full caption.)

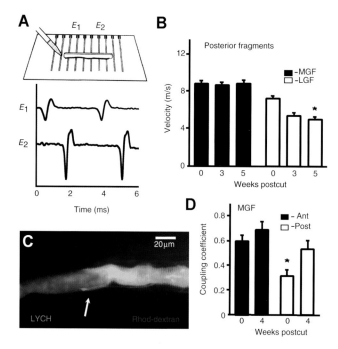

Figure 10.5 Changes in GF conduction and connectivity during neural morphallaxis. (See text for full caption.)

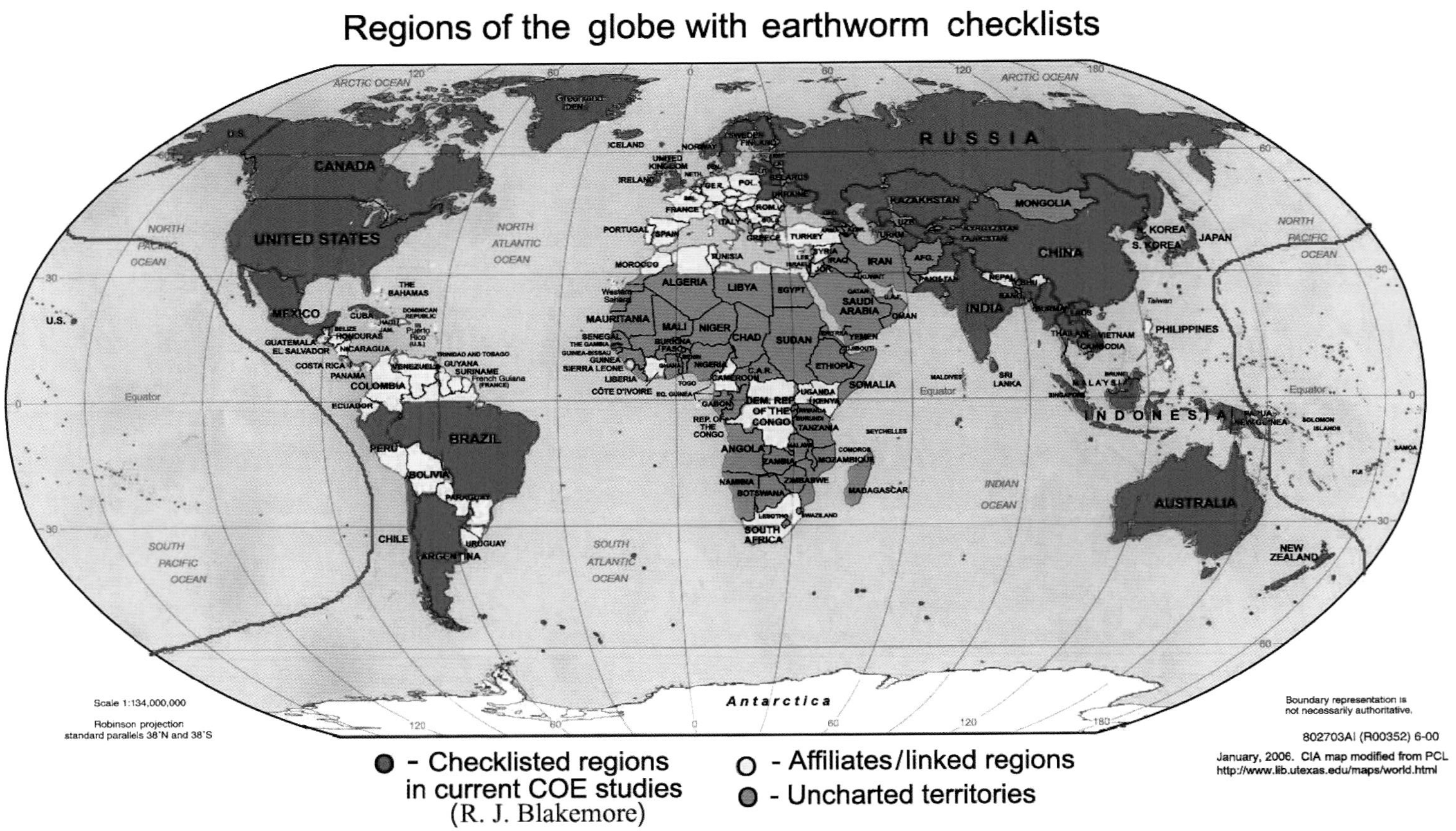

Regions of the globe with earthworm checklists

● – Checklisted regions in current COE studies (R. J. Blakemore)

○ – Affiliates/linked regions

◐ – Uncharted territories

Scale 1:134,000,000
Robinson projection
standard parallels 38°N and 38°S

Boundary representation is not necessarily authoritative.

January, 2006, CIA map modified from PCL
http://www.lib.utexas.edu/maps/world.html

802703AI (R00352) 6-00

Figure 14.1 Regions with current checklists of earthworm species.

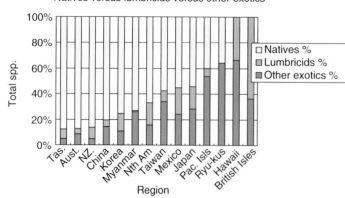

Natives versus lumbricids versus other exotics

Figure 14.2 Lumbricidae as a proportion of total species from selected regions (only British Isles have >40%). Abbreviations: Tas = Tasmania; Aust = Mainland Australia; N.Z. = New Zealand; Nth Am = North America (Rio Grande to Arctic); Pac Isls = Pacific Isles (not Hawaii); China = China + Hainan.

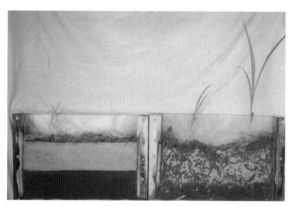

Figure 14.3 Demonstration of effect of adding earthworms (*Pontoscolex corethrurus*) on soil and litter layers, and plant (*Sorghum bicolor*) growth after 2 weeks: lhs = no worms; rhs + five worms.
Source: Photo courtesy of Dr. Les Robertson.

Figure 15.1 *Alvinella pompejana* specimen out of its tube (a) and on a smoker wall after an anthropogenic disturbance (b) (worm sampling by Nautile).

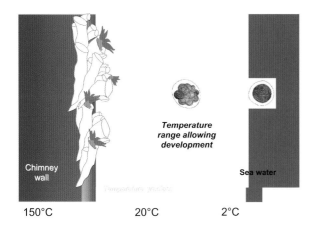

Figure 15.3 Diagram of thermal conditions *in situ* allowing *Alvinella pompejana* embryonic development. Above 20 °C, near adult tubes, development is prevented by high temperatures. At the abyssal temperature (2 °C), development is arrested.

Chimney wall

Temperature range allowing development

Sea water

150°C   20°C   2°C

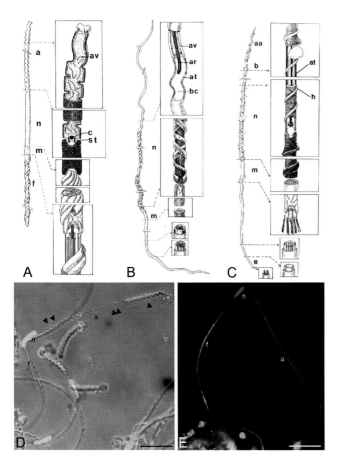

Figure 17.5 Spermatozoa of branchiobdellids, acanthobdellids and hirudineans. (See text for full caption.)

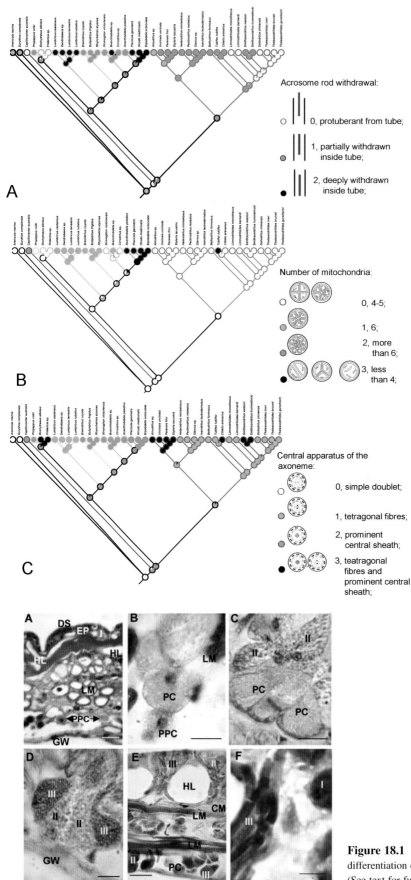

Figure 17.7 Maximum likelihood spermatozoal character state reconstructions on a corrected combined data Bayesian topology using the ancestral state reconstruction packages as implemented in Mesquite version 1.12 (Maddison and Maddison 2006). (See text for full caption.)

Figure 18.1 Temporal development and differentiation of type II and type III cells. (See text for full caption.)

not been explored. However, neither T nor P mechanosensory cells are among the neurons that express these RPTPs, and thus are unlikely to participate in the tiling of their sensory fields.

Among P cells in leech, the sensory tile boundaries between segmental homologues appear to be established in a different way than P_D and P_V cells. Segmental homologues overlap slightly at the segmental margins and, at least for P_D cells, extend into adjacent segments if the neighboring P_D cell is deleted, suggesting a competitive interaction for sensory space between homologues (Gan and Macagno 1995a, b). The boundaries between P_D and P_V cells at the lateral midline, however, are much sharper and appear to be established through a different mechanism; deletion of a P_D cell does not result in the adjacent P_V cell expanding into its territory. A possible explanation for this phenomenon is that the target regions of the two P-cell classes, dorsal and ventral territories, express different markers that regulate the ability of P cells to establish arbors. A candidate for such a marker is leech netrin, which is expressed in the body wall by ventral but not dorsal longitudinal muscle cells (Fig. 9.2A, B; Gan et al. 1999; Aisemberg et al. 2001). This hypothesis is supported by the fact that P_D cells express the netrin receptor unc-5, whereas P_V cells do not (M.W. Baker and O. Shefi, unpublished observations). We are currently using a micro-capillary gene gun to shoot gold particles coated with netrin siRNA into small areas of the ventral body wall to knock down netrin expression, and gold particles coated with a DNA plasmid containing the full-length leech netrin gene into dorsal territory to knock in netrin expression where it is normally absent (Shefi et al. 2006). Efficient ectopic gene expression in leech has been made possible by the cloning and characterization of several strong leech promoters (Baker and Macagno 2006).

An interesting question arises when the tiling pattern of epidermal sensory innervation by the P (pressure-sensitive) cells is compared to that of the T (touch-sensitive) neurons. Three pairs of T cells are present in each segmental ganglion (T_V, T_L and T_D) compared to two pairs of P cells. Hence, T cells innervate the skin in a pattern of three tiles per side, with dorsoventral boundaries located at positions different from the boundary between P cells. This difference has spurred a search for factors whose distributions might reflect the T-cell tiling pattern using imaging mass spectrometry, a novel approach discussed below.

9.3.5 Formation of Central Arbors by Afferent Projections of Peripheral Sensory Cells

A subset of peripheral sensory cells in the body wall of *H. medicinalis* resides in seven bilateral pairs of sensilla that are located in the central annulus of each segment. In the embryo, afferent projections of these sensory neurons travel along common nerve tracts to the CNS, where they defasciculate, branch and arborize into separate, modality-specific regions of the neuropil, forming synaptic connections with specific target neurons. Observations reported in earlier studies by Zipser and collaborators (Song and Zipser 1995; Tai and Zipser 1998, 1999, 2002) indicate that these stages of growth and differentiation are accompanied by specific changes in molecules expressed by the sensory afferents, tractin and leech CAM (related to vertebrate adhesion molecules L1 and NCAM, respectively). In particular, a shift occurs from constitutive mannosidic glycosylation along the ectodomains of these transmembrane proteins to modality-specific galactosidic glycosylations as axons grow into the ganglion and differentiate their terminal arbors, which segregate into different areas within the neuropil. As proposed by Tai and Zipser (2002), constitutive mannosidic glycosylation promotes dynamic growth, while developmentally regulated galactosidic modifications of the same cell adhesion molecules promote tissue stability. Both types of neutral glycans persist in adult leech nervous systems, and may function in synaptic plasticity during habituation and learning (Tai and Zipser 2002).

To gain further insight into the dynamics of afferent growth into ganglia, Baker et al. (2003) examined these projections by time-lapse morphological observations of single, dye-filled sensory afferents using two-photon laser scanning microscopy of intact developing embryos. Images of individual sensory projections recorded at 3- to 30-min intervals over several hours and at different developmental stages revealed highly dynamic and age-dependent growth patterns (Fig. 9.3). Upon entering the CNS, the peripherally compact growth cones expanded and sprouted numerous long filopodial processes, many of which repeated cycles of extension and retraction. With time, the growth cone changed into a terminal arbor through bifurcation into secondary branches that extended within the ganglionic neuropil in a stereotypic lateral position, and out of the ganglia via the lateral connectives, anteriorly and posteriorly. Numerous tertiary and quaternary processes grew from these branches and also displayed cycles of extension and retraction. The motility of these higher-order branches changed with age, with younger afferent terminals displaying higher branch densities and greater motility than older, more mature arborizations. Branches remained ipsilateral, infrequently crossing to the opposite side of the neuropil. Finally, coincident with a reduction in the numbers and densities of higher-order branches, concavolar structures appeared on secondary processes. Rows of these indentations suggest the formation of presynaptic en passant specializations accompanying the developmental onset of synapse formation. Thus, the maturation of sensory afferent terminal arbors followed a stereotyped series of events beginning with an exploratory phase and extensive branching, and ending with a simpler arbor and synaptic contacts with undetermined targets in the ganglionic neuropil. A causal relationship between the shifts in glycosylation and these structural changes is being explored.

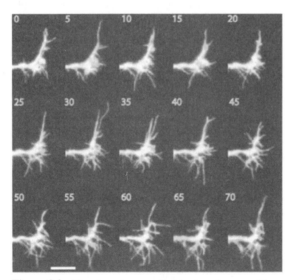

Figure 9.3   Time-lapse images of a late stage 1 neuron. Consecutive images were taken 5 min apart. Dynamic filopodial motility was observed throughout the arbor during the 70-min imaging session, with most of the long filopodial/neuritic extensions appearing along an anterior-projecting trajectory. Scale bar = 10 μm.
Source: Modified from Baker et al. 2003, with permission from John Wiley & Sons, Inc.

9.3.6   Abnormal Central Projections Following Disruption of a Peripheral Target

Retrograde signaling from target tissues has been shown to influence many aspects of neuronal development in a number of developmental systems, including leech. For example, studies by Gao and Macagno (1988) in *H. medicinalis* showed that depriving some leech embryonic motor neurons (AE and AP cells) of their peripheral targets by cutting the local nerve roots led to the retention and overextension of interganglionic projections that were normally retracted, and to the establishment of peripheral connections through nerve roots of adjacent ganglia. Johnson et al. (2000) reexamined this question but with a different set of motor neurons. Their studies in *H. medicinalis* embryos focused on cell 3, a motor neuron that normally innervates dorsal longitudinal muscle fibers. Unlike the AP and AE motor neurons, embryonic cell 3 does not normally extend into the interganglionic connectives. In the protocol used by Johnson et al. (2000), the embryonic DP nerve (the route taken by the peripheral projection of cell 3) was cut at one of two locations within the body wall: proximal or distal to the ganglion and in embryos of different ages. The response was an extensive overgrowth of cell 3's central arbor, with some branches extending far into the connective nerves, though without reaching neighboring ganglia. When cell 3's peripheral axon was cut distally relatively early in development, its overgrown central branches eventually retracted. However, cells that were disrupted later in development retained their overextended branches into adulthood. Additionally, when the axon was cut proximal to the ganglion early in development, depriving the cell of contact with any dorsal tissues, the central branches failed to retract and were instead retained into adulthood. Moreover, cutting the root nerves in adults did not produce central overgrowth of cell 3. A possible explanation for these different results, as pointed out by Johnson et al. (2000), is that different potentials exist for the regeneration of the peripheral axons as a consequence of when and where the growing projection is severed. In contrast to cell 3, cell 1, an inhibitory motor neuron (iMN) with the same peripheral target muscles as cell 3, remained morphologically unchanged following the same surgical protocols. The authors note that an important difference between these two cells might explain the different responses: while severing the peripheral projection of cell 3 abolishes all of its synaptic outputs, the same intervention of cell 1 only deletes a fraction of this cell's outputs, since its strong synaptic connections to motor neurons and interneurons within the ganglion are maintained.

These experimental results on cells 3 and 1 complement earlier results on AE and AP cells mentioned above, and show that neurons that innervate peripheral targets, as do the motor neurons, respond in different ways to being deprived of their targets. Obtaining mechanistic explanations for this range of responses will require knowledge of the genetic programs expressed normally by each of these identified cells, how they change when they contact their targets and when these contacts are abolished. Efforts along these lines are considered below.

9.3.7   Development of Neuronal Circuits and Behaviors

The accessibility of the neuronal circuitry responsible for generating specific behaviors is one of the great advantages of *H. medicinalis* as a model system for studying behavioral mechanisms (for a recent review, see Kristan et al. 2005). This advantage also extends to the study of the genesis of behaviors and the development of the corresponding neuronal circuitry, an important area that has been fruitfully addressed by French, Kristan and collaborators (Kristan et al. 2000 and earlier work referenced therein; French et al. 2005; Marin-Burgin et al. 2005, 2006). Spontaneous and evoked behaviors become evident in leech embryos in a particular order, with some occurring transiently in embryonic stages, while others continuing throughout life (Reynolds et al. 1998). Recent reports have examined the development of two important adult behaviors, local bending and swimming.

Local bending is a behavioral response to a localized mechanosensory stimulus consisting of the local shortening of the stimulated region, about one segment wide, resulting in a movement away from the stimulus. Neurons responsible for generating this response were first identified in adults, and consist of three layers in each ganglion: a sensory layer comprising four pressure-sensitive cells (Ps), a complement of ~17 interneurons [local bend interneurons (LBIs)] that receive inputs from Ps, and ~16 pairs of excitatory motor neurons (eMNs) and iMNs that receive inputs from LBIs, and are also interconnected via electrotonic as well as inhibitory chemical synapses (Reynolds et al. 1998; see Fig. 9.4). The iMNs, which are GABAergic (Cline et al. 1985), synapse centrally onto an antago-

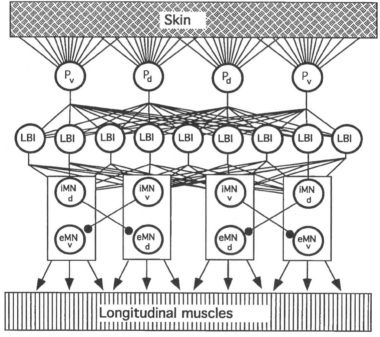

Figure 9.4 Diagram of the segmental circuitry responsible for local bending. A total of four pressure-sensitive mechanoreceptors (P cells) innervate the circumference of the body wall in overlapping receptive fields. The receptive field of each P cell is indicated by a subscript; P_D innervates the dorsal body wall, P_V the ventral. All four P cells excite all 17 identified LBIs in a graded manner (9 of the 17 identified LBIs are shown). LBIs excite the eMNs and iMNs in such a way that P-cell activation most strongly excites eMNs innervating the same region of the body wall. For example, P_D cells activate eMNd and P_V cells activate eMNv. Activating a P cell generates a local bend on the side of the animal that was touched. Motor neurons with the same input (e.g. iMNd and eMNv) are enclosed in the same box. Connections indicated by lines are excitatory; connections indicated by lines ending in closed circles are inhibitory.

Source: Modified from Kristan et al. 2000, with permission from Elsevier Limited.

nistic set of eMNs (e.g. dorsal iMNs onto ventral eMNs) and peripherally onto the muscle cells innervated by this set of eMNs.[1]

During embryogenesis, local bending appears to supplant a similar but bilaterally symmetric embryonic behavior (i.e. circumferential indentation), a change that is temporally correlated with the establishment of chemical synapses. As in many other developing nervous systems, electrical connections are established among sets of neurons before chemical connections; in leech embryos cultured under standard conditions (~23 °C; Kristan et al. 2000), this delay is 1–2 days. Since electrical connections are necessarily excitatory, circumferential indentation might be elicited by a localized mechanosensory stimulus initiating an excitatory response in one P cell, which then spreads excitation bilaterally across the LBIs and MNs. These considerations led Kristan et al. (2000) to the hypothesis that the delayed establishment of inhibitory connections from iMNs to eMNs, which allow the animal to contract on one side while relaxing the opposite side, might be responsible for the disappearance of circumferential indentation and the emergence of local bending (Cacciatore et al. 2000).

This hypothesis was tested by Marin-Burgin et al. (2005) in an elegant set of experiments. To assay whether GABAergic synaptic inhibition was responsible for the switch to local bending, they applied a GABA blocker (bicuculline methiodide) to older embryos that responded to a standard stimulus with local bending, and observed that afterwards, the same stimulus elicited only circumferential indentation. Using a voltage-sensitive dye to detect membrane potentials and intracellular microelectrodes to measure or elicit electrical activity, they studied the electrical activity of neuron populations in individual ganglia from early embryos to adults. Their observations demonstrated that electrical connections were present at early stages, and that at these stages (before the onset of local bending), stimulating a single neuron produced only in-phase neuronal activity in a set of follower neurons on both sides of the ganglion. At later embryonic stages (after the onset of local bending) and in adult ganglia, the follower neurons divided into two groups, with one group (located mostly ipsilateral to the driven neuron) showing in-phase activity, and the other group (bilateral) displaying out-of-phase responses (phase difference more than 90°). Data from these experiments lead to two important conclusions: first, electrical synapses are present early and produce a bilateral circumferential indentation, and second, synaptic inhibition appears later, changing the existing synaptic network and producing the unilateral adult local bending behavior (Marin-Burgin et al. 2005).

In a subsequent paper, Marin-Burgin et al. (2006) reported on the onset of gap junctional coupling between embryonic leech neurons involved in the local bending behavior by injecting identified neurons in this circuit with a tracer, neurobiotin, which passes through leech GJs. By injecting the tracer at different developmental stages, they were able to determine an invariant sequence in the formation of GJs within and among the layers of neurons that form the local bending circuit. The first evidence of dye coupling occurred among the MNs, followed after a 1 day or longer interval by dye coupling of the P cells to the LBIs, and of the LBIs to the MNs. Unlike the case for MNs, however, no dye passage was observed from a P to other Ps, or from an LBI to other LBIs. Furthermore, dye passed from a P cell to LBIs, but not from an LBI to Ps, and from an LBI to MNs,

---

[1] The synaptic connections of the LB reflex are more complex than this simplified description presents. In fact, every P cell excites every LBI and every LBI excites every MN, but the sizes of the postsynaptic responses vary greatly among the connections made by each cell, with some of the connections being much stronger than the rest for each cell. This pattern of distributed and variable-strength connectivity can be modeled with a population coding algorithm, which determines the motor output that will produce the appropriate direction of the response.

but not in the opposite direction. The order in which the LB circuit neurons become coupled shows an interesting correlation with the order in which behaviors are first observed. As in many other systems, the first movements that a leech embryo makes are spontaneous (they may be neurogenic or myogenic) and without any apparent response to mechanical stimuli, a property that is acquired later. Spontaneous movements start at about the time that the MN pool becomes electrically interconnected, and evoked responses appear when P cells and LBIs are added to the circuit (Marin-Burgin et al. 2006).

Marin-Burgin et al. (2006) also found that some electrical connections in the embryo were transient and were not observed in the adult nervous system. For example, dye injection into a young P cell shows dye passage to neuropil glia, a coupling that disappears when the secondary branches of the P cell grow in the neuropil. Some electrotonic connections between embryonic neurons (e.g. Ps to LBIs) were absent in the adult, but the majority persisted (e.g. LBIs to MNs). One reason for the transient nature of some electrical synapses may be their replacement by chemical junctions, which could occur in the transient P to LBI electrical connections. However, the transient coupling of the sensory cell to the very large neuropil glia is not a prelude to the formation of chemical synapses, and a different explanation is required. An interesting attribute of these two cell types is that they express different innexins from early developmental stages and into the adult (Dykes et al. 2004; Dykes and Macagno 2006). Ongoing RNAi experiments (A. Sanchez and E.R. Macagno, pers. comm.) that knock down innexin expression in ganglionic glia may yield some clues regarding the function of this transient neuron–glia connection.

In contrast to the segmental nature of the local bending behavior, swimming is a behavior that requires the coordination of muscular contractions over the whole body of the leech. The neuronal circuitry and the patterns of electrical activity that generate the traveling wave propelling the leech through water are well-known in adult leeches (Kristan et al. 2005), but only the kinematics of swimming behavior has been explored in embryonic stages (French et al. 2005). An interesting result of these studies is that embryos initially generate body shapes (e.g. ribbon-like flattening, full-body flexions) and movements that do not produce forward motion. With time, the body shape shifts toward that of the adult (one full wave) and forward progress increases (French et al. 2005). These observations provide the background for complementary studies on the development of the swimming neuronal circuitry underlying this complex behavior.

Complex behaviors like swimming involve the coordination of the full nervous system, including the larger ganglia at the head and the tail, which have not been studied as thoroughly as ganglia in the 21 body segments. The head ganglion, in particular, is thought to be a nexus for regulatory and modulatory functions, such as initiation or termination of complex activities (Cornford et al. 2006). It comprises a supraesophageal ring of nonmetameric origin, and a SubEG comprising four bilateral neuromeres that fuse early in development (i.e. they do not form connective nerves between them). To examine its functions, Cornford et al. (2006) fully or partially ablated the SubEG late in the first half of embryonic development, thus disconnecting the head ganglion from the rest of the nerve cord. Assays of 17 different behaviors were carried out a few weeks later in young juveniles. Behavioral changes readily observed included a much larger fraction of time spent either being inactive or swimming spontaneously. This was at the expense of other activities, such as crawling and probing, which became less frequent. From these results, they concluded that the head ganglion normally provides both general excitation to the rest of the CNS and selective inhibition of some behaviors. They also assayed the relative frequency of responses elicited by gentle touch of the skin in anterior, middle and posterior areas. The data showed a number of relatively small but significant changes in the frequencies of several behaviors with respect to controls. Interestingly, these were much smaller than the differences in the frequencies of behaviors elicited by touching different regions in the controls. The authors concluded that the head ganglion has an important role in the organization of the behavioral repertoire of the leech (Cornford et al. 2006).

9.3.8 Expression of Leech Innexins in Development

Intercellular communication through GJs is a hallmark of early developmental interactions that establish neuronal properties, particularly in the formation of synaptic circuits. Our ability to carry out mechanistic studies of the developmental functions of gap junctional communication in invertebrates has been greatly enhanced by the discovery that invertebrates use a different family of proteins, the innexins, to make GJs (Phelan et al. 1998). Like vertebrate connexins, the innexins form channels by assembling in cell membranes as six-monomer ring structures that couple with similar structures on an opposing membrane. Despite the lack of amino acid sequence similarity between families, both innexin and connexin proteins have four transmembrane domains, with both N and C termini within the cell. Physiologically, vertebrate and invertebrate GJs have similar properties, allowing the passage of ions as well as small signaling molecules such as nucleotides.

Cloning of leech innexins in *H. medicinalis* was first reported by Dykes et al. (2004), who used sequence homology with known innexins in other invertebrates to identify *Hm-inx1* and *Hm-inx2*, and demonstrated their expression in adult neurons and giant glia, respectively. Expression of *Hm-inx1* and *Hm-inx2* in frog oocytes showed that they can form characteristic GJs when two expressing oocytes were pushed together. More recently, Dykes and Macagno (2006) reported the cloning of 10 additional leech innexins, helped in part by the creation of *H. medicinalis* EST libraries (discussed below). They also reported the embryonic expression patterns of all 12 known leech innexins, obtained by *in situ* hybridization of whole-mounted embryos at different developmental stages. Most leech innexins were expressed in more than one tissue, with *Hm-inx1, Hm-inx5, Hm-inx6* and *Hm-inx8* expressed by central neurons and *Hm-inx2* and *Hm-inx3* expressed in the giant glia of ganglia and connective nerves. *Hm-inx1* mRNA was also detected in clusters of peripheral neurons, probably the sensilla. Ongoing studies by the Macagno group suggest at least four to six additional leech innexins. Thus, the probable number of *H. medicinalis* innexins falls within the range defined by *Drosophila melanogaster* (eight innexins) and *Caenorhabditis elegans* (25 innexins), numbers derived from complete genome sequences. An accurate count of innexin genes for the leech *Helobdella robusta* will be available soon from the completed sequencing of its genome.

The availability of cloned leech innexin genes opens the possibility of assaying function by silencing (using RNAi) or by ectopic expression (using plasmids carrying innexin genes). The ability to modulate gene expression using these approaches in single identified neurons in an embryo that is otherwise normal is perhaps one of the important experimental advantages of the leech, as discussed below.

9.4 NEURONAL REGENERATION AND REPAIR

An important property of leeches is their capacity to regenerate neurites and synaptic connections in the adult CNS. Neurites that have been damaged or severed can sprout, establish de novo growth cones, and extend and reconnect specifically with normal targets (Nicholls and Hernandez 1989). Possible explanations for this useful attribute include a continued presence of embryonic factors that are required for neuronal growth and maturation, along with the ability to induce expression or repression of critical factors in response to signals released by damaged tissues. Preliminary molecular analyses of changes in gene expression modulation provoked by damage are considered below. Another aspect that may be important is that leech central neurons continue to expand their central and peripheral dendritic and terminal arbors throughout the life of the animal, suggesting that the machinery for growth and addition of synaptic coupling may never be turned off completely in this invertebrate group.

In mammals, by contrast, not only are many embryonic growth-promoting molecules and their receptors apparently no longer present in the adult, but the adult CNS produces various growth-inhibiting molecules that are not present in the embryo or neonate (Koeberle and Bahr 2004). The limited level of neuronal regeneration following trauma in the adult mammalian CNS has made it difficult to assay possible mechanisms for successful axonal regeneration and whether the possibility exists for restoring specific, functional synaptic connections. The relatively recent demonstration of adult neurogenesis in some restricted brain regions and the incorporation of new neurons into existing circuits, however, may permit an entry point into repair mechanisms that might be accessed therapeutically. Information gained from model systems such as the leech CNS could help focus on what mechanisms should be sought and explored.

Several recent reports examine important aspects of neural regeneration in the adult leech (for reviews of relevant earlier work, see (Drapeau et al. 1995; Fernandez-de-Miguel and Drapeau 1995). Since neurons, glia and even microglia in leech resemble their mammalian counterparts in their physiological and molecular properties, and in their assembly as networks during development, we may expect that this new knowledge of regeneration in leech will bear upon the important question of how successful central regeneration might be produced in the mammalian CNS (Duan et al. 2005).

9.4.1 Recovery of Learning Following Regeneration of a Specific Synapse

Neuronal regeneration leading to full recovery of normal function requires not just reconnection to targets but also reestablishment of complex behaviors mediated by regenerated parts. This has been tested at the single cell level by Burrell et al. (2003) in an elegant experiment. A leech escape reflex—body shortening—can be modified by nonassociative conditioning, one result of which is sensitization, an enhancement of the response. The S-cell system is critical for this simple form of learning, and capacity for sensitization is lost when a single S interneuron is ablated or the nerve carrying the S-cell axon is cut (Sahley et al. 1994; Modney et al. 1997), although the capacity for the reflex itself continues (Shaw and Kristan 1999). Instead of cell or nerve ablation, Burrell et al. (2003) cut a single S-cell axon with a laser without disrupting other axons in the nerve. They observed that the experimental preparations initially failed to show sensitization of the shortening reflex. By ~4 weeks postaxotomy, however, full recovery of sensitization was achieved in those preparations that had regenerated the S-cell axons and had reestablished synaptic contacts with neighboring S cells. Interestingly, full sensitization does not occur immediately upon synapse regeneration, which can occur in 2–3 weeks, suggesting that it must require additional changes. One likely possibility is that axotomy may downregulate some factor in the S cell that can only be restored once the cell has recovered its synaptic connections with other S cells (Burrell et al. 2003).

9.4.2 Response to Nerve Injury: Nitric Oxide (NO) Efflux and Migration of Microglial Cells

Early stages of leech CNS regeneration following a mechanical lesion are characterized by two events that appear to be crucial for successful repair: one is the increased activity of epithelial nitric oxide synthase (NOS) in the area of the lesion and the generation and diffusion of NO, and the second is the induced migration of microglia toward, and their accumulation at, the injury site (McGlade-McCulloh and Muller 1989; Shafer et al. 1998). Further analyses of these phenomena have been reported in the past few years and are summarized below (Duan et al. 2003, 2005).

To obtain an accurate measurement of the dynamics of NO generated following a lesion, Kumar et al. (2001) used standard citrulline assays to show that NO is generated within 30 min after the nerve cord has been injured. To obtain higher temporal and spatial resolution, they developed a polarographic NO-selective microsensor. With this probe, they were able to detect, immediately following a nerve crush, significant efflux of NO from the lesion site. This efflux peaked within minutes after the crush and then decreased exponentially (time constant ~120 s) to a steady, low value known to follow injury. Inhibition of NOS activity significantly reduced the magnitude of NO efflux, indicating that NO efflux was indeed a result of the injury. The significance of the transient peak of NO is unknown, but Kumar et al. (2001) suggest that transient NO diffusing away from the lesion might activate microglia and enable their migratory behavior.

To assay directly for a role of NO on microglial accumulation at the injury site, Chen et al. (2000) modulated NO levels in several ways. As demonstrated by NOS immunoreactivity, a large increase in NOS occurs at the crush site within 5 min of injury, and this high level persists for at least 24 h. Microglial accumulation at the lesion, however, is not detectable at 5 min but is quite strong after a few hours, and peaks at ~24 h. Inhibition of NO synthesis by the prior application of the NOS inhibitor N^G-nitro-L-arginine methyl ester (L-NAME) effectively blocks microglial accumulation, while the presence of its inactive enantiomer N^G-nitro-D-arginine methyl ester (D-NAME) has little or no effect. Interestingly, increasing NO levels with the NO donor spermine NONOate (SPNO) also inhibits accumulation of microglia at the crush, but not in the presence of the NO scavenger carboxy-2-phenyl-4,4,5,5-tetramethyl-imidazoline-1-oxyl-3-oxide (cPTIO). Examination of microglial kinetics in living nerve cords shows that the effect of SPNO application occurs by the reduction of average microglial migratory speeds, even to no movement. Thus, NO is clearly implicated as a modulator of microglial movement, and indeed appears to function as a stop signal at high levels, leading to higher densities of these cells at the injury site.

What might be the mechanisms by which NO affects microglial motility? One possibility is suggested by the observation that damage also induces an increase in cGMP immunoreactivity at the lesion, in a pattern coextensive with that of NOS immunoreactivity and with microglial cell accumulation, as reported by Duan et al. (2003). Blocking the increase in cGMP at the lesion with the guanylate cyclase inhibitor methylene blue (MB) also abolishes microglial migration and accumulation at the injury site, suggesting a possible role in the leech's NO signaling pathway for this nucleotide, as demonstrated in other systems. Interestingly, time-lapse analysis of microglial movement in living nerve cords shows that while MB does not affect microglial speeds, it does affect direction, with significantly more cells moving away from the lesion or reversing direction, and fewer cells moving toward the lesion than is observed following a nerve crush. Thus, assuming cGMP is downstream of NO, these data support a model that has NO produced at a lesion affecting the direction of microglial migration and stopping microglial movement, both of which are effects required for microglia to accumulate at the injury site (Chen et al. 2000; Duan et al. 2003). A remaining uncertainty is the role, if any, of NO production by the microglia themselves. Answering this and other mechanistic questions regarding microglial function in regeneration of the nervous system will require attaining a more complete understanding of the molecular response to injury, an important objective currently being pursued (see below).

9.4.3 Identification of Genes Regulated in Response to CNS Injury

Over the past decade, Blackshaw and collaborators have implemented a differential screening strategy to assay for changes in gene expression that accompany neuronal regeneration in leech (Korneev et al. 1997; Emes et al. 2003; Blackshaw et al. 2004; Wang et al. 2005). Their approach is based on the use of subtractive probes, constructed by hybridizing cDNAs from regenerating and nonregenerating central ganglia, and selecting those sequences enriched either in the regenerating sample (upregulated genes) or in the nonregenerating sample (downregulated genes). These probes were then used to screen cDNA libraries constructed from whole leech CNS or from identified microdissected neurons (i.e. Retzius cells, see Fig. 9.5; Blackshaw et al. 2004; Wang et al. 2005). Among sequences upregulated 24 h following axotomy are leech homologues of mammalian genes with established functions, such as actin, tubulin and protein 4.1, thioredoxin (TRX), rough endoplasmic reticulum protein 1 (RER-1) and the neuron-specific protein synapsin. Other sequences, like the cysteine-rich intestinal protein (CRIP), have been shown to be expressed in developing mammalian intestinal cells but not in adult regenerating nerve cells (Blackshaw et al. 2004). Other genes regulated by injury in leech have counterparts in the mammalian genome but are not known in regeneration processes. Two additional genes, myohemerythrin (Vergote et al. 2004) and the novel protein ReN3, are expressed exclusively in invertebrates. Still other regulated genes have no known homologues in vertebrate

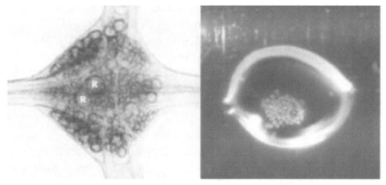

Figure 9.5 Whole ganglia from the leech ventral nerve cord and microdissected neuron cell bodies from which cDNA libraries were constructed. (A) An isolated ganglion contains ~400 nerve cells including primary sensory and motor neurons, modulatory neurons and interneurons, macroglia and microglia, cells comprising the external connective tissue sheath, and the internal capsule, a structure that separates cell bodies from the central neuropile. (B) Isolated Retzius cell bodies microdissected from successive ganglia in the ventral nerve cord, collected into a drop of fluid on the wall of an Eppendorf tube, and used to construct a Retzius neuron cDNA library. Abbreviation: R = Retzius cells.

Source: Modified from Blackshaw et al. 2004, with permission from Blackwell Publishing, Ltd.

genomes, and these invertebrate-specific sequences are interesting in view of the different capacity for CNS repair in invertebrates such as the leech (Blackshaw et al. 2004). To investigate the key role of such genes in regeneration, knockdown studies such as siRNA have to be undertaken. Single cell studies employing siRNA for protein 4.1 have demonstrated their feasibility in leech (Wang et al. 2005). Further analyses, particularly those examining the time course of injury-evoked changes in gene expression, have the potential to yield interesting information at a systems level on the genetic programs underlying nervous system repair.

As important as these studies of modulation of gene expression at the mRNA level are, so are studies of the changes in protein levels and protein modifications, as some of the critical effects of injury may be in the regulation of translation and posttranslational processing. Ongoing proteomic studies of the leech CNS have shown that some molecules detected by transcriptomic approaches, e.g. cytoskeletal and metabolic proteins, foldases, calcium sensors, kinases and neurohemerythrin (reflecting specific cytoskeletal rearrangements linked to cell migration), vesicular trafficking as well as the modulation of synaptic activity, are also observed in excised adult ganglia challenged by bacterial toxins (Vergote et al. 2006; see below). *In vitro* studies will add an important dimension, and the demonstration that 3-D culture of adult and embryonic tissues in collagen gels supports the regenerative outgrowth of axon tracts as well as the migration of microglial cells (Babington et al. 2005) is a useful advance. The *in vitro* approach will offer the possibility of assaying the nature of soluble factors produced in the environment of regenerating tissue using mass spectrometry strategies.

Molecular studies of leech CNS regeneration are still at an early stage but should advance rapidly in the near future. Transcriptomic, proteomic and functional studies should provide the basic knowledge required for a mechanistic understanding of the integrative physiological processes involved in leech CNS regeneration, from which we expect to obtain insights into such phenomena in mammals.

9.5 NEUROIMMUNE RESPONSES

Perhaps reflecting their long lives (more than 10 years for some species), leeches have evolved the capacity to respond to microbial infections not only by peripheral tissues but also by the CNS. For example, a differential display analysis with 2-D gel electrophoresis coupled to mass spectrometry yielded evidence that the leech CNS responds to bacterial infection by modulating the expression of at least 16 proteins. These proteins appear between 1 and 24 h after bacterial challenge and have been assigned to the immune response because they are not induced by exposure to control sterile medium (Table 9.1). These immune response-induced proteins include cytoskeletal and metabolic proteins, foldases, calcium sensors, kinases and neurohemerythrin, reflecting specific cytoskeletal rearrangements linked to cell migration, vesicular trafficking and/or phagocytosis, as well as the modulation of synaptic activity. Interestingly, several of these upregulated proteins, such as gliarin (Xu et al. 1999) and neurohemerythrin (Fig. 9.6; Vergote et al. 2004), are expressed specifically in glial and microglial cells, suggesting a key role for these cells in the immune response of the leech nervous system, similar to what has been observed in vertebrates (Vergote et al. 2006). Gliarin upregulation, in particular, could thus serve as a new marker of proliferation and maturation of leech glial cells, its upregulation reflecting glial activation in response to immune challenge. The accompanying cytoskeletal rearrangements might result from morphological changes associated with phagocytosis or cell migration. Indeed, leech microglial cells are able to migrate within the ganglionic chain

Table 9.1   List of proteins modulated after 1, 6, 12 or 24h of bacterial challenge in the leech nervous system 42

Spot Number	MM (kDa)	pI	Access Number[a]	Protein Family Identification	Bacteria/control[b] Spot Representativity			
					1 h	6 h	12 h	24 h
1,049	55–60	5.0–5.2	AJ005808	PDI	—	—	—	1.82**
1,155	50–55	6.0	AAN69196	Aminoacyl dehydrogenase	—	0.63*	—	0.60*
1,164	50	5.1–5.2	Q25117	β Subunit ATP synthase	2.10*	—	—	—
1,528	36–38	4.8	P42637	Tropomyosin	0.62**	1.89***	—	0.48***
1,585	35–37	4.8–4.9	Q9U0U9	Acetyl transferase	0.29***	3.53***	—	—
1,749	33–36	6.9–7.0	NP_492651	NCS	app	—	—	dis
1,759	33–35	5.9	AAD29248	Gliarin	2.11*	—	—	—
2,220	22–24	6.8–6.9	U92087	Cyclophilin	—	1.65*	—	—
2,802	14–15	5.6	AY521548	Neurohemerythrin	—	1.40*	—	—
1,726	33–36	5.8–5.9	—	Unknown	—	3.43**	—	—
1,917	28–32	7.0	—	Unknown	—	1.81*	—	—
1,932	28–32	6.8–6.9	—	Unknown	1.81*	—	—	—
2,090	25	7.5–8.0	—	Unknown	—	—	0.27***	—
2,144	23–25	5.8	—	Unknown	—	1.86***	—	—
2,256	21–23	5.3	—	Unknown	—	—	2.32***	—

Spot numbers, experimental molecular masses (MM) and pI, protein identifications and modulation ratios of spot between bacterial-challenged and control leech nervous system are indicated.

[a]NCBI GenBank accession numbers.

[b]Ratio of bacterial to control conditions.

*$p < 0.05$;

**$p < 0.01$;

***$p < 0.001$.

Abbreviations: — = no significant change; app = spot present in stimulated but absent in control nervous systems.

to a lesion site and to play a phagocytic role (Morgese et al. 1983). By contrast, the expected properties of neurohemerythrin have led to the proposal of several putative functions for this protein in the responses of leech nervous tissue: (1) a role as an oxygen supplier for metabolism; (2) a role as a trap for reactive oxygen species and NO, protecting cells from cell death; and (3) a role as an antibacterial factor depriving bacteria of iron (Deloffre et al. 2003).

Data from these initial studies of bacterial toxins show that the leech CNS is able to respond in an intrinsic manner to a septic stimulus, mounting a "neuroimmune" response. Careful study of identified proteins will be essential to understand the mechanisms involved, but the role of some candidates can be proposed based on the proteins identified thus far: (1) cytoskeletal rearrangements potentially responsible for morphological changes, cell migration, vesicular trafficking and/or phagocytosis; (2) modulation of synaptic activity; (3) calcium signaling; and (4) unfolded protein response controlling the functionality of proteins affected by stress-generated sepsis. The involvement in innate immunity of some proteins or protein families has been described in transcriptomic studies, but at the peripheral level and not within the nervous system (Table 9.2; Lefebvre and Salzet 2003). Moreover, the protein families involved in the immune response of the medicinal leech nervous system appear to also be involved in nerve regeneration, as shown by Blackshaw et al. (2004) in the medicinal leech (discussed above) and by Perlson et al. (2004) in *Lymnaea stagnalis* (Table 9.2).

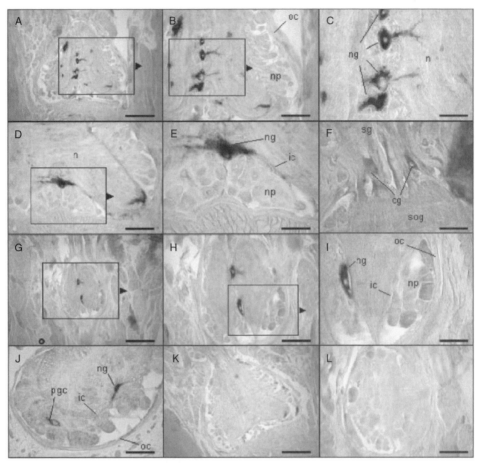

Figure 9.6 Detection of hemerythrin message in the CNS of *Hirudo medicinalis*. Sections were hybridized with digoxygenin-labeled RNA probes, followed by immunodetection. (A–E) Antisense hemerythrin probe labeling in head ganglia. (F) Antisense hemerythrin probe labeling in connectives between segmental ganglia. (G–J) Antisense hemerythrin probe labeling in segmental ganglia. (K) Sense hemerythrin probe labeling in head ganglion. (L) Sense hemerythrin probe labeling in segmental ganglia. Scale bars = 40 µm (A, G and K), 20 µm (B, D, F, J and L) and 10 µm (C, E and I). Abbreviations: cg = connective glia; ic = inner capsule; n = neuropile; ng = neuropile glia; np = neuronal packet; oc = outer capsule; pgc = packet glial cell; sg = segmental ganglia; sog = suboesophageal ganglion.

Source: From Vergote et al. 2004, with permission from the American Society for Biochemistry and Molecular Biology.

These observations suggest a certain parallelism between CNS defense mechanisms and CNS regeneration, suggesting the involvement of neuroinflammation in such phenomena. Modulation of neuroinflammation after a crush seems necessary but is highly regulated in the CNS during regeneration. Involvement of molecules implicated in NO regulation as well as some immunomodulators (including lipids) may be an essential part of this biological process.

Table 9.2 Similarities between proteins involved in innate immunity within invertebrate and those involved in CNS regeneration within Lophotrochozoa

Immunity		Leech CNS	Leech Periphery	Insect Periphery
Cytoskeletal	Microfilaments	Tropomyosin	Tropomyosin-2	Myosin II regulatory light-chain
	IF	Gliarin	Actin-2	Actin 5C
Calcium	Calcium sensor	NCS-2/neurocalcin		Calmodulin Sarcoplasmic CaBP1
Metabolism	AA/nt metabolism	AA dehydrogenase	Aldehyde dehydrogenase	Aldehyde dehydrogenase
	Energy	ATP synthase β subunit		ATP synthase β subunit
	Others	Acetyl transferase		
Hsp and chaperones	Cyclophilin/PPI	Cyclophilin	PPI	FK506-BP-PPI, cyclophilin
	PDI	PDI		PDI (Erp60), CaBP1-PDI
Metal oxidation	Respiratory molecule	Neurohemerythrin		Hemocyanin
	Others			TRX, transferrin, ferritin

Regeneration		Leech CNS		Mollusk CNS
Cytoskeletal	Microfilaments	Protein 4.1		Tropomyosin
	IF	Synapsin		Intermediate filament
	Microtubules	α- and β-tubulin		Tubulin
Calcium	Calcium sensor	Calmodulin-like		Calmodulin, calbindin
	Others			Calpain
Metabolism	AA/nt metabolism			Glutamine synthase
	Energy	ATP synthase inhibitor		ATP synthase
Hsp and Chaperones	Cyclophilin/PPI			Cyclophilin
	PDI			PDI
	Others	Hsp90		Hsp60, 14-3-3
Metal oxidation	Respiratory molecule	Myohemerythrin		
	Others	COX I		Peroxiredoxin, ferritin

All are identified by proteomic or transcriptomic approaches.

9.6 CELLULAR AND HUMORAL IMMUNE MECHANISMS: A LEECH INNATE IMMUNE RESPONSE

Since a coelom is absent in most hirudinid species, the cellular immune response has been less investigated compared to the humoral response. However, differential CD [CD11a+, CD45RA+, CD45RO+, CDw49b+, CD54+, β_2-m+ and Thy-1+ (CD90)] staining at different time points following grafting experiments indicate the presence of two distinct cell types, namely, natural killer (NK)-like and CD8+ cells (generally classified as cytotoxic lymphocytes in vertebrates) in the medicinal leech (de Eguileor et al. 2000a, b). These coelomocytes are selectively stimulated by different antigens (de Eguileor et al. 2000b) and are involved in inflammatory responses (de Eguileor et al. 2000a) as well as in angiogenesis (Tettamanti et al. 2003). Like other annelids, leech leukocytes phagocytose and encapsulate foreign material selectively, depending upon size—wounding in leeches is accompanied by "leucopoiesis," as it is in vertebrates (de Eguileor et al. 2003).

A coelom is present in the glossiphoniid leech *T. tessulatum*, and three types of circulating immune cells have been detected, one of them similar to macrophages in their ability to phagocytose foreign factors (Lefebvre and Salzet 2003). A gene tandem cystatin B (*Tt-cysb*)–cathepsin L (*Tt-catl*) that belongs to the cysteine protease inhibitor family has been identified in *T. tessulatum* (Lefebvre et al. 2004), and is the first cystatin B protein isolated in invertebrates. *Tt-catl* displays 68% sequence identity with *D. melanogaster* cathepsin L, and is only expressed in one of the circulating coelomic cell populations. Bacterial challenge leads to an increase of the *Tt-cysb* transcript exclusively in these cells. *Tt-catl* is expressed in the same cells, suggesting that both *Tt*-cyst and *Tt-catl* gene products might play a role in immune regulation through these circulating coelomic cells. These data are in line with observations in mammals and *D. melanogaster*, in which cathepsin L is present in small granules within hemocytes and might play a role in phagocytic events (Tryselius and Hultmark 1997; Lefebvre et al. 2004).

Considering the humoral response, three antimicrobial peptides have been isolated recently from the coelomic fluid of *T. tessulatum*: theromacin, theromyzin and peptide B (Tasiemski et al. 2000a, b; Salzet 2001b, 2002a; Tasiemski et al. 2003). Theromacin and theromyzin are processed from a larger precursor comprising a signal peptide directly followed by the active peptide. Interestingly, peptide B is processed from the C terminal part of proenkephalin A (PEA), the major precursor of enkephalins. Peptide B and theromyzin possess bacteriostatic activity, in contrast with theromacin, which exhibits potent bactericidal activity toward gram-positive bacteria. Theromyzin is a histidine-rich peptide sharing a His-x-x-Asn divalent ion active site, similar to histatin in vertebrates (Oppenheim et al. 1988). Analysis of theromacin and theromyzin gene expression demonstrated both a developmental and a septic injury induction in large fat cells (LFCs), which constitute a specific tissue that seems to be a functional equivalent of the insect fat body. Both peptides seem to exert their antimicrobial activity by a systemic action. All these data suggest the presence of an innate immune response in a lophotrochozoan comparable to the antibacterial response of an ecdysozoan (e.g. *D. melanogaster*). Consequently, *T. tessulatum* is an original invertebrate model that has developed two modes of fighting infections by antimicrobial peptides: (1) storage of antibacterial peptide derived from PEA, particularly in coelomocytes and/or in nervous system, and release of the peptide into the coelomic fluid after immune challenge; and (2) induction after septic injury of genes coding for more classical antimicrobial peptides, mainly in LFC, and rapid release of antibiotic peptides into the body fluid (Fig. 9.7).

Besides these antimicrobial peptides, an intrinsic lysozyme-like activity was demonstrated for destabilase (Baskova and Zavalova 2001; Baskova et al. 2001). Several isoforms of destabilase constitute a protein family with at least two members characterized by lysozyme activity (Baskova and Zavalova 2001; Baskova et al. 2001). The corresponding gene family implies an ancient evolutionary history of the genes, although the function of various lysozymes in leech remains unclear. Differences in primary structures of the destabilase family members and members of known lysozyme families allow one to assign the former to a new family of lysozymes. New proteins homologous to destabilase have

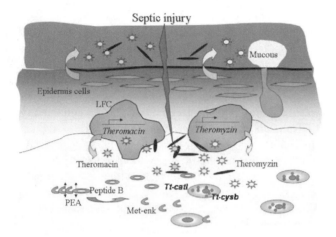

Figure 9.7   Schematic of putative innate immune response of leeches involving antibacterial peptides. Septic injury provokes an important production of mucous that would trap bacteria in the external environment of the leech; bacteria would be killed by theromacin and theromyzin present in the mucous. At the same time, injury would induce theromacin and theromyzin gene expression and the secretion of gene products from the LFC into the body fluid (Tasiemski et al. 2003). Thus, theromacin and theromyzin would exert their antimicrobial property through a systemic action. Moreover, antibacterial peptide (peptide B) as well as immune activator (Met-enk) derived from PEA, particularly in coelomocytes and/or in nervous system, are released into the coelomic fluid after immune challenge (Tasiemski et al. 2000). This would lead, in conjunction with the phagocytic action of coelomocytes containing *Tt-cysb* and *Tt-catl*, to the microbial killing in the body fluid. Abbreviation: Met-enk = methionine enkephalin. (See color insert.)

recently been described in *C. elegans* and bivalve mollusks, suggesting that the new lysozyme family is widely distributed among invertebrates (Baskova and Zavalova 2001; Baskova et al. 2001). Another protein sharing a bacteriostatic activity and belonging to the hemerythrin family has been found in fat cells (Coutte et al. 2001) as well as in glial cells of leeches (Vergote et al. 2004).

In sum, the leech innate immune response is not restricted to a systemic response but is also central, as in vertebrates. The nervous system controls all systems involved in the organism's homeostasis, including immunity. Circulating levels of proinflammatory cytokines (tumor necrosis factor-α, interleukin-1β and interleukin-6) that increase upon systemic endotoxemia have been detected in vertebrates. Systemic inflammation is detected by brain areas directly in contact with vascular vessels, such as circumventricular organs, and then spreads to other brain areas. Furthermore, without having direct access to the brain parenchyma, pathogens trigger an innate immune reaction throughout cerebral tissue leading to the concept of "neuroimmunity." This inflammation is followed by double-edged consequences. Inflammation in the brain leads to nuclear factor κB signaling and transcriptional activation of molecules that engage and control the innate immune response for pathogen elimination. Although most bacterial infections are outside of the brain, when bacteria penetrate the brain tissue, the release of toxic compounds (e.g. reactive oxygen species) induces necrosis and apoptosis of both neurons and glial cells. The data summarized here point out that an innate immune response occurs in the brain during pathogen infections and is a conserved physiological process across the Metazoa (e.g. leech, human).

9.7 CONCLUSIONS AND FUTURE DIRECTIONS

The studies we have reviewed in this chapter attest to the broad range of inquiry, from structure to behavior, and to the breadth of techniques currently employed to study the genesis, repair and maintenance of the leech nervous system. Clearly, the increasing application of biochemical and molecular genetic tools is beginning to yield insights into the nature of the molecular mechanisms responsible for these processes. However, while progress is being made, there is also a strong need to accelerate the implementation and application of genomic, transcriptomic and proteomic tools in leech. The leech nervous system has many important advantages, perhaps the most important being the ability to bridge the physiological properties of identified neurons to behavior, in the context of having detailed knowledge about all the neurons in the CNS and thus an unparalleled level of completeness. In most other invertebrate systems as well as in all vertebrate systems under current study, only a small fraction of their neuronal populations can be included in physiological models, because many of their neurons are inaccessible and only average population properties are attainable. Only the *C. elegans* CNS has been described in more detail structurally, but many of its neurons are not accessible to physiological tools. However, some other systems, including *C. elegans, D. melanogaster,* zebra fish and some rodents, greatly benefit from the possibility of conducting standard genetic analyses of neuronal properties and from the availability of their complete genome sequences (though this is about to change, since the complete genome sequence of the glossiphoniid leech *H. robusta* will soon be assembled and published). Nonetheless, the ability to conduct systems-level functional studies, the possibility of relating physiological programs to genetic programs that encode the underlying circuitry and its properties, the capacity of the system to regenerate and repair, and the availability of reverse genetic tools to study gene expression and regulation all strongly justify continuing and enhancing our efforts to understand neurogenesis, neuroregeneration and neuroimmune responses in leech.

In this review of recent studies of the leech CNS, we endeavored in particular to include molecular-level approaches, as we see this as a necessary direction for future advancement. The application of genomic, transcriptomic and proteomic tools in parallel with functional studies will open the door to a deeper understanding of integrative physiological processes. An effort in this direction is an ongoing collaboration with several laboratories that has resulted in the generation of a *H. medicinalis* transcriptomic database of ~130,000 EST sequences from two libraries, one from whole embryos and the second from adult nerve cords. Clustering the raw data yields ~45,000 contigs and singletons, providing good though incomplete representation of the *H. medicinalis* transcriptome. The database has already provided access to specific genes of interest, for example, several new innexins (Dykes and Macagno 2006). About 400 EST clones, representing genes expected to be involved in neurogenesis, neuroregeneration and neuroimmune response, have been selected for use in an initial microarray experiment to assay for changes in gene expression under different experimental conditions, such as silencing specific genes with RNAi (Baker and Macagno 2000a). Finally, coupling transcriptomic studies with peptidomic ones using matrix-assisted laser desorption/ionization (MALDI) imaging (Lemaire et al. 2006) will yield a more complete understanding of molecular aspects and mechanisms that participate in neurogenesis and neuroregeneration. For example, an ongoing collaboration between our laboratories is exploring the possibility of obtaining peptide/protein distributions for whole leech embryos, opened and fixed flat but otherwise with tissues in their normal relative locations (Fig. 9.8). With this systems-level approach, one can select

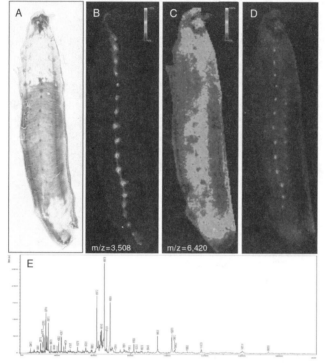

Figure 9.8   Matrix-assisted laser desorption/ionization time-of-flight (MALDI-TOF) imaging of an opened
E12 leech embryo in whole mount. Panel (A) shows a low-power image of a lightly fixed embryo before
covering with a molecular matrix required for this procedure. Mass spectra were obtained for ~20,000
locations in a raster covering the whole embryo. Panels (B) and (C) display distributions of peptides of mass/
charge (m/z) = 3,508 and 6,420 daltons, respectively, derived from mass spectra for each point in the image.
The peptide in (B) is most strongly present in the CNS, in segmental ganglia and interganglionic connectives,
while the peptide in (C) is more widely distributed but with maxima associated with the nephridiopores. An
example of the mass spectra recorded is shown in panel (E). Relative abundances are color coded [color bars,
upper right in panels (B) and (C)], with red–white being high and blue–black low. Panel (D) superposes the
images of panels (A) and (B), but is now coded in green (m/z = 3,508) and red (m/z = 6,420).
(See color insert.)

distributions of unknown molecules that are of particular interest (e.g. molecules expressed
in patterns that reflect the tiling of T-cell sensory arbors, as netrin does for P cells), before
proceeding to the arduous task of obtaining the amino acid sequence to identify and functionally characterize a particular protein. Applying these new technologies to leech will
make a critical contribution toward exploiting the advantages of this annelid system, and
will provide insights on the relationship between genetic and physiological programs
underlying the genesis, repair and maintenance of the leech CNS.

ACKNOWLEDGMENTS

Research in the authors' laboratories is supported by grants from the Centre National de
la Recherche Scientifique (CNRS), Ministère de L'Education Nationale, de L'Enseignement
Supérieur et de la Recherche, from the National Institutes of Health and from the National
Science Foundation. The authors would like to thank Genoscope for the adult *H. medicinalis* CNS EST sequencing.

REFERENCES

AISEMBERG, G.O., KUHN, J., MACAGNO, E.R. 2001. Netrin signal is produced in leech embryos by segmentally iterated sets of central neurons and longitudinal muscle cells. Dev. Genes Evol. 211, 589–596.

BABINGTON, E.J., VATANPARAST, J., VERRALL, J., BLACKSHAW, S.E. 2005. Three-dimensional culture of leech and snail ganglia for studies of neural repair. Invert. Neurosci. 5, 173–182.

BAKER, M.W., MACAGNO, E.R. 2000a. RNAi of the receptor tyrosine phosphatase HmLAR2 in a single cell of an intact leech embryo leads to growth-cone collapse. Curr. Biol. 10, 1071–1074.

BAKER, M.W., MACAGNO, E.R. 2000b. The role of a LAR-like receptor tyrosine phosphatase in growth cone collapse and mutual-avoidance by sibling processes. J. Neurobiol. 44, 194–203.

BAKER, M.W., MACAGNO, E.R. 2001. Neuronal growth and target recognition: lessons from the leech. Can. J. Zool. 79, 204–217.

BAKER, M.W., MACAGNO, E.R. 2006. Characterizations of *Hirudo medicinalis* DNA promoters for targeted gene expression. J. Neurosci. Methods 156, 145–153.

BAKER, M.W., MACAGNO, E.R. 2007. In vivo imaging of growth cone and filopodial dynamics: evidence for contact-mediated retraction of filopodia leading to the tiling of sibling processes. J. Comp. Neurol. 500, 850–862.

BAKER, M.W., RAUTH, S.J., MACAGNO, E.R. 2000. Possible role of the receptor protein tyrosine phosphatase HmLAR2 in interbranch repulsion in a leech embryonic cell. J. Neurobiol. 45, 47–60.

BAKER, M.W., KAUFFMAN, B., MACAGNO, E.R., ZIPSER, B. 2003. In vivo dynamics of CNS sensory arbor formation: a time-lapse study in the embryonic leech. J. Neurobiol. 56, 41–53.

BAKER, M.W., PETERSON, S.M., MACAGNO, E.R. 2008. The receptor phosphatase HmLAR2 sheds its ectodomain and collaborates with focal adhesion proteins in filopodial tips to control growth cone morphology. Dev. Biol. 215–225.

BAPTISTA, C.A., GERSHON, T.R., MACAGNO, E.R. 1990. Peripheral organs control central neurogenesis in the leech. Nature 346, 855–858.

BASKOVA, I.P., ZAVALOVA, L.L. 2001. Proteinase inhibitors from the medicinal leech *Hirudo medicinalis*. Biochemistry (Mosc) 66, 703–714.

BASKOVA, I.P., ZAVALOVA, L.L., BASANOVA, A.V., SASS, A.V. 2001. Separation of monomerizing and lysozyme activities of destabilase from medicinal leech salivary gland secretion. Biochemistry (Mosc) 66, 1368–1373.

BISWAS, S.C., DUTT, A., BAKER, M.W., MACAGNO, E.R. 2002. Association of LAR-like receptor protein tyrosine phosphatases with an enabled homolog in *Hirudo medicinalis*. Mol. Cell. Neurosci. 21, 657–670.

BLACKSHAW, S.E., NICHOLLS, J.G. 1995. Neurobiology and development of the leech. J. Neurobiol. 27, 267–276.

BLACKSHAW, S.E., BABINGTON, E.J., EMES, R.D., MALEK, J., WANG, W.-Z. 2004. Identifying genes for neuron survival and axon outgrowth in *Hirudo medicinalis*. J. Anat. 204, 13–24.

BRIGGMAN, K.L., KRISTAN, W.B., JR. 2006. Imaging dedicated and multifunctional neural circuits generating distinct behaviors. J. Neurosci. 26, 10925–10933.

BURRELL, B.D., SAHLEY, C.L., MULLER, K.J. 2003. Progressive recovery of learning during regeneration of a single synapse in the medicinal leech. J. Comp. Neurol. 457, 67–74.

CACCIATORE, T.W., ROZENSHTEYN, R., KRISTAN, W.B., JR. 2000. Kinematics and modeling of leech crawling: evidence for an oscillatory behavior produced by propagating waves of excitation. J. Neurosci. 20, 1643–1655.

CHEN, A., KUMAR, S.M., SAHLEY, C.L., MULLER, K.J. 2000. Nitric oxide influences injury-induced microglial migration and accumulation in the leech CNS. J. Neurosci. 20, 1036–1043.

CLINE, H.T., NUSBAUM, M.P., KRISTAN, W.B., JR. 1985. Identified GABAergic inhibitory motor neurons in the leech central nervous system take up GABA. Brain Res. 348, 359–362.

CORNFORD, A., KRISTAN, W.B., III, MALNOVE, S., KRISTAN, W.B., JR., FRENCH, K.A. 2006. Functions of the subesophageal ganglion in the medicinal leech revealed by ablation of neuromeres in embryos. J. Exp. Biol. 209, 493–503.

COUTTE, L., SLOMIANNY, M.C., MALECHA, J., BAERT, J.L. 2001. Cloning and expression analysis of a cDNA that encodes a leech hemerythrin. Biochim. Biophys. Acta 1518, 282–286.

DELOFFRE, L., SALZET, B., VIEAU, D., ANDRIES, J.C., SALZET, M. 2003. Antibacterial properties of hemerythrin of the sand worm *Nereis diversicolor*. Neuro Endocrinol. Lett. 24, 39–45.

DRAPEAU, P., CATARSI, S., MERZ, D.C. 1995. Signalling synapse formation between identified neurons. J. Physiol. Paris 89, 115–123.

DUAN, Y., HAUGABOOK, S.J., SAHLEY, C.L., MULLER, K.J. 2003. Methylene blue blocks cGMP production and disrupts directed migration of microglia to nerve lesions in the leech CNS. J. Neurobiol. 57, 183–192.

DUAN, Y., PANOFF, J., BURRELL, B.D., SAHLEY, C.L., MULLER, K.J. 2005. Repair and regeneration of functional synaptic connections: cellular and molecular interactions in the leech. Cell. Mol. Neurobiol. 25, 441–450.

DYKES, I.M., MACAGNO, E.R. 2006. Molecular characterization and embryonic expression of innexins in the leech *Hirudo medicinalis*. Dev. Genes Evol. 216, 185–197.

DYKES, I.M., FREEMAN, F.M., BACON, J.P., DAVIES, J.A. 2004. Molecular basis of gap junctional communication in the CNS of the leech *Hirudo medicinalis*. J. Neurosci. 24, 886–894.

DE EGUILEOR, M., GRIMALDI, A., TETTAMANTI, G., VALVASSORI, R., COOPER, E.L., LANZAVECCHIA, G. 2000a.

Different types of response to foreign antigens by leech leukocytes. Tissue Cell 32, 40–48.

DE EGUILEOR, M., GRIMALDI, A., TETTAMANTI, G., VALVASSORI, R., COOPER, E.L., LANZAVECCHIA, G. 2000b. Lipopolysaccharide-dependent induction of leech leukocytes that cross-react with vertebrate cellular differentiation markers. Tissue Cell 32, 437–445.

DE EGUILEOR, M., TETTAMANTI, G., GRIMALDI, A., CONGIU, T., FERRARESE, R., PERLETTI, G., VALVASSORI, R., COOPER, E.L., LANZAVECCHIA, G. 2003. Leeches: immune response, angiogenesis and biomedical applications. Curr. Pharm. Des. 9, 133–147.

EMES, R.D., WANG, W.Z., LANARY, K., BLACKSHAW, S.E. 2003. HmCRIP, a cysteine-rich intestinal protein, is expressed by an identified regenerating nerve cell. FEBS Lett. 533, 124–128.

FERNANDEZ-DE-MIGUEL, F., DRAPEAU, P. 1995. Synapse formation and function: insights from identified leech neurons in culture. J. Neurobiol. 27, 367–379.

FRENCH, K.A., KRISTAN, W.B., JR. 1994. Cell-cell interactions that modulate neuronal development in the leech. J. Neurobiol. 25, 640–651.

FRENCH, K.A., CHANG, J., REYNOLDS, S., GONZALEZ, R., KRISTAN, W.B., III, KRISTAN, W.B., JR. 2005. Development of swimming in the medicinal leech, the gradual acquisition of a behavior. J. Comp. Physiol. A Neuroethol. Sens. Neural Behav. Physiol. 191, 813–821.

GAN, W.B., MACAGNO, E.R. 1995a. Developing neurons use a putative pioneer's peripheral arbor to establish their terminal fields. J. Neurosci. 15, 3254–3262.

GAN, W.B., MACAGNO, E.R. 1995b. Interactions between segmental homologs and between isoneuronal branches guide the formation of sensory terminal fields. J. Neurosci. 15, 3243–3253.

GAN, W.B., WONG, V.Y., PHILLIPS, A., MA, C., GERSHON, T.R., MACAGNO, E.R. 1999. Cellular expression of a leech netrin suggests roles in the formation of longitudinal nerve tracts and in regional innervation of peripheral targets. J. Neurobiol. 40, 103–115.

GAO, W.Q., MACAGNO, E.R. 1988. Axon extension and retraction by leech neurons: severing early projections to peripheral targets prevents normal retraction of other projections. Neuron 1, 269–277.

GERSHON, T.R., BAKER, M.W., NITABACH, M., WU, P., MACAGNO, E.R. 1998. Two receptor tyrosine phosphatases of the LAR family are expressed in the developing leech by specific central neurons as well as select peripheral neurons, muscles, and other cells. J. Neurosci. 18, 2991–3002.

HUANG, Y., JELLIES, J., JOHANSEN, K.M., JOHANSEN, J. 1998. Development and pathway formation of peripheral neurons during leech embryogenesis. J. Comp. Neurol. 397, 394–402.

HUGHES, M.E., BORTNICK, R., TSUBOUCHI, A., BAUMER, P., KONDO, M., UEMURA, T., SCHMUCKER, D. 2007. Homophilic Dscam interactions control complex dendrite morphogenesis. Neuron 54, 417–427.

JELLIES, J., JOHANSEN, J. 1995. Multiple strategies for directed growth cone extension and navigation of peripheral neurons. J. Neurobiol. 27, 310–325.

JELLIES, J., KOPP, D.M., JOHANSEN, K.M., JOHANSEN, J. 1996. Initial formation and secondary condensation of nerve pathways in the medicinal leech. J. Comp. Neurol. 373, 1–10.

JELLIES, J., JOHANSEN, K.M., JOHANSEN, J. 2000. Ectopic CNS projections guide peripheral neuron axons along novel pathways in leech embryos. Dev. Biol. 218, 137–145.

JOHANSEN, K.M., KOPP, D.M., JELLIES, J., JOHANSEN, J. 1992. Tract formation and axon fasciculation of molecularly distinct peripheral neuron subpopulations during leech embryogenesis. Neuron 8, 559–572.

JOHNSON, L.A., KRISTAN, W.B., JELLIES, J., FRENCH, K.A. 2000. Disruption of peripheral target contact influences the development of identified central dendritic branches in a leech motor neuron in vivo. J. Neurobiol. 43, 365–378.

KOEBERLE, P.D., BAHR, M. 2004. Growth and guidance cues for regenerating axons: where have they gone? J. Neurobiol. 59, 162–180.

KORNEEV, S., FEDOROV, A., COLLINS, R., BLACKSHAW, S.E., DAVIES, J.A. 1997. A subtractive cDNA library from an identified regenerating neuron is enriched in sequences up-regulated during nerve regeneration. Invert. Neurosci. 3, 185–192.

KRISTAN, W.B., JR., EISENHART, F.J., JOHNSON, L.A., FRENCH, K.A. 2000. Development of neuronal circuits and behaviors in the medicinal leech. Brain Res. Bull. 53, 561–570.

KRISTAN, W.B., JR., CALABRESE, R.L., FRIESEN, W.O. 2005. Neuronal control of leech behavior. Prog. Neurobiol. 76, 279–327.

KUMAR, S.M., PORTERFIELD, D.M., MULLER, K.J., SMITH, P.J., SAHLEY, C.L. 2001. Nerve injury induces a rapid efflux of nitric oxide (NO) detected with a novel NO microsensor. J. Neurosci. 21, 215–220.

KUWADA, J.Y. 1985. Pioneering and pathfinding by an identified neuron in the embryonic leech. J. Embryol. Exp. Morphol. 86, 155–167.

LEFEBVRE, C., SALZET, M. 2003. Annelid neuroimmune system. Curr. Pharm. Des. 9, 149–158.

LEFEBVRE, C., COCQUERELLE, C., VANDENBULCKE, F., HOT, D., HUOT, L., LEMOINE, Y., SALZET, M. 2004. Transcriptomic analysis in the leech *Theromyzon tessulatum*: involvement of cystatin B in innate immunity. Biochem. J. 380, 617–625.

LEMAIRE, R., TABET, J.C., DUCOROY, P., HENDRA, J.B., SALZET, M., FOURNIER, I. 2006. Solid ionic matrixes for direct tissue analysis and MALDI imaging. Anal. Chem. 78, 809–819.

MACAGNO, E.R. 1980. Number and distribution of neurons in leech segmental ganglia. J. Comp. Neurol. 190, 283–302.

MCGLADE-MCCULLOH, E., MULLER, K.J. 1989. Developing axons continue to grow at their tip after synapsing with their appropriate target. Neuron 2, 1063–1068.

MARIN-BURGIN, A., EISENHART, F.J., BACA, S.M., KRISTAN, W.B., JR., FRENCH, K.A. 2005. Sequential development of electrical and chemical synaptic connections generates a specific behavioral circuit in the leech. J. Neurosci. 25, 2478–2489.

MARIN-BURGIN, A., EISENHART, F.J., KRISTAN, W.B., JR., FRENCH, K.A. 2006. Embryonic electrical connections appear to pre-figure a behavioral circuit in the leech CNS. J. Comp. Physiol. A Neuroethol. Sens. Neural. Behav. Physiol. 192, 123–133.

MATTHEWS, B.J., KIM, M.E., FLANAGAN, J.J., HATTORI, D., CLEMENS, J.C., ZIPURSKY, S.L., GRUEBER, W.B. 2007. Dendrite self-avoidance is controlled by Dscam. Cell 129, 593–604.

MODNEY, B.K., SAHLEY, C.L., MULLER, K.J. 1997. Regeneration of a central synapse restores nonassociative learning. J. Neurosci. 17, 6478–6482.

MORGESE, V.J., ELLIOTT, E.J., MULLER, K.J. 1983. Microglial movement to sites of nerve lesion in the leech CNS. Brain Res. 272, 166–170.

NICHOLLS, J.G., HERNANDEZ, U.G. 1989. Growth and synapse formation by identified leech neurones in culture: a review. Q. J. Exp. Physiol. 74, 965–973.

OPPENHEIM, F.G., XU, T., MCMILLIAN, F.M., LEVITZ, S.M., DIAMOND, R.D., OFFNER, G.D., TROXLER, R.F. 1988. Histatins, a novel family of histidine-rich proteins in human parotid secretion. Isolation, characterization, primary structure, and fungistatic effects on Candida albicans. J. Biol. Chem. 263, 7472–7477.

PERLSON, E., MEDZIHRADSZKY, K.F., DARULA, Z., MUNNO, D.W., SYED, N.I., BURLINGAME, A.L., FAINZILBER, M. 2004. Differential proteomics reveals multiple components in retrogradely transported axoplasm after nerve injury. Mol. Cell. Proteomics 3, 510–520.

PHELAN, P., STEBBINGS, L.A., BAINES, R.A., BACON, J.P., DAVIES, J.A., FORD, C. 1998. Drosophila Shaking-B protein forms gap junctions in paired Xenopus oocytes. Nature 391, 181–184.

REYNOLDS, S.A., FRENCH, K.A., BAADER, A., KRISTAN, W.B., JR. 1998. Development of spontaneous and evoked behaviors in the medicinal leech. J. Comp. Neurol. 402, 168–180.

SAHLEY, C.L., MODNEY, B.K., BOULIS, N.M., MULLER, K.J. 1994. The S cell: an interneuron essential for sensitization and full dishabituation of leech shortening. J. Neurosci. 14, 6715–6721.

SALZET, M. 2001a. Neuroimmunology of opioids from invertebrates to human. Neuro Endocrinol. Lett. 22, 467–474.

SALZET, M. 2001b. Vertebrate innate immunity resembles a mosaic of invertebrate immune responses. Trends Immunol. 22, 285–288.

SALZET, M. 2002a. Antimicrobial peptides are signaling molecules. Trends Immunol. 23, 283–284.

SALZET, M. 2002b. Immune cells express endocrine markers. Neuro Endocrinol. Lett. 23, 8–9.

SALZET, M., TASIEMSKI, A., COOPER, E. 2006. Innate immunity in lophotrochozoans: the annelids. Curr. Pharm. Des. 12, 3043–3050.

SAWYER, R.T. 1986. Leech Biology and Behaviour. Oxford: Oxford University Press.

SHAFER, O.T., CHEN, A., KUMAR, S.M., MULLER, K.J., SAHLEY, C.L. 1998. Injury-induced expression of endothelial nitric oxide synthase by glial and microglial cells in the leech central nervous system within minutes after injury. Proc. Biol. Sci. 265, 2171–2175.

SHAIN, D.H., STUART, D.K., HUANG, F.Z., WEISBLAT, D.A. 2004. Cell interactions that affect axonogenesis in the leech Theromyzon rude. Development 131, 4143–4153.

SHAW, B.K., KRISTAN, W.B., JR. 1999. Relative roles of the S cell network and parallel interneuronal pathways in the whole-body shortening reflex of the medicinal leech. J. Neurophysiol. 82, 1114–1123.

SHEFI, O., SIMONNET, C., BAKER, M.W., GLASS, J.R., MACAGNO, E.R., GROISMAN, A. 2006. Microtargeted gene silencing and ectopic expression in live embryos using biolistic delivery with a pneumatic capillary gun. J. Neurosci. 26, 6119–6123.

SOBA, P., ZHU, S., EMOTO, K., YOUNGER, S., YANG, S.J., YU, H.H., LEE, T., JAN, L.Y., JAN, Y.N. 2007. Drosophila sensory neurons require Dscam for dendritic self-avoidance and proper dendritic field organization. Neuron 54, 403–416.

SONG, J., ZIPSER, B. 1995. Kinetics of the inhibition of axonal defasciculation and arborization mediated by carbohydrate markers in the embryonic leech. Dev. Biol. 168, 319–331.

TAI, M.H., ZIPSER, B. 1998. Mannose-specific recognition mediates two aspects of synaptic growth of leech sensory afferents: collateral branching and proliferation of synaptic vesicle clusters. Dev. Biol. 201, 154–166.

TAI, M.H., ZIPSER, B. 1999. Sequential steps in synaptic targeting of sensory afferents are mediated by constitutive and developmentally regulated glycosylations of CAMs. Dev. Biol. 214, 258–276.

TAI, M.H., ZIPSER, B. 2002. Sequential steps of carbohydrate signaling mediate sensory afferent differentiation. J. Neurocytol. 31, 743–754.

TASIEMSKI, A., SALZET, M., BENSON, H., FRICCHIONE, G.L., BILFINGER, T.V., GOUMON, Y., METZ-BOUTIGUE, M.H., AUNIS, D., STEFANO, G.B. 2000a. The presence of antibacterial and opioid peptides in human plasma during coronary artery bypass surgery. J. Neuroimmunol. 109, 228–235.

TASIEMSKI, A., VERGER-BOCQUET, M., CADET, M., GOUMON, Y., METZ-BOUTIGUE, M.H., AUNIS, D., STEFANO, G.B., SALZET, M. 2000b. Proenkephalin A-derived peptides in invertebrate innate immune processes. Brain Res. Mol. Brain Res. 76, 237–252.

TASIEMSKI, A., VIZIOLI, J., VANDENBULCKE, F., SAUTIERE, P.E., LEMOINE, J., SALZET, M. 2003. Inducible expression of theromacin and theromyzin, two novel antibacterial peptides in leech Theromyzon tessulatum. 9th International Congress of I.S.D.C.I., St. Andrews, Scotland.

TETTAMANTI, G., GRIMALDI, A., FERRARESE, R., PALAZZI, M., PERLETTI, G., VALVASSORI, R., COOPER, E.L., LANZAVECCHIA, G., de EGUILEOR, M. 2003. Leech responses to tissue transplantation. Tissue Cell 35, 199–212.

THOREY, I.S., ZIPSER, B. 1991. The segmentation of the leech nervous system is prefigured by myogenic cells at the embryonic midline expressing a muscle-specific matrix protein. J. Neurosci. 11, 1786–1799.

TRYSELIUS, Y., HULTMARK, D. 1997. Cysteine proteinase 1 (CP1), a cathepsin L-like enzyme expressed in the *Drosophila melanogaster* haemocyte cell line mbn-2. Insect Mol. Biol. 6, 173–181.

VENKITARAMANI, D.V., WANG, D., JI, Y., XU, Y.Z., PONGUTA, L., BOCK, K., ZIPSER, B., JELLIES, J., JOHANSEN, K.M., JOHANSEN, J. 2004. Leech filamin and Tractin: markers for muscle development and nerve formation. J. Neurobiol. 60, 369–380.

VERGOTE, D., SAUTIERE, P.E., VANDENBULCKE, F., VIEAU, D., MITTA, G., MACAGNO, E.R., SALZET, M. 2004. Up-regulation of neurohemerythrin expression in the central nervous system of the medicinal leech, *Hirudo medicinalis*, following septic injury. J. Biol. Chem. 279, 43828–43837.

VERGOTE, D., MACAGNO, E.R., SALZET, M., SAUTIERE, P.E. 2006. Proteome modifications of the medicinal leech nervous system under bacterial challenge. Proteomics 6, 4817–4825.

WANG, W.Z., EMES, R.D., CHRISTOFFERS, K., VERRALL, J., BLACKSHAW, S.E. 2005. *Hirudo medicinalis*: a platform for investigating genes in neural repair. Cell. Mol. Neurobiol. 25, 427–440.

WONG, V.Y., AISEMBERG, G.O., GAN, W.B., MACAGNO, E.R. 1995. The leech homeobox gene Lox4 may determine segmental differentiation of identified neurons. J. Neurosci. 15, 5551–5559.

XU, Y., BOLTON, B., ZIPSER, B., JELLIES, J., JOHANSEN, K.M., JOHANSEN, J. 1999. Gliarin and macrolin, two novel intermediate filament proteins specifically expressed in sets and subsets of glial cells in leech central nervous system. J. Neurobiol. 40, 244–253.

ZIPSER, B., MCKAY, R. 1981. Monoclonal antibodies distinguish identifiable neurones in the leech. Nature 289, 549–554.

Chapter 10

Lumbriculus variegatus and the Need for Speed: A Model System for Rapid Escape, Regeneration and Asexual Reproduction

Mark J. Zoran* and Veronica G. Martinez†

*Department of Biology, Texas A&M University, College Station, TX
†Department of Biology, University of the Incarnate Word, San Antonio, TX

10.1 INTRODUCTION

Many traditional models of regeneration (e.g. *Hydra*, planarians, amphibians) have garnered renewed interest as the application of emerging cellular, molecular and genetic approaches allow more fruitful exploitation of the long-appreciated capabilities of these animals (Sanchez Alvarado and Tsonis 2006; Gurley et al. 2008; Petersen and Reddien 2008). Along with these organisms, several oligochaete species have been the basis for new contributions to our understanding of regenerative mechanisms (Bely and Wray 2001; Martinez et al. 2005; Myohara et al. 2006). Annelids have a reputation for impressive regenerative abilities, which is certainly justified, even though this ability varies widely. Leeches, for example, are incapable of regenerating lost segments, while other groups can regenerate an entire new individual from a single body segment (Morgulis 1907; Okada 1929; Berrill 1952). Many annelid worms are limited in their ability to regenerate anterior body parts, whereas posterior segment regeneration is much more common and is likely an ancestral character of the phylum (Bely 2006). Since some species capable of regeneration are closely related to others that are nonregenerating, Bely (2006) has argued that mechanisms of regeneration have been independently lost in certain annelid groups, with some of these losses being evolutionarily recent. In addition to their ability to regenerate body segments, annelids generally have a marked capacity for wound healing. For example, even though leeches do not regenerate lost segments, they generate an effective response to injury by assembling an extracellular scaffold of proteins that facilitates the restoration of traumatized structures (Tettamanti et al. 2005).

Annelids in Modern Biology, Edited by Daniel H. Shain
Copyright © 2009 John Wiley & Sons, Inc.

In those annelid worms that replace lost body parts, regenerative capabilities often vary according to the position of the fragmentation site along the anterior–posterior axis (Hyman 1940; Berrill 1952; Herlant-Meewis 1964). For example, in *Eisenia fetida* (Lumbricidae), regeneration of anterior structures is drastically reduced as more segments are removed. Consequently, *E. fetida* is not capable of regenerating its head when >20 anterior segments are removed (Moment 1950; Berrill 1952). In contrast, *Lumbriculus variegatus* (Lumbriculidae) regenerates a stereotyped number of seven to eight head segments regardless of the fragmentation site along the animal's body axis (Morgulis 1907; Drewes and Fourtner 1990). Some annelids, including *L. variegatus*, possess the ability to regenerate a complete adult animal following reduction to as little as three segments (Morgulis 1907; Berrill 1952). Lumbriculids have long been known as the most favorable oligochaetes for experimental studies of regeneration, not only on account of their high capacity for regeneration, but also their large body size, low mortality, and the fact that they exhibit qualitative and quantitative axial differences (Hyman 1916).

The remarkable ability of some annelid worms to reconstruct virtually their entire body requires coordinated activation of multiple developmental, regenerative and wound healing processes in response to injury. Wound healing employs very different mechanisms from the cell proliferation-dependent processes used in segmental regeneration. Supporting this idea, wound healing occurs following amputation of anterior segments and prolonged exposure to x-radiation, but mitosis and head regeneration is abolished (Turner 1934). Head and tail bud formation is called epimorphic regeneration, i.e. the dedifferentiation of extant tissues and the activation of stem cell populations to form a blastema and subsequent segmental bud (Morgan 1901; Goss 1969). Thus, epimorphosis is a process that requires ongoing cell division and differentiation. This process of epimorphic segmental regeneration is accompanied by morphallactic regeneration in the old (original) segments of some annelids, including the polychaete species *Sabella pavonina* (Berrill 1978). Morphallaxis, on the other hand, is a pattern of regeneration that involves the transformation of existing tissues with no significant differentiation of stem cells (Holstein et al. 2003). Thus, body segments that survive injury and transform into newly organized segments via morphallaxis often have new positional identities along the anterior–posterior axis (Drewes and Fourtner 1990; Martinez et al. 2005). In two well-defined examples, the regeneration of body segments in *L. variegatus* (Drewes and Fourtner 1990; Martinez et al. 2005) and *Enchytraeus japonensis* (Myohara et al. 2006) involves both epimorphosis and morphallaxis.

The emergence of new methods in cellular and molecular biology has permitted the exploration of genes and gene products that mediate animal regeneration (Sanchez Alvarado and Tsonis 2006). Although little is currently understood about the genes and proteins key to regeneration in annelids, some genes involved in nervous system repair have been determined in the leech *Hirudo medicinalis* (Blackshaw et al. 2004). In oligochaetes, several recent studies indicate that such knowledge may soon be obtained for this group. For example, Bely and Wray (2001) have investigated regulatory gene expression during regeneration in *Pristina leidyi* (Naididae). These authors suggest that genetic mechanisms employed in embryonic development are redeployed during regeneration and asexual fission. Myohara et al. (2006) performed suppression subtractive hybridization analyses using cDNAs from regenerating and nonregenerating oligochaetes, *E. japonensis*. Expression levels of many regeneration-associated genes were highest during blastema formation, and a total of 165 unique sequences were discovered to be upregulated during regeneration. These studies, taken together with others (Dupin et al. 1991; Martinez et al. 2005), are beginning to uncover the molecular players involved in oligochaete regeneration, processes that for decades were only tractable at an organismal level.

Morphallaxis is a rare, or at least rarely reported, form of annelid regeneration. Its presence, together with several other important life history traits in *L. variegatus*, makes this species an ideal model for examining behavioral, physiological, cellular and molecular mechanisms of development, regeneration and systems-level plasticity that underlie the transformation of body segments following fragmentation due to injury, asexual fission or autotomy. In this chapter, we present a review of recent investigations into the regenerative capabilities of oligochaete worms, with primary emphasis on morphallaxis of the *L. variegatus* nervous system. We begin with an overview of the organization of rapid escape behaviors in oligochaetes and the neural correlates of these reflexive events. We continue with how these neurobehaviors are transformed during morphallactic regeneration to quickly retool the animal for its changing needs for speed. Finally, we describe how this simple model system is ideal as a template for science education in the K-12 classroom and beyond.

10.2 NEURAL REGENERATION IN OLIGOCHAETES

Annelids, like most invertebrate animals, regenerate their nervous systems rapidly and accurately after neural trauma (Moffet 1996). Studies of regeneration in earthworms have focused largely on the injured central nervous system (CNS) and axonal regeneration, particularly the recovery of function in giant interneuronal pathways (Birse and Bittner 1981; Drewes et al. 1988; Lyckman et al. 1992). For example, in *Lumbricus terrestris*, regenerating giant fibers (GFs) show a high degree of specificity in establishing new contacts with their appropriate partners (Birse and Bittner 1976, 1981). This recovery of function is also rapid. Intersegmental connectivity along the giant axons of *E. fetida* is reestablished 24 h after ventral nerve cord (VNC) crush (Fig. 10.1A, B). These neuroregenerative capabilities of oligochaetes include the rapid reconstruction of the VNC following crush, transection or ablation. Additionally, in all terrestrial oligochaetes studied, functional recovery is seen following removal and transplantation of large sections of nerve cord (Drewes et al. 1988). Because the CNS of oligochaetes has been the focus of most neural regeneration studies, a review of some basic oligochaete CNS features is necessary.

10.2.1 Oligochaete CNS

The CNS of oligochaetes, like those of other annelids, comprises a brain and nerve cord. As most biology students who have dissected an earthworm might recall, an oligochaete brain is a fused pair of supraesophageal ganglia connected via two circumesophageal connectives to a pair of subesophageal ganglia, with the location of the brain in specific anterior segments depending on the species (Stephenson 1930; Bullock 1965; Jamieson 1981). The VNC extends down the midline of the animal and gives rise to multiple pairs of nerves within each segment. These segmental nerves extend laterally around the body wall and contain the axons of sensory (input) and motor (output) pathways to and from the CNS. Within the brain and the VNC, the processes and synapses between sensory neurons, interneurons and motor neurons form a core or neuropile, which is the site of neural network integration that underlies behavioral regulation (Bullock 1965; Jamieson 1981). Some interneurons within the VNC are responsible for intersegmental communication and for the coordination of neural networks responsible for multisegmental behaviors,

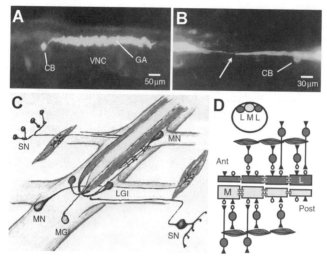

Figure 10.1   Oligochaete escape reflex anatomy. Rapid escape in oligochaete worms is mediated by segmentally arranged giant axons that form through-conducting GF systems. (A) In the VNC of *Eisenia fetida*, for example, a single cell body (CB) within the VNC gives rise to a small process that expands in the dorsal part of the cord to form a giant axon (GA). Each segmental GA is electrically coupled at synaptic contacts with neighboring axons forming the GF pathway. (B) Within 24 h of axonal crush of the *E. fetida* VNC, fine projections from the GAs in neighboring segments meet precisely at their tips (arrow; adapted from Drewes et al. 1988, with permission from the American Society of Zoologists). (C) The dorsal region of the VNC of most oligochaete worms contains three giant nerve fibers: one medial and one pair of laterals. Each MGF axon (yellow) is connected to one medial giant interneuron (MGI) CB. Each of the two lateral giant interneuron (LGI) CBs gives rise to an axonal process that, after crossing the VNC midline, expands to form the LGF pathway (blue). Each GF possesses multiple, ventrally projecting collaterals, the sites of synaptic communication between sensory neuron (SN) inputs and motor neuron (MN) outputs. In each segment (except a few in the head region), the VNC gives rise to pairs of segmental nerves containing sensory and motor fibers. (D) The neural substrate of rapid escape reflexes in oligochaetes involves sensory inputs (green circles) relatively restricted to anterior segments for the MGF and posterior segments for the LGF. Motor outputs (brown circles) activated by both systems are present along most of the animal, although the MGF outputs are more efficacious in anterior segments and elicit head withdrawal, while LGF outputs are more effective in posterior segments and mediate tail withdrawal. The MGF pathway (yellow) is large in diameter in anterior segments and possesses stronger gap junction coupling there. In contrast, the LGF pathway (blue) is small in diameter in anterior segments and gradually increases in size posteriorly. The inset shows the spatial configuration of the MGF (M) and LGF (L) pathways of the VNC in cross section. (See color insert.)

such as swimming and crawling. Also among these behaviors are rapid escape reflexes, mediated by giant interneuronal pathways (Drewes et al. 1983; Zoran and Drewes 1987). The regeneration of head segments and their complex structures, such as brain ganglia, might very well present an oligochaete with a formidable task, that is, more formidable than replacing lost posterior segments that are largely metameric repeats. Nonetheless, we shall see later that lumbriculid worms possess rather marked abilities for head replacement and also novel regenerative mechanisms for transforming neural connections in posterior segments into anterior synaptic networks.

A relatively common feature of the oligochaete lifestyle is the need for rapid behavioral responses to potential predatory attacks. Consequently, giant axons that traverse the worm's longitudinal axis and regulate rapid withdrawals are ubiquitous. Although a marked diversity in the number, size and arrangement of GFs within polychaetes exists (Bullock

1948; Nicol 1948; Bullock 1965), a conserved feature of most oligochaetes is the presence of three GFs (Fig. 10.1C, D; Zoran and Drewes 1987). Oligochaetes predominately possess one medial and one pair of lateral giant axons. These three giant axons are each derived segmentally from a single interneuron (Fig. 10.1C; Zoran and Drewes 1987). In turn, each giant axon is connected to partner axons in neighboring segments by gap junctions, forming through-conducting pathways along the animal's anterior–posterior axis (Fig. 10.1D). In contrast, polychaete GFs are highly elaborated structures with diverse developmental and evolutionary origins. This diversity in GF structure encompasses giant axons derived from the fusion of numerous anterior interneurons to a simple pair of GFs derived from two cell bodies (Nicol 1948; Bullock 1965). Although GF pathways in polychaetes have certainly evolved independently many times within this taxon, the conserved pathway configuration and development (i.e. segmentally arranged cell bodies) of giant axons in oligochaetes suggests a single evolutionary lineage.

10.2.2 Rapid Escape Reflexes in Oligochaetes

The giant interneuronal pathways constructed from chains of segmental giant axons are the central basis of two independent escape circuits: the medial giant fiber (MGF) pathway that regulates anterior shortening (head withdrawal) and the lateral giant fiber (LGF) pathway that regulates posterior shortening (tail withdrawal). Although both pathways traverse the length of the worm, the MGF axonal diameter is largest in anterior segments and decreases in size posteriorly. In an opposing fashion, the LGFs are larger in posterior segments (Fig. 10.1D). Similarly, sensory inputs that activate the GFs and elicit rapid withdrawal reflexes are restricted to anterior segments for MGF activation and posterior segments for LGF activation (Fig. 10.1D). Because speed is critical for these escape reflexes, the axonal elements of the GF pathways are electrically coupled via gap junctions, allowing rapid and uninterrupted through conduction of nerve impulses along the GF system (Mulloney 1970; Brink and Ramanan 1985). In most oligochaetes, the two LGFs are connected by cross-bridges within each segment. These interconnections are undoubtedly the basis for electrotonic coupling between the LGFs and the resulting bilateral synchronization of LGF action potentials during spike propagation (Drewes 1984). In addition to being large in diameter, which facilitates spike conduction velocity, most oligochaete giant axons are ensheathed by glial cell membranes, resulting in myelin-like sheaths (Günther 1976; Roots and Lane 1983; Zoran et al. 1988), which are also thought to increase conduction velocity along GF pathways (Zoran et al. 1988; Drewes and Fourtner 1990). We shall see that the GF pathways of *L. variegatus* undergo rapid transformations in both axonal size and gap junctional coupling during CNS morphallactic regeneration.

Each segmental axon of an oligochaete GF pathway has 2–4 ventrally projecting collaterals that are the synaptic sites of sensory inputs and motor outputs (Fig. 10.1C, D; Drewes 1984). These collaterals are also critical for salutatory conduction of GF spikes, as they are high-density sites of voltage-gated ion channels and analogous structures to the nodes of Ranvier in vertebrates (Zoran et al. 1988). Although only subsets of anterior and posterior segments activate MGF and LGF responses, the motor outputs of both systems extend largely along the length of the animal. Still, the efficacy of neuromuscular transmission varies along anterior–posterior gradients (Fig. 10.1D). We will discuss in the next section how sensory and motor elements of *L. variegatus* rapid escape reflexes transform during segmental regeneration and asexual fission.

10.2.3 A Natural Experiment

Freshwater oligochaetes have long been thought to have evolved from ancestral groups of terrestrial worms. Interestingly, terrestrial oligochaetes generally live in burrows and emerge head first from their earthly dwelling places. In contrast, most aquatic oligochaetes live with their heads burrowed into the substrate and their tails extending up into the water column of the lake, stream or marsh they inhabit. This habit likely evolved as tail segments became specially adapted for gas exchange (Fig. 10.2). Thus, earthworms are especially vulnerable to predatory attacks directed toward anterior segments. Many tubificid and lumbriculid groups, however, have likely been subject to millions of years of predation pressure directed at the tail, as posterior segments modified for respiration were exposed to attacks from fish, diving ducks, shore birds and leeches. Therefore, nature has provided an experiment whereby a conserved escape neural circuit, constructed of anterior- and posterior-specific pathways, has been subjected to predatory pressures at opposite ends, consequently imposing opposing needs for speed.

When escape reflex properties between terrestrial and aquatic species are examined using behavioral, anatomical and physiological approaches, a strikingly clear set of differences is observed (Drewes et al. 1983; Zoran and Drewes 1987). Many terrestrial worms, for example, have MGF sensory fields (inputs that activate head withdrawal) along two-thirds of their segments extending back from the prostomium or first body segment. In contrast, aquatic oligochaetes have LGF sensory fields from the tips of their tails forward over two-thirds of their body segments. Likewise, LGFs are larger and faster conducting in the tails of freshwater species, whereas MGFs are largest along most of the body segments of terrestrial species. Among aquatic oligochaetes, tail withdrawal reflexes have evolved remarkable traits to assure speedy escape. For example, the tubificid worm *Branchiura sowerbyi* needs to generate only a single GF spike to elicit a complete tail withdrawal reflex, which from stimulus to shortening is 7 ms (Zoran and Drewes 1988). In contrast, most terrestrial species require a closely spaced train of GF spikes to activate a complete withdrawal response (Drewes 1984), perhaps since a life of crawling through hard substrates brings with it frequent GF activation not associated with predatory attack.

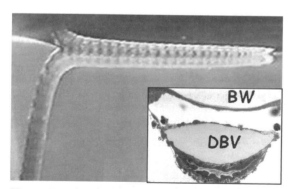

Figure 10.2 Respiratory adaptations of *Lumbriculus variegatus*. *L. variegatus* is common in habitats that include layers of decomposing leaves, and its head probes the sediments during foraging behavior. Its tail, specialized for gas exchange, projects upward into the water column, just breaking the air/water interface where it forms a right angle bend (background image). This posture facilitates gas exchange between the air and the pulsating dorsal blood vessel (DBV, inset), which lies just beneath the modified body wall (BW) devoid of musculature in the last 30 segments of tail. Photos courtesy of C.D. Drewes.

In *L. variegatus*, the posteriormost 30 segments are unique in that, along with respiratory adaptations (Fig. 10.2), a large population of photoreceptor cells exists in the body wall (Jamieson 1981; Drewes and Fourtner 1989). These photoreceptors are connected via sensory inputs to the LGF escape system and thereby provide an early withdrawal activation system, detecting the shadows of approaching predatory threats (Drewes and Fourtner 1989, 1990; Drewes and Zoran 1989; Drewes 1990).

Given this insight into the evolution of rapid escape reflexes and selective pressures associated with vulnerability to attack, we might expect anterior and posterior segmental regeneration to be superior in terrestrial and aquatic oligochaetes, respectively. However, this is not the case; rather, both groups exhibit generally limited capacity for head regeneration. The impressive abilities for head regeneration in *L. variegatus* make this annelid species an exception to this limitation. Bely (2006) suggests that the evolution of rapid escape reflexes, presumably providing effective avoidance of body damage, may have been associated with a loss of regenerative capabilities in some species. This does not seem to be supported by comparative studies of escape reflexes—at least among oligochaetes—where if any trend exists, worms with some of the most rapid escape responses, like *B. sowerbyi* (Zoran and Drewes 1987; Zoran et al. 1988), are among the most efficient regenerators of lost body parts.

10.3 *LUMBRICULUS VARIEGATUS*, A MODEL SYSTEM FOR REGENERATION AND ASEXUAL REPRODUCTION

Lumbriculus variegatus is a freshwater oligochaete of the order Lumbriculida (Brinkhurst and Jamieson 1971; Jamieson 1981; Brinkhurst and Gelder 1991) and is commonly called the California blackworm or mudworm. Composed primarily of mud-dwelling species, lumbriculids are distinct from other freshwater oligochaetes (e.g. tubificid worms) and are thought to have diverged as an early lineage from the oligochaete branch of the annelid phylogenetic tree (Brinkhurst and Jamieson 1971; Jamieson 1981). *L. variegatus* is a hermaphroditic species ranging in length from ~5 to 10 cm (100–250 segments) depending upon developmental state, and can be collected during the spring and summer months in shallow ponds, lakes and marshes. Sexually mature worms are typically large, ~1.5 mm in diameter (Drewes and Brinkhurst 1990) and produce cocoons containing 4–11 fertilized eggs during the sexual reproductive phase of their life cycle (Fig. 10.3). Worms raised in the laboratory are usually smaller (100–150 segments or 4–6 cm in length) and rarely reach sexual maturity. Rather, these animals reproduce via asexual reproduction, as they do throughout the summer and fall months in temperate climates of North America. Asexual reproduction, or architomic fission, involves the production of body fragments, or zooids, that regenerate new segments to become clones of the fragmented adult (Stephenson 1930; Berrill 1952; Brusca and Brusca 1990). Therefore, architomy in *L. variegatus* involves fission and subsequent replacement of lost segments via epimorphosis. Additionally, these lumbriculids are able to self-amputate by autotomy in response to noxious stimulation (Lesiuk and Drewes 1999a). Autotomy, although involving fragmentation, should be noted as being fundamentally distinct from the regulated reproductive process of architomy. Thus, *L. variegatus*, besides having a life cycle that includes sexual reproduction via cocoons, has several alternative reproductive mechanisms involving zooid clones: asexual fission, injury-induced regeneration and self-fragmentation by autotomy (Fig. 10.3).

The ease of collecting lumbriculid worms in most areas of North America and Europe, together with their straightforward culture and manipulation, makes them a favorable

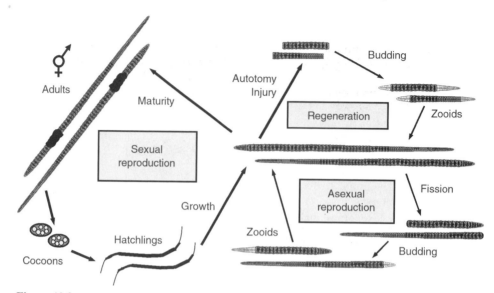

Figure 10.3 Natural history of *Lumbriculus variegatus*. The life cycle of *L. variegatus* involves both sexual and asexual reproduction. Hermaphroditic adults generate cocoons that grow and mature into new adult animals. Alternatively, mature or immature worms generate new individuals from fragments of body segments by both asexual reproduction and regeneration. During asexual reproduction, zooids produced by architomic fission give rise to buds that expand to regenerate clones. During regeneration following injury, autotomy or experimental transaction, fragments form new buds by epimorphosis at the transection site.

model organism for many biological studies. *L. variegatus* is used extensively for environmental toxicology (ASTM 1995; Phipps et al. 1995; Ding et al. 2001; Mino et al. 2005) and developmental biology studies (Lesiuk and Drewes 2001a, b; Martinez et al. 2005, 2006).

10.4 NEURAL MORPHALLAXIS

Lumbriculus variegatus exhibits anterior–posterior gradients in behavior that are easily monitored (Drewes and Fourtner 1990; Lesiuk and Drewes 2001a). Stimuli (touch or vibration) detected by sensory systems in the anterior 1/3 region of the worm's body (i.e. the MGF sensory field; Fig. 10.4A) result in quick anterior shortening or head withdrawal (Drewes 1984; Zoran and Drewes 1987; Drewes and Fourtner 1990). Conversely, stimulation of segments in the posterior 2/3 region of the worm's body (i.e. the LGF sensory field) results in posterior shortening or tail withdrawal (Drewes 1984; Zoran and Drewes 1987; Drewes and Fourtner 1989, 1990). In addition to prompting head and tail shortening, stimulation of anterior segments can result in a 180° body turn or reversal response, which directs the animal's locomotion away from aversive stimuli (Drewes 1999a). In contrast, stimulation of posterior segments can elicit undulatory swimming movements (Drewes 1999a). These behaviors, activated by sensory inputs to LGF and MGF sensory fields, are body region specific and are therefore subject to transformation during certain instances of neural morphallaxis.

As mentioned, *L. variegatus* body fragments regenerate constant numbers of head segments and varying lengths of tail segments following amputation or autotomy (Berrill

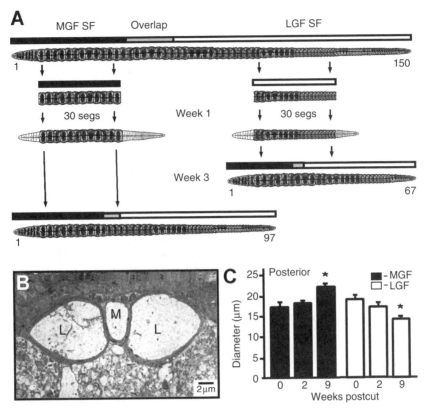

Figure 10.4 Neural morphallaxis in *Lumbriculus variegatus*. (A) Amputated fragments (segments ~8–38) from animals of ~150 segments regenerate seven to eight head segments and variable numbers of tail segments (~30–60). Posterior fragments (segments ~100–130) also generate constant head segment numbers (~7–8) and thus acquire a more anterior position following segmental regeneration. Head and tail buds form in 1 week, and neural morphallaxis is largely complete by 3 weeks postamputation. Sensory field (SF) maps (rectangular boxes above worm diagrams) illustrate that stimulation of the anterior 1/3 region of the body wall activates only MGF spikes, resulting in head withdrawal (black box). Stimulation of the posterior 2/3 region of the body wall activates only LGF spiking and tail withdrawal (white box). An area of both MGF and LGF spike activation (sensory field overlap, indicated by the gray box) is common. Following 3 weeks of regeneration, sensory fields of posterior fragment gain MGF and overlap sensory fields previously absent. Numbers represent segmental identity, with no. 1 being the first head segment (prostomium) and no. 150 the last tail segment. (B) Photomicrograph of a cross section through a posterior body fragment illustrating that lateral giant axons (L) are relatively larger than medial giant axons (M) in tail segments (photo by C.D. Drewes). (C) In experiments comparing control (nonregenerating) and experimental (regenerating) worms, both MGF (solid bars) and LGF (open bars) pathways showed marked changes in axonal diameter after 9 weeks of regeneration (*$p < 0.02$, $n = 9$; adapted from Martinez et al. 2006, with permission from Springer).

1952; Drewes and Fourtner 1990; Martinez et al. 2006). This characteristic regeneration of limited numbers of head segments results in positional transformations of fragments removed from the posteriormost body regions. Thus, following regeneration, posterior fragments often become more anteriorly located (Fig. 10.4A; Drewes and Fourtner 1990; Martinez et al. 2006). This change in positional identity coincides with anatomical and physiological changes in these segments as they come to occupy a more anterior position along the new body. Changes in CNS structure and function associated with this regenerative transformation have been defined as neural morphallaxis (Drewes and Fourtner 1990; Martinez et al. 2006).

Evidence of morphallactic regeneration has been described in at least three oligochaete species: *L. variegatus* (Drewes and Fourtner 1990; Martinez et al. 2006), *E. fetida* (Chapron 1970) and *E. japonensis* (Myohara et al. 1999; Yoshida-Noro et al. 2000; Müller 2004; Myohara 2004). Here, morphallaxis is defined as a transformation of existing body parts or tissues into newly organized structures without the necessity of new cell proliferation (Morgan 1901; Gilbert 2000). Drewes and Fourtner (1990) studied amputated body fragments of *L. variegatus* and described morphallaxis of the MGF and LGF systems. These studies, together with experiments from other groups, have defined some of the neurobehavioral properties that transform during neural morphallaxis. First, sensory fields are reorganized based on their new body position (Fig. 10.4A). Specifically, tail escape responses in a regenerating posterior fragment, originally governed by the LGF system, gradually transform to head withdrawal behaviors mediated by MGF pathways (Drewes and Fourtner 1990). This behavioral morphallaxis is also accompanied by structural and functional changes within the GF pathways. Medial and lateral giant axon diameters in regenerating posterior fragments, quantified in histological sections (Fig. 10.4B) or whole mounts of dye-filled fibers (Fig. 10.5C), are significantly increased and decreased, respectively, over a 9-week period following amputation (Fig. 10.4C; Martinez et al. 2006). Noninvasive electrophysiological recordings of GF spikes, where large action potentials can be detected through the body wall, demonstrate changes in GF conduction velocity associated with altered giant axon size (Fig. 10.5A, B).

Although GF size is tightly correlated with conduction velocity along axons, and clearly contributes to the changes observed during *L. variegatus* neural morphallaxis, changes in giant axon electrical coupling have also been demonstrated. GF coupling has been analyzed following injection with Lucifer yellow, a fluorescent dye small enough to pass through the gap junctions that link giant axons at their segmental boundaries. Dye coupling coefficients, calculated for posterior fragments, demonstrate that gap junctional communication along the MGF pathway is significantly increased during regeneration (Fig. 10.5C, D). These dye fills also demonstrate that the GF pathways of *L. variegatus* are septate, or are separated by a membranous septum, as opposed to syncytial. That is, each segment of giant axon is separated from its neighbors by cellular divisions that are not lost during neural development. Taken together, neural morphallaxis in this species involves transformations in neurobehavioral substrates as fragments acquire new identities. Note that by virtue of strictly regenerating short heads and long tails (Fig. 10.4A), *L. variegatus* segments generally take on more anterior identities, although posterior transformation is possible in certain circumstances. Additionally, multiple cellular and molecular mechanisms are clearly involved in neural morphallaxis of a fragment as it gains the specific kind of speed it will soon need.

Whether morphallaxis among oligochaetes is rare or, as with epimorphic regeneration of body segments (Bely 2006), has simply been underreported due to inadequate investigations across diverse groups remains unclear. Nevertheless, significant changes in neural function occur immediately following asexual fragmentation in some oligochaetes (e.g. *Dero digitata*, Naididae), including rapid switching of sensory-to-giant pathways in rostral segments of posterior zooids (Drewes and Fourtner 1991). In *L. variegatus*, any body segment along the anterior–posterior axis, with the possible exception of some head segments, possesses the ability to change its positional identity via morpallactic mechanisms (Drewes and Fourtner 1990; Lesiuk and Drewes 2001a; Martinez et al. 2006). Likewise, using an environmental shift protocol to induce asexual fission (Martinez et al. 2006),

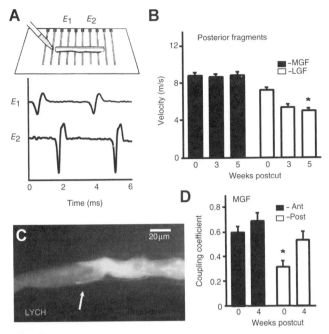

Figure 10.5 Changes in GF conduction and connectivity during neural morphallaxis. (A) Noninvasive recordings of GF conduction velocity are possible using electrode grids connected to a digital data recorder (top). E_1 and E_2 represent two recording sites. (B) Conduction velocity (calculated as the conduction distance divided by the conduction time) of LGF spikes (open bars) in posterior fragments decreases over a 5-week period following amputation (*$p < 0.05$; adapted from Martinez et al. 2005, with permission from Elsevier). (C) *Lumbriculus variegatus* GFs are septate in nature as demonstrated by injection of a mixture of Lucifer yellow CH (LYCH) and rhodamine-dextran (Rhod-dextran) into a MGF. Rhodamine-dextran, a molecule of high molecular weight (10,000 daltons), does not pass through gap junctions at septal boundaries between axons (arrow). Thus, both red and green dyes are present in the injected axon, giving it a yellow-orange appearance in this merged image. The small LYCH molecules (>500 daltons), in contrast, exhibit dye coupling between giant axons. (D) Dye-coupling coefficients calculated for posterior fragments (open bars) before and after fragmentation demonstrate an increase in MGF dye coupling during neural morphallaxis and the acquisition of a more anterior positional identity ($n = 5$). (See color insert.)

virtually every worm in an experimental population will fragment within 3 weeks after the inductive temperature shift (Fig. 10.6A, B). Interestingly, only a single fission plane is formed en route to generating two zooids. This fragmentation stereotypically occurs at the animal's 1/3–2/3 anterior–posterior boundary. That is, in animals of 150 segments in length, architomic fission occurs at segment number 48 ± 10, producing an anterior fragment of about 50 segments including the original head, and a posterior fragment of approximately 100 segments including the original tail (Fig. 10.6C, D). Thus, a predictable fission site that is located along an identical segmental range to that of the zone of MGF/ LGF sensory field overlap is produced during asexual reproduction (Fig. 10.6D).

Neural morphallactic changes in sensory field maps involve an expansion of the MGF sensory field and a retraction of the LGF sensory field posterior to the architomy site. Development of this exclusively MGF sensory field (in ~15 segments) occurs prior to the actual fragmentation event. Immunohistochemical analyses have revealed the expression of a lumbriculid protein associated with neural morphallaxis. This protein, detected with an antibody called Lan 3-2, is expressed in neural structures, including the GF system,

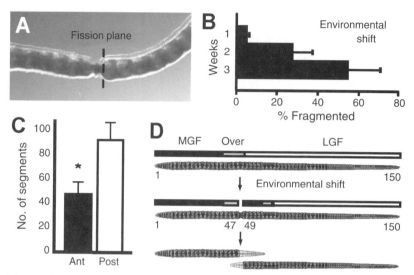

Figure 10.6   Asexual reproduction by architomic fission in *Lumbriculus variegatus* involves neural morphallaxis. (A) *L. variegatus* reproduces asexually through the formation of architomic fission zones that progress to sites of fission plane (dashed line) development prior to fragmentation. No new segmental differentiation is observed in segments adjacent to the fission site, a feature characteristic of architomy. (B) Asexual reproduction can be induced in whole animals by an environmental shift in culture conditions. The solid bars represent the percentage of animals (noncumulative) that fragmented in each of 3 weeks after temperature shift. Three weeks following the shift, fragmentation occurred in over 90% of animals in a population (adapted from Martinez et al. 2005, with permission from Elsevier Limited). (C) Architomy sites form consistently in *L. variegatus* at approximately segment 50 in animals of 150 segments; *$p < 0.05$ (adapted from Martinez et al. 2006, with permission from Springer). (D) During worms' asexual reproduction, fission zones are formed 2 weeks following an environmental shift. Prior to fragmentation, GF sensory fields expand such that the zone of MGF/LGF sensory field overlap is disrupted by newly emerging MGF sensory inputs (black rectangular box above worm illustration).

and is induced during morphallaxis (Martinez et al. 2005). We have named this protein "morphallaxis protein 66" (MP66); however, its molecular identity remains undetermined. Interestingly, the Lan 3-2 epitope is enriched on GF axonal collaterals (Martinez et al. 2005), the synaptic sites of sensory input and motor output along oligochaete giant axons (Günther and Walther 1971; Zoran and Drewes 1987) and the likely sites of synaptic plasticity correlated with morphallactic changes in sensory fields.

The evolutionary origins and potential phylogenic relationships between asexual reproduction and regeneration in annelids remain unclear (Bely 1999; Sanchez Alvarado 2000; Bely and Wray 2001; Bely 2006). The cellular mechanisms underlying neural morphallaxis appear to have been co-opted in *L. variegatus* to mediate both of these developmental events. Reorganization of the neural circuits that occurs during fission and regeneration exhibits similar temporal and spatial patterns of MP66 expression. However, during segmental regeneration, Lan 3-2 epitope expression is upregulated several weeks after injury-induced fragmentation (Fig. 10.7A), whereas the epitope is upregulated several weeks prior to asexual fragmentation (Fig. 10.7B). In both situations, the induction of MP66 correlates with the observed changes in escape reflex anatomy and physiology. Thus, although the cellular and molecular substrates of neural morphallaxis are developed in anticipation of architomy and in compensation for injury or autotomy, these mechanisms

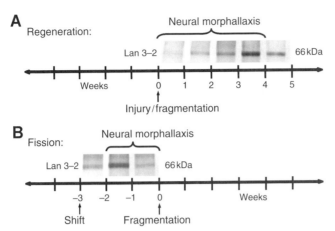

Figure 10.7 Molecular marker of neural morphallaxis common to both injury-induced regeneration and asexual reproduction. (A) Neural morphallaxis, including changes in neuroanatomy, physiology and behavior, occur during a 5-week period following injury-induced fragmentation. (B) In contrast, neural morphallactic changes associated with asexual fission occur 1–2 weeks prior to architomic fragmentation. Lan 3-2 epitope expression associated with a morphallaxis-induced protein of 66 kDa (Western blot above time lines) was differentially upregulated prior to architomy, but following transection. Induction of this neural glycoepitope correlates with both compensatory (A) and anticipatory (B) neural morphallaxis.

are clearly shared elements of both life history events. Consequently, future "evo-devo" studies involving the genetic identification of MP66 and other regeneration- or reproduction-induced proteins may eventually shed light on the evolutionary relationships between regeneration and asexual reproduction in oligochaetes.

Lesiuk and Drewes (2001b) first demonstrated that removal of part of the VNC in *L. variegatus* could produce, in a percentage of cases, ectopic heads (Fig. 10.8A, B). These extra head segments are generated following loss of several segments of VNC, but not following ablation of dorsal body wall and blood vessel (Martinez et al. 2008). Interestingly, ectopic heads form with greater efficacy in the same body segments that correspond to zones of GF sensory field overlap and architomic fission plane formation. Changes in GF sensory fields (Fig. 10.7A; Lesiuk and Drewes 2001b) and the expression of the Lan 3-2 epitope marker of neural morphallaxis (Fig. 10.8C) occur simultaneously following the induction of ectopic head formation (Martinez et al. 2008). These findings not only indicate cellular and molecular links between the mechanisms governing regeneration and reproduction, as others have suggested (Bely and Wray 2001), but they also support a role for the CNS in orchestrating these events.

Although the VNC in annelids is widely thought to be involved in establishing the dorsoventral polarity of a regenerating fragment (Salo and Baguna 2002), rigorous data to support this idea are sparse. In the lumbriculid species *Rhynchelmis limosella*, removal of the nerve cord leads to failure of posterior segment regeneration (Zhinkin 1936). Interestingly, only once in hundreds of cord ablations that lead to ectopic head formation have we seen an ectopic tail form in *L. variegatus* (V.G. Martinez and M.J. Zoran, unpublished data). Head or tail bud development on inappropriate ends of a fragmented annelid is rare, but has been observed in the polychaete worm *Sabella melanostigma* (Fitzharris and Lesh 1969), in the aquatic oligochaete *Dero limosa* (Hyman 1916) and in the terrestrial oligochaete *E. japonensis* (Kawamoto et al. 2005). Bipolar head regeneration (i.e. a head formed posteriorly) in *E. japonensis* appears to result from the combined effects of artificial

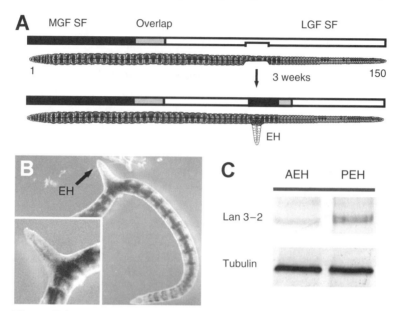

Figure 10.8 Nervous system lesion is necessary for induction of neural morphallaxis. (A) Ventrally protruding ectopic heads typically form following removal of five segments of ventral body wall and nerve cord from posterior segments of *Lumbriculus variegatus*. (B) Ectopic heads (EHs) generally contain five to eight body segments, including a well-defined prostomium. Emergence of MGF sensory fields (black box above illustration in A) occurs within 3 weeks of nerve cord ablation (adapted from Martinez et al. 2008, with permission from The Company of Biologists). (C) Western blot analyses of protein extracts from ventrally lesioned animals indicate that the Lan 3-2-positive molecular marker of neural morphallaxis is upregulated in segments posterior to forming ectopic heads (PEH). Lan 3-2 expression is much less abundant in segments anterior to ectopic heads (AEH).

amputation and anesthesia, although the specific causes remain undefined. No bipolar tails were observed in these experiments (Kawamoto et al. 2005). Like *L. variegatus* and *E. japonensis*, appropriate epimorphic regeneration at damaged segments appears to be a stable oligochaete trait, and is only rarely disrupted in very small fragments or with chemical perturbation. In summary, mechanisms exist that rigidly regulate the formation of head and tail buds along the anterior–posterior axis, and the VNC may play a critical role in that regulation.

10.5 ACCESSIBLE MODEL FOR LIFE SCIENCE EDUCATION

As outlined in this chapter, *L. variegatus* has proven to be an excellent model system for studies of neural plasticity, rapid escape reflexes and regeneration. These features, together with the fact that animals can be easily collected from freshwater ponds, purchased at affordable prices from biological supply houses and local pet stores (e.g. Flinn Scientific and Carolina Biological Supply), and maintained with basic laboratory supplies, make *L. variegatus* an especially useful educational tool for K-12 biology classrooms. The late Professor Charles Drewes of Iowa State University successfully combined much of his decades of research experience in annelid neurophysiology with his love of educational outreach to create a series of original "hands-on" classroom applications using these worms for elementary, high school and college-level biological experiments. For example, in one laboratory exercise, students simply count segments regenerated by body fragments and discover the relationship between original segment size and the number of new segments produced (Drewes 1996a). In another, students learn about experimental design, careful technique, data quantification, proper graphing of scientific data, statistical correlation and the interpretation of results. Most importantly, they learn through personal discovery. Largely as a result of Dr. Drewes' efforts, California blackworms (*L. variegatus*) are now commonly used in school science laboratories, at-home science projects and in research university settings. Their ease of culturing makes them highly amenable for both structured and open-ended student investigations, such as studies of segmental regeneration, circulatory physiology, locomotion, neurophysiology and toxicology (Drewes 1996a, b, 1999a, b; Drewes and Cain 1999; Lesiuk and Drewes 1999b).

Teachers have found lumbriculids to be highly versatile worms. Their varied life history traits can be integrated into course units on ecology, behavior, development, physiology and biodiversity. This flexibility as a teaching tool makes *L. variegatus* an ideal species for inquiry-based curricula and student research experiences. For example, these worms exhibit unique forms of locomotor reflexes (helical swimming and body reversal), which are easily quantified. Moreover, blood pulsations and effects of drugs and toxins on oxygen consumption and photic responses can be determined. Additionally, all-or-none action potentials from giant nerve fibers are easily recorded from intact, freely moving worms. Finally, these worms are readily eaten by a wide array of invertebrates and therefore are useful in simple predator–prey exercises. In the authors' research laboratories, high school students, undergraduate and graduate students use *L. variegatus* in a range of experiments, from biological clocks to learning and memory. Therefore, lumbriculid worms represent an emerging annelid model system that is a relatively unconstrained platform for discovery in the science laboratory, determined largely by the creativity of the student investigator.

ACKNOWLEDGMENTS

Charlie Drewes, on many levels, contributed to the publication of this work and to the emergence of *L. variegatus* as a model system in general, through his many years of patient mentoring and expert assistance to students and peers alike. We wish he were still with us to share in our studies and the expanding importance of *L. variegatus* as both a research and an educational tool. We thank the many undergraduate student researchers in the Department of Biology at Texas A&M University who have contributed to the work reviewed here. Some of this work was supported by an American Psychological Association Minority Fellowship Grant (T32 MH-18882 to V.G.M) and by a National Institutes of Health National Institute of Neurological Disorders and Stroke grant (PO1 NS-39546 to M.J.Z.).

REFERENCES

American Society for Testing and Materials (ASTM). 1995. Standard guide for conducting sediment toxicity tests with freshwater invertebrates. In *Annual Book of ASTM Standards*, vol. 11.05, E1393-94a. Philadelphia: ASTM International, pp. 802–834.

BELY, A.E. 1999. Decoupling of fission and regeneration capabilities in an asexual oligochaete. Hydrobiologia 406, 243–251.

BELY, A.E. 2006. Distribution of segment regeneration ability in the Annelida. Int. Comp. Biol. 46, 508–518.

BELY, A.E., WRAY, G.A. 2001. Evolution of regeneration and fission in annelids: insights from engrailed- and orthodonticle-class gene expression. Development 128, 2781–2791.

BERRILL, N.J. 1952. Regeneration and budding in worms. Biol. Rev. 27, 401–438.

BERRILL, N.J. 1978. Induced segmental organization in sabellid worms. J. Embryol. Exp. Morphol. 47, 85–96.

BIRSE, S.C., BITTNER, G.D. 1976. Regeneration of giant axons in earthworms. Brain Res. 113, 575–581.

BIRSE, S.C., BITTNER, G.D. 1981. Regeneration of earthworm giant axons following transection or ablation. J. Neurophysiol. 45, 724–742.

BLACKSHAW, S.E., BABINGTON, E.J., EMES, R.D., MALEK, J., WANG, W-Z. 2004. Identifying genes for neuron survival and axon outgrowth in *Hirudo medicinalis*. J. Anat. 204, 13–24.

BRINK, P.R., RAMANAN, S.V. 1985. A model for the diffusion of fluorescent probes in the septate axon of earthworm. Axoplasmic diffusion and junctional membrane permeability. Biophys. J. 48, 299–309.

BRINKHURST, R.O., GELDER, S.R. 1991. Annelida: Oligochaeta and Branchiobdellida. In *Ecology and Classification of North American Freshwater Invertebrates*, edited by J.H. Thorp and A.P. Covich. New York: Academic Press, pp 431–463.

BRINKHURST, R.O., JAMIESON, B.G.M. 1971. *Aquatic Oligochaeta of the World*. Toronto, ON: University of Toronto Press.

BRUSCA, R.C., BRUSCA, G.J. 1990. *Invertebrates*. Sunderland, MA: Sinauer.

BULLOCK, T.H. 1948. Physiological mapping of giant nerve fiber systems in polychaete annelids. Physiol. Comp. Oecol. 1, 1–14.

BULLOCK, T.H. 1965. Annelida. In *Structure and Function in the Nervous System of Invertebrates*, vol. 1, edited by T.H. Bullock and G.A. Horridge. San Francisco, CA: W.H. Freeman and Co, pp. 661–790.

CHAPRON, C. 1970. Study in oligochaetous *Eisenia foetida* of morphallaxis phenomena which are demonstrated in the previous digestive duct during cephalic regeneration. C. R. Acad. Sci. Hebd. Seances Acad. Sci. D 270, 1362–1364.

DING, J., DREWES, C.D., HSU, W.H. 2001. Behavioral effects of ivermectin in a freshwater oligochaete, *Lumbriculus variegatus*. Environ. Toxicol. Chem. 20, 1584–1590.

DREWES, C.D. 1984. Escape reflexes in earthworms and other annelids. In *Neural Mechanisms of Startle Behavior*, edited by R.C. Eaton. New York: Plenum Press, pp. 43–91.

DREWES, C.D. 1990. Tell-tail adaptations for respiration and rapid escape in a freshwater oligochaete (*Lumbriculus variegatus* Müll.). J. Iowa Acad. Sci. 97 (4), 112–114.

DREWES, C.D. 1996a. Heads or tails: patterns of segmental regeneration in a freshwater oligochaete. In *Tested Studies for Laboratory Teaching*, vol. 17, edited by J.C. Glase. Toronto, Canada: Association for Biology Laboratory Education (ABLE), pp. 23–34.

DREWES, C.D. 1996b. Those wonderful worms. Carolina Tips 59, 17–20.

DREWES, C.D. 1999a. Helical swimming and body reversal behaviors in *Lumbriculus variegatus* (family Lumbriculidae). Hydrobiologia 406, 263–269.

DREWES, C.D. 1999b. Electromyography: recording electrical signals from human muscle. In *Tested Studies for Laboratory Teaching*, vol. 21, edited by S.J. Karcher. Toronto, Canada: Association for Biology Laboratory Education (ABLE), pp. 248–270.

DREWES, C.D., BRINKHURST, R.O. 1990. Giant nerve fibers and rapid escape reflexes in newly hatched aquatic oligochaetes, *Lumbriculus variegatus* (family Lumbriculidae). Invertebr. Reprod. Dev. 17, 91–95.

DREWES, C.D., CAIN, K. 1999. As the worm turns: locomotion in a freshwater oligochaete worm. Am. Biol. Teach. 51, 438–442.

DREWES, C.D., FOURTNER, C.R. 1989. Hindsight and rapid escape in a freshwater oligochaete. Biol. Bull. 177, 363–371.

DREWES, C.D., FOURTNER, C.R. 1990. Morphallaxis in an aquatic oligochaete, *Lumbriculus variegatus*: reorganization of escape reflexes in regenerating body fragments. Dev. Biol. 138, 94–103.

DREWES, C.D., FOURTNER, C.R. 1991. Reorganization of escape reflexes during asexual fission in an aquatic oligochaete, *Dero digitata*. J. Exp. Zool. 260, 170–180.

DREWES, C.D., ZORAN, M.J. 1989. Neurobehavioral specializations for respiratory movements and rapid escape from predators in posterior segments of the tubificid *Branchiura sowerbyi*. Hydrobiologia 180, 65–71.

DREWES, C.D., CALLAHAN, C.A., FENDER, W.M. 1983. Species specificity of giant nerve fiber conduction velocity in oligochaetes. Can. J. Zool. 61, 2688–2694.

DREWES, C.D., VINING, E.P., ZORAN, M.J. 1988. Regeneration of rapid escape reflex pathways in earthworms. Am. Zool. 28, 1077–1089.

DUPIN, F., COULON, J., LE PARCO, Y., FONTES, M., THOUVENY, Y. 1991. Formation of the extracellular matrix during the epimorphic anterior regeneration of *Owenia fusiformis*: autoradiographical and in situ hybridization studies. Int. J. Dev. Biol. 35, 109–119.

FITZHARRIS, T.P., LESH, G.E. 1969. Gut and nerve–cord interaction in sabellid regeneration. J. Embryol. Exp. Morphol. 22, 279–293.

GILBERT, S.F. 2000. *Developmental Biology*. Sunderland, MA: Sinauer Press.

GOSS, R.J. 1969. *Principles of Regeneration*. New York: Academic Press.

GÜNTHER, J. 1976. Impulse conduction in the myelinated giant fibers of the earthworm. Structure and function of the dorsal nodes in the median giant fiber. J. Comp. Neurol. 168, 505–531.

GÜNTHER, J., WALTHER, J.B. 1971. Funktionelle Anatomie der dorsalen Riesenfaser-Systeme von *Lumbricus terrestris*. Z. Morphol. Tiere 70, 253–280.

GURLEY, K.A., RINK, J.C., SÁNCHEZ ALVARADO, A. 2008. Beta-catenin defines head versus tail identity during planarian regeneration and homeostasis. Science 319, 323–327.

HERLANT-MEEWIS, H. 1964. Regeneration in annelids. Adv. Morphog. 4, 155–215.

HOLSTEIN, T.W., HOBMAYER, E., TECHNAU, U. 2003. Cnidarians: an evolutionarily conserved model system for regeneration? Dev. Dyn. 226, 257–267.

HYMAN, L.H. 1916. An analysis of the process of regeneration in certain microdrilous oligocheates. J. Exp. Zool. 20, 99–163.

HYMAN, L.H. 1940. Aspects of regeneration in annelids. Am. Nat. 74, 513–527.

JAMIESON, B.G.M. 1981. *The Ultrastructure of the Oligochaeta*. New York: Academic Press.

KAWAMOTO, S., YOSHIDA-NORO, C., TOCHINAI, S. 2005. Bipolar head regeneration induced by artificial amputation in *Enchytraeus japonensis* (Annelida, Oligochaeta). J. Exp. Zool. A Comp. Exp. Biol. 303, 615–627.

LESIUK, N.M., DREWES, C.D. 1999a. Autotomy reflex in a freshwater oligochaete, *Lumbriculus variegatus* (Clitellata: Lumbriculidae). Hydrobiologia 406, 253–261.

LESIUK, N.M., DREWES, C.D. 1999b. Blackworms, blood pulsations, and drug effects. Am. Biol. Teach. 51, 48–53.

LESIUK, N.M., DREWES, C.D. 2001a. Behavioral plasticity and central regeneration of locomotor reflexes in the freshwater oligochaete, *Lumbriculus variegatus*. I. Transection studies. Invertebr. Biol. 120, 248–258.

LESIUK NM, DREWES CD (2001b) Behavioral plasticity and central regeneration of locomotor reflexes in the freshwater oligochaete, *Lumbriculus variegatus*. II. Ablation studies. Invertebr. Biol. 120, 259–268.

LYCKMAN, A.W., HEIDELBAUGH, S.M., BITTNER, G.D. 1992. Analysis of neuritic outgrowth from severed giant axons in *Lumbricus terrestris*. J. Comp. Neurol. 318, 426–438.

MARTINEZ, V.G., MENGER, G.J., III, ZORAN, M.J. 2005. Regeneration and asexual reproduction share common molecular changes: upregulation of a neural glycoepitope during morphallaxis in *Lumbriculus*. Mech. Dev. 122, 721–732.

MARTINEZ, V.G., REDDY, P.K., ZORAN, M.J. 2006. Asexual reproduction and segmental regeneration, but not morphallaxis, are inhibited by boric acid in *Lumbriculus variegatus* (Annelida: Clitellata: Lumbriculidae). Hydrobiologia 564, 73–86.

MARTINEZ, V.G., MANSON, J.M.B., ZORAN, M.J. 2008. Effects of nerve injury and segmental regeneration on the cellular correlates of neural morphallaxis. J. Exp. Zool. B Mol. Dev. Evol. 310B, 520–533.

MINO, L.A., FOLCO, S., PECHEN DE ANGELO, A.M., GUERRERO, N.R.V. 2005. Modeling lead bioavailability and bioaccumulation by Lumbriculus variegatus using artificial particles. Potential use in chemical remediation processes. Chemosphere 63, 261–268.

MOFFET, S.B. 1996. *Nervous System Regeneration in the Invertebrates*. New York: Springer-Verlag.

MOMENT, G.B. 1950. A contribution to the anatomy of growth in earthworms. J. Morphol. 86, 59–72.

MORGAN, T.H. 1901. *Regeneration*. New York: The Macmillan Co.

MORGULIS, S. 1907. Observations and experiments on regeneration in *Lumbriculus*. J. Exp. Zool. 4, 549–574.

MÜLLER, M.C.M. 2004. Nerve development, growth and differentiation during regeneration in *Enchytraeus fragmentosus* and *Stylaria lacustris* (Oligochaeta). Dev. Growth Differ. 46, 471–478.

MULLONEY, B. 1970. Structure of the giant fibers of earthworms. Science 168, 994–996.

MYOHARA, M. 2004. Differential tissue development during embryogenesis and regeneration in an Annelid. Dev. Dyn. 231, 349–358.

MYOHARA, M., YOSHIDA-NORO, C., KOBARI, F., TOCHINAI, S. 1999. Fragmenting oligochaete *Enchytraeus japonensis*: a new material for regeneration study. Dev. Growth Differ. 41, 549–555.

MYOHARA, M., NIVA, C.C., LEE, J.M. 2006. Molecular approach to annelid regeneration: cDNA subtraction cloning reveals various novel genes that are upregulated during the large-scale regeneration of the oligochaete, *Enchytraeus japonensis*. Dev. Dyn. 235, 2051–2070.

NICOL, J.A.C. 1948. The giant axons in annelids. Q. Rev. Biol. 23, 291–319.

OKADA, Y.K. 1929. Regeneration and fragmentation in the syllidian polychaetes. Roux Arch. Entw. Mech. Organ. 115, 542–600.

PETERSEN, C.P., REDDIEN, P.W. 2008. Smed-betacatenin-1 is required for anteroposterior blastema polarity in planarian regeneration. Science 319, 327–330.

PHIPPS, G.L., WELLS, P.G., McCARTY, L.S. 1995. Introduction to aquatic toxicology. In *Fundamentals of Aquatic Toxicology*, 2nd edn., edited by G.M. Rand. London: Francis & Taylor, pp. 3–67.

ROOTS, B.I., LANE, N.J. 1983. Myelinating glia of earthworm giant axons: thermally-induced intramembrane changes. Tissue Cell 15, 695–709.

SALO, E., BAGUNA, J. 2002. Regeneration in planarians and other worms: new findings, new tools, and new perspectives. J. Exp. Zool. 292, 528–539.

SANCHEZ ALVARADO, A. 2000. Regeneration in the metazoans: why does it happen? Bioessays 22, 578–590.

SANCHEZ ALVARADO, A., TSONIS, P. 2006. Bridging the regeneration gap: genetic insights from diverse animal models. Nat. Rev. Genet. 7, 873–884.

STEPHENSON, J. 1930. *The Oligochaeta*. Oxford: Claredon Press.

TETTAMANTI, G., GRIMALDI, A., CONGIU, T., PERLETTI, G., RASPANTI, M., VALVASSORI, R., DE EGUILEOR, M. 2005. Collagen reorganization in leech wound healing. Biol. Cell 97, 557–568.

TURNER, C.D. 1934. The effects of x-rays on posterior regeneration in *Lumbriculus inconstans*. J. Exp. Zool. 68, 95–115.

YOSHIDA-NORO, C., MYOHARA, M., KOBARI, F., TOCHINAI, S. 2000. Nervous system dynamics during fragmentation and regeneration in *Enchytraeus japonensis* (Oligochaeta, Annelida). Dev. Genes Evol. 210, 311–319.

ZHINKIN, L. 1936. The influence of the nervous system on regeneration in *Rhynchelmis limosella*. Hof. J. Exp. Zool. 73, 43–65.

ZORAN, M.J., DREWES, C.D. 1987. Rapid escape reflexes in aquatic oligochaetes: variations in design and function of evolutionary conserved giant fiber systems. J. Comp. Physiol. [A] 161, 729–738.

ZORAN, M.J., DREWES, C.D. 1988. The rapid tail withdrawal reflex of the tubificid worm, *Branchiura sowerbyi*. J. Exp. Biol. 137, 487–500.

ZORAN, M.J., DREWES, C.D., FOURTNER, C.R., SIEGEL, A.J. 1988. The lateral giant fibers of the tubificid worm, *Branchiura sowerbyi*: structural and functional asymmetry in a paired interneuronal system. J. Comp. Neurol. 275, 76–86.

Part IV

Environmental and Ecological Studies

Chapter 11

Polychaetes in Environmental Studies

Victoria Díaz-Castañeda* and Donald J. Reish†

*Departamento de Ecología, CICESE, Baja California, Mexico
†Department of Biological Sciences, California State University, Long Beach, CA

11.1 INTRODUCTION

Polychaetes are segmented worms common throughout the marine environment, from estuarine muds to coral and rocky reefs to deep-sea sediments, and in the pelagic water column where they contribute to meroplankton and holoplankton communities. Polychaetes are especially common in soft sediments where they constitute 35–70% of the macroinvertebrate population (Knox 1977), but they are also found in hard substrates and in algal holdfasts. Several polychaete species have symbiotic associations, existing as nonboring commensals (where an association is clearly advantageous—not necessarily from a trophic point of view—to one of the members without seriously affecting or harming the other), nonboring parasites and symbiotic borers (Martin and Britayev 1998). To date, 292 species of commensal polychaetes belonging to 28 families are reported to be involved in 713 different commensal relationships with organisms including cirripeds, sponges, bivalves, gastropods, decapods, echinoids, holothuroids, cnidarians, asteroids, sipunculids, nudibranchs, ophiuroids and other polychaetes.

The number of recognized polychaete families ranges between 84 (Day 1967; Fauchald 1977) and 87 (Pettibone 1982), with more than 16,000 described species (Blake 1994). Commensals are thus present in just over 31% of known polychaete families and comprise ~1.8% of known species (Martin and Britayev 1998). Polychaetes play an important role in the burial of organic matter, controlling how much organic carbon is moved out of global oceans and the atmosphere. They exhibit extensive morphological variations and symbiotic relationships, reflecting a great diversity of lifestyles.

Polychaetes are present in all oceans where they float or swim in the water column, burrow or wander on sediment surfaces, or live in tubes that they construct. Some species occur in either hypersaline or hyposaline waters. Species belonging to families Capitellidae, Paraonidae and Cirratulidae impact the biogeochemistry of the organically enriched

Annelids in Modern Biology, Edited by Daniel H. Shain
Copyright © 2009 John Wiley & Sons, Inc.

sediments by burrowing and ingesting sediments, influencing particle distribution and affecting redox conditions.

The ability of polychaetes to survive in a wide range of environmental parameters allows them to occupy a variety of habitats. Some polychaetes, for example, tolerate large changes in temperature and salinity. The mesopsammal polychaete *Trilobodrilus axi* is highly resistant to cold temperatures, showing no signs of damage after 5h at −12°C (Westheide and von Basse 2004), while two species of intertidal tube-dwelling polychaetes are resistant to 42.5°C (*Diopatra cuprea*) and 40.5°C (*Clymenella torquata*) (Kenny 1969). The salinity resistance of *Alitta virens* (Nereididae) in the White Sea increases during ontogenesis; successful fertilization and larval development occur at 22–34‰, while early nektochaetes normally develop at 12–45‰ (Ushakova and Saranchova 2003). The nektochaeta larvae of *Harmothoe imbricata* (Polynoidae) show the greatest salinity resistance, tolerating salinities of up to 14‰. The larvae of *Spirorbis spirorbis* die in 8–14 days at 10‰ (Ushakova and Saranchova 2003). Large numbers of *Nereis succinea*, *Streblospio benedicti* associated with barnacles and amphipods live in the Salton Sea, California, which has a salinity of ~47‰ and is still rising (Oglesby 2005).

Polychaetes are important zoobenthos in oxygen minimum zones (OMZs). OMZs generally form in areas where upwelling leads to high surface productivity that sinks and degrades, with the largest OMZs occurring at bathyal depths in the eastern Pacific Ocean. Polychaetes inhabiting OMZ maximize oxygen uptake through morphological, physiological and architectural adaptations (Levin 2003). For example, some sessile species construct compacted mud dwellings to obtain stability in soupy sediments. These dwellings take the form of mudballs in the cirratulid polychaete *Monticellina* sp. (Levin and Edesa 1997) and tubes in spionid polychaetes (Levin et al. 2000). Sediments in the Santa Catalina Basin, where O_2 is 0.4 mL/L, harbor similar mud-walled cirratulids (e.g. *Tharyx luticastellus*; Smith 1986). Elongated, proliferated and numerous branchiae appear to be adaptations to permanent hypoxia in spionid, dorvilleid, cirratulid and lumbrinerid polychaetes in OMZ sediments. Cossurid polychaetes within the Oman margin OMZ have an exceptionally long median antenna that is thought to aid in respiration (Lamont and Gage 2000).

A few polychaete species occur in nonmarine habitats. Some Spionida species live in freshwater lakes; for example, the feather duster *Manayunkia speciosa* is found in the Great Lakes and in other lakes in North America (Kleem 1985). Other species (e.g. *Hrabeiella periglandulata* and *Parergodrilus heideri*) inhabit terrestrial environments and have many features in common with Clitellata taxa, including a prostomium lacking appendages, simple and short chaetae, absence of parapodia, epidermis without kinocilia, sensory cells with cilia that project only slightly beyond the cuticle or not at all, direct development, direct sperm transfer and eggs laid in cocoons (Purschke 1999). Two terrestrial polychaete species, *P. heideri* and *H. periglandulata*, are known in Europe, and have been found in Polish oak-hornbeam stands (Dumnicka and Rożen 2002). The intertidal polychaete *Stygocapitella subterranean* occurs in sandy beaches between marine and terrestrial realms. A few species, primarily Nereididae (e.g. *Neanthes lighti)*, are known from the supralittoral zone in central California.

11.2 ESTUARINE OCCURRENCE

Estuaries are biologically diverse ecosystems, especially in the spatial relationships between macrobenthos and their food sources. They are also harsh ecosystems subject to highly variable environmental conditions, sites of major industries and repositories for effluents. Organisms able to colonize estuaries have evolved mechanisms that allow them to survive fluctuations in water salinity, temperature, long periods of immersion and emersion and, in some instances, pollution (Reish 1979). Polychaetes have developed different strategies for adapting to estuarine habitats, varying in morphology, feeding types and reproductive modes. Species found in estuaries are unique to that environment; the benthic estuarine community is dominated by polychaetes that burrow into sediments or live within tubes or burrows that extend above the sediment surface.

11.3 INTERTIDAL OCCURRENCE

Intertidal polychaetes are exposed to highly dynamic environments, as related to tidal state and amplitude, current velocity and sediment morphodynamics (Magni et al. 2006). Intertidal areas are under increasing pressure from urbanization as 50% of the world's population lives near the coast. Polychaetes in the intertidal zone possess a variety of adaptations to cope with the effect of physical variables, including changes caused by moving sediment and feeding behavior (Tamaki 1987). Resuspension and deposition of organic matter from the overlying water plays an important role in the settlement and composition of these populations. Low tide exposes animals to higher temperatures and desiccation, presenting challenges to their metabolism (Sokolova and Pörtner 2001).

Polychaetes living in high intertidal zones may have increased thermal resistance including heat stability of metabolic enzymes and expression of heat shock proteins. Polychaetes living in intertidal mud flats are often the prey of shore birds, fishes and arthropods. For example, the stomachs of 73 shorebird specimens representing seven species were analyzed from Bahia San Quintin, Baja California. Polychaetes were the main food items for the marbled godwit and dunlin, and to a lesser extent the willet, sandpiper and short-billed dowitcher (Reish and Barnard 1991). In Argentina, fire ants *Solenopsis richteri* prey on polychaetes during the summer, building intertidal galleries perpendicular to the shoreline that are used to transport polychaetes to their nest (Palomo et al. 2003).

Polychaetes gain protection from desiccation and potential predators in burrows, tubes and galleries. Burrowing morphologies vary, but subsurface deposit feeders (e.g. *Cossura* sp., *Arenicola marina*, *Capitella* sp., *Sternaspis scutata*, *Euclymene* sp., *Mediomastus* sp.) have increased protection from predation. Sediment-dwelling polychaetes can increase the percolation of water through pumping or by the bioturbation associated with deposit feeding or burrowing. Burrowing depletes oxygen because mud can be anoxic even at shallow depths. Many polychaetes, such as *Chaetopterus variopedatus*, pump oxygenated water through their parchment tube. Anoxic conditions cause changes in an animal's metabolism, including glycogen content. The glycogen content of the whole body or muscle of 14 polychaete species was determined, and the substrate affinity of glycogen synthetase from four of the species was measured (Augenfeld 1978). Species-average glycogen concentrations correlated with the risk of anoxia each species faced in the field. Polychaetes most likely to experience anoxia had the highest proportion of active glycogen synthetase and, in such species, glycogen synthetase had the highest affinity for substrate. Augenfeld (1978) concluded that evolutionary adaptation to oxygen-poor habitats occurs through qualitative changes in enzyme properties rather than through quantitative changes in enzyme production.

11.4 MUSSEL BEDS

Mussels attach and grow on rocks, pilings, boat floats, ships, and on intertidal and subtidal mud flats, providing refuge and habitats for a broad range of associated organisms including polychaetes (Rius Viladomiu 2004). Tokeshi (1995) analyzed patterns of poly-chaete abundance and dispersal associated with mussel beds in wave-exposed rocky intertidal habitats on the Pacific coast of South America. Total polychaete densities in the lower intertidal habitat ranged from 18,800 to 36,300 ind/m^2, comparable to some high values recorded for polychaete communities in soft sediments. Populations of *Pseudone-reis, Halosydna, Lumbrineris, Scoloplos, Mediomastus, Typosyllis* and other Syllidae genera were abundant, indicating that mussel bed environments on hard substrata are a major polychaete habitat.

A seasonal study over 2.7 years of polychaetes associated with a *Mytilus edulis* (=*Mytilus galloprovincialis*) community on floating docks in a southern California bay showed that their occurrence was related to temperature, with a greater number of species taken during summer months (Reish 1964). The most common species, *Halosydna john-soni* and *Platynereis bicanaliculata*, settled during spring months, while *Hydroides nor-vegica* (=*Hydroides pacifica*) and *Polydora ligni* polychaetes were abundant during the summer; no single species was present throughout the year (Reish 1964). Polychaetes associated with these mussel beds feed on particulate matter present on shells or between mussel shells. Nereids and polynoids feed on associated amphipods and other species of polychaetes (Reish 1964), in contrast to fauna associated with mussel beds on soft bottoms that contain species feeding on material accumulated by mussels, such as feces and pseu-dofeces. Thiel and Ullrich (2002) concluded that mussels on hard substrate provide a substratum for associated fauna, whereas mussels on soft bottoms provide both substratum and food resources. Reducing sediments caused by the accumulation of biodeposits in soft sediment mussel beds may result in lowering the redox potential values in calm waters, making it possible for the polychaete *Capitella capitata*, an indicator for organic enrich-ment and other opportunists, to flourish.

11.5 SEA GRASSES

Sea grasses play an essential role for the coastal benthos due to their structural complexity, which strongly influences the local environment and provides the spatial structure and resources for rich associated communities including many species of polychaetes and peracarids.

Sea grasses are a conspicuous part of intertidal and shallow subtidal marine communi-ties; their roots accumulate sediment and are a haven for many polychaete species and other invertebrates. Eel grasses are found in protected waters and surf grasses on open coastal rocks. Crouch (1991) studied the polychaete occurrence in surf grass roots present in Southern California over 1 year. She recorded 91 polychaete species, six of which comprised over 50% of the population and were present throughout the year: *Fabricia berkeleyi, Exogone lourei, Mooreonuphis stigmatis, Branchiomaldane vincentii, Parapionosyllis* sp. and *Typosyllis* sp. The largest number of species and specimens were collected in December. The dominant polychaetes were deposit feeders and a combination of filter/deposit feeders, reflecting the importance of sediment accumulation around the roots.

Gambi et al. (2003) described polychaete borers from the family Eunicidae within sheaths of the sea grass *Thalassia testudinum*, in the Mexican Caribbean coast. The poly-chaetes *Lysidice ninetta, Lysidice collaris* and *Nematonereis unicornis* bore into dead sheath tissues and are considered detritivores.

11.6 SABELLARID AND SERPULID REEFS

Reefs are constructed by colonies of polychaetes (Gosner 1978) that live in individual tubes made from cemented sand grains (sabellarids) or calcareous secretions (serpulids). Worms attach their tubes to rocks or to their neighbor's tubes, forming large colonies that grow into massive reefs found in intertidal and shallow subtidal rocky shores. Sabellarids are filter feeders that feed on algae and other microbes, encrusting sand and shell fragments, and planktonic microorganisms such as diatoms and foraminiferans. The crevices of tubes provide habitat protection for many species of polychaetes and microcrustaceans. These reefs have a wide distribution in modern seas, but their existence has been traced back to the early Quaternary. Arenaceous tubes have been found in rocks dating back to the Cambrian era [see Hartman (1944) for photographs of sabellarid reefs in Peru].

Serpulids construct calcareous tubes that can form reefs up to 1 km in length and 1–2 m thickness (Ten Hove and van den Hurk 1993), occurring mostly in shallow, temperate subtidal and lagoonal locations. Factors contributing to these mass aggregations include a short larval period, high habitat selectivity, larval gregariousness and lack of competition. Bianchi and Morri (2001) reported that two serpulid species, *Hydroides dianthus* and *Ficopomatus enigmaticus*, build reefs in the Lagoon of Orbetello, a poly-hyperhaline coastal pond. The latter species reached higher densities and built the most conspicuous reefs; the former species, however, had tubes with more robust biomechanical properties (e.g. size, thickness, bulk density and porosity).

The calcareous serpulid worm *F. enigmaticus* is also prominent in the coastal lagoon of Albufera, Portugal, where it encrusts hard substrates. Its tubes differ in shape depending on location, substratum and water level. On the bottom of the lagoon, *F. enigmaticus* develop microatolls and patch reefs measuring over 2 m in height and 4 m in diameter. The rapid growth of these serpulids can change ecological dynamics that result in the lagoon being progressively filled by reefs and empty tubes (Fornos et al. 1997). In cold regions, the serpulid *Serpula narconensis* was reported by Ramos and San Martín (1999) from the South Georgia Island shelf. This particular mass occurrence is the second serpulid "reef" reported in the Antarctic region. Its habitat, at 91- to 105-m depth, is very different from typical serpulid reefs usually present at intertidal zones or shallow depths, and may be one of the largest serpulid reefs in the world.

11.7 BENTHIC COMMUNITY STRUCTURE

Polychaetes play a dominant role in the establishment and maintenance of subtidal benthic communities (Díaz-Castañeda and Safran, 1988; Snelgrove et al. 1997; Díaz-Castañeda 2000). Reish (1961) followed the initial settlement of animals over a 3-year period in a newly dredged marina in southern California and found that 7 weeks after the bottom was dredged, the initial inhabitants were dominated by polychaetes in both species diversity and specimens, and these continued to increase over the next year. The community structure was similar to other areas of Alamitos Bay, California. Díaz-Castañeda et al. (1989) studied the colonization process of macrofauna in the North and Mediterranean Seas and found that polychaetes are important pioneer species arriving rapidly to defaunated sediments. *Capitella* sp., *Spio* sp., *Cirratulus cirratus*, *Pygospio elegans* and *Polydora ciliata* were among the early colonizers after sediment disturbance.

Predatory polychaetes also influence the structure of soft-bottom communities. In a series of field experiments, Ambrose (1984) observed that *Glycera dibranchiata* consumes

Nereis virens, allowing *P. ligni, S. benedicti, Scoloplos robustus* and several phyllodocids to increase in number. Biologically generated refuges involve an organism's utilization of a refuge that is the product of another organism(s), and this can influence the structure of soft-bottom communities. Woodin (1978) demonstrated the effectiveness of several refuge types, particularly generated by polychaetes. The refuge-forming onuphid *D. cuprea* inhabits shallow water, medium-grained sand flats. The abundance and species richness of associated infauna were positively correlated with the presence of *Diopatra* tubes in the immediate vicinity.

The processes influencing patterns of community composition and diversity in marine habitats include those that operate before and after colonization. Postcolonization processes include abiotic disturbance, predation and competition. Differences in polychaete larval and juvenile supply as well as larval and juvenile habitat selection are colonization-related processes that could influence benthic patterns. Larval supply is generally accepted as playing an important role for some benthic species at multiple temporal and spatial scales (Snelgrove et al. 2001).

Some marine invertebrates live in habitats such as hydrothermal vents, hydrocarbon seeps, coastal mudflats and marshes, where they are regularly exposed to sulfide that can reach millimolar concentrations (Hance et al. 2008). *Glycera dibranchiata* was tested and was unaffected at sulfide concentrations up to 0.25 mM/L; at higher concentrations, animals showed decreased erythrocyte count indicating cell loss. Hance et al. (2008) concluded that at least some sulfide-tolerant polychaetes experience cellular injury and impaired tissue proliferation when exposed to environmentally relevant sulfide concentrations.

11.8   UNUSUAL BENTHIC HABITATS

Polychaetes are found in a broad range of marine environments and have been encountered in some unusual habitats. Examples of some of these unique environments are described below.

11.8.1   Shell-Boring Polychaetes

Some polychaetes, especially members of the family Spionidae (e.g. *Polydora commensalis, Polydora concharum, Polydora socialis, Polydora websleri, Boccardia hamata, Polydora* sp.) penetrate the shells of mollusks. Some have been found in the oyster *Crassostrea virginica*, which reduces the oyster's ability to accumulate nutritional reserves (Wargo and Ford 1993). The cirratulid *Dodecaceria* sp. and the sabellid *Pseudopolamilla reniformis*, are commonly found in shells of *Placopeclen magellanicits* in Maine waters (Blake 1969). The sabellid *Terebrasabella heterouncinata* has been found infesting cultured abalones (*Haliotis rufescens*) in California, causing abnormal shell formation and rendering the abalone unsuitable as an appetizer. This polychaete is not host-specific since many other California native gastropods are also infested (Cáceres-Martinez et al. 1999; Kuris and Culver 1999).

11.8.2   Skeletons of Marine Mammals

Polychaetes that feed on the fatty tissues within bones of dead whales have been found. The endemic genus *Osedax* has been described from whale-fall communities (Glover et al. 2005). *Osedax* species lack a digestive tube as an adult but grow a system of root-like tubes into freshly exposed whale bone using lipids broken down by intracellular symbionts inside the worm. Bacteria in the roots help digest fats in the bone and transfer nutrients to the worms. The largest female recovered was ~5 cm long and 1 cm in diameter. Evidence from molecular genetics suggests that *Osedax* species arose at least 40 million years ago, in the late Eocene (Rouse et al. 2004).

11.8.3 Hydrothermal Vents (Also See Chapter 15)

At seafloor spreading centers, cold water penetrates into the seafloor through fissures where it is heated near the roof of shallow magmatic chambers beneath the ridge axis. The heated seawater, enriched with metals and other substances leached from surrounding rock, is then expelled at the seafloor in highly localized sites known as hydrothermal vents.

The discovery of life at hydrothermal vents in 1977 opened the door to a new realm of life without sunlight and photosynthesis. Vent organisms rely on chemical energy from the oxidation of hydrogen sulfide to fuel life processes. Hydrothermal vent regions have extremes in temperature, areas of very low oxygen, and toxic hydrogen sulfide and heavy metals. Polychaetes are a large component of deep-sea hydrothermal vent invertebrate faunas, tolerating darkness, great pressures and hot temperatures near the vent itself. Nearly 25 years after the discovery of the first hydrothermal vent system, one of the most thermotolerant animal on earth was identified, *Alvinella pompejana*, the Pompeii worm. Pompeii worms are deep-sea polychaetes that reside in tubes buried into the sides of hydrothermal vents along the seafloor. They can reach up to 5 in. in length and are pale gray with red tentacle-like gills on their heads (Van Dover 2000). Perhaps most fascinating is that their pygidium often rests in temperatures as high as 80 °C, while their feather-like head sticks out of their tubes into much cooler water (22 °C). The rich community of epibiotic bacteria that coat its dorsal surface appear to shield worms from intermittent blasts of hot, metal-rich water. Living in a symbiotic relationship, these polychaetes secrete mucous from dorsal glands to feed the bacteria, and in return they are protected by some degree of insulation. Several polychaete families have been found in deep-sea vents including Nereididae, Ampharetidae, Amphinomidae, Eunicidae, Polynoidae, Serpulidae, Siboglinidae and Spionidae (Shank et al. 1998).

11.8.4 Cold Methane Environments

Ecosystems known as cold seeps are found where reduced sulfur and methane emerge from seafloor sediments without an appreciable temperature rise. The first cold seep was found ~20 years ago on the Florida Escarpment in the Gulf of Mexico (Paull et al. 1984). Several polychaete families have been found in this marine habitat including Maldanidae, Polynoidae, Phyllodocidae, Hesionidae, Sabellidae, Ampharetidae, Chaetopteridae, Capitellidae, Dorvilleidae (especially the genus *Ophryotrocha*), Nerillidae, Paraonidae and Serpulidae, as well as vestimentiferan worms (e.g. *Seepiophyla jonesi, Lamellibranchia* cf. *luymesi*).

Levin et al. (2006) studied macrofauna colonization at methane seeps on the northern California margin (525 m) using colonization trays. They examined the influence of sulfide on recruitment and survival by deploying sediments with and without sulfide added. Within seep patches, annelids comprised five of the top 10 dominant taxa; the polychaetes *Mediomastus* sp., *Aphelochaeta* sp., *Paraonidae* sp. and *Nerillidae* sp. exhibited significantly higher densities in sulfide additions. Six dorvilleid polychaete species were collected; four of them occurred exclusively in trays with sulfide added.

Gas hydrate mounds in the Gulf of Mexico provide a specialized substratum for the methane iceworm *Hesiocaeca methanicola*, which burrows into the deposits. These polychaetes inhabit the surface of the gas hydrate at densities up to 3,000 ind/m^2. To what extent the worm colonies use hydrate mounds for protection or nutrition is not yet known, but they are the only animals known to inhabit this unique habitat. Carbon isotope signatures were consistent with consumption of methane-oxidizing archaea by some dorvilleid polychaetes, and with grazing on filamentous sulfur bacteria by gastropods and polychaetes from Oregon and California seeps (Levin and Michener 2002). *H. methanicola* appears to feed heterotrophically on bacteria associated with the hydrates (Levin 2005). In the northern Gulf of Mexico, MacDonald et al. (2003) found hydrate deposits also colonized by the polychaete *H. methanicola*. Seep mussel beds support a diverse group of smaller macrofauna that include chaetopterid, maldanid, capitellid and nautiliniellid (mussel symbiont) polychaetes.

In seep sediments from northern California and Oregon, 17 species of dorvilleid polychaetes were found, with 10 species belonging to the genus *Ophryotrocha*. Other polychaetes frequently encountered in seep sediments include ampharetid, phyllodocid, polynoid, chaetopterid, capitellids, maldanids, serpulid and hesionid polychaetes (Van Dover et al. 2003; Levin et al. 2006).

11.8.5 Plankton

Some polychaetes are holopelagic, completing their entire life cycle in the water column and forming part of the pelagic ecosystem. These polychaetes comprise six families: Tomopteridae, Alciopidae, Lopadorhynchidae, Typhloscolecidae, Iospilidae and Pontodoridae (Fauchald 1977). Rouse and Pleijel (2001) also include the family Poeobiidae as planktonic. Fernández-Alamo (2003) have found an important species richness of pelagic polychaete assemblages in some areas along the California current system, partially due to the high primary productivity in this region.

Many benthic polychaetes leave the benthos and swim to the surface to spawn. Most polychaetes are gonochoristic, having separate sexes; however, some families undergo epitoky, a form of differentiation for sexual reproduction. Epitokes may reach critical abundances, as in members of the family Nereididae, or reproductive stolons, as in members of the family Syllidae. Such behavior occurs at dusk or early evening and may be related to phases of the moon. These worms typically die after spawning or are eaten by fish.

Macroscopic marine particles of detritus, living organisms and inorganic matter known as marine snow are an important component of the environment for many polychaete larvae. Marine snow aggregates contain rich microbial communities and adaptations to life in these microhabitats may have played a role in the evolution of polychaete larval traits. Shanks and Carmen (1997) collected 12 polychaete families from plankton, 10 of which were associated with marine snow, suggesting that precompetent polychaete larvae may spend ~5 h/day visiting marine snow aggregates and may visit ~90 aggregates per day. Competent larval polychaetes may spend >19 h/day on aggregates.

11.9 FEEDING GUILDS

Some studies have tested the applicability of polychaete feeding guilds to ecological and environmental assessments (Pagliosa 2005). Diverse feeding structures can be grouped into different functional categories that are useful for ecological purposes (e.g. raptorial feeding, nonselective deposit feeders, selective deposit feeders and filter feeders). The most common feeding schemes generally divide polychaete assemblages into five or six trophic categories: omnivores, herbivores, carnivores, filter feeders, surface deposit feeders and burrowers. These characteristics are related to the size and composition of food particles, as presented in the Fauchald and Jumars (1979) scheme of polychaete feeding guilds. According to these authors, two types of feeding (macrophagous and microphagous) can be differentiated, which in turn can be subdivided into five submodes and 12 subgroups formed according to the feeding morphology of each mode. These subgroups are combined with the three components of feeding motility to form 22 feeding guilds that are biologically acceptable.

Penry and Jumars (1990) analyzed gut architectures of 42 marine polychaete species in terms of their anatomically distinct compartments. They classified polychaetes into four groups: carnivores with tubular guts, deposit feeders with tubular guts, deposit feeders with three gut compartments and deposit feeders with four or five gut compartments. Tubular guts are common among deposit feeders and may allow relatively rapid ingestion rates. Median gut volume per unit body volume in deposit feeders (31%) is twice that of carnivores (15%) and up to 83% more in one deep-sea polychaete species (*T. luticastellus*). Deep-sea deposit feeders tend to have relatively larger and longer guts than closely related nearshore and shelf species. Guts of a number of deep-sea deposit feeders and nearshore and shelf deposit feeders from muddy environments are relatively longer and narrower as body size increases, suggesting that digestive diffusion limitations may be important (Penry and Jumars 1990).

Polychaetes are abundant within sediments and are present in all oceanic regions of the world. Deposit-feeding polychaetes constitute the dominant macrofauna in environments that tend to be depositional centers for organic matter and contaminants (Jumars et al. 1990). Deposit-feeding invertebrates dominate metazoan life in marine sediments, both in terms of abundance and biomass (Pearson and Rosenberg 1978). Polychaetes are major controllers of sediment ecosystems, and their mixing of sediment particles is an important driving force behind chemical reactions and transport of organic matter in marine sediments (Levinton 1995). Typical "conveyor-belt" feeders, such as the polychaetes *C. torquata* and *Heteromastus filiformis*, are intense perturbers. They live head-down in relatively permanent vertical burrows, feeding on deposits at some depth. Anoxic material is transported as fecal pellets to the sediment surface. Surface deposit-feeding polychaetes (e.g. *Nereis* spp. and *Amphitrite* spp.) rework the sediment more by burrowing than by feeding, because they mainly depend on fresh detritus at the surface as a food source.

The quantity and the nutritional quality of the organic matter present in marine sediment are believed to affect the biomass and trophic structure of macrofaunal communities. Marine sediments comprise only a few percent organic matter, and even organic-rich sediments may be 95% mineral phase. Because most sedimentary organic matter is refractory, deposit feeders survive on a remarkably poor food source (Jumars and Wheatcroft 1989) and thus meet their nutritional requirements by ingesting large amounts of bulk sediment. Species like the common lugworm *A. marina* ingest up to 20 times their body weight of wet sediment per day (Cadée 1976).

11.10 ALGAL "GARDENING" BEHAVIOR

Some animals engage in behaviors that result in the maintenance of food patches that are predictable and available (Plagányi and Branch 2000). Such behaviors include those of some polychaetes (e.g. Nereididae, Onuphidae) that increase the biomass or productivity of an algal community (Branch et al. 1992).

Benthic animals provide anchorages to tufts of algae, and these in turn provide shelter and food for mobile benthos. Nereid polychaetes *Nereis vexillosa* and *P. bicanaliculata* attach pieces of drift algae to their tube surfaces. The presence of permanent algal cover increases the predictability of the food supply for herbivores (including nereids) and deposit feeders, and modulates temperature and salinity stresses of the marine intertidal soft-bottom environment. The attachment behavior of nereid polychaetes probably increases the ability of algae to colonize these habitats (Woodin 1977). Schories and Reise (1993) suggest that primary and secondary attachment of *Enteromorpha* filaments provided by benthic fauna (e.g. *A. marina*) is an essential step in the development of green algal mats on sedimentary tidal flats. Associated with shoots of seagrass (*Zostera noltii*) are the tubes and burrows of several polychaetes: *Pygospio elegans, Pectinaria koreni, Nereia diversicolor, Neanthes virens* and *A. marina*. Olivier et al. (1996) revealed a gardening type of activity of *Nereia diversicolor* on *Enteromorpha intestinalis* where polychaetes attach fragments of algae near their tube entrance and ingest parts of algae and aquatic macrophytes.

11.11 POLYCHAETES AS ENVIRONMENTAL INDICATORS AND REMEDIATORS

Benthic organisms have been especially useful in applied research and are good indicators of environmental stress for several reasons. As stated by Montagna (2005): "… sediments are the 'memory' of the ecosystem. Because they live and feed in sediments, benthic organisms are usually the first organisms affected by pollution or stress."

Benthic macroinvertebrates are important components of marine ecosystems, serving essential functions in wide-ranging capacities; for example, they are known to alter sediment properties by feeding, irrigating, and by constructing tubes and burrows, which cause changes in sediment parameters such as porosity, particle size, fluxes and diffusion rates (Aller 2001). Polychaetes play a major role in benthic communities, in terms of recycling and reworking benthic sediments, bioturbating sediments and in the burial of organic matter. Many species living within or on the surface of sediments provide the prey base for fish and other marine predators, while others decompose organic material as a crucial step in nutrient cycling. Polychaetes are often dominant in both species diversity and abundance in macrobenthic communities; their trophic flexibility and life history traits are considered preadaptive to conditions of disturbed habitats (Giangrande 1997).

Polychaetes, by their burrowing and feeding activity, may enhance various sedimentary processes. Gunnarsson et al. (2000) has shown that bioturbation enhances the release of hydrophobic organic contaminants from sediment to overlying water. Measures of community condition have long been used for water quality assessments because polychaetes tend to be more sedentary and thus more reliable site indicators over time, compared to nekton and plankton. Ventilation of burrows and tubes is a major factor controlling biogeochemical processes occurring in sediments (Kristensen 1989). Polychaetes buried in the sediment generally pump water through their burrows by active ventilation. Such renewal of burrow water serves several important purposes including gas exchange, food transport and metabolite removal.

From a management perspective, polychaetes are useful organisms for identifying problem sites and for assessing problem severity. They respond to disturbance induced by different kinds of pollution and exhibit quantitative changes in assemblage distributions (Díaz-Castañeda et al. 1989). Polychaetes can also be used as indicators of benthic environment recovery from perturbation because in most cases, they are major elements of the recolonization process in hard and soft bottoms, and are important to initial community structuring. In colonization studies in the Mediterranean, North Sea, Caribbean and Pacific Ocean, polychaetes have been pioneering species, rapidly colonizing soft and hard bottoms (Díaz-Castañeda et al. 1989, 1993; Díaz-Castañeda and Almeda 1999; Díaz-Castañeda 2000).

Giangrande et al. (2005) proposed that the filter feeder sabellid *Sabella spallanzanii* is a biofilter useful to treat wastes from intensive aquaculture. Laboratory experiments indicate a positive action of its feeding activity on solid waste removal from the water column coupled with an interesting microbiological activity. Stabili et al. (2006, 2007) identified a bioremediation activity of *S. spallanzanii* in the Mediterranean Sea based on the ability of this species to feed and accumulate bacteria from water samples. This ability enables this polychaete to be used as a bioindicator in detecting and monitoring microbial pollution, as well as a potential bioremediator by reducing pollution in the water column. Stabili et al. (2006) determined that bacterial densities were usually orders of magnitude higher in polychaete homogenates than in the corresponding seawater, confirming the ability of *S. spallanzanii* to accumulate the microbial pollution. Licciano et al. (2005) estimated clearance rates and retention efficiencies of two polychaetes using the bacterial species *Vibrio alginolyticus* at a concentration of 25×10^3 cells/mL for *S. spallanzanii* and 5×10^3 cells/mL for *Branchiomma luctuosum*. The retention efficiency was 98% for *B. luctuosum* and 70% for *S. spallanzanii*. Both species were efficient at removing *V. alginolyticus* from seawater, indicating their possible use as biofilters of microbially contaminated waters. In most cases, polychaete assemblages demonstrate the same distribution patterns as the benthic fauna taken in its entirety (Fauchald 1973). Due to time, cost and taxonomic expertise necessary to identify aquatic fauna at a species level, some pollution studies have used polychaetes as surrogates because of their sufficiency to detect community responses to pollution or disturbance without important loss of information (Warwick 1988; Chapman 1998).

Dorgan et al. (2006) correlated behaviors and anatomies of some burrowing animals, including polychaetes, with physical constraints of the sediment environment. These recent studies increase our knowledge about feeding ecology in polychaetes and modify previous ideas. For example, Dorgan et al. (2005) describe a previously unknown mechanism that burrowing polychaetes use to progress through sediments, namely, crack propagation, in which an alternating "anchor" system of burrowing serves as a wedge to extend the crack-shaped burrow. Burrowing polychaetes have developed different mechanisms that act as wedges to create cracks and to elongate their burrows. Animals with subterminal expansions likely used them as wedges for crack propagation: many cirratulids have expansions at both ends for burrowing in either direction; some terebellids have wedges of unknown function, and some trichobranchids carry highly ornate wedges whose functions remain unknown. Magelonids have an apparently wedge-shaped prostomium that was thought to facilitate burrowing, but others have classified them as surface deposit feeders (Dorgan et al. 2005).

Members of the family Sternaspidae have been described as active burrowers that use their anterior region to dig into sediments. Fauchald and Jumars (1979) classified them as subsurface deposit feeders, and recent observations by Dorgan et al. (2005) have shown *Sternaspis* as a vigorous subsurface burrower that rarely comes to the sediment surface. Cossurids are surface deposit feeders that use their ciliated buccal tentacles by opening their buccal cavity and placing them on the sediment surface (Jumars 1978; Tzetlin 1994). Dorgan et al. (2005) found that they feed below the sediment–water interface in cracks of their own manufacture. Cirratulids are considered surface deposit feeders that use their palps for food collection and in doing so can accumulate toxins such as heavy metals and arsenic (Rouse and Pleijel 2001). *Chaetozone setosa* was found to be an early colonizer of sediments that had been defaunated and could reach densities up to 10,000/m^2 (Hily 1987). Terebellid, trichobranchid and cirratulid polychaetes generally are considered surface deposit feeders (Fauchald and Jumars 1979); however, Nowell et al. (1989) note that at least some terebellids, trichobranchids and bipalpate cirratulids rarely appear at the sediment surface. Burrowing by liquefaction in sandy sediments has been described among terebellids, although the mechanism is not clear (Nowell et al. 1989).

Maldanids are generally considered head-down, conveyor-belt feeders that consume deep sediments and defecate at the surface. They are capable of moving surface deposits into their feeding cavities deep in the sediment, resulting in downward mixing of organic matter (Dobbs and Whitlatch 1982). Based on tracer experiments, Levin et al. (1997) proposed that subduction of organic matter by maldanids may serve a keystone function in bathyal sediments, promoting microbial activity, enhancing organic matter diagenetic processes and providing organic-rich food to deep-dwelling infauna. Mayer et al. (1997, 2001) filmed representative species from several phyla in the process of burrowing and developed image-analytic computer techniques to analyze the process. Worms using peristaltic mechanisms also propagate cracks, and the threshold concentrations of sedimentary food at which marine benthic invertebrates can obtain nutrition and ingest sediment have been determined. They also reassessed feeding and movement modes of various deposit feeders to establish feeding and movement patterns that can be used in bioturbation models.

Burrowing deposit feeders are of both global and local importance. They enhance the burial of organic matter, which assists in controlling how much organic carbon is moved out of contact with the global oceans and atmosphere. An important fraction of the nutrients used in coastal ocean primary production arises from nutrient cycling within the deposit feeder-modulated seabed, and evidence is growing that this feedback from deposited organic particles is a strongly positive one that may "run away," causing coastal eutrophication and "dead zones" (Donard 1996; Grall and Chauvaud 2002).

Infaunal macrobenthos are known to modify sediment properties by feeding, irrigating, and constructing tubes and burrows, which cause changes in sediment parameters such as porosity, particle size, fluxes and diffusion rates (Krantzberg 1985). Most burrow constructors maintain contact with the overlying water by ventilating water through their burrow system, thus increasing the transport of ions and gases (e.g. O_2) across the sediment–water interface (Kristensen 1989). In the same way, burrowing activity may also alter pollutant concentrations in sediments, causing pollutants to become buried with long-lasting effects on the marine ecosystem. For example, bioturbation by both small and large polychaetes (e.g. *C. capitata* and *A. marina*) is known to bury polycyclic aromatic hydrocarbons (PAHs) into anoxic zones of the sediment (Madsen et al. 1997). This has implications for the fate of PAHs because microbial degradation of these compounds is redox sensitive. Reworking of sediment by polychaetes allows aerobic degradation of PAHs by distributing oxygen deeper in the sediment (Fenchel 1996), and repetitive oscillating redox conditions relative to stable conditions have been shown to stimulate organic matter degradation (Aller 1994).

The benthic landscape is characterized by mounds and depressions produced by the burrowing and feeding activities of polychaetes. Mounds created by subsurface feeders have been shown to influence sediment erosion and transport (Grant 1983). For example, cirratulid species in the genus *Monticellina* construct mudballs in OMZs. The mudballs are cigar shaped, 4.5–25.0 mm long, and positioned vertically so as to project several millimeters above the sediment–water interface, and can reach densities of $16,000/m^2$. This is important because mudballs appear to represent a source of heterogeneity that can influence macrofaunal community structure in deep-sea sediments (Levin and Edesa 1997).

Analyses in different polychaete assemblages and their responses to habitat conditions reflect the biological effects of marine pollution and habitat disturbance. Sediments act as a sink and eventually as a source of contaminants. Thus, the search for benthic organisms sensitive to chemical contaminants has led to intensive research in the past decades in order to assess pollution effects. Chemical measurements are often used to assess the presence of contaminants in sediment, but they do not offer indications of bioavailability and biological effects.

Due to conservation demands, reliable rapid-assessment methods for mapping biodiversity are currently needed. One approach is to use surrogate species, i.e. quantities that correlate strongly with the number of species and diversity but are easier to obtain—preferably a taxonomically well-defined group with long-lived, large species that are easy to sort from the sediment. Some authors (e.g. Olsgard et al. 2003) have tested the potential for using polychaetes to assess the overall diversity and community structure of the macrobenthos. Polychaetes in the order Terebellida have been considered a good indicator of polychaete species richness and to a lesser extent of whole benthic assemblages. This family is promising as a proxy for species richness in marine biodiversity studies (Olsgard et al. 2003).

Many studies use soft-bottom communities, particularly polychaetes, to construct biotic indices, because macrobenthic animals are relatively sedentary (and therefore cannot avoid pollution), have relatively long life spans (thus indicate and integrate water/sediment quality conditions over time), comprise different species with different tolerances to stress, and have an important role in cycling nutrients and materials between the underlying sediments and overlying water column (Borja et al. 2000). Based on their contribution, a marine biotic index (BI) was designed to establish the ecological quality of European coasts (Borja et al. 2003). This index explores the response of soft-bottom communities to natural and human-induced changes in water quality, integrating long-term environmental conditions.

11.12 BIOMONITORING

The object of biomonitoring is to assess the impact of man-made changes, such as organic pollution, introduction of toxic chemicals, dredging and construction, on the marine environment. Two types of biomonitoring are recognized. One is surveillance before and after a project is complete or a toxic substance enters the water, and the other is to have compliance with regulations or to maintain water and sediment quality. Biomonitoring involves the use of indicator species or communities, and can be achieved by (1) examining biota that have been affected by organic and/or chemical pollution at species, population or community levels, (2) using toxicological techniques to examine chemical effects on representative organisms and (3) examining the body burdens of chemical residues in selected organisms. Polychaetes are used in each of these methods of biomonitoring. The discussion below demonstrates their wide uses as marine environmental quality indicators on a global scale.

Polychaetes are used with other species in faunal analyses at species, population and community levels, in pollution gradient studies, as part of multiparameter monitoring programs, as test organisms, and as field monitors of chemical elements and organic compounds. By using polychaetes in benthos, the status of marine environments subjected to increasing loads of pollution can be assessed (Soule 1988). To facilitate these activities, (1) polychaete analyses should be included in all marine environmental quality monitoring and assessment programs; (2) evaluation of the comparative toxicology of polychaetes and other invertebrates (especially crustaceans and mollusks) living in sediments should be conducted; and (3) taxonomic keys and atlases for the polychaetes of coastal waters should be prepared and published (e.g. Pocklington and Wells 1992).

The use of polychaetes in assessing the effects of marine pollution has a long history but has not been recognized by ecologists and public officials until recently. Wilhelmi (1916) found that *C. capitata* played a similar role in the marine waters of Germany as the oligochaete *Tubifex tubifex* does in freshwater streams. The benthic communities around the outfall sewers in Copenhagen Harbor were divided into three zones by Blegvad (1932) based on polychaete species and other invertebrates. No further studies on the role of polychaetes in marine studies were carried out until Hartman (1952), Reish (1955) and Filice (1954) examined the benthos of Los Angeles/Long Beach harbors and divided them into five zones based on polychaete populations. One zone, which received oil refinery wastes, was devoid of marine life, and *C. capitata* was present in areas contaminated with fish cannery wastes or domestic pollution. Similar findings were made around the sewage discharge in Marseille, France (Bellan 1967).

Pearson and Rosenberg (1978) described a general model for the effects of organic enrichment on benthic organisms. Based on this model, a number of methods have been proposed to describe and quantify the effects of pollution (e.g. Gray 1981; Lambshead et al. 1983; Ferraro and Cole 1990; Gray et al. 1990; Clarke 1993; Warwick 1993; Elliott 1994). Pearson and Rosenberg (1978) found that the first detectable change in population structure was an increase in species number, followed by an increase in biomass and subsequently an increase in abundance. Abundance increased dramatically at relatively high organic matter load, generating a peak of so-called opportunistic species that were mainly polychaetes.

Species that can respond rapidly to open or unexploited habitats have been called opportunistic (Hutchinson 1967), fugitive (Hutchinson 1951) or colonizing (Lewontin 1965). Some of their characteristics include a lack of equilibrium population size, density-independent mortality, the ability to increase population size rapidly, high birth rate, poor competitive ability, high dispersal ability (Grantham et al. 2003) and high proportion of resources devoted to reproduction. None of these features alone is adequate to define an opportunist, and thus Grassle and Grassle (1974) proposed an opportunist as a species adapted for life in a short-lived, unpredictable habitat by relying on a high r (rate of increase in the exponential population growth curve) to make use of ephemeral resources. In a new habitat, such a species would discover the habitat quickly, reproduce rapidly to use up resources before other competing species could exploit the habitat and disperse in search of other new habitats as the existing one grows unfavorable.

Grassle and Grassle (1974) suggested that genetic variation and high mortality, which results in intense selection, were important components of adaptation to unpredictable environments. Pearson and Rosenberg (1978) suggested four major stages of change in response to organic enrichment of the benthos: normal, transitory, polluted and grossly polluted along a gradient of organic enrichment. In reality, a smooth transition from one stage to the next and classification into four stages is simply a convenient abstraction of the data. These stages can be used to assess the state of eutrophication in benthic systems and are based on the species composition determined from grab samples and species identification. Normal is characterized by the presence of large, deep-burrowing species such as decapods and echinoids, with a deep redox potential discontinuity (RPD) layer, while transitory is characterized by the presence of smaller organisms, usually deposit-feeding species that replace the large, deep-burrowing species. Polluted is characterized by a shallow RPD layer and strong dominance of small tube-building polychaetes, indicating severe eutrophication, and grossly polluted is characterized by sulfide patches at the sediment surface and no macrofauna (only nematodes survive).

Pearson and Rosenberg (1978) also ranked macrobenthic species (99 species from literature data) in order of their occurence on a gradient of decreasing organic enrichment, and the top 58 species were polychaetes. Polychaetes in this group included the genera *Capitella, Polydora, Streblospio, Scolelepis, Nereis, Platynereis, Neanthes, Dorvillea, Heteromastus, Mediomastus, Eteone, Eumida, Cirriformia, Anaitides, Prionospio, Goniada, Pygospio, Lumbrineris* and *Nepthys*. The dominance of specific polychaetes in sediments can indicate the condition or health of the benthic environment. Of particular interest are members of the families Capitellidae and Spionidae; when found in large numbers, these have become accepted as indicators of pollution (Pocklington and Wells 1992). Other authors suggested that the quality of a marine environment can also be assessed by the absence of some so-called sensitive or intolerant species. For example, Rygg (1985) studied 150 stations of soft-bottom sediments in Norway affected in part by industrial pollution and classified 100 common species according to their distribution along a diversity gradient. About 45 species were absent from the low diversity stations: 25 polychaetes, 5 crustaceans, 7 mollusks and 5 echinoderms.

Licciano et al. (2005) examined the filter-feeding ability of two sabellid species, *S. spallanzanii* and *B. luctuosum*, to accumulate microorganisms under natural and experimental conditions. The authors suggest their employment as biofilters of microbially contaminated waters in intensive aquaculture. As a consequence of their filter-feeding activity, sabellids may accumulate and concentrate many pollutants in seawater, particularly those that are particulate or associated with particles (Stabili et al. 2006; Licciano et al. 2007). This ability to accumulate materials may facilitate the detection and measurement of pollutants at very low environmental concentrations. Sabellids and other polychaetes with efficient filtering capacities may affect the global amount of suspended abiotic and biotic particles, thus contributing to the energy flow from plankton to the benthos.

11.13 TOXICOLOGICAL TESTS

The use of polychaetes in toxicological testing has been reviewed by Reish and Gerlinger (1997). A total of 48 species from 20 polychaete families have been employed as test organisms. Initial tests were aquatic and determined the lethal concentration of a metal chloride for a specified time period, often 96 h. Prior to dredging, tests were initiated to determine if the sediments were toxic for disposal elsewhere. Growth, as measured by body weight, was the criterion used to determine the disposal method of contaminated sediment.

Mercury and copper were the most toxic elements to the polychaete species tested, followed by chromium and cadmium; zinc and lead were the least toxic. No one species

Table 11.1   The 96-h LC50 of selected metals to polychaetes (mg/L)

Species	Cd	Cr	Cu	Pb	Hg	Zn
Capitella capitata	0.6–7.5	4.1–8.0	0.02–0.2	3.0–6.8	0.014–<1.0	2.5–4.9
Cirriformia luxuriosa	15.0		0.9			15.0
Ctenodrilus serratus	4.3		0.3	7.2	0.04	7.1
Dinophilus gyrociliatus	0.8		0.026	2.5		0.2
Ophryotrocha diadema		4.2	0.16	14.0	0.09	1.4
Namanereis merukensis			0.55	3.75	0.04	
Neanthes virens	9.3–<40.0	2.0			0.01	8.1
Neanthes arenaceodentata	12.0–14.0	1.0–4.3	0.08–0.57	10.0	0.15–0.2	1.5–1.8
Nereis diversicolor	84.0–>87.0	0.7–80.0	0.3–0.5	>100.0	0.3–0.75	6.0–49.0
Nereis grubei	4.7	1.7	0.1	>5.0	0.09	1.8
Pectinaria californiensis	2.6	>10.0	0.2	>5.0	0.06	2.8
Halosydna johnsoni	13.8	1.2	0.15	6.3	0.3	6.0

Data modified from Reish and Gerlinger (1997), with permission from the Rosenstiel School of Marine and Atmospheric Science.

was the most sensitive to all metals tested (Table 11.1). For example, *Neanthes arenaceodentata* was sensitive to chromium but was tolerant to lead. Other criteria have been employed to measure aquatic toxicity in polychaetes; for instance, the 96-h cadmium LC50 for *N. arenaceodentata* was 12–14 mg/L, and growth was significantly reduced (Reish and Gerlinger 1984). The minute species *Dinophilus gyrociliatus*, which has a 7-day life cycle, was used by Nipper and Carr (2003) to measure a spiked toxicant in pore water. *D. gyrociliatus* reproduction was consistently among the most sensitive end points when compared to early-life-stage tests with several other marine species, including macroalgae, sea urchins and fish. Other species with short life cycles have been employed to measure toxicity levels based on their reproduction (e.g. number of offspring), including *Ophryotrocha* spp., *Ctenodrilus serratus* and *C. capitata* (=*Capitella* sp. I), and bifurcated trochophore larvae were induced by sublethal concentrations of chromium (Reish 1977). The effects of other aquatic toxicants, including petroleum hydrocarbons, detergents/dispersants, chlorinated hydrocarbons, tributyltin and ionizing hydrocarbons have been tested using polychaetes (Reish and Gerlinger 1997). Johns and Ginn (1990) developed the contaminated sediment growth test using *N. arenaceodentata*. While the survival was >90% in contaminated test containers, growth (measured by dry weight) was one order of magnitude less than controls.

Data on the effects of organic compounds on polychaetes are limited. The 28-day LC50 to dichloro-diphenyl trichloroethane (DDT) was 0.1 mg/L in *N. arenaceodentata* (Reish 1980). Using the same species, Murdoch et al. (1997) did not find an acute effect by either DDT or polychlorinated biphenyls (PCBs) in field- or laboratory-spiked sediments; however, chronic exposure to DDT resulted in a significant reduction in reproductive fitness, as measured by the number of emerging juveniles at concentrations of 173 and 553 mg/kg dry weight. The effect of petroleum hydrocarbons depends on whether they are crude or refined; specifically, the toxicity of southern Louisiana crude oil and refined no. 2 fuel oil to *N. arenaceodentata* was 12.5 and 2.7 mg/L, respectively (Rossi et al. 1976).

Bioaccumulation, or body burden of polychaetes, has been measured both in field and laboratory tests. Bocchetti et al. (2004) measured three trace metals on *S. spallanzanii* at four locations of the Adriatic and Tyrrhenian Sea, Italy. Chromium ranged from 0.5 to 1.3 μg/g, lead from 0.65 to 1.96 μg/g and nickel from 1.13 to 2.6 μg/g. Christensen et al. (2002) demonstrated that the polychaetes *N. diversicolor* and *A. marina* were both capable of rapidly accumulating and eliminating pyrene. Steady-state concentrations of pyrene were established in both species within 5 days of exposure to contaminated sediments, with *A. marina* having 5–10 times higher bioaccumulation factors than *N. diversicolor*. Data on the accumulation of elements and organic compounds in vertebrates and invertebrates, including polychaetes, in both field and laboratory studies, are summarized yearly by Water Environment Research (Mearns et al. 2006).

11.14 ECONOMIC IMPORTANCE OF POLYCHAETES

In the early 1970s, it became apparent that contaminated sediments in marine environments resulted in the bioaccumulation of toxicants, which affected reproduction, death rates and the prevention of repopulating benthic animals. The United States Environmental Protection Agency/US Army Corps of Engineers (US EPA/US ACE 1977) produced a manual for conducting sediment tests with polychaetes and other animals, which is periodically updated (e.g. Reish et al. 2005). A laboratory population of *N. arenaceodentata* was established in 1964 for research by D.J. Reish and students, which became the source of polychaetes for aquatic and sediment toxicological analyses that led to the development of polychaete aquaculture.

Undoubtedly, polychaetes have been used as fish bait for centuries. Polychaetes *G. dibranchiata* and *Neanthes virens* are harvested in estuarine mud flats of Maine and the Canadian Atlantic maritime provinces and are shipped to North American localities. Commercial aquaculture of *N. virens* began in England and The Netherlands in the mid-1980s, establishing alternative sources of this animal as a means of supplying existing markets with marine polychaetes to be used as bait. Other species harvested for bait include *Marphysa sanguinea, Marphysa leidi, Onuphis teres* and *Perinereis nuntia* (Olive 1994). Fishing industries have shown a sustained growth pattern (Olive 1999) and have generated techniques for the cryopreservation of marine polychaete larvae, photoperiodic manipulation of breeding time cycles and optimization of growth processes. Many of these polychaete species are also used as a nutrient for stimulating gonad maturation and for spawning in hatchery-reared species of shrimp and fish, such as *Solea vulgaris, Solea senegalensis, Penaeus kerathurus* and *Penaeus vannani*.

To a limited extent, polychaetes have also been used as human food. The annual swarming of the eunicid *Palola viridis* occurs for 3 days during the third quarter of the October–November moon. Male and female worms cast off their posterior ends, which subsequently swim to the surface to spawn in Samoa, Fiji and the Solomon Islands. The natives know the timing well and consider the worms a delicacy (Caspers 1984). Likewise, people in Fujian, Guangdong and Guangxi coastal provinces (China), Japan and southeast Asia are fond of eating nereids, especially the nereids bearing genital glands (Wu et al. 1985).

11.15 CONCLUSIONS

Studies of soft-bottom macrofauna are important because most marine species are benthic, and sedimentary habitats cover most of the ocean bottom. Polychaetes occupy most parts of the marine ecosystem, but are especially abundant in the littoral zone, where they have been used as pollution-indicator species. They constitute a dominant functional component of macrobenthic communities and reveal a wide range of adaptability to different marine and coastal habitats. Analyses in different polychaete assemblages and their responses to habitat conditions reflect the biological effects of marine pollution and habitat disturbance. Healthy benthic communities are characterized by high biomass and high species richness dominated by polychaetes, and by relatively long-lived and often deep-dwelling species. The movements and deposit-feeding mode of some polychaete species can enhance bioturbation, decompose organic matter and recycle nutrients, all of which affect an ecosystem's dynamics. Polychaetes can also modify the redox status of the surficial sediments.

Many polychaete species show high physiological tolerance to extreme variations in environments, and can grow and reproduce in different sediment types and in stressful habitats. The ability of some polychaetes to feed on bacteria provides the possibility of using them as bioindicators to monitor microbial pollution in marine environments. They can also be used in bioremediation as they can reduce microbial pollution in the water column. Recent studies (e.g. Giangrande et al. 2005; Licciano et al. 2005) suggest that sabellids and other polychaetes with high filtering capacities may be used as biofilters in microbially contaminated waters. Clearly, polychaetes should be included in all marine environmental quality monitoring programs.

The importance of polychaetes in the marine environment, especially the subtidal benthos, was largely overlooked until the 1960s due to collection methods and sample processing. Subtidal collections were done by trawling, which focused on larger animals such as mollusks, crustaceans and echinoderms. While quantitative benthic sampling began in the Danish seas in the late 1890s by Petersen (1918), the method of processing samples was essentially the same. The mesh size of screens used to remove sediment from samples was large, resulting in smaller animals, including polychaetes, being washed back to sea. Hartman (1952, 1955) reported on polychaetes collected in Los Angeles/Long Beach harbors and in locations beyond these harbors in which a screen with a 1-mm mesh size was used. Polychaetes were a dominant member of the subtidal benthos, as summarized later by Knox (1977). Reports by Hartman (1952, 1955) and similar studies by Reish and Winter (1954) and Reish (1959) stimulated others to use fine mesh sieves in processing benthic samples. Later studies used sieves measuring 0.5 and 0.3 mm.

Although polychaetes are useful in environmental studies, their potential is currently underestimated. By monitoring the benthos in general, and polychaetes in particular, the status of marine ecosystems subjected to increasing loads of pollution, climatic change and alien species can be assessed. However, increased monitoring of the marine environment has put pressure on polychaete taxonomists. In the past, polychaete systematists were versed in the entire phylum, but increases in benthic studies have led to specialization at the family level. Now a polychaetologist typically focuses on a specific family or on groups of families having similar characteristics. From where will the next generation of workers capable of identifying polychaetes come? Systematics is no longer emphasized at most universities. The late J. Laurens Barnard (pers. comm.) estimated that it takes an experienced taxonomist ~40 h to describe a new marine invertebrate species. Over 300 new species of polychaetes are known to exist off the California coast alone. Thousands must be elsewhere. Who will describe them? Careful descriptions of newly discovered species are critical to catalog and understand the diversity of life on our planet.

REFERENCES

ALLER, R.C. 1994. Bioturbation and remineralization of sedimentary organic matter: effects of redox oscillation. Chem. Geol. 114, 331–345.

ALLER, R.C. 2001. Transport and reactions in the bioirrigated zone. In *The Benthic Boundary Layer. Transport Processes and Biogeochemistry*, edited by B. Boudreau and B. Jørgensen. Oxford: Oxford University Press, pp. 269–301.

AMBROSE, W. 1984. Influences of predatory polychaetes and epibenthic predators on the structure of a soft-bottom community in a Maine estuary. J. Exp. Mar. Biol. Ecol. 81, 115–145.

AUGENFELD, J. 1978. Relation of habitat to glycogen concentration and glycogen synthetase in polychaetes. Mar. Biol. 48, 57–62.

BELLAN, G. 1967. Pollution et peuplements benthiques sur substrat meuble dans la region de Marseille. Premiére Partie. Rev. Intern. Océanogr. Méd. 6/7, 53–87.

BIANCHI, C., MORRI, C. 2001. The battle is not to the strong : serpulid reefs in the lagoon of Orbetello (Tuscany, Italy). Estuar. Coast. Shelf Sci. 53, 215–220.

BLAKE, J. 1969. Systematics and ecology of shell-boring polychaetes from New England. Am. Zool. 9, 813–820.

BLAKE, J.A. 1994. Introduction to the Polychaeta. In *Taxonomic Atlas of the Benthic Fauna of the Santa Maria Basin and Western Santa Barbara Channel*, edited by J.A. Blake and B. Hilbig. Santa Barbara: Santa Barbara Museum of Natural History, vol. 4, pp. 39–113.

BLEGVAD, H. 1932. Investigations of the bottom fauna at outfalls of drains in the Sound. Danish Biol. Stn. 37, 1–20.

BOCCHETTI, R., FATTORINI, D., GAMBI, M.C., REGOLI, F. 2004. Trace metal concentrations and susceptibility to oxidative stress in the polychaete *Sabella spallanzanii* (Gmelin) (Sabellidae): potential role of antioxidants in revealing stressful environmental conditions in the Mediterranean. Arch. Environ. Contam. Toxicol. 46, 353.

BORJA, A., FRANCO, J., PÉREZ, V. 2000. A marine biotic index to establish the ecological quality of soft-bottom benthos within European estuarine and coastal environments. Mar. Pollut. Bull. 40, 1100–1114.

BORJA, A., MUXIKA, I., FRANCO, J. 2003. The application of a Marine Biotic Index to different impact sources affecting soft-bottom benthic communities along European coasts. Mar. Pollut. Bull. 46, 835–845.

BRANCH, G., HARRIS, J., PARKINS, C., BUSTAMANTE, R., EEKHOUT, S. 1992. Algal gardening by grazers: a comparison of the ecological effects of territorial fish and limpets. In *Systematics Association Special: Plant-Animal Interactions in the Marine Benthos*, vol. 46, edited by D. John, S. Hawkins and J. Price. Oxford: Clarendon Press, pp. 405–423.

CÁCERES-MARTINEZ, J., TINICO, G., BUSTAMENTE, M., GÓMEZ-HUMARAN, I.M. 1999. Relationship between the burrowing worm *Polydora* sp. and the black clam *Chione fluctifraga* Sowerby. J. Shellfish Res. 19, 85–89.

CADÉE, G.C. 1976. Sediment reworking by *Arenicola marina* on tidal flats in the Dutch Wadden Sea. Neth. J. Sea Res. 10, 440–460.

CASPERS, H. 1984. Spawning periodicity and habitat of the palolo worm *Eunice viridis* in the Samoan Islands. Mar. Biol. 79, 229–239.

CHAPMAN, M. 1998. Relationships between spatial patterns of benthic assemblages in a mangrove forest using different levels of taxonomic resolution. Mar. Ecol. Prog. Ser. 162, 71–78.

CHRISTENSEN, M., ANDERSEN, O., BANTA, G. 2002. Metabolism of pyrene by the polychaetes *Nereis diversicolor* and *Arenicola marina*. Aquat. Toxicol. 58, 15–25.

CLARKE, K. 1993. Non-parametric multivariate analysis on changes in community structure. Aust. J. Ecol. 18, 117–143.

CROUCH, C.A. 1991. Infaunal polychaetes of a rocky intertidal surfgrass bed in southern California. Bull. Mar. Sci. 48, 386–394.

DAY, J.H. 1967. *A Monograph on the Polychaeta of Southern Africa*. London: British Museum of Natural History.

DÍAZ-CASTAÑEDA, V. 2000. The early establishment and development of a polychaete community settled on artificial substrata at Todos Santos Bay, Baja California (Mexico). Bull. Mar. Sci. 67, 321–335.

DÍAZ-CASTAÑEDA, V., ALMEDA, C. 1999. Early benthic organism colonization on a Caribbean coral reef (Barbados, West Indies): a plate experimental approach. Mar. Ecol. 20, 197–220.

DÍAZ-CASTAÑEDA, V., SAFRAN, P. 1988. Dynamique de la colonization par les Annélides Polychetes de sédiments defaunés par la pollution dans des enceintes experimentales en Rade de Toulon. Oceanol. Acta 11, 285–297.

DÍAZ-CASTAÑEDA, V., RICHARD, A., FRONTIER, S. 1989. Preliminary results on colonisation, recovery and succession in a polluted area of the southern North Sea. Proceedings of the 22nd EMBS, Barcelona, Spain. Topics in Marine Biology. J. Ros. Ed. Scient. Mar. 53, 705–716.

DÍAZ-CASTAÑEDA, V., FRONTIER, S., ARENAS, V. 1993. Experimental re-establishment of a soft-bottom community following defaunation by pollution. Utilization of multivariate analyses to characterize different benthic recruitments. Estuar. Coast. Shelf Sci. 37, 387–402.

DOBBS, F., WHITLATCH, R.B. 1982. Aspects of deposit-feeding by the polychaete *Clymenella torquata*. Ophelia 21, 159–166.

DONARD, O. 1996. Bioavailability of sedimentary contaminants subject to deposit-feeder digestion. Environ. Sci. Technol. 30, 2641–2645.

DORGAN, K., JUMARS, P. 2006. Macrofaunal burrowing: the medium is the message. Oceanogr. Mar. Biol. Ann. Rev. 44, 85–121.

DORGAN, K., JUMARS, P., JOHNSON, B., BOUDREA, B., LANDIS, E. 2005. Burrowing by crack propagation through muddy sediment. Nature 433, 475.

DORGAN, K.M., JUMARS, P.A., JOHNSON, B.D., BOUDREAU, B.P. 2006. Macrofaunal burrowing: the medium is the message. Ocean. Mar. Biol. 44, 85–141.

DUMNICKA, E., ROŻEN, A. 2002. The first record of the terrestrial polychaete *Hrabeiella periglandulata* Pižl et Chalupský, 1984, in Poland, with a note on anatomy and ecology. Fragm. Faun. 45, 1–7.

ELLIOTT, M. 1994. The analysis of macrobenthic community data. Mar. Pollut. Bull. 28, 62–64.

FAUCHALD, K. 1973. Polychaetes from Central American sandy beaches. Bull. South Calif. Acad. Sci. 72, 19–31.

FAUCHALD, K. 1977. The polychaete worms: definitions and keys to the orders, families and genera. Nat. Hist. Mus. Los Angeles County Sci. Ser. 28, 1–188.

FAUCHALD, K., JUMARS, P. 1979. The diet of worms: a study of polychaete feeding guilds. Oceanogr. Mar. Biol. Ann. Rev. 17, 193–284.

FENCHEL, T. 1996. Worm burrows and oxic microniches in marine sediments. 1. Spatial and temporal scales. Mar. Biol. 127, 289–295.

FERNÁNDEZ-ALAMO, M. 2003. Distribution of holoplanktonic typhloscolecids (Annelida-Polychaeta) in the eastern tropical Pacific Ocean. J. Plankton Research 24, 647–657.

FERRARO, S., COLE, F. 1990. Taxonomic level and sample size sufficient for assessing pollution impacts on the Southern California Bight macrobenthos. Mar. Ecol. Prog. Ser. 67, 251–262.

FILICE, F.P. 1954. An ecological survey of the Castro Creek area in San Pablo Bay. Wasmann J. Biol. 12, 1–24.

FORNOS, J., CORTEZA, V., MARTÍNEZ-TABERNER, A. 1997. Modern polychaete reefs in western Mediterranean lagoons: *Ficopomatus enigmaticus* (Fauvel) in the Albufera of Menorca, Balearic Islands. Paleogeog, Paleoclimat. Paleoecol. 128, 75–186.

GAMBI, M.C., TUSSENBROEK, B., BREARLEY, A. 2003. Mesofaunal borers in seagrasses: world-wide occurrence and a new record of boring polychaetes in the Mexican Caribbean. Aquat. Bot. 76, 65–77.

GIANGRANDE, A. 1997. Polychaete reproductive patterns, life cycles and life histories: an overview. Oceanogr. Mar. Biol. Ann. Rev. 35, 323–386.

GIANGRANDE, A., CAVALLO, A., LICCIANO, M., MOLA PIERRE, C., TRIANNI, L. 2005. The utilization of the filter feeder polychaete *Sabella spallanzanii* (Sabellidae) as bioremediator in aquaculture. Aquac. Int. 13, 129–136.

GLOVER, A., KALLSTROM, B., SMITH, C., DAHLGREN, T. 2005. World-wide whale worms? A new species of *Osedax* from the shallow north Atlantic. Proc. R. Soc. B 272, 2587–2592.

GOSNER, K.L. 1978. *Peterson Field Guides: Atlantic Seashore*. Boston, MA: Houghton-Mifflin.

GRALL, J., CHAUVAUD, L. 2002. Marine eutrophication and benthos: the need for new approaches and concepts. Glob. Chang. Biol. 8, 813–830.

GRANT, J. 1983. The relative magnitude of biological and physical sediment reworking in an intertidal community. J. Mar. Res. 14, 673–689.

GRANTHAM, G., ECKERT, E., SHANKS, A. 2003. Dispersal potential of marine invertebrates in diverse habitats. Ecol. Appl. 13, 108–116.

GRASSLE, J.F., GRASSLE, J.P. 1974. Opportunistic life histories and genetic systems in marine benthic polychaetes. J. Mar. Res. 32, 253–284.

GRAY, J.S. 1981. *The Ecology of Marine Sediments. An Introduction to the Structure and Function of Benthic Communities*. Cambridge: Cambridge University Press.

GRAY, J.S., CLARKE, K., WARWICK, R.M., HOBBS, G. 1990. Detection of initial effects of pollution on marine benthos: an example from the Ekofisk and Eldfisk oilfields, North Sea. Mar. Ecol. Prog. Ser. 66, 185–299.

GUNNARSSON, J., BJÖRK, M., GILEK, M., GRANBERG, M., ROSENBERG, R. 2000. Effects of eutrophication on contaminant cycling in marine benthic systems. Ambio 29, 252–259.

HANCE, J.J., ANDRZEJEWSKI, B., PREDMORE, K., DUNLAP, K., MISIAK, K.L., JULIAN, D. 2008. Cytotoxicity from sulfide exposure in a sulfide-tolerant marine invertebrate. J. Exp. Mar. Biol. Ecol. 359, 102–109.

HARTMAN, O. 1944. Polychaetous annelids. Pt. 6. Paraonidae, Magelonidae, Longosomidae, Ctenodrilidae and Sabellariidae. Allan Hancock Pacific Expedition 10, 311–389.

HARTMAN, O. 1952. *Identification of Marine Animals in Los Angeles—Long Beach Harbor Pollution Study*. Appendix IV, No. 4. Los Angeles, CA: Los Angeles Regional Water Pollution Control Board, p. 41.

HARTMAN, O. 1955. Endemism in the North Pacific Ocean, with emphasis on the distribution of marine annelids, and descriptions of new or little known species. In *Essays in the Natural Sciences in Honor of Captain A. Hancock*. Los Angeles, CA: Hancock Foundation, University of Southern California, pp. 39–60.

HILY, C. 1987. Spatio-temporal variability of *Chaetozone setosa* (Malmgren) populations on an organic gradient in the Bay of Brest, France. J. Exp. Mar. Biol. Ecol. 112, 201–216.

HUTCHINSON, G.E. 1951. Copepodology for the ornithologist. Ecology 32, 571–577.

HUTCHINSON, G.E. 1967. *A Treatise on Limnology*. New York: John Wiley & Sons.

JOHNS, D.M., GINN, T. 1990. *Development of a Neanthes Sediment Bioassay for Use in Puget Sound*. EPA 910/9-90-005. Seattle, WA: US Environmental Protection Laboratory, Region 10.

JUMARS, P. 1978. Spatial autocorrelation with RUM (remote underwater manipulator): vertical and horizontal structure of a bathyal benthic community. Deep-Sea Res. 25, 589–604.

JUMARS, P., WHEATCROFT, R.A. 1989. Responses of benthos to changing food quality and quantity, with a focus on deposit feeding and bioturbation. In *Productivity of the Ocean: Present and Past*, edited by W.H. Berger, V.S. Smetacek and G. Wefer. London: John Wiley and Sons Limited, pp. 235–253.

JUMARS, P., MAYER, L.M., DEMING, J.W., BAROSS, J.A., WHEATCROFT, R.A. 1990. Deep-sea deposit-feeding strategies suggested by environmental and feeding constraints. Philos. Trans. Royal Soc. Lond. 331, 85–101.

KENNY, R. 1969. Effects of temperature, salinity and substrate on distribution of Clymenella torquata (Leidy), *Polychaeta*. Ecology 50 (4), 624–631.

KLEMM, D.J. 1985. *A Guide to the Freshwater Annelida (Polychaeta, Naididae and Tubificid Oligochaeta, and Hirudinea) of North America*. Dubuque, IA: Kendall/Hunt.

KNOX, C. 1977. The role of polychaetes in benthic soft-bottom communities. In *Essays on Polychaetous Annelids in Memory of Dr. O. Hartman*, edited by D.J. Reish and K. Fauchald. Los Angeles, CA: Allan Hancock Foundation, pp. 547–604.

KRANTZBERG, G. 1985. The influence of bioturbation on physical, chemical and biological parameters in aquatic environments: a review. Environ. Pollut. 39, 99–122.

KRISTENSEN, E. 1989. Oxygen and carbon dioxide exchange in the polychaete *Nereis virens*: influence of ventilation activity and starvation. Mar. Biol. 101, 381–388.

KURIS, M., CULVER, C. 1999. An introduced sabellid polychaete pest infesting cultured abalones and its potential spread to other California gastropods. Invertebr. Biol. 118, 391–403.

LAMBSHEAD, P., PLATT, J.H., SHAW, K. 1983. Detection of differences among assemblages of benthic species based on an assessment of dominance and diversity. J. Nat. Hist., Lond. 17, 859–874.

LAMONT, P.A., GAGE, J.D. 2000. Morphological responses of macrobenthic polychaetes to low oxygen on the Oman continental slope, NW Arabian Sea. Deep-Sea Res. 47, 9–24.

LEVIN, L. 2003. Oxygen minimum zone benthos: adaptation and community response to hypoxia. Oceanogr. Mar. Biol. Ann. Rev. 41, 1–45.

LEVIN, L. 2005. Ecology of cold seep sediments: interactions of fauna with flow, chemistry and microbes. Oceanogr. Mar. Biol. Ann. Rev. 43, 1–46.

LEVIN, L., EDESA, S. 1997. The ecology of cirratulid mudballs on the Oman margin, northwest Arabian Sea. Mar. Biol. 128, 671–678.

LEVIN, L., MICHENER, R. 2002. Isotopic evidence for chemosynthesis-based nutrition of macrobenthos: the lightness of being at Pacific methane seeps. Limnol. Oceanogr. 47, 1336–1345.

LEVIN, L.A., BLAIR, N., DEMASTER, D.J., PLAIA, G., FORNES, W., MARTIN, C., THOMAS, C. 1997. Rapid subduction of organic matter by maldanid polychaetes on the North Carolina slope. J. Mar. Res. 55, 595–611.

LEVIN, L., GAGE, J., MARTIN, C., LAMONT, P. 2000. Macrobenthic community structure within and beneath the oxygen minimum zone, NW Arabian Sea. Deep Sea Res. Part II Top. Stud. Oceanogr. 47, 189–226.

LEVIN, L.W., ZIEBIS, G., MENDOZA, V., GROWNEY, V., WALTHER, S. 2006. Recruitment response of methane-seep macrofauna to sulfide-rich sediments: an *in situ* experiment. J. Exp. Mar. Biol. Ecol. 330, 132–150.

LEVINTON, J. 1995. Bioturbators as ecosystem engineers: control of the sediment fabric, interindividual interactions, and material fuxes. In *Linking Species and Ecosystems*, edited by C. Jones and J.H. Lawton. New York: Chapman & Hall, pp. 29–38.

LEWONTIN, R.C. 1965. Selection for colonizing ability. In *The Genetics of Colonizing Species*, edited by H. Baker and G. Stebbins. New York: Academic Press, pp. 79–94.

LICCIANO, M.L., STABILI, L., GIANGRANDE, A. 2005. Clearance rates of *Sabella spallanzanii* and *Branchiomma luctuosum* (Annelida: Polychaeta) on a pure culture of *Vibrio alginolyticus*. Water Res. 139, 4375–4384.

LICCIANO, M.L., SATBILI, L., GIANGRANDE, A., CAVALLO, R. 2007. Bacterial accumulation by *Branchiomma luctuosum* (Annelida Polychaeta): a tool for biomonitoring marine systems and restoring polluted waters. Mar. Environ. Res. 63, 291–302.

MACDONALD, I., SAGAR, W., PECCINI, M. 2003. Gas hydrate and chemosynthetic biota in mounded bathymetry at mid-slope hydrocarbon seeps: Northern Gulf of Mexico. Mar. Geol. 198, 133–158.

MADSEN, S., FORBES, T., FORBES, V. 1997. Particle mixing by the polychaete *Capitella* species 1: coupling fate and effect of a particle-bound organic contaminant (fluoranthene) in a marine sediment. Mar. Ecol. Prog. Ser. 147, 129–142.

MAGNI, P., COMO, S., MONTANI, S., TSUTSUMI, H. 2006. Interlinked temporal changes in environmental conditions, chemical characteristics of sediments and macrofaunal assemblages in an estuarine intertidal sandflat (Seto Inland Sea), Japan. Mar. Biol. 149, 1185–1197.

MARTIN, D., BRITAYEV, T. 1998. Symbiotic polychaetes: review of known species. Oceanogr. Mar. Biol. Ann. Rev. 36, 217–340.

MAYER, L.M., SCHICK, L., SELF, R., JUMARS, P., FINDLAY, R., CHEN, Z., SAMSON, S. 1997. Digestive environments of benthic macroinvertebrate guts: enzymes, surfactants and dissolved organic matter. J. Mar. Res. 55, 785–812.

MAYER, L.M., JUMARS, P., BOCK, M., VETTER, Y., SCHMIDT, J. 2001. Two roads to Sparagmos: extracellular digestion of sedimentary food by bacterial infection vs. deposit feeding. In *Organism-Sediment Interactions*, edited by J.Y. Aller, S. Woodin and R. Aller. Columbia, SC: University of South Carolina Press, pp. 335–347.

MEARNS, A.J., REISH, D., OSHIDA, P., BUCHMAN, M., GINN, T. 2006. Effects of pollution on marine organisms. Water Environ. Res. 78, 2033–2086.

MONTAGNA, P.A. 2005. Measuring marine benthic community response. In *Practical Biological Indicators of Human Impacts in Antarctica*, Vol. 2, COMNAP Secretariat.

MURDOCH, M.H., CHAPMAN, P., JOHNS, D., PAINE, M. 1997. Chronic effects of organochlorine exposure in sediment to the marine polychaete *Neanthes arenaceodentata*. Environ. Toxicol. Chem. 16, 1494–1505.

NIPPER, M., CARR, R.S. 2003. Recent advances in the use of meiofaunal polychaetes for ecotoxicological assessments. Hydrobiologia 496, 347–353.

NOWELL, A., JUMARS, P., SELF, R., SOUTHWARD, A. 1989. The effects of sediment transport and deposition on infauna: results obtained in a specially designed flume. In *Ecology of Marine Deposit Feeders*, vol. 31, edited by G. Lopez, G. Taghon and J. Levinton. New York: Springer, pp. 247–268.

OGLESBY, L.C. 2005. The Salton Sea. Geology, history, potential problems, politics, and possible futures of an unnatural desert salt lake. Mem. Calif. Acad. Sci. 10, 240.

OLIVE, P. 1994. Polychaeta as a world resource: a review of patterns of exploitation as sea angling baits and the potential for aquaculture based production. Mem. Mus. Natl. Hist. Nat. 162, 603–610.

OLIVE, P. 1999. Polychaete aquaculture and polychaete science: a mutual synergism. Hydrobiologia 402, 175–183.

OLIVIER, M., DESROSIERS, G., CARON, A., RETIERE, C., CAILLOU, A. 1996. Juvenile growth of *Nereis diversicolor* (O.F. Muller) feeding on a range of marine vascular and macroalgal plant sources under experimental conditions. J. Exp. Mar. Biol. Ecol. 208, 1–12.

OLSGARD, F., BRATTEGARD, T., HOLTHE, T. 2003. Polychaetes as surrogates for marine biodiversity: lower taxonomic resolution and indicator groups. Biodivers. Conserv. 12, 1033–1049.

PAGLIOSA, P. 2005. Another diet of worms: the applicability of polychaete feeding guilds as a useful conceptual framework and biological variable. Mar. Ecol. 26, 246–254.

PALOMO, G., MARTINETTO, P., PEREZ, C., IRIBARNE, O. 2003. Ant predation on intertidal polychaetes in a SW Atlantic estuary. Mar. Ecol. Prog. Ser. 253, 165–173.

PAULL, C.K., HECKER, B., COMMEAU, R., FREEMAN-LYNDE, R., NEUMANN, C., CORSO, W., GOLUBIC, S., HOOK, J., SIKES, E., CURRAY, J. 1984. Biological communities at the Florida Escarpment resemble hydrothermal vent taxa. Science 226, 965–967.

PEARSON, T.H., ROSENBERG, R. 1978. Macrobenthic succession in relation to organic enrichment and pollution of the marine environment. Oceanogr. Mar. Biol. Ann. Rev. 16, 229–311.

PENRY, D., JUMARS, P. 1990. Gut architecture, digestive constraints and feeding ecology of deposit-feeding and carnivorous polychaetes. Oecologia 82, 1–11.

PETERSEN, C.J. 1918. The sea bottom and its production of fishfood. Danish Biol. Statin Repts. 25, 1–62.

PETTIBONE, M.H. 1982. Annelida. In *Synopsis and Classification of Living Organisms*, 2 vols., edited by S.P. Parker. New York: McGraw-Hill, pp. 1–43.

PLAGÁNYI, E., BRANCH, G. 2000. Does the limpet Patella cochlear fertilize its own algal garden? Mar. Ecol. Prog. Ser. 194, 113–122.

POCKLINGTON, P., WELLS, P.G. 1992. Polychaetes: key taxa for marine environmental quality monitoring. Mar. Pollut. Bull. 24, 593–598.

PURSCHKE, G. 1999. Terrestrial polychaetes models for the evolution of the Clitellata (Annelida)? Hydrobiologia 406, 87–99.

RAMOS, A., SAN MARTÍN, G. 1999. On the finding of a mass occurrence of *Serpula narconensis* Baird, 1885 (Polychaeta, Serpulidae) in South Georgia (Antarctica). Polar Biol. 22, 379–383.

REISH, D.J. 1955. The relation of polychaetous annelids to harbor pollution. Public Health Rep. 70, 1168–1174.

REISH, D.J. 1959. *The Ecological Study of Pollution in Los Angeles–Long Beach Harbors, California*. Occasional Paper No. 22. Allan Hancock Foundation, University of Southern California, Los Angeles.

REISH, D.J. 1961. A study of succession in a recently constructed boat harbor in Southern California. Ecology 42, 84–91.

REISH, D.J. 1964. Studies on the *Mytilus edulis* community in Alamitos Bay, California: II. Population variations and discussion of the associated species. Veliger 6, 202–207.

REISH, D.J. 1977. Effects of chromium on the life history of *Capitella capitata* (Annelida: Polychaeta). In *Physiological Responses of Marine Biota to Pollutants*, edited by F.J. Vernberg, A. Calabrese, F. Thurberg and W. Verngerg. New York: Academic Press, pp. 199–207.

REISH, D.J. 1979. Bristle worms (Annelida:Polychaeta). In *Pollution Ecology of Estuarine Invertebrates*, edited by C.W. Hart and S.L.H. Fuller. New York: Academic Press, pp. 77–125.

REISH, D.J. 1980. Use of polychaetous annelids as test organisms for marine bioassay experiments. In *Aquatic Invertebrate Bioassays. STP 715*, edited by A.L. Buikema, Jr. and J. Cairns. Phildelphia, PA: American Society for Testing and Materials, pp. 140–154.

REISH, D.J., BARNARD, J.L. 1991. Marine invertebrates as food for the shorebirds of Bahia de San Quitin, Baja California. *Proceedings of the International Marine Biology VIII Symposium*, June 4–8, 1990. Ensenada, Baja, California, pp. 1–6.

REISH, D.J., GERLINGER, T. 1984. The effects of cadmium and zinc on survival and reproduction in polychaetous annelid *N. arenaceodentata* (Nereididae). In *Proceedings of the First International Polychaete Conference, Sydney, Australia*, edited by P.A. Hutchings. Sydney: Linnean Society of New South Wales, pp. 383–389.

REISH, D.J., GERLINGER, T.V. 1997. A review of the toxicological studies with polychaetous annelids. Bull. Mar. Sci. 60, 584–607.

REISH, D.J., WINTER, H.A. 1954. The ecology of Alamitos Bay, California, with special reference to pollution. Calif. Fish Game 40, 105–121.

REISH, D.J., CHAPMAN, P.M., INGERSOLL, C., MOORE, D., PESCH, C. 2005. Annelids. In *Standard Methods for Examination of Water and Wastewater*, 21st edn., edited by A.E. Eaton, L. Clesceri, E.W. Rice and A.E. Greenberg. Washington, DC: American Public Health Association, American Water Works Association and Water Environment Federation, pp. 880–892.

RIUS VILADOMIU, M. 2004. *The effects of the invasive mussel Mytilus galloprovincialis and human exploitation on the indigenous mussel Perna perna on the South Coast of South Africa*. Masters Thesis, Rhodes University.

ROSSI, S., ANDERSON, J., WARD, G. 1976. Toxicity of water soluble fractions of four test oils on growth and reproduction in *Neanthes arenaceodentata* and *Capitella capitata* (Polychaeta: Annelida). Environ. Pollut. 10, 9–18.

Rouse, G., Pleijel, F. 2001. *Polychaetes.* New York: Oxford University Press.

Rouse, G., Goffredi, S., Vrijenhoek, R. 2004. *Osedax*: bone-eating marine worms with dwarf males. Science 305, 668–671.

Rygg, B. 1985. Distribution of species along pollution-induced diversity gradients in benthic communities in Norwegian fjords. Mar. Pollut. Bull. 16, 469–474.

Schories, D., Reise, K. 1993. Germination and anchorage of *Enteromorpha* spp. in sediments of the Wadden Sea. Helgoland Mar. Res. 47, 275–285.

Shank, T., Fornari, D., Von Damm, K., Lilley, M., Haymon, R., Lutz, R. 1998. Temporal and spatial patterns of biological community development at nascent deep-sea hydrothermal vents (9° 50′ N, East Pacific Rise). Deep-Sea Res. II 45, 465–515.

Shanks, A., Carmen, K. 1997. Larval polychaetes are strongly associated with marine snow. Mar. Ecol. Prog. Ser. 154, 211–221.

Smith, C.R. 1986. Nekton falls. Low-intensity disturbance and community structure of infaunal benthos in the deep sea. J. Mar. Res. 44, 567–600.

Snelgrove, P., Blackburn, T.H., Hutchings, P., Alongi, D.M., Grassle, J.F., Hummel, H., King, G., Koike, I., Lambshead, P.J.D., Ramsing, N.B., SolisWeiss, V. 1997. The importance of marine sediment biodiversity in ecosystem processes. Ambio 26, 578–583.

Snelgrove, P., Grassle, F., Grassle, J., Petrecca, R., Stocks, K. 2001. The role of colonization in establishing patterns of community composition and diversity in shallow-water sedimentary communities. J. Mar. Res. 59, 813–831.

Sokolova, I., Pörtner, H. 2001. Physiological adaptations to high intertidal life involve improved water conservation abilities and metabolic rate depression in *Littorina saxatilis*. Mar. Ecol. Prog. Ser. 224, 171–186.

Soule, D. 1988. Marine organisms as indicators: reality or wishful thinking? In *Marine Organisms as Indicators*, edited by D. Soule and G. Kleppel. New York: Springer-Verlag, pp. 1–11.

Stabili, L., Licciano, M., Giangrande, A., Fanelli, G., Cavallo, R. 2006. *Sabella spallanzanii* filter-feeding on bacterial community: ecological implications and applications. Mar. Environ. Res. 61, 74–92.

Stabili, L., Licciano, M., Giangrande, A., Fanelli, G., Cavallo, R. 2007. Bacterial accumulation by *Branchiomma luctuosum* (Annelida: Polychaeta): a tool for biomonitoring marine systems and restoring polluted waters. Mar. Environ. Res. 63, 291–302.

Tamaki, A. 1987. Comparison of resistivity to transport by wave action in several polychaete species on an intertidal sand flat. Mar. Ecol. Prog. Ser. 37, 181–189.

Ten Hove, H., van den Hurk, P. 1993. A review of recent and fossil serpulid "reefs"; actuopaleontology and the "Upper Malm" serpulid limestones in NW Germany. Geologie en Mijnbouw 72, 23–67.

Thiel, M., Ullrich, N. 2002. Hard rock versus soft bottom: the fauna associated with intertidal mussel beds on hard bottoms along the coast of Chile, and considerations on the functional role of mussel beds. Helgoland Mar. Res. 56, 21–30.

Tokeshi, M. 1995. Polychaete abundance and dispersion patterns in mussel beds: a non-trivial "infaunal" assemblage on a Pacific South American rocky shore. Mar. Ecol. Prog. Ser. 125, 137–147.

Tzetlin, A. 1994. Fine morphology of the feeding apparatus of *Cossura* sp. (Polychaeta, Cossuridae) from the White Sea. Mem. Mus. Natl. Hist. Nat. 162, 137–143.

United States Environmental Protection Agency/ US Army Corps of Engineers (US EPA/US ACE). 1977. *Ecological Evaluation of Proposed Discharge of Dredged Material into Ocean Waters.* Tech. Committee on Criteria for Dredged and Fill Material. Environmental Effects: Laboratory. Vicksburg, MI: US Army Waterways Environment Station.

Ushakova, O., Saranchova, O. 2003. Changes in resistance to salinity and temperature in some polychaetes from the White Sea. Russ. J. Mar. Biol. 29, 236–241.

Van Dover, C. 2000. *The Ecology of Hydrothermal Vents.* Princeton, NJ: Princeton University Press.

Van Dover, C., Aharon, P., Bernhard, J., Doerries, M., Flickinger, W., Gilhooly, W., Knick, K., Macko, S., Rapoport, S., Ruppel, C., Salerno, J., Seitz, R., Sen Gupta, B.K., Shank, T.M., Turnipseed, M., Vrijenhoek, R., Watkins, E. 2003. Blake Ridge methane seeps: characterization of a soft-sediment, chemosynthetically based ecosystem. Deep-Sea Res. 50, 281–300.

Wargo, R., Ford, S. 1993. The effect of shell infestation by *Polydora* sp. and infection by *Haplosporidium nelsoni* on the tissue condition of oysters *Crassostrea virginica*. Estuaries 16, 229–234.

Warwick, R. 1988. The level of taxonomic discrimination required to detect pollution effects on marine benthic communities. Mar. Pollut. Bull. 19, 259–268.

Warwick, R. 1993. Environmental impact studies on marine communities: pragmatical considerations. Aust. J. Ecol. 18, 63–80.

Westheide, W., von Basse, M. 2004. Chilling and freezing resistance of two polychaetes from a sandy tidal beach. Oecologia 33, 45–54.

Wilhelmi, J. 1916. Übersicht über die biologische Beurteilung Wassers. Ges Naturf Freunde Berlin Sitzber 297–306.

Woodin, S. 1977. Algal "gardening" behavior by Nereid polychaetes: effects on soft-bottom community structure. Mar. Biol. 44, 39–42.

Woodin, S. 1978. Refuges, disturbance, and community structure: a marine soft-bottom example. Ecology 59, 274–284.

Wu, B.L., Ruiping, S., Deijian, Y. 1985. *Nereidae (polychaetous annelids) of the Chinese coast.* Berlin: Springer-Verlag.

Chapter 12

Oligochaete Worms for Ecotoxicological Assessment of Soils and Sediments

Jörg Römbke and Philipp Egeler

ECT Oekotoxikologie GmbH, Flörsheim, Germany

12.1 INTRODUCTION

The ecological services of lumbricid earthworms in soil are the best-known example of the important environmental role of oligochaetes (Petersen and Luxton 1982; Edwards and Bohlen 1996; Lavelle et al. 1997). Other oligochaetes including soil enchytraeids and lumbriculids or tubificids in sediments, however, can have similar ecological roles, but their importance is usually much less acknowledged. In fact, these often inconspicuous worms have been chosen as test species for the ecotoxicological risk assessment of soils and sediments for good reasons. First, they play an important ecological role in soils and sediments, as exemplified by high biomass and high diversity (Brinkhurst and Jamieson 1971; Didden 1993; Edwards and Bohlen 1996). Some species act as "ecosystem engineers," such as the lumbricid *Lumbricus terrestris*, which inhabits temperate meadow and forest soils (Milcu et al. 2006), and the enchytraeid *Cognettia sphagnetorum*, which inhabits acid soils of European coniferous forests (Abrahamsen 1990).

Second, they are practical—some species like the compost worm *Eisenia fetida* (Jänsch et al. 2005) or the sludge worm *Lumbriculus variegatus* (ASTM International 2000, 2002) are easy to breed in the laboratory. In several cases, the same or related enchytraeid species can be used for various test levels (Moser et al. 2004).

Third, several oligochaete species have shown medium to high sensitivity to chemicals (Edwards and Bohlen 1992; Chapman 2001; Jensen et al. 2001), partly because oligochaetes are soft-bodied organisms; i.e. they are exposed to chemicals dissolved in water (in soil: pore water). Additionally, through feeding, they are also exposed to chemicals adsorbed to clay and/or organic particles occurring in soils and sediments.

Fourth, international standardized and validated test methods are available from the Organisation for Economic Co-operation and Development (OECD) and the International Organization for Standardization (ISO) (comparable guidelines have also been published by national organizations like ASTM), as described below. Finally, the general importance of oligochaete worms is well understood by the public; specifically, worms have been highly valued since the times of Charles Darwin, and the services provided by worms can be explained by referring to common earthworms like *L. terrestris*. For these reasons, oligochaetes have been the first choice as invertebrate test species for soils and second only to chironomids for sediments.

12.2 PRINCIPLES OF ENVIRONMENTAL RISK ASSESSMENT

Environmental risk assessment (ERA), a concept developed in the USA during the late 1970s (Fava et al. 1987), can be defined as "a systematic means of developing a scientific basis for regulatory decision making" (Barnthouse et al. 1992). It was first used for anthropogenic stress factors with potential environmental impacts. Then in the 1980s, the risk of industrial chemicals was assessed (US EPA 1992). Shortly thereafter, this concept was also used by European authorities, initially for harmonizing the registration of pesticides according to Council Directive 91/414 and its accompanying documents (e.g. "Directive concerning the placing of plant protection products on the market"; EU 1991) to ensure a safe and efficient use of these chemicals. Later, ERA was used for other classes of chemicals as well (Leeuwen and Hermens 2001).

Any ERA can be divided into four steps (Fig. 12.1). The aim of hazard identification is to define whether (and where) a pesticide may occur in the environment. In the exposure assessment, the predicted environmental concentration (PEC) is determined (in mg/kg soil, dry weight). During the effect assessment, the toxic concentration (TC) is measured in standardized tests (also in mg/kg soil, dry weight), and the potential pesticide risk is assessed by calculating the toxic exposure ratio (TER): TC/PEC. Risk is present if TER is ≤1, in which case the result can be refined by using test methods more relevant for the field (e.g. semifield approaches) or by remodeling exposure by using actual values from the area where the test substance will be used. Using these new TC and PEC data, the risk is repeatedly assessed (refined ERA) until measures to decrease the risk are deemed necessary or not. If risks cannot be avoided by safety measures, the pesticide may be banned.

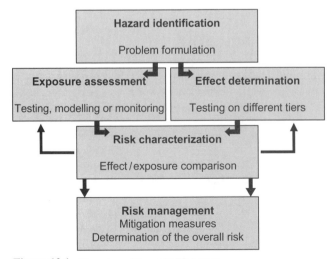

Figure 12.1 Overview of the main ERA steps.

The PEC for a pesticide is determined by using complex models based on the physicochemical properties of the substance, how and when they reach the environment (i.e. use pattern, application rate) and, to a certain extent, environmental variables. For example, chemicals with high log P_{ow} values can accumulate in fat tissues and along the food chain that may result in secondary poisoning of species at higher trophic levels. The determination of TC is usually performed by measuring the chemical's effects on mortality, growth or reproduction of individual species in laboratory tests, semifield or field tests. Using safety or assessment factors, each effect concentration determined in these tests is extrapolated to the concentration expected to have no effect in the field (Leeuwen and Hermens 2001).

ERAs must be performed separately for each environmental compartment (i.e. surface water, ground water, soil, sediment and air). Only soil and sediment compartments will be considered here. Organisms to be tested for pesticide ERA in soil include microorganisms, plants, collembolans, beetles and oligochaetes, while mainly insects and oligochaetes are tested in sediment. A diversity of organisms is necessary since no one species is "most sensitive" to all chemicals (Cairns 1986). All tests used for ERA must be standardized and validated by the OECD or the ISO.

The ERA for soil is usually performed by testing earthworms (Lumbricidae) as a representative soil invertebrate. Earthworm tests are an obligatory component of any chemical ecotoxicological assessment (including biocides, pesticides or veterinary pharmaceuticals) in the European Union as well as for the assessment of potentially contaminated land. An alternative or addition to the ERA for soil is the testing of enchytraeids. Although these tests are not yet integral for pesticide ERA, they are consistently used with other chemical groups (e.g. soil quality assessment).

For sediment ERA, two tests with the midge *Chironomus riparius* (Chironomidae) or related species are used (OECD 2004a, b). However, since midge larvae inhabit the boundary between sediment and water, feeding primarily on the surface of (and not within) sediment (e.g. Rasmussen 1984), they may not be fully representative for evaluating the effects of chemicals on sediment-dwelling organisms. On the other hand, benthic oligochaetes (Tubificidae, Lumbriculidae) are "true" sediment inhabitants since they are exposed to sediment contaminants by all uptake routes, including ingestion of contaminated sediment (e.g. Rodriguez and Reynoldson 1999; Chapman 2001). Consequently, these worms are used for sediment ERA as an alternative or in addition to chironomids.

12.3 SOIL TESTS WITH LUMBRICIDAE

Standardized tests with lumbricid earthworms are available for the laboratory as well as the field (for a recent overview on the status of earthworm tests, see Spurgeon et al. 2003). The most widely used earthworm and Enchytraeidae tests are described here, including tests currently under development. Tabular descriptions of these tests can be found in ISO (2005), Römbke (2006), and Egeler and Römbke (2008).

12.3.1 Earthworm Laboratory Tests

The acute toxicity test with the compost worm *E. fetida* (or its sister species *Eisenia andrei*) in artificial soil (OECD 1984) is the most important test with soil organisms to date. Despite its limited sensitivity in comparison to chronic earthworm tests, it is usually the initial step when assessing the toxicity of chemicals in soil. Its main end point is mortality, but various attempts have been made to include biomass development as a second (and often more sensitive) end point. In a slightly modified version published by the ASTM International (2004; No. E 1676), soil quality assessment and single chemical testing are described within one document, resulting in a general text. More recently, Environment Canada published an updated version suitable for the assessment of contaminated land. In other countries, the OECD version has simply been translated into the respective language as a national guidance document.

The chronic earthworm reproduction test, despite being based on the OECD acute test, was standardized first by the ISO (1998; No. 11268-2) and 5 years later by the OECD (2004d; No. 222). Generally speaking, it is an extended version—56 days instead of 14—of the acute test (i.e. using the same species, test substrate and also the same test conditions) while adding the endpoint reproduction (number of juveniles). Ecologically, this end point is highly relevant and is also very sensitive toward different stressors. For example, in a recent guideline describing the requirements for the environmental risk of pharmaceuticals, the earthworm reproduction test instead of the acute test is used as a starting point (VICH 2005). This change is due to the required sensitivity, since the drugs to be tested occur only at very low concentrations.

Currently, only two soil field tests have been standardized: the litterbag test (Römbke et al. 2003) and the earthworm field test that uses a natural earthworm community (ISO 1999). Recently, however, recommendations have been made to improve the applicability and practicality of the latter test (Kula et al. 2006). In the European Union, for example, the earthworm field test is regularly required for pesticides that show risks in laboratory tests. However, since the test outcome can be complex (e.g. effects can vary considerably between individual sampling times or individual species), additional guidance is provided by the authorities responsible for evaluating test results (De Jong et al. 2006). Nevertheless, useful comparisons between pesticide laboratory and field effects have been published by Heimbach (1997) and Jones and Hart (1998).

Data from the aforementioned three tests (acute and chronic laboratory tests, as well as a field study) and from a newly developed avoidance test (see below) were used to demonstrate their pros and cons (Table 12.1). Since many earthworm tests were performed for the registration of pesticides, their results were never published. The presented example from Garcia (2004) is representative of pesticide studies in Europe and North America.

As expected, the two pesticides used in these tests (the insecticide lambda-cyhalothrin and the fungicide carbendazim) showed lower toxicity in acute (14 days) mortality tests compared to chronic (56 days) reproduction tests. For carbendazim, the difference is

Table 12.1 Comparison of standardized earthworm tests performed under tropical conditions with the fungicide carbendazim and the insecticide lambda-cyhalothrin spiked into tropical artificial soil (Garcia 2004)

Earthworm Test	End Point	Carbendazim	Lambda-cyhalothrin
Laboratory: acute mortality	LC_{50}	>1,000	23.9
Laboratory: chronic reproduction	NOEC	3.16	6.2
Laboratory: avoidance	NOEC	<1.0	<0.32
Field: abundance of the native species *Andiorrhinus amazonicus*	LOEC	1.33	?

Numbers are given in mg a.s. / kg soil for the acute earthworm test (as lethal concentration = LC50), the chronic and the avoidance earthworm test [no-observed-effect concentration (NOEC), lowest-observed-effect concentration (LOEC)] and an earthworm field study (modified from ISO 1999).
? = Not known.

unusually high (>300-fold; typically factors of ~10 are observed); for example, in tests under temperate conditions, the factor is ~60 (Garcia 2004). Considering the mode of action of lambda-cyhalothrin (an insecticide), the observation that earthworm reproduction was impacted at concentrations >6.2 mg/kg is surprising. Interestingly, the short-term avoidance test (2 days) shows higher sensitivity than the chronic tests for both pesticides. While this test is usually more sensitive than the acute mortality test, other experiments [e.g. with the biocide pentachlorophenol (PCP) or the heavy metal copper] often show toxicity results in the same order of magnitude as the chronic test (Hund-Rinke et al. 2005). In field tests performed in an abandoned rubber tree plantation, no effects of these two pesticides were found on the earthworm community in general, but the abundant species *Andiorrhinus amazonicus* (Glossoscolecidae) reacted negatively at concentrations of carbendazim comparable to those impacting reproduction or behavior in laboratory tests. No consistent effects on earthworms were found in the field for lambda-cyhalothrin.

12.3.2 Earthworm Bioaccumulation Test

Despite the large number of chemical bioaccumulation studies in earthworms (including the development of bioaccumulation models; Sample et al. 1998), no standardized test is currently available (Belfroid et al. 1995). In the USA and in Europe, however, draft methods using oligochaetes have recently been proposed (Bruns et al. 2001a, b; ASTM International 2004), and the European version is now considered as an OECD standard. The need for this test is based on the observation that current practices to estimate bioaccumulation using fish data or other models are not sufficient (e.g. lindane was not eliminated from earthworm tissue as expected; Bruns et al. 2001b). In North America, only lumbricid earthworms are proposed as test species, while the OECD version also includes enchytraeids. Here only earthworms are highlighted: test species are the compost worms *E. fetida* or *E. andrei* as well as the common epigeic species *Lumbricus rubellus*. One adult earthworm per test vessel is used, and the duration of both phases is 21 days. The test is invalid if the control mortality is >10%.

12.3.3 Earthworm Tests under Development

In the mid-1990s, avoidance behavior tests were developed to evaluate the effect of contaminated substrate on earthworms (Yeardley et al. 1996). Based mainly on experiences in Canada and in Germany, a modified method is now under standardization by the ISO (2006b). The sensitivity of this test is comparable to the much longer (but ecologically more relevant) reproduction test discussed previously. Most likely, behavioral tests will be employed regularly as a screening method, e.g. as part of a battery of biological on-site tests when evaluating the potential toxicity of contaminated soil.

12.4 SOIL TESTS WITH ENCHYTRAEIDAE

Currently, chronic laboratory tests (OECD 2004c; OECD Guideline 220) with selected species, a long-term semifield test with natural enchytraeid communities (Knacker et al. 2004; Moser et al. 2004) and a laboratory bioaccumulation test (Bruns et al. 2001a, b) are available. A recent overview of enchytraeid tests is provided by Römbke et al. (2005), while examples of toxicity results are given by Didden and Römbke (2001).

12.4.1 Enchytraeid Laboratory Tests

The enchytraeid reproduction test (ERT) was standardized in three versions: the OECD (OECD 2004d) focuses on testing individual chemicals (particularly pesticides); the ISO covers retrospective sample testing from contaminated sites (ISO 2003), and the ASTM has a broad approach that covers enchytraeids and also earthworms (ASTM 2004). In the ERT, acute mortality and chronic reproduction parameters are the most common end points. This test has often been used for testing pesticides (Römbke 2003; Amorim et al. 2005b), and currently is listed in several legal documents as an alternative to earthworm testing (EPPO 2003; VICH 2005; ISO 2006a). Usually, the species *Enchytraeus albidus* is the test organism, but *Enchytraeus crypticus* is becoming popular due to its broader ecological range and practicality (e.g. shorter test duration, higher juvenile numbers; Kuperman et al. 2006).

Below we describe the effects of the fungicide carbendazim on *E. albidus* in laboratory tests (Römbke and Moser 2002). These data are from an international ring test, and thus variability in the data is partly caused by the experience level of participants (e.g. mortality; Fig. 12.2). Nevertheless, a large difference in sensitivity between acute and chronic end points was found: the acute tests resulted in an LC_{50} of >10 mg/kg soil, while the EC_{50} in the chronic test was 2.8–3.7 mg/kg soil.

12.4.2 Enchytraeidae Bioaccumulation Test

The oligochaete bioaccumulation test (OBT) is currently under validation in an international ring test and will be published as an OECD guideline in the near future. This test can be conducted with different enchytraeids as well as earthworm species (see above). Until recently, the use of enchytraeids in bioaccumulation studies has been limited. As an

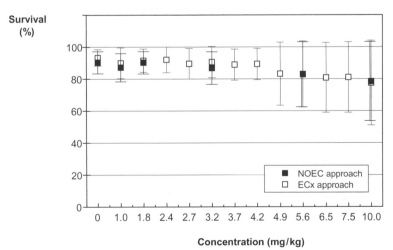

Figure 12.2 Survival of adult *Enchytraeus albidus* individuals in a laboratory test after exposure to the fungicide derosal (a.i. carbendazim), using two different test designs. Data from an international ring test (Römbke and Moser 2002), with permission from Elsevier Limited.

Table 12.2 Bioaccumulation and tissue residues at the end of elimination using two pesticides (lindane and HCB) in two soils (OECD artificial soil and LUFA 2.2 standard field soil) and two enchytraeid species (*Enchytraeus albidus* and *Enchytraeus luxuriosus*)

Pesticides	Soil Type	BAF + SE [kg Soil / kg Worm (ww)]	BSAF (g OC / g Lipid)	Tissue Residues at End of Elimination (% of Accumulated Residues)
Lindane				
E. albidus	OECD	12.0 ± 0.7	9.5	50.1
	LUFA 2.2	22.0 ± 0.4	7.7	8.6
E. luxuriosus	OECD	12.0 ± 1.1	9.4	2.6
	LUFA 2.2	36.0 ± 1.1	12.7	4.1
HCB				
E. albidus	OECD	14.0 ± 4.4	11.0	14.3
	LUFA 2.2	28.0 ± 0.8	9.7	10.0
E. luxuriosus	OECD	27.0 ± 0.9	21.2	0.02
	LUFA 2.2	35.0 ± 0.7	12.1	8.4

BAF = bioaccumulation factor based on concentrations in wet worm and soil; BSAF = biota-soil accumulation factor normalized for concentrations in lipid and OC; SE = standard error; ww = wet weight; OC = organic carbon. (Reprinted with permission from Umweltbundesamt Dessau.)

example, the bioaccumulation factors (BAFs) of two organic substances [lindane and hexachlorobenzene (HCB)] mixed into two soils (OECD artificial soil and LUFA St. 2.2 standard field soil) are presented in Table 12.2 (Bruns et al. 2001a). Two enchytraeid species, *E. albidus* and *Enchytraeus luxuriosus*, accumulated both substances in each soil type. While the BAF values of lindane were two to three times higher in the LUFA 2.2 field soil compared with OECD soil, the BAFs for HCB are not that different between the two soils (OECD = 1.5, LUFA = 2.0). This difference is probably influenced by different organic matter contents between the two soils (i.e. soil organic matter is about five times higher in OECD than in LUFA soil). With the exception of lindane in OECD soils, BAFs are always higher for the smaller species, *E. luxuriosus*, than for *E. albidus*. This difference may be caused by different volume : surface ratios (i.e. small-bodied species take up higher amounts of test chemicals from pore water). This hypothesis is supported by earthworm tests performed in parallel in which BAF values were always lower than those of enchytraeids.

The importance of BAF values determined in the uptake phase of the bioaccumulation test is a well-established part of ERAs, since the protection of species at higher trophic levels has historically been an important topic in ecotoxicology (e.g. the effects of DDT on birds of prey; Wiemeyer et al. 1975). The rationale for investigating the elimination phase is illustrated by noneliminated tissue residues (Table 12.2). HCB is almost completely eliminated by both species in each soil at the end of the test. However, 50% of the accumulated lindane is still present in *E. albidus* (but not in *E. luxuriosus*), which may imply a risk of secondary poisoning for species foraging on *E. albidus* (e.g. carabid beetles, centipedes or gamasid mites). Interestingly, such species-specific elimination behavior was also found in earthworms (Füll and Nagel 1994), which can be interpreted as follows: for lindane, a risk of secondary predator poisoning is proven, while HCB seems not to be a problem after short-term exposure to enchytraeids.

12.4.3 Enchytraeid Tests under Development

In the mid-1990s, an avoidance test using enchytraeids was proposed by Achazi et al. (1996). While a comparable test using earthworms is already standardized by ISO (2006b), little experience concerning the use of enchytraeid avoidance behavior is available. The fungicides benomyl and carbendazim, however, seem to cause effects only within the range known from acute tests (Amorim et al. 2005a). Behavioral tests seem to be a useful complement to existing acute and chronic tests, since the toxicity of a substance or oil sample can be evaluated within a few hours or days. Nonetheless, more tests with different chemicals and additional enchytraeid species must be made before ISO can standardize an enchytraeid avoidance test guideline.

12.5 SEDIMENT TESTS WITH LUMBRICULIDAE AND TUBIFICIDAE

12.5.1 Chronic Laboratory Tests with Lumbriculidae

Several chronic sediment toxicity tests with aquatic oligochaetes (e.g. Reynoldson et al. 1991; ASTM International 2002; OECD 2006) as well as bioaccumulation tests (e.g. ASTM International 2000; Egeler et al. 2006; Egeler and Römbke 2008) are available. The main goal of a newly standardized toxicity method using *L. variegatus* (OECD 2006) is to determine the effects of a test substance on oligochaete reproduction and biomass. The methodology is closely related to chironomid tests described by OECD (2004c) but also takes into account experiences reported in US guidelines (mainly ASTM International 2002). Artificial sediment, which is similar to artificial soil (see above), was used as test substrate, and the test method was validated in an international ring test within the OECD Test Guidelines Programme (Egeler et al. 2005).

Some results of PCP toxicity of *L. variegatus* were derived from the report of an international ring test (Egeler et al. 2005). Figure 12.3 shows a steep concentration–response relationship of worms after a 28-day exposure to PCP, demonstrating that the test method is suitable for assessing effects of the sediment-spiked test substance with *L. variegatus*. An ERA was not performed with these data because the model substance PCP was banned in the European Union many years ago.

12.5.2 Bioaccumulation Tests with Lumbriculidae and Tubificidae

The methodology of the sediment bioaccumulation test, like the enchytraeid bioaccumulation test, was recently standardized on the basis of an international ring test (Egeler et al. 2006). In fact, both tests share similar features, e.g. artificial sediment is spiked with the test chemical; however, the bioaccumulation test not only uses *L. variegatus* but also the sludge worm, *Tubifex tubifex*. This recommendation is based on studies performed by Egeler et al. (1997). A third species, *Branchiura sowerbyi*, can also be used in this test. Although not validated in a ring test, the option to include *B. sowerbyi* is based on relevant literature (e.g. Marchese and Brinkhurst 1996; Roghair et al. 1996). Since *B. sowerbyi* is often found in tropical regions, its inclusion extends the applicability of this guideline considerably (the inclusion of nontemperate species into OECD test guidelines will likely become a general issue in the near future).

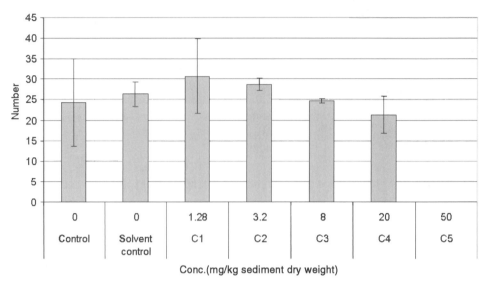

Figure 12.3   Total number of worms (*Lumbriculus variegatus*) per treatment (mean values and standard deviation, *n* = 3) after 28 days of exposure to PCP-spiked artificial sediment.

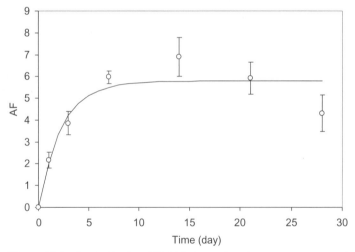

Figure 12.4   Uptake kinetics of ^{14}C-HCB in *Lumbriculus variegatus*; accumulation factor (AF): BAF as a ratio of dpm (disintegrations per minute)/kg worm dry weight and dpm/kg sediment dry weight (mean values, *n* = 3; error bars are standard deviation); data from Egeler et al. (2006). Data points represent the mean AF as the ratio of concentration in the worms and the sediment of each replicate at each sampling date.

Bioaccumulation of the highly persistent chemical HCB in *L. variegatus* is used as an example of this test. The uptake kinetics of HCB in worms during the exposure phase is shown in Fig. 12.4. The accumulation of radioactivity occurred rapidly during the initial part of the uptake phase, with the accumulation factor (AF) reaching a plateau after approximately 2 weeks. Following the uptake phase, the remaining worms were transferred to vessels containing clean water and sediment to eliminate accumulated residues. Interestingly, worms almost completely lost accumulated HCB during the elimination phase. According to Franke et al. (1994), the noneliminated residues remaining after 10 days provide useful information for evaluating the bioaccumulation behavior of test chemicals in risk assessment schemes.

12.6   OLIGOCHAETES IN ECOTOXICOLOGY

12.6.1   Earthworms in Soil Ecotoxicology

Some issues must be improved to employ earthworm tests more efficiently. For example, since the onset of earthworm testing, the selection of compost worms as test organisms has been scrutinized. In particular, these species were thought to live in almost purely organic matter (i.e. not representative of typical soil-inhabiting worms), and it was unclear whether *E. fetida* and/or *E. andrei* were as sensitive as common field-relevant species (e.g. *L. terrestris* or *Aporrectodea caliginosa*).

In general, lumbricid earthworm species do not differ in sensitivity toward many pesticides (Heimbach 1985), but of course some species react differently to selected substances (Bauer and Römbke 1997). When comparing results of laboratory tests between *E. fetida* and other species, differences in LC_{50} values (lethal concentration for 50% of test animals) were nearly always within a factor of 10. One noteworthy exception is the insecticide propoxur, in which the LC_{50} of *E. fetida* was 72 times higher than that of *Aporrectodea longa* (Jones and Hart 1998). Some species, however, are more exposed to chemicals because of the worm's lifestyle (Edwards 1983). In particular, *L. terrestris* can be affected by direct spray, feeding on contaminated leaves or exposure to aqueous pesticide solutions washed into its burrows (Edwards et al. 1995).

On laboratory and field levels, standardized tests exist or are under development, but further work is necessary at the semifield level. In particular, the question of whether earthworms can reliably be used in terrestrial model ecosystems (TMEs) or similar systems (Knacker et al. 2004) must be clarified. Although preliminary results are promising (Römbke et al. 2004), further studies are necessary to determine whether or not earthworms are too large for the relatively small TMEs proposed (diameter 17 cm).

12.6.2   Enchytraeidae in Soil Ecotoxicology

The use of enchytraeids in soil ecotoxicology is hampered in the area of pesticide risk assessment because clear legal requirements for such tests are lacking. In other areas, e.g. the assessment of contaminated land, such requirements have been formulated recently (e.g. ISO 2005, 2006a). *E. crypticus* seems to be easier to handle (e.g. the mean number of juveniles is higher and more or less independent from soil properties) than the current standard species, *E. albidus*; thus, the former will likely be the main test species in the near future. Species closely related to *E. crypticus* are common in many temperate soils, but no higher-tier field test (including an evaluation scheme) with enchytraeids has been standardized. The most promising method proposed is to include enchytraeids when using TMEs (Moser et al. 2004). This semifield method has intact soil cores that are kept under controlled conditions (e.g. in a greenhouse). Pesticides can be applied on the soil surface before or directly on the soil core after their extraction. Typically, after a test duration of 16 weeks, samples are obtained using a soil corer to determine the abundance and species composition of potworms. Based on experiences with the fungicide carbendazim, enchytraeids may well be suited for the risk assessment of pesticides (Moser et al. 2007).

12.6.3 Oligochaetes in Sediment Ecotoxicology

Although aquatic oligochaete tests are not yet implemented formally in pesticide risk assessment, they are increasingly used for testing other chemical groups (e.g. industrial chemicals), where an exposure of benthic organisms to contaminated sediments is to be represented. Additionally, aquatic oligochaetes are evaluated as a component of aquatic outdoor mesocosms (e.g. Warren et al. 1998; Verdonschot and Ter Braak 2004). As the acceptance of such higher-tier tests is growing, they are becoming a regular part of the pesticide registration process. Experiences and guidance information are more elaborated for these aquatic mesocosms than for soil model ecosystems, particularly with guidance on test design [see the Higher-Tier Aquatic Risk Assessment for Pesticides (HARAP) workshop; Campbell et al. 1999] and multivariate evaluation tools (Maltby et al. 2005). The presence and evaluation of oligochaetes in aquatic mesocosms can provide valuable information for risk assessment—especially for substances that tend to associate with sediments—as these organisms are an integral part of benthic communities.

12.7 CONCLUSIONS

Oligochaetes are the most important organisms in soil ecotoxicology due to a variety of factors (e.g. ecological niche, practicality, sensitivity, standardized test methods). For sediment assessments, oligochaetes are currently second only to midge larvae. However, standardized test methods have been developed with lumbriculids and tubificids for both toxicity and bioaccumulation; their relevance for ERA is due to an endobenthic lifestyle and, consequently, exposure to sediment-associated chemicals. In current ecotoxicology, the importance of higher-tier (semifield and field) tests is growing because ecologically relevant test conditions as well as site-specific investigations are increasingly required. Thus, the use of oligochaetes in ecotoxicology will surely increase since these organisms can be studied on all experimental investigation levels. In this context, however, more research is needed to extend the advances made with relatively few "standard" species to field-relevant species as well.

ACKNOWLEDGMENTS

We thank the Federal Environmental Agency (Umweltbundesamt, UBA, Dessau, Germany) for supporting the development and standardization of oligochaete tests. We appreciate the help of Daniel Shain and Sonia Krutzke for editorial comments.

REFERENCES

ABRAHAMSEN, G. 1990. Influence of *Cognettia sphagnetorum* (Oligochaeta: Enchytraeidae) on nitrogen mineralization in homogenized mor humus. Biol. Fertil. Soils 9, 159–162.

ACHAZI, R.K., CHROSZCZ, G., PILZ, B., ROTHE, B., STEUDEL, I., THROL, C. 1996. Der Einfluss des pH-Werts und von PCB52 auf Reproduktion und Besiedlungsaktivität von terrestrischen Enchytraeen in PAK-, PCB- und schwermetallbelasteten Rieselfeldböden. Verh. Ges. Oekol. 26, 37–42.

AMORIM, M.J.B., RÖMBKE, J., SOARES, A.M.V.M. 2005a. Avoidance behaviour of *Enchytraeus albidus*: effects of benomyl, carbendazim, phenmedipham and different soil types. Chemosphere 59, 501–510.

AMORIM, M.J.B., RÖMBKE, J., SCHEFFCZYK, A., SOARES, A.M.V.M. 2005b. Effect of different soil types on the enchytraeids *Enchytraeus albidus* and *Enchytraeus luxuriosus* using the herbicide Phenmedipham. Chemosphere 61, 1102–1114.

ASTM INTERNATIONAL. 2000. Standard guide for the determination of the bioaccumulation of sediment-associated contaminants by benthic invertebrates, E 1688-00a. In *ASTM International 2004 Annual Book of Standards*, vol. 11.05. Biological Effects and Environmental Fate; Biotechnology; Pesticides. West Conshohocken, PA: ASTM International, pp. 1–26.

ASTM INTERNATIONAL. 2002. Standard test method for measuring the toxicity of sediment-associated contaminants with freshwater invertebrates, E1706-00. In *ASTM International 2004 Annual Book of Standards*, vol. 11.05. Biological Effects and Environmental Fate; Biotechnology; Pesticides. West Conshohocken, PA: ASTM International, pp. 1–118.

ASTM INTERNATIONAL. 2004. Standard guide for conducting laboratory soil toxicity or bioaccumulation tests with the lumbricid earthworm *Eisenia fetida* and the enchytraeid potworm *Enchytraeus albidus*, E1676–1697. In *ASTM International 2004 Annual Book of Standards*, vol. 11.05. Biological Effects and Environmental Fate; Biotechnology; Pesticides. West Conshohocken, PA: ASTM International, pp. 1–26.

BARNTHOUSE, L.W., SUTER, G.W., ROSEN, A.E., BEAUCHAMP, J.J. 1992. Estimating responses of fish populations to toxic contaminants. Environ. Toxicol. Chem. 6, 811–824.

BAUER, C., RÖMBKE, J. 1997. Factors influencing the toxicity of two pesticides on three lumbricid species in laboratory tests. Soil Biol. Biochem. 29, 705–708.

BELFROID, A., SEINE, W., Van GESTEL, C.A.M., HERMENS, J.L., Van LEEUWEN, K.J. 1995. Modelling the accumulation of hydrophobic organic chemicals in earthworms—application of the equilibrium partitioning theory. Environ. Sci. Pollut. Res. 2, 5–15.

BRINKHURST, R.O., JAMIESON, B.G.M. 1971. *Aquatic Oligochaeta*. Edinburgh: Oliver & Boyd.

BRUNS, E., EGELER, PH, RÖMBKE, J., SCHEFFCZYK, A., SPÖRLEIN, P. 2001a. Bioaccumulation of lindane and hexachlorobenzene by the oligochaetes *Enchytraeus luxuriosus* and *Enchytraeus albidus* (Enchytraeidae, Oligochaeta, Annelida). Hydrobiologia 463, 185–197.

BRUNS, E., EGELER, PH, MOSER, T., RÖMBKE, J., SCHEFFCZYK, A., SPÖRLEIN, P. 2001b. *Standardisierung und validierung eines bioakkumulationstests mit terrestrischen oligochaeten*. R&D No. 298 64 416. Report to the Federal Environmental Agency (Umweltbundesamt Berlin), 158 pp.

CAIRNS, J. 1986. The myth of the most sensitive species. Bioscience 36, 670–672.

CAMPBELL, P.J., ARNOLD, D.J.S., BROCK, T.C.M., GRANDY, N.J., HEGER, W., HEIMBACH, F., MAUND, S.J., STRELOKE, M. 1999. *Guidance Document on Higher-Tier Aquatic Risk Assessment for Pesticides (HARAP)*. Brussels, Belgium: SETAC-Europe.

CHAPMAN, P.M. 2001. Utility and relevance of aquatic oligochaetes in ecological risk assessment. Hydrobiologia 463, 149–169.

DE JONG, F.M.W., VAN BEELEN, P., SMIT, C.E., MONTFORTS, M.H.M.M. 2006. *Guidance for Summarising Earthworm Field Studies*. RIVM Report 601506006. Bilthoven, The Netherlands, 46 pp.

DIDDEN, W.A.M. 1993. Ecology of terrestrial Enchytraeidae. Pedobiologia 37, 2–29.

DIDDEN, W.A.M., RÖMBKE, J. 2001. Enchytraeids as indicator organisms for chemical stress in terrestrial ecosystems. Ecotoxicol. Environ. Saf. 50, 25–43.

EDWARDS, C.A. 1983. Earthworm ecology in cultivated soils. In *Earthworm Ecology - from Darwin to Vermiculture*, edited by J.E. Satchell. London: Chapman & Hall, pp. 123–137.

EDWARDS, C.A., BOHLEN, P.J. 1992. The effects of toxic chemicals on earthworms. Rev. Environ. Contam. Toxicol. 125, 23–99.

EDWARDS, C.A., BOHLEN, P.J. 1996. *Biology of Earthworms*, 3rd edn. London: Chapman and Hall.

EDWARDS, C.A., BOHLEN, P.J., LINDEN, D.R., SUBLER, S. 1995. Earthworms in agroecosystems. In *Earthworm Ecology and Biogeography in North America*, edited by P.F. Hendrix. Boca Raton, FL: Lewis Publishers, pp. 185–215.

EGELER, P., RÖMBKE, J. 2008. Oligochaete (microdrile) worms in ecotoxicology, in particular their use for the environmental risk assessment of pesticides in the European Union. Acta Hydrobiologica Sinica (in press).

EGELER, PH, RÖMBKE, J., MELLER, M., KNACKER, TH, FRANKE, C., STUDINGER, G., NAGEL, R. 1997. Bioaccumulation of lindane and hexachlorobenzene by tubificid sludgeworms (Oligochaeta) under standardised laboratory conditions. Chemosphere 35, 835–852.

EGELER, PH, MELLER, M., SCHALLNASS, H.J., GILBERG, D. 2005. *Validation of a Sediment Toxicity Test with the Endobenthic Aquatic Oligochaete Lumbriculus variegatus by an International Ring Test*. In co-operation with R. Nagel and B. Karaoglan. R&D No. 202 67 429. Report to the Federal Environmental Agency (Umweltbundesamt Berlin), 87 pp.

EGELER, PH, MELLER, M., SCHALLNASS, H.J., GILBERG, D. 2006. Validation of a Sediment Bioaccumulation Test with Endobenthic Aquatic Oligochaetes by an International Ring Test. R&D No. 202 67 437. Report to the Federal Environmental Agency (Umweltbundesamt Dessau), 150 pp.

EUROPEAN PLANT PROTECTION ORGANISATION (EPPO). 2003. EPPO standards. Environmental risk assessment scheme for plant protection products. Chapter 8: soil organisms and functions. EPPO Bull. 33, 195–209.

EUROPEAN UNION (EU). 1991. *Council Directive Concerning the Placing of Plant Protection Products on the Market*. No. 91/414/EEC. Brussels, Belgium.

FAVA, J.A., ADAMS, W.J., LARSON, R.J., DICKSON, G.W., DICKSON, K.L., BISHOP, W.E. 1987. *Research Priorities in Environmental Risk Assessment.* SETAC Workshop Report. Breckenridge, Colorado (SETAC).

FRANKE, C., STUDINGER, G., BERGER, G., BÖHLING, S., BRUCKMANN, U., COHORS-FRESENBORG, D., JÖHNCKE, U. 1994. The assessment of bioaccumulation. Chemosphere 29, 1501–1514.

FÜLL, C., NAGEL, R. 1994. Bioaccumulation of lindane (γ-HCH) and Dichlorprop (2,4-DP) by earthworms (*Lumbricus rubellus*). *Third European Conference on Ecotoxicology*, Zürich, Switzerland, 1994 (Abstract Book, p. 4.03).

GARCIA, M. 2004. Effects of pesticides on soil fauna: development of ecotoxicological test methods for tropical regions. Bonn (Zentrum für Entwicklungsforschung, Universität Bonn). Ecol. Dev. Ser. No. 19, 1–285.

HEIMBACH, F. 1985. Comparison of laboratory methods, using *Eisenia foetida* and *Lumbricus terrestris*, for the assessment of the hazard of chemicals to earthworms. Z. Pflanzenkr. Pflanzenschutz 92, 186–193.

HEIMBACH, F. 1997. Field tests on the side effects of pesticides on earthworms: influence of plot size and cultivation practices. Soil Biol. Biochem. 29, 671–676.

HUND-RINKE, K., LINDERMANN, M., SIMON, M. 2005. Experiences with novel approaches in earthworm testing alternatives. J. Soils Sediments 5, 233–239.

INTERNATIONAL COOPERATION ON HARMONISATION OF TECHNICAL REQUIREMENTS FOR REGISTRATION OF VETERINARY MEDICINAL PRODUCTS (VICH). 2005. Environmental Impact Assessment for Veterinary Medicinal Products—Phase II. Guidance, vol. 38. VICH GL.

ISO (INTERNATIONAL ORGANIZATION FOR STANDARDIZATION). 1998. Soil Quality—Effects of Pollutants on Earthworms (*Eisenia fetida*)—Part 2: Method for the Determination of Effects on Reproduction. ISO 11268-2. Genève, Switzerland.

ISO (INTERNATIONAL ORGANIZATION FOR STANDARDIZATION). 1999. Soil Quality—Effects of Pollutants on Earthworms—Part 3: Guidance on the Determination of Effects in Field Situations. ISO No. 11268-3. Genève, Switzerland.

ISO (INTERNATIONAL ORGANIZATION FOR STANDARDIZATION). 2003. Soil Quality—Effects of Pollutants on Enchytraeidae (*Enchytraeus* sp.). Determination of Effects on Reproduction. ISO No. 16387. Genève, Switzerland.

ISO (INTERNATIONAL ORGANIZATION FOR STANDARDIZATION). 2005. Soil Quality—Guidance on the Ecotoxicological Characterization of Soils and Soil Materials. ISO/CD No. 15799. Genève, Switzerland.

ISO (INTERNATIONAL ORGANIZATION FOR STANDARDIZATION). 2006a. Soil Quality – Guidance for the Choice and Evaluation of Bioassays for Ecotoxicological Characterization of Soils and Soil Materials. ISO 17616. Genève, Switzerland.

ISO (INTERNATIONAL ORGANIZATION FOR STANDARDIZATION). 2006b. Soil Quality—Avoidance Test for Testing the Quality of Soils and Effects of Chemicals on Behaviour—Part 1: Test with Earthworms (*Eisenia fetida* and *Eisenia andrei*). ISO 17512-1. Genève, Switzerland.

JÄNSCH, S., AMORIM, M.J.B., RÖMBKE, J. 2005. Identification of the ecological requirements of important terrestrial ecotoxicological test species. Environ. Rev. 13, 51–83.

JENSEN, J., LØKKE, H., HOLMSTRUP, M., KROGH, P-H., ELSGAARD, L. 2001. Effects and risk assessment of linear alkylbenzene sulfonates in agricultural soil. 5. Probabilistic risk assessment of linear alkylbenzene sulfonates in sludge-amended soils. Environ. Toxicol. Chem. 20, 1690–1697.

JONES, A., HART, A.D.M 1998. Comparison of laboratory toxicity tests for pesticides with field effects on earthworm populations: a review. In *Advances in Earthworm Ecotoxicology*, edited by S.C. Sheppard, J.D. Bembridge, M. Holmstrup and L. Posthuma. Pensacola, FL: SETAC Press, pp. 247–267.

KNACKER, T., Van GESTEL, C.A.M., JONES, S.E., SOARES, A.M.V.M., SCHALLNASS, H-J., FÖRSTER, B., EDWARDS, C.A. 2004. Ring-testing and field-validation of a terrestrial model ecosystem (TME)—an instrument for testing potentially harmful substances: conceptual approach and study design. Ecotoxicology 13, 5–23.

KULA, C., HEIMBACH, F., RIEPERT, F., RÖMBKE, J. 2006. Technical recommendations for the update of the ISO earthworm field test guideline (ISO 11268-3). J. Soils Sediments 6, 182–186.

KUPERMAN, R.K., AMORIM, M.J.B., RÖMBKE, J., LANNO, R., CHECKAI, R.T., DODARD, S.G., SUNAHARA, G.I., SCHEFFCZYK, A. 2006. Adaptation of the Enchytraeid toxicity test for use with natural soil types. Eur. J. Soil Biol. 42, S234–S243.

LAVELLE, P., BIGNELL, D., LEPAGE, M., WOLTERS, V., ROGER, P., INESON, P., HEAL, O.W., DHILLION, S. 1997. Soil function in a changing world: the role of invertebrate ecosystem engineers. Eur. J. Soil Biol. 33, 159–193.

LEEUWEN, C.J., HERMENS, J.L.M. 2001. *Risk Assessment of Chemicals: an Introduction.* Dordrecht, Holland: Kluwer Academic Publishers.

MALTBY, L., BLAKE, N., BROCK, T.C.M., Van den BRINK, P.J. 2005. Insecticide species sensitivity distributions: importance of test species selection and relevance to aquatic ecosystems. Environ. Toxicol. Chem. 24, 379–388.

MARCHESE, M.R., BRINKHURST, R.O. 1996. A comparison of two tubificid species as candidates for sublethal bioassay tests relevant to subtropical and tropical regions. Hydrobiologia 334, 163–168.

MILCU, A., SCHUMACHER, J., SCHEU, S. 2006. Earthworms (*Lumbricus terrestris*) affect plant seedling recruitment and microhabitat heterogeneity. Funct. Ecol. 20, 261–268.

MOSER, T., Van GESTEL, C.A.M., JONES, S.E., KOOLHAAS, J.E., RODRIGUES, J.M.L., RÖMBKE, J. 2004. Ring-testing and field-validation of a terrestrial model ecosystem (TME) – an instrument for testing potentially harmful substances: effects of carbendazim on enchytraeids. Ecotoxicology 13, 85–99.

MOSER, T., RÖMBKE, J., SCHALLNASS, H-J., Van GESTEL, C.A.M. 2007. The use of the multivariate principal response curve (PRC) for community level analysis: a case study on the effects of carbendazim on enchytraeids in terrestrial model ecosystems (TME). Ecotoxicology 16, 573–583.

OECD (ORGANISATION FOR ECONOMIC CO-OPERATION AND DEVELOPMENT). 1984. *OECD-Guideline for Testing of Chemicals No. 207. Earthworm Acute Toxicity Test.* Paris, France: Organisation for Economic Co-operation and Development.

OECD (ORGANISATION FOR ECONOMIC CO-OPERATION AND DEVELOPMENT). 2004a. *OECD-Guideline for Testing of Chemicals No. 218. Sediment-Water Chironomid Toxicity Test Using Spiked Sediment.* Paris, France: Organisation for Economic Co-operation and Development.

OECD (ORGANISATION FOR ECONOMIC CO-OPERATION AND DEVELOPMENT). 2004b. *OECD-Guideline for Testing of Chemicals No. 219. Sediment-Water Chironomid Toxicity Test Using Spiked Water.* Paris, France: Organisation for Economic Co-operation and Development.

OECD (ORGANISATION FOR ECONOMIC CO-OPERATION AND DEVELOPMENT). 2004c. *OECD-Guideline for Testing of Chemicals No. 220. Enchytraeidae Reproduction Test.* Paris, France: Organisation for Economic Co-operation and Development.

OECD (ORGANISATION FOR ECONOMIC CO-OPERATION AND DEVELOPMENT). 2004d. *Guideline for Testing of Chemicals No. 222. Earthworm Reproduction Test.* Paris, France: Organisation for Economic Co-operation and Development.

OECD (ORGANISATION FOR ECONOMIC CO-OPERATION AND DEVELOPMENT). 2006. *Guideline for Testing of Chemicals No. 225: Sediment-water Lumbriculus Toxicity Test Using Spiked Sediment.* Adopted October 2007. Paris, France: Organisation for Economic Co-operation and Development.

PETERSEN, H., LUXTON, M. 1982. A comparative analysis of soil fauna populations and their role in decomposition processes. Oikos 39, 287–388.

RASMUSSEN, J.B. 1984. Comparison of gut contents and assimilation efficiency of fourth instar larvae of two coexisting chironomids, *Chironomus riparius* Meigen and *Glyptotendipes paripes* (Edwards). Can. J. Zool. 62, 1022–1026.

REYNOLDSON, T.B., THOMPSON, S.P., BAMSEY, J.L. 1991. A sediment bioassay using the tubificid oligochaete worm *Tubifex tubifex.* Environ.Toxicol. Chem. 10, 1061–1072.

RODRIGUEZ, P., REYNOLDSON, T.B. 1999. Laboratory methods and criteria for sediment bioassessment. In *Manual of Bioassessment of Aquatic Sediment Quality,* edited by A. Mudroch, J.M. Azcue and P. Mudroch. Boca Raton, FL: CRC Press LLC Lewis Publishers, pp. 33–83.

ROGHAIR, C.J., BUIJZE, A., HUYS, M.P.A., WOLTERS-BALK, M.A.H., YEDEMA, E.S.E., HERMENS, J.L.M. 1996. *Toxicity and Toxicokinetics for Benthic Organisms; II: QSAR for Base-Line Toxicity to the Midge Chironomus riparius and the Tubificid Oligochaete Worm Branchiura*

sowerbyi. RIVM Report 719101026. Bilthoven, The Netherlands, 50 pp.

RÖMBKE, J. 2003. Ecotoxicological laboratory tests with enchytraeids: a review. Pedobiologia 47, 607–616.

RÖMBKE, J. 2006. Tools and techniques for the assessment of ecotoxicological impacts of contaminants in the terrestrial environment. Human Environ. Risk Assess. 12, 84–101.

RÖMBKE, J., MOSER, T. 2002. Validating the enchytraeid reproduction test: organisation and results of an international ring test. Chemosphere 46, 1117–1140.

RÖMBKE, J., HEIMBACH, F., HOY, S., KULA, C., SCOTT-FORDSMAND, J., SOUSA, P., STEPHENSON, G., WEEKS, J. 2003. *Effects of Plant Protection Products on Functional Endpoints in Soil (EPFES Lisbon 2002).* Pensacola, FL: SETAC Publications.

RÖMBKE, J., Van GESTEL, C.A.M., JONES, S.E., KOOLHAAS, J.E., RODRIGUES, J.M.L., MOSER, T. 2004. Ring-testing and field-validation of a terrestrial model ecosystem (TME) – an instrument for testing potentially harmful substances: effects of carbendazim on earthworms. Ecotoxicology 13, 101–114.

RÖMBKE, J., JÄNSCH, S., MOSER, T. 2005. State-of-the-art: the use of Enchytraeidae as test and indicator organisms in ecotoxicology. Proc. Estonian Acad. Sci. Biol. Ecol. (Special Issue on Enchytraeidae—Newsletter on Enchytraeidae No. 9) 54, 342–346.

SAMPLE, B.E., BEAUCHAMP, J.J., EFROYMSON, R.A., SUTER, G.W., ASHWOOD, T.L. 1998. *Development and Validation of Bioaccumulation Models for Earthworms.* Report ES/ER/TM-200. Lockheed Martin Energy Systems, Inc. Oak Ridge National Laboratory, 45 pp.

SPURGEON, D.J., WEEKS, J.M., Van GESTEL, C.A.M. 2003. A summary of eleven years progress in earthworm ecotoxicology. Pedobiologia 47, 588–604.

UNITED STATES ENVIRONMENTAL PROTECTION AGENCY (US EPA). 1992. *Framework for Ecological Risk Assessment.* EPA/630/R-92/001. Washington.

VERDONSCHOT, P.F.M., TER BRAAK, C.J.F. 2004. An experimental manipulation of oligochaete communities in mesocosms treated with chlorpyrifos or nutrient additions: multivariate analyses with Monte Carlo permutation tests. Hydrobiologia 278, 251–266.

WARREN, L.A., TESSIER, A., HARE, L. 1998. Modelling cadmium accumulation by benthic invertebrates in situ: the relative contributions of sediment and overlying water reservoirs to organism cadmium concentrations. Limnol. Oceanogr. 43, 1442–1454.

WIEMEYER, S.N., SPITZER, P.R., KRANTZ, W.C., LAMONT, T.G., CROMARTIE, E. 1975. Effects of environmental pollutants on Connecticut and Maryland ospreys. J. Wildl. Manage. 1, 124–139.

YEARDLEY, R.B., LAZORCHAK, J.M., GAST, L.C. 1996. The potential of an earthworm avoidance test for evaluation of hazardous waste sites. Environ. Toxicol. Chem. 15, 1532–1537.

Chapter 13

Evolution and Ecology of *Ophryotrocha* (Dorvilleidae, Eunicida)

Daniel J. Thornhill,* Thomas G. Dahlgren,† and Kenneth M. Halanych*

*Department of Biological Sciences, Auburn University, Auburn, AL
†Göteborgs Universitet, Zoologiska Institutionen, Systematik och Biodiversitet, Göteborg, Sweden

13.1 INTRODUCTION

Members of *Ophryotrocha* are small, opportunistic benthic worms that inhabit nutrient-rich environments ranging from polluted harbors to deep-sea sediments. These worms are typically small and nondescript with a rounded or blunt prostomium and well-delineated setigers. Most studies on the group have focused on approximately 20 relatively shallow-water species that have been maintained in culture (largely due to Bertil Åkesson's efforts, e.g. Åkesson 1967, 1973a, b). A wide variety of reproductive modes are found in species of this genus, including gonochorism, simultaneous hermaphrodism and sequential hermaphrodism. As a result, these organisms have been ideal models for study of reproductive strategies. They occur from the oceans' depths at hydrothermal vent environments to continental slopes to littoral regions. *Ophryotrocha* spp. are typically found as infauna or at the sediment surface, and in general appear to graze food (e.g. bacteria, eukaryotic microbes, detritus) from their substrate. Parasitic forms are also known (Martin et al. 1991). The purpose of this chapter is to broadly review the evolution and ecology of *Ophryotrocha* spp., to assess the current state of knowledge for the group, and to provide some future research directions.

13.2 GENERAL MORPHOLOGY

Ophryotrocha spp. are typically small (generally between 1.5 and 3.0 mm long, but parasitic taxa approach 110 mm), fairly simple, cylindrical and nondescript. For most species, the protostomium ranges from a rounded to a broad, spade-shaped structure that may contain two antennae and palps, which occur in a variety of forms but are reported never to be moniliform (i.e. beaded). Branchiae are usually absent; parapodia are uniramous, and ventral and dorsal cirri are usually lacking. Pleijel and Eide (1996) report these losses as apomorphies for *Ophryotrocha* spp.; however, their conclusions were based only on a limited taxon sample. For example, *Ophryotrocha labidion* clearly possess a ventral parapodial cirrus and *Ophryotrocha mandibulata* has a short dorsal cirrus (Hilbig and Blake 1991). Most species are ciliated or have clear bands or tufts of cilia. Retention of cilia is most likely due to the worm's small size, allowing cilia to have a locomotory function.

Perhaps the most conspicuous structures in *Ophryotrocha* spp. are jaws (a feature generally shared with other Eunicida), which have been described as ctenognaths (Orensanz 1990; Struck et al. 2006). Jaws usually have maxillae comprising eight elements: seven symmetrical sets of anterior free denticles and a relatively large ice-tong-shaped basal maxillae. Whether or not basal maxillae constitute "carrier" structures homologous to other Eunicida has been debated (see Paxton 2004). Jaw morphologies are referred to as either P or K type. P types (from German "primitiv") are found in all species, whereas K types ("kompliziert"), which occur later ontologically, are missing in several taxa (Ockelmann and Åkesson 1990; Dahlgren et al. 2001). The ice tong shape of the mandible is usually visible in mature specimens. Paxton (2004) gives a particularly thorough discussion of jaw growth and replacement in *Ophryotrocha labronica* LaGreca and Bacci 1962. In particular, she points out that an ontogenetic series of jaws are replaced by a molting-like process, and that clear morphological differences between sets of P-type jaws exist. This observation suggests that, at least for some species, the P/K dichotomy is an oversimplification (Paxton and Åkesson 2007).

13.3 TAXONOMIC AND PHYLOGENETIC CONSIDERATIONS

Ophryotrocha was first described by Claparède and Mecznikow (1869) and over 40 species have since been described (Table 13.1), with numerous additional species awaiting description (Table 13.2). Fossil evidence suggests that *Ophryotrocha* is an old lineage, spanning back to the Cretaceous (Eriksson and Lindström 2000). Although species within the group have been studied since shortly after their discovery (e.g. Bonnier 1893; Bergh 1895; Bergmann 1903; Meek 1912), the taxonomy and evolutionary history of the group have received little attention until recently. Both Hilbig and Blake (1991) and Pleijel and Eide (1996) have provided redescriptions of the genus *Ophryotrocha*. Based on the taxa employed, Pleijel and Eide (1996) provided diagnostic apomorphies for the group. They list five characters as supporting *Ophryotrocha* monophyly, but three of these are problematic: one is homoplastic and two (dorsal and ventral cirri) represent losses of structures. The remaining putative apomorphies includes a single pair of maxillary rows and eight maxillary plates or elements (other non-*Ophryotrocha* taxa have more pairs of rows and plates). Within *Ophryotrocha*, the overall structure of the jaws varies considerably. Interestingly, much of the group's taxonomy has focused on the jaw apparatus and cephalic structures (i.e. antennae and palps). As jaws are involved in feeding, the evolution of jaw morphology is presumably influenced by diet, about which we know little. Thus, the taxonomic reliability of these structures deserves further evaluation.

Ophryotrocha has traditionally been placed within Dorvilleidae, Annelida (Fauchald 1977; Hilbig and Blake 1991; Pleijel and Eide 1996), but Orensanz (1990) asserted that it should be placed with *Iphitime* in Iphitimidae, a phylogenetic hypothesis not supported by molecular data (see below). Phylogenetic studies of dorvilleid relationships have been limited. Notably, Eibye-Jacobsen and Kristensen (1994) conducted a morphological

Table 13.1 List of described *Ophryotrocha* species with their authority information and original collection localities

Ophryotrocha Species and Authority	Location and Habitat
Ophryotrocha adherens Paavo et al. 2000	Cyprus and Hawaii, littoral
Ophryotrocha akessoni Blake 1985	Galapagos Rift, East Pacific Basin, deep sea
Ophryotrocha atlantica Hilbig and Blake 1991	NW Atlantic, slope depths
Ophryotrocha baccii Parenti 1961	Roscoff, France, littoral
Ophryotrocha bifida Hilbig and Blake 1991	NW Atlantic, slope depths
Ophryotrocha claparedii Studer 1878	Kerguelen, littoral
Ophryotrocha costlowi Åkesson 1978	Beaufort, North Carolina, littoral
Ophryotrocha cosmetandra Oug 1990	Northern Norway, littoral
Ophryotrocha diadema Åkesson 1976	Los Angeles har bor, littoral
Ophryotrocha dimorphica Zavarzina and Tzetlin 1986	Peter the Great Bay, littoral
Ophryotrocha dubia Hartmann-Schröder 1974	North Sea (off Scotland), 68 m
Ophryotrocha gerlachi Hartmann-Schröder 1974	North Sea (off Denmark) 52 m
Ophryotrocha geryonicola (Esmark 1874)	Skagerack, Kattegat, sublittoral
Ophryotrocha globopalpata Blake and Hilbig 1990	Juan de Fuca Ridge, deep sea
Ophryotrocha gracilis Huth 1933	Helgoland, Germany, littoral
Ophryotrocha hadalis Jumars 1974	Aleutian Trench, deep sea
Ophryotrocha hartmanni Huth 1933	NE Atlantic, littoral
Ophryotrocha irinae Tzetlin 1980b	Kandalaksha Bay, White Sea, littoral
Ophryotrocha kagoshimaensis Miura 1997	Kagoshima Bay, Japan, 197 m
Ophryotrocha labidion Hilbig and Blake 1991	NW Atlantic, slope depths
Ophryotrocha labronica LaGreca and Bacci 1962	Naples, Italy, littoral
Ophryotrocha lipscombae Lu and Fauchald 2000	NW Atlantic, slope depths
Ophryotrocha littoralis (Levinsen 1879)	Egesminde, Greenland, littoral
Ophryotrocha lobifera Oug 1978	West Norway, in mud, 50 m
Ophryotrocha longidentata Josefson 1975	Skagerack, Kattegat, 50–100 m
Ophryotrocha maciolekae Hilbig and Blake 1991	NW Atlantic, slope depths
Ophryotrocha maculata Åkesson 1973b	Skagerack, Kattegat, 25 m
Ophryotrocha mandibulata Hilbig and Blake 1991	NW Atlantic, slope depths
Ophryotrocha mediterranea Martin et al. 1991	Mediterranean, parasitic on *Geryon longipes*, 600–1,800 m
Ophryotrocha minuta Lévi 1954	Roscoff, France, littoral
Ophryotrocha natans Pfannenstiel 1975b	Red Sea, littoral
Ophryotrocha notialis (Ehlers 1908)	Southern South America, sublittoral
Ophryotrocha notoglandulata Pfannenstiel 1972	Japan, littoral
Ophryotrocha obtusa Hilbig and Blake 1991	NW Atlantic, slope depths
Ophryotrocha pachysoma Hilbig and Blake 1991	NW Atlantic, slope depths
Ophryotrocha paralbidion Hilbig and Blake 1991	NW Atlantic, slope depths
Ophryotrocha paragerlachi Brito and Núñez 2003	Canary Islands, *Cymodcea nodosa* meadows, 6–15 m
Ophryotrocha platykephale Blake 1985	Guayamas Basin, hydrothermal vents, deep sea
Ophryotrocha puerilis puerilis Claparède and Mecznikow 1869	Naples, Italy, littoral
Ophryotrocha puerilis siberti (McIntosh 1885)	Plymouth, England, littoral
Ophryotrocha scarlatoi Averincev 1989	Franz Josefs Land, littoral
Ophryotrocha schubravyi Tzetlin 1980a	Marine aquarium in Moscow, Russia
Ophryotrocha socialis Ockelmann and Åkesson 1990	Marine aquarium in Helsingör, Denmark
Ophryotrocha spatula Fournier and Conlan 1994	Arctic Canada, littoral
Ophryotrocha splendida Brito and Núñez 2003	Canary Islands, *Cymodcea nodosa* meadows, 6–15 m
Ophryotrocha vivipara Banse 1963	San Juan Archipelago, USA, 22 m
Ophryotrocha wubaolingi Miura 1997	Kagoshima Bay, Japan, 200 m

Source: Adapted from Dahlgren et al. (2001; with permission from the Marine Biological Laboratory) with additions of new *Ophryotrocha* species.

Table 13.2 Undescribed *Ophryotrocha* species, including distribution and reference information

Putative *Ophryotrocha* Species	Location	Source
Ophryotrocha alborana nom. nud.	Mediterranean	Dahlgren et al. (2001)
Ophryotrocha sp. Eilat-Hurghada	Red Sea	Dahlgren et al. (2001)
Ophryotrocha japonica nom. nud.	Japan, California, Mediterranean	Dahlgren et al. (2001)
Ophryotrocha macrovifera nom. nud.	Mediterranean, NE and NW Atlantic	Dahlgren et al. (2001)
Ophryotrocha obscura nom. nud.	Public aquarium, Gothenburg	Dahlgren et al. (2001)
Ophryotrocha olympica nom. nud.	Olympic peninsula, WA, USA	Pleijel and Eide (1996)
Ophryotrocha permanni nom. nud.	Florida, Okinawa, China	Dahlgren et al. (2001)
Ophryotrocha prolifica nom. nud.	Indian River, FL, USA	Pleijel and Eide (1996)
Ophryotrocha robusta nom. nud.	Mediterranean	Dahlgren et al. (2001)
Ophryotrocha rubra nom. nud.	Mediterranean	Pleijel and Eide (1996)
Ophryotrocha sp. Qingdao	China	Dahlgren et al. (2001)
Ophryotrocha sp. Sanya 2	South Hainan, China	Dahlgren et al. (2001)
Ophryotrocha sativa nom. nud.	Public aquarium, Gothenburg	Pleijel and Eide (1996)

All species listed above occur in either intertidal or shallow-water environments. Names pending formal description are followed by "nom. nud." to indicate their status as nomina nuda (i.e. name not available).

phylogenetic analysis, including several dorvilleids, in an explicit attempt to use objective methods to understand the evolution of this group. However, many small dorvilleids have limited morphology that may be reduced during progenetic evolutionary processes and thus confound phylogenetic interpretation (Struck et al. 2005, 2006). In terms of molecular data, the 18S gene is the most extensively sampled for Dorvilleidea relationships. Although dorvilleids appear not to be monophyletic, *Ophryotrocha* is nested within the main "Dorvilleidae" clade (Struck et al. 2005, 2006). Even though molecular data corroborate the traditional morphological view, many dorvilleid taxa, including *Ophrotrocha*, have long-branch lengths (see Struck et al. 2006), an issue known to be potentially problematic for phylogenetic reconstruction (Felsenstein 1978).

Based on our current knowledge, *Ophryotrocha* appears to be within Dorvilleidae, closely related to *Pseudophryotrocha, Parophryotrocha, Parapodrilus, Exallopus, Parougia, Ougia, Microdorvillea* and some of *Dorvillea* (the latter may be polyphyletic; Hilbig and Blake 1991; Pleijel and Eide 1996; Struck et al. 2006). However, monophyly of *Ophryotrocha* is currently debatable. Using cytochrome c oxidase subunit I (CO1) mitochondrial gene data, Heggøy et al. (2007) shows that *Iphitime paguri* Fage and Legendre 1934 is deeply nested within *Ophryotrocha* as a sister clade to an *Ophryotrocha hartmanni/Ophryotrocha gracilis* clade. Høisæter and Samuelsen (2006) discuss the similarities between *Iphitime* and *Ophryotrocha* and suggest the two genera are indistinguishable. Also, unpublished data (D.J. Thornhill and K.M. Halanych) based on the 16S and cytochrome b mitochondrial genes indicate that *Exallopus* is within *Ophryotrocha*.

Three studies have focused mainly on phylogenetic relationships within *Ophryotrocha*, and all three have taken advantage of the numerous species held in culture by Bertil Åkesson. In the first, Pleijel and Eide (1996; Fig. 13.1a) conducted a morphological cladistic analysis that also included electrophoretic loci for 20 *Ophryotrocha* taxa (note that allozymes are typically used for intra- but not interspecific questions). Their analysis was important in providing a phylogenetic hypothesis of *Ophryotrocha* evolution that could be explicitly tested. Dahlgren et al. (2001) expanded this work using data from the 16S ribosomal mitochondrial gene in 18 cultured taxa (Fig. 13.1b), many of which were also

Figure 13.1 Phylogenetic reconstruction of the evolutionary history of *Ophryotrocha* spp. based on (a) Pleijel and Eide (1996; with permission from Taylor & Francis), (b) Dahlgren et al. (2001; with permission from the Marine Biological Laboratory) and (c) Heggøy et al. (2007; with permission from Taylor & Francis).

examined in the previous morphological analysis. In the third analysis, Heggøy et al. (2007) expanded the 16S dataset in Dahlgren et al. (2001) to the CO1 data for 17 *Ophryotrocha* taxa from cultures (Fig. 13.1c).

Prior to these studies, *Ophryotrocha* species were clustered into three main groups, *labronica, hartmanni* and *gracilis*, based on reproductive biology and jaw morphology (Åkesson 1973b, 1984). Pleijel and Eide (1996) suggested that the *hartmanni* group was not a monophyletic clade, but they found some phylogenetic support for the *labronica* and *gracilis* groups. Results of molecular analyses do not support *hartmanni* and *gracilis* groups, but generally support the idea of a *labronica* group, whose members are gonochoristic (other *Ophryotrocha* are hermaphroditic; Dahlgren et al. 2001; Heggøy et al. 2007). However, relationships with the *labronica* group vary greatly between the morphological analysis and molecular studies. Similarly, all three analyses support a close relationship between *Ophryotrocha diadema* Åkesson 1976 and *Ophryotrocha alborana* nom. nud., but relationships among other *Ophryotrocha* species are inconsistent between morphology and molecules.

In contrast, considerable congruence exists between the two molecular studies, which employ similar taxa and gene fragments. Mitochondrial sequence data support a close relationship between *O. hartmanni* Huth 1933 and *O. gracilis* Huth 1933, which are simultaneous hermaphrodites, but differ in jaw type (*O. gracilis* only has P-type jaws; *O. hartmanni* has both) and shape of the egg mass (*O. gracilis* lays irregular-shaped masses; *O. hartmanni* lays fusiform-shaped masses). One of the major shortcomings of phylogenetic analyses to date is the lack of representatives from the deep-sea or specialized habitats (Jumars 1974; Hilbig and Blake 1991; Levin et al. 2003). Unpublished data from methane seep (K.M. Halanych and D.J. Thornhill) and whale-fall (T.G. Dahlgren) *Ophryotrocha* show a greater diversity of this group.

13.4 REPRODUCTIVE BIOLOGY

A variety of reproductive modes are found in *Ophryotrocha* spp., including gonochorism, simultaneous hermaphrodism and sequential hermaphrodism. As a result, *Ophryotrocha* spp. offer an interesting opportunity to explore hypotheses about sexual selection, sexual conflict and the evolution of mating systems in a related group of organisms. Additionally, many of these polychaetes are readily maintained in cultures, have short generation times, are small in size, breed semicontinuously, and have easily observable behavior and quantifiable reproductive success. These characteristics make *Ophryotrocha* spp. ideal organisms for controlled laboratory experiments on sexual strategy (Berglund 1991; Sella and Ramella 1999). Table 13.3 summarizes published data on reproductive characteristics of well-known *Ophryotrocha* spp.

In most studied *Ophryotrocha* species, reproductive individuals generally produce a network of mucous-lined tubes, which, along with corresponding mucous trails, presumably assist individuals in finding a mate. Once a potential mate is located, individuals engage in a lengthy (approximately 4.5 h) courtship and subsequent mating via close physical contact followed by external fertilization, known as pseudocopulation (Rolando 1981; Sella 1985, 1988, 1990, 1991; Premoli and Sella 1995). The resulting brood is generally cared for by one or both parents and is housed in a protective tube or cocoon of mucous, parchment-like material, or loose jelly (Table 13.3).

Although details of reproduction are poorly known in many *Ophryotrocha*, gonochorism is the most common reproductive strategy found in the genus (Sella and Ramella

Table 13.3 Data on sexuality, reproductive output, fertilization mode, parental care and egg protection from species of *Ophryotrocha*

Ophryotrocha Species	Form of Sexuality	Mean No. of Eggs per Spawning	Egg Diameter (μm)	Fertilization Mode	Parental Care	Brood Protection
Ophryotrocha labronica labronica	Gonochorism, sex dimorphism	120–130	120	Pseudocopulation aflagellate sperm	Maternal	Mucuous tube
Ophryotrocha labronica pacifica	Gonochorism, sex dimorphism	90	120	Pseudocopulation aflagellate sperm	Maternal	Mucuous tube
Ophryotrocha costlowi	Gonochorism, sex dimorphism	100	125	Pseudocopulation aflagellate sperm	Maternal	Mucuous tube
Ophryotrocha notoglandulata	Gonochorism, sex dimorphism	115	120–130	Pseudocopulation aflagellate sperm	Maternal	Mucuous tube
Ophryotrocha robusta	Gonochorism, strong sex dimorphism	150–200	125	Pseudocopulation aflagellate sperm	Biparental	Mucuous cocoon
Ophryotrocha cosmetandra	Gonochorism, strong sex dimorphism	30–60	270–350	Pseudocopulation aflagellate sperm	Maternal	Parchment-like tube
Ophryotrocha macrovifera	Gonochorism, sex dimorphism	80–100	180	Pseudocopulation aflagellate sperm	Maternal	Mucuous tube
Ophryotrocha diadema	Simultaneous hermaphrodism	25	180	Pseudocopulation aflagellate sperm	Biparental	Mucuous tube
Ophryotrocha gracilis	Simultaneous hermaphrodism	11	200–275	Pseudocopulation aflagellate sperm	Biparental	Mucuous cocoon
Ophryotrocha hartmanni	Simultaneous hermaphrodism	30	175	Pseudocopulation aflagellate sperm	Maternal	Mucuous cocoon
Ophryotrocha bacci	Simultaneous hermaphrodism	15–35	250	Pseudocopulation	Biparental	Mucuous cocoon
Ophryotrocha maculata	Simultaneous hermaphrodism	20–40	125	Pseudocopulation	Absent	Loose jelly
Ophryotrocha socialis	Simultaneous hermaphrodism	50	150	Pseudocopulation aflagellate sperm	Communal care	Parchment-like open long tube
Ophryotrocha puerilis puerilis	Sequential hermaphrodism	200–300	100–110	Pseudocopulation aflagellate sperm	Biparental	Loose jelly

Source: Adapted from Sella and Ramella (1999), with permission from Springer.

1999). In *O. labronica*, the most studied species, body size and fecundity are correlated for females, but not for males when mating pairs are placed in isolation (Berglund 1991). However, when multiple males are present, larger male *O. labronica* have increased reproductive success resulting from either male–male competition, female choice or some combination of these two factors (Berglund 1991). Sex ratios in some *Ophryotrocha* spp. have also been examined: *O. labronica* are reported to be 1:1 or female biased, with shifts in ratio possible within a lineage (Åkesson 1972, 1975; Sella and Zambaldi 1985). On the other hand, sex ratio has also been reported to be heritable and unchanged after several generations (Åkesson and Paxton 2005). Adjustments in sex ratio possibly relate to population density, with female-dominated populations favored at low densities (Hamilton 1967; Sella and Ramella 1999). For many gonochoristic *Ophryotrocha* spp., parental care is generally maternal and broods comprise a relatively large clutch of small-diameter eggs (Table 13.3).

Simultaneous hermaphrodism is present in at least eight *Ophryotrocha* species (Sella and Ramella 1999). Hermaphrodism is hypothesized to be a successful reproductive strategy for low-density populations, as it increases the likelihood of encountering a suitable potential mate (Westheide 1984; Ghiselin 1987; Sella and Ramella 1999; Puurtinen and Kaitala 2002). Simultaneous hermaphrodites have an initial protandrous phase, but do not typically reproduce until becoming fully hermaphroditic (Sella et al. 1996). These *Ophryotrocha* species exhibit a mating behavior known as egg trading (Fischer 1980), in which partners regularly switch between male and female roles, and only small parcels of eggs are released at any one time, thereby insuring an opportunity for each to reciprocally fertilize its partner's eggs (Sella 1985, 1988; Sella et al. 1996, 1997). Egg trading is thought to be a protection against "cheating," wherein one partner could exclusively assume the male role producing only energetically cheaper sperm. When male–male competition is lacking, this behavior results in biased allocation of ovarian tissue and female gametes (~80% in *O. diadema* and *O. gracilis*; Sella 1990; Sella et al. 1996), as only enough sperm to fertilize available eggs is required. However, presence of reproductive competitors causes individuals to drastically increase their sex allocation in male function (up to 100% in *O. diadema*; Lorenzi et al. 2005). Partner fidelity also appears dependent on availability of ovigerous mates (higher availability results in higher rates of desertion; Sella 1990; Sella et al. 1996, 1997; Sella and Lorenzi 2000). Generally, simultaneous hermaphrodites produce fewer, larger eggs, and spawn more frequently than do congeneric gonochoristic species (Table 13.3; Sella et al. 1996; Prevedelli et al. 2006). Eggs of simultaneous hermaphrodites are typically cared for by both parents, which may be a by-product of pair formation (Table 13.3; Sella 1991; Sella and Lorenzi 2000).

Ophryotrocha puerilis Claparède and Mecznikow 1869, for which multiple lineages have been identified (e.g. Paxton and Åkesson 2007), is the only known sequential hermaphrodite in the genus *Ophryotrocha*. In this species, smaller, younger individuals tend to be males, whereas larger, older individuals tend to be females. However, individuals can alternate between sexes several times throughout their lives, and females commonly change sex after production of energetically costly eggs and corresponding biomass loss (Berglund 1986, 1990; Premoli and Sella 1995). Females express preference for smaller males, perhaps because small males are less likely to change sex (Berglund 1986, 1990). Natural populations of *O. puerilis* tend to have a male-biased sex ratio, likely due to higher mortality among females and the ability of large females to hormonally inhibit the production of eggs in other individuals (Pfannenstiel 1975a; Grothe and Pfannenstiel 1986; Premoli and Sella 1995; Prevedelli et al. 2006). *O. puerilis* produces large broods of small eggs that are cared for by both parents (Table 13.3).

Whether the last common ancestor of extant *Ophryotrocha* was gonochoristic or hermaphroditic remains an open question (Sella and Ramella 1999). Results of Dahlgren et al. (2001) indicate that the change from one reproductive strategy to another has taken place only once during *Ophryotrocha* evolution. Theoretically, the transition from hermaphrodism to gonochorism is more evolutionarily likely (see Sella and Ramella 1999). The occasional presence of hermaphroditic characters in gonochoristic species also indicates that gonochorism may be derived (Pfannenstiel 1976; Bacci et al. 1979; Rolando 1982; Åkesson 1984; Premoli et al. 1996). Recently available phylogenetic evidence supports this conjecture; for example, Heggøy et al. (2007; Fig. 13.1c) and unpublished data (D.J. Thornhill and K.M. Halanych) suggest that gonochorism is indeed derived from simultaneous hermaphrodism in *Ophryotrocha*. Likewise, the observation of an occasional simultaneous hermaphrodite in *O. puerilis* is consistent with phylogenetic placement of this species as being derived from simultaneous hermaphrodites (Fig. 13.1b; Dahlgren et al. 2001).

13.5 ECOLOGY

Ophryotrocha largely consists of infaunal species (Hilbig and Blake 1991), which are globally distributed from tropics to poles in a wide variety of soft-sediment habitats (Tables 13.1 and 13.2). The best-studied *Ophryotrocha* species occur interstitially, at relatively low density in shallow, nutrient-rich waters (La Greca and Bacci 1962; Åkesson 1976). However, a number of species are opportunistic or stress tolerant, reaching high abundance in environments that are inhospitable to many other metazoans (Levin et al. 2003). These include organically enriched environments (Grassle and Morse-Porteous 1987; Lenihan et al. 1990; Hall et al. 1997; Tzetlin et al. 1997; Conlan et al. 2004), organic patches throughout the deep sea (Grassle and Morse-Porteous 1987; Miura 1997; Lu and Fauchald 2000), polluted harbors (Simonini and Prevedelli 2003), eutrophied areas associated with mariculture (Brooks et al. 2003; Hall-Spencer et al. 2006; Lee et al. 2006), pulp mill and sewage outfalls (Hall et al. 1997; Paavo et al. 2000), fouling organisms on the hulls of boats (Bonnier 1893) and poorly maintained aquaria (Åkesson 1973b, 1984), commensally inside the branchial chambers of deep-water crabs (Martin et al. 1991) and shipworm siphons (H. Wiklund, unpublished observation), Arctic (Conlan and Kvitek 2005) and Antarctic ice scours (Richardson and Hedgpeth 1977; Lenihan and Oliver 1995; Bromberg et al. 2000; Varella-Petti et al. 2006) and oxygen minimum zone (OMZ) sediments (Levin et al. 1991). *Ophryotrocha* species are also a dominant fauna at highly reductive and sulfidic environments, including east Pacific hydrothermal vents (Blake 1985; Grassle et al. 1985; Petrecca and Grassle 1987; Mullineaux et al. 2003), methane seeps (Sahling et al. 2002; Levin et al. 2003; Robinson et al. 2004) and whale-fall environments (Smith et al. 1998; Dahlgren et al. 2006). Outside of biogeographic and habitat associations, and laboratory studies of demography (e.g. Prevedelli and Simonini 2003; Simonini and Prevedelli 2003), relatively little is known about the ecology of this group, mostly due to the challenges of studying small, well-dispersed marine organisms.

The ability of certain *Ophryotrocha* species to tolerate environments that are inhospitable to most other eukaryotes likely releases it from biological controls including interspecific competition and predation (Levin et al. 2003). As a result, *Ophryotrocha* spp. in marginal environments are often found in areas with very low species richness, but at extremely high abundance (e.g. Lenihan and Oliver 1995; Hall et al. 1997; Paavo et al.

2000; Levin et al. 2003). Outside of the marginal environments described above, abundance of these *Ophryotrocha* species often decreases dramatically.

Physiological and biochemical mechanisms that underlie various *Ophryotrocha* spp. resistance to sulfides, hydrocarbons, heavy metals and other toxins are not well understood. Possibly, this resistance results from general insensitivity and / or from well-developed detoxification mechanisms (Grieshaber and Volkel 1998; Levin et al. 2003). In either case, the presence of certain *Ophryotrocha* species may be indicative of impacted or marginal habitats: *Ophryotrocha claparedii* Studer 1878 associated with heavy human impacts at McMurdo Station, Antarctica (Lenihan et al. 1990); *O. hartmanni* associated with sewage effluent in the Tyne Estuary, England (Hall et al. 1997); *Ophryotrocha* cf. *vivipara* Banse 1963 around British Columbia salmon farms (Brooks et al. 2003); and *Ophryotrocha adherens* Paavo et al. 2000 associated with Pacific threadfin mariculture and sewage outfalls in Hawaii (Paavo et al. 2000; Lee et al. 2006), among others. Therefore, the distribution and abundance of these and other *Ophryotrocha* spp. may be used as bioindicators to assay environmental quality.

Several reports suggest that certain *Ophryotrocha* spp. are early successional species that are competitively displaced over time. For instance, Mullineaux et al. (2003) experimentally tested temporal successional patterns at East Pacific Rise hydrothermal vents. They found that *Ophryotrocha akessoni* Blake 1985 was most abundant during the early-to midsuccessional period at these vents (up to 8 months), with its abundance dropping significantly in subsequent periods. Similarly, *Ophryotrocha spatula* Fournier and Conlan 1994 is present as an early successional species in Arctic ice scours of northern Canada, with its numbers declining precipitously as the scours age (Fournier and Conlan 1994; Conlan and Kvitek 2005), and *O. adherens* is an early- to midsuccessional species in a Hawaiian mariculture operation (Lee et al. 2006). Despite this, many *Ophryotrocha* taxa, such as *Ophryotrocha platykephale* Blake 1985 at methane seeps (Levin et al. 2003), appear to exist as stable populations in constant, high-stress environments.

Typically, only one *Ophryotrocha* sp. is found in a particular habitat or region; however, instances of sympatric *Ophryotrocha* species have been noted. For example, Prevedelli et al. (2005) investigated patterns of seasonal abundance in different members of a mixed *Ophryotrocha* spp. community in La Spezia Harbor, Italy. Species exhibited differential abundance throughout the year with the most abundant taxa changing seasonally (Prevedelli et al. 2005). Specifically, *O. puerilis* dominated during the spring; *Dinophilus gyrociliatus* Schmidt 1857 (traditionally considered a dorvilleid, but see Struck et al. 2005) dominated during the summer; *O. labronica* dominated during the autumn, and *Schistomeringos rudolphii* (Delle Chiaje 1828) (Dorvilleidae) along with an undescribed *Ophryotrocha* sp. were most abundant during the winter (Prevedelli et al. 2005). *O. hartmanni* was also present at very low abundance during all months sampled (Prevedelli et al. 2005). This pattern indicates that some *Ophryotrocha* species are able to coexist by specializing under different environmental conditions, in this case associated with seasonal environmental change. Other studies have suggested that similar phenomena explain the coexistence of several congeneric *Ophryotrocha* species. Levin et al. (2003) found that at least three *Ophryotrocha* species, and likely many more, coexisted at methane seeps along the California and Oregon coast. They postulated that the geochemical gradients and unusual microbial food sources at methane seeps facilitate evolutionary specialization on different microhabitats. Thus, several *Ophryotrocha* species may coexist by resource partitioning (Levin et al. 2003). Preliminary analysis of gut contents and stable isotope ratios supports this hypothesis (Levin et al. 2003). Similarly, at an experimentally implanted whale carcass in the shallow North Atlantic, five species have been found associated with the bacterial mat covering bones and reduced sediment (T.G. Dahlgren, unpublished data).

At least two species of *Ophryotrocha* are suspected of being introduced, occurring outside their endogenous habitat due to anthropogenic interference. *O. labronica* is considered a cosmopolitan species that occurs in the Mediterranean Sea as well as in the Pacific and Atlantic Oceans. Åkesson and Paxton (2005) recently suggested that this wide distribution could be a result of anthropogenic introduction. *Ophryotrocha* species that have been reported as ship fouling organisms (e.g. Bonnier 1893) could in theory be carried in ballast water tanks and are commonly reported in eutrophic harbors (Simonini and Prevedelli 2003). Simonini (2002) reported the recent appearance of *Ophryotrocha* sp. *japonica* in Italian harbors, which had previously been known only from the waters of Japan and California. Increasing international commerce, coupled with the hardy life history of this group, makes future introductions of nonnative *Ophryotrocha* spp. progressively more likely.

13.6 FUTURE RESEARCH

Ophryotrocha is relatively well studied in comparison to most annelid and marine invertebrate genera. Efforts to culture multiple species of these animals have permitted numerous excellent comparative studies (many discussed above). Nonetheless, we have a limited understanding of various aspects of their evolution and biology. Thus, to promote additional research on species of *Ophryotrocha*, several unresolved, important questions remain: What is their phylogenetic diversity? Why, from a physiological perspective, are they so successful in inhospitable environments? What do they eat (besides flaked spinach)? What reproductive strategies are found in *Ophryotrocha* species from extreme and/or ephemeral environments such as whale falls, methane seeps and hydrothermal vents? What is the genetic basis for differences in reproductive biology? How different are cultured populations from their wild counterparts? How much do developmental patterns and ontological timing vary between species? How do they disperse to new habitats?

If examined in a rigorous manner, these topics of interest should yield valuable insight on not only the biology of *Ophryotrocha*, but into annelid and marine invertebrate systems in general.

ACKNOWLEDGMENTS

We thank Dan Shain for the opportunity to contribute this work, which was made possible with support from the National Science Foundation (EAR-0120646, OCE-0425060) to K.M.H. and the Swedish Research Council to T.G.D., Contribution No. 22 to the Auburn University Marine Biology Program.

REFERENCES

ÅKESSON, B. 1967. On the biology and larval morphology of *Ophryotrocha puerilis* Claparède and Metschnikov (Polychaeta). Ophelia 4, 111–119.

ÅKESSON, B. 1972. Incipient reproductive isolation between geographic populations of *Ophryotrocha labronica* (Polychaeta, Dorvilleidae). Zool. Scr. 1, 207–210.

ÅKESSON, B. 1973a. Reproduction and larval morphology of five *Ophryotrocha* species (Polychaeta, Dorvilleidae). Zool. Scr. 2, 145–155.

ÅKESSON, B. 1973b. Morphology and life history of *Ophryotrocha maculata* sp. n. (Polychaeta, Dorvilleidae). Zool. Scr. 2, 141–144.

ÅKESSON, B. 1975. Reproduction in the genus *Ophryotrocha* (Polychaeta, Dorvilleidae). Pubbl. Staz. zool. Napoli. (Suppl.) 39, 377–398.

ÅKESSON, B. 1976. Morphology and life cycle of *Ophryotrocha diadema*, a new polychaete species from California. Ophelia 15, 23–35.

ÅKESSON, B. 1978. A new *Ophryotrocha* species of the *Labronica* group (Polychaeta, Dorvilleidae) revealed in crossbreeding experiments. In *NATO Conference Series (Marine Science)*, edited by B. Battaglia and J. Beardmore. New York: Plenum Publishing, pp. 573–590.

ÅKESSON, B. 1984. Speciation in the genus *Ophryotrocha* (Polychaeta, Dorvilleidae). Fortschr. Zool. 29, 299–316.

ÅKESSON, B., PAXTON, H. 2005. Biogeography and incipient speciation in *Ophryotrocha labronica* (Polychaeta, Dorvilleidae). Mar. Biol. Res. 1, 127–139.

AVERINCEV, V.G. 1989. The seasonal dynamics of Polychaeta in high-arctic coastal ecosystems of Franz-Josef Land (Errantia). Akademiya Nauk SSSR, 1–78.

BACCI, G., LANFRANCO, M., MANTELLO, I., TOMBA, M. 1979. New pattern of hermaphroditism (inducible hermaphroditism) in populations of *Ophryotrocha labronica* (Annelida Polychaeta). Experientia 35, 605–606.

BANSE, K. 1963. Polychaetous annelids from Puget Sound and the San Juan Archipelago, Washington. Proc. Biol. Soc. Wash. 76, 197–208.

BERGH, R.S. 1895. Neue Untersuchungen über *Ophryotrocha* und über Anneliden-larven. Zool. Centralblatt. Leipsig. 2, 257–263.

BERGLUND, A. 1986. Sex change in a polychaete: effects of social and reproductive costs. Ecology 67, 837–845.

BERGLUND, A. 1990. Sequential hermaphroditism and the size-advantage hypothesis: an experimental test. Anim. Behav. 39, 426–433.

BERGLUND, A. 1991. To change or not to change sex: a comparison between two *Ophryotrocha* species (Polychaeta). Evol. Ecol. 5, 128–135.

BERGMANN, W. 1903. Untersuchungen über die Eibildung bei Anneliden und Cephalopoden. Z. Wiss. Zool. 73, 278–301.

BLAKE, J.A. 1985. Polychaeta from the vicinity of deep-sea geothermal vents in the Eastern Pacific. I. Euphrosinidae, Phyllodocidae, Hesionidae, Nereididae, Glyceridae, Dorvilleidae, Orbiniidae, and Maldanidae. Bull. Biol. Soc. Wash. 6, 67–101.

BLAKE, J.A., HILBIG, B. 1990. Polychaeta from the vicinity of deep-sea hydrothermal vents in the Eastern Pacific Ocean. II. New species and records from the Juan de Fuca and Explorer Ridge systems. Pac. Sci. 44, 219–253.

BONNIER, J. 1893. Notes sur les annélides du Boulonnais. I. L' *Ophryotrocha puerilis* (Claparède et Metschnikoff) et son appareil maxillaire. Bull. Biol. Fr. Belg. 25, 198–226.

BRITO, M.D., NÚÑEZ, J. 2003. Three new interstitial dorvilleids (Annelida: Polychaeta) from the *Cymodocea nodosa* meadows of the Canary Islands. Hydrobiologia 496, 27–34.

BROMBERG, S., NONATO, E.F., CORBISIER, T.N., VARELLA-PETTI, M.A. 2000. Polychaete distribution in the nearshore zone of Martel Inlet, Admiralty Bay (King George Island, Antarctica). Bull. Mar. Sci. 67, 175–188.

BROOKS, K.M., STIERNS, A.R., MAHNKEN, C.V.W., BLACKBURN, D.B. 2003. Chemical and biological remediation of the benthos near Atlantic salmon farms. Aquaculture 219, 355–377.

CLAPARÈDE, E., MECZNIKOW, E. 1869. Beiträge zur Kenntnis der Entwicklungsgeschichte der Chaetopoden. Z. Wiss. Zool. Abt. A. 19, 163–205.

CONLAN, K.E., KVITEK, R.G. 2005. Recolonization of soft-sediment ice scours on an exposed Arctic coast. Mar. Ecol. Prog. Ser. 286, 21–42.

CONLAN, K.E., KIM, S.L., LENIHAN, H.S., OLIVER, J.S. 2004. Benthic changes during 10 years of organic enrichment by McMurdo Station, Antarctica. Mar. Pollut. Bull. 49, 43–60.

DAHLGREN, T.G., ÅKESSON, B., SCHANDER, C., HALANYCH, K.M., SUNDBERG, P. 2001. Molecular phylogeny of the model annelid *Ophryotrocha*. Biol. Bull. 201, 193–203.

DAHLGREN, T.G., WIKLUND, H., KÄLLSTRÖM, B., LUNDÄLV, T., SMITH, C.R., GLOVER, A. 2006. A shallow-water whale-fall experiment in the north Atlantic. Cah. Biol. Mar. 47, 385–389.

EHLERS, E. 1908. Die Bodensaessigen Anneliden aus den Sammlungen der deutschen Tiefsee-Expedition. In *Wissenschaftliche Ergebnisse der deutschen Tiefsee-Expedition auf dem Dampfer 'Valdivia' 1898–1899*, edited by C. Chun. Jena, Germany: Gustav Fischer, vol. 16, pp. 1–168.

EIBYE-JACOBSEN, D., KRISTENSEN, R.M. 1994. A new genus and species of Dorvilleidae (Annelida, Polychaeta) from Bermuda, with a phylogenetic analysis of Dorvilleidae, Iphitimidae and Dinophilidae. Zool. Scr. 23, 107–131.

ERIKSSON, M., LINDSTRÖM, S. 2000. *Ophryotrocha* sp., the first report of a jawed polychaete from the Cretaceous of Skåne, Sweden. Acta Palaeontol. Pol. 45, 311–315.

ESMARK, L. 1874. Eteonopsis geryonicola. Forh. VidenskSelsk. Christiania. 1873, 497–498.

FAGE, L., LEGENDRE, R. 1934. Les annélides polychètes du genre *Iphitime*. A propos d'une espece nouvelle commensale des pagures, *Iphitime paguri* n. sp. Bull. Soc. Zool. France 58, 299–305.

FAUCHALD, K. 1977. The polychaete worms. Definitions and keys to the orders, families and genera. Nat. Hist. Museum Los Angeles County. Sci. Ser. 28, 1–188.

FELSENSTEIN, J. 1978. Cases in which parsimony or compatibility methods will be positively misleading. Syst. Zool. 27, 401–410.

FISCHER, E.A. 1980. The relationship between mating system and simultaneous hermaphroditism in the coral reef fish, *Hypoplectrus nigricans* (Serranidae). Anim. Behav. 28, 620–633.

FOURNIER, J.A., CONLAN, K.E. 1994. A new species of *Ophryotrocha* (Polychaeta, Dorvilleidae) associated with ice scours in the Canadian Arctic Archipelago. In *Actes de la 4ème conférence internationale des polychètes,* edited by J.C. Dauvin, L. Laubier, D.J. Reish. Mem. Mus. Natl. Hist. Nat. 162, 185–190.

GHISELIN, M.T. 1987. Evolutionary aspects of marine invertebrate reproduction. In *Reproduction of Marine Invertebrates*, edited by C.A. Giese, J.S. Pearse and V.B. Pearse. Palo Alto, CA: Blackwell Scientific Publications, pp. 609–666.

GRASSLE, J.F., MORSE-PORTEOUS, L.S. 1987. Macrofaunal colonization of disturbed deap-sea environments and the structure of deep-sea benthic communities. Deep Sea Res. 34, 1911–1950.

GRASSLE, J.F., BROWN-LEGER, S., MORSE-PORTEOUS, L.S., PETRECCA, R., WILLIAMS, I. 1985. Deep-sea fauna of sediments in the vicinity of hydrothermal vents. Bull. Biol. Soc. Wash. 6, 443–452.

GRIESHABER, M.K., VOLKEL, S. 1998. Animal adaptations for tolerance and exploitation of poisonous sulfide. Annu. Rev. Physiol. 60, 33–53.

GROTHE, C., PFANNENSTIEL, H.D. 1986. Cytophysiological study of neurosecretory and pheromonal influences on sexual development in *Ophryotrocha puerilis* (Polychaeta, Dorvilleidae). Int. J. Invertebr. Rep. Dev. 10, 227–239.

HALL, J.A., FRID, C.L.J., GILL, M.E. 1997. The response of estuarine fish and benthos to an increasing discharge of sewage effluent. Mar. Pollut. Bull. 34, 527–535.

HALL-SPENCER, J., WHITE, N., GILLESPIE, E., GILLHAM, K., FOGGO, A. 2006. Impact of fish farms on maerl beds in strongly tidal areas. Mar. Ecol. Prog. Ser. 326, 1–9.

HAMILTON, W.D. 1967. Extraordinary sex ratios. Science 156, 477–488.

HARTMANN-SCHRÖDER, G. 1974. Polychaeten von Expeditionen der "Anton Dohrn" in Nordsee und Skagerrak. Veröff. Inst. Meeresforsch. Bremerh. 14, 169–274.

HEGGØY, K.K., SCHANDER, C., ÅKESSON, B. 2007. The phylogeny of the annelid genus *Ophryotrocha* (Dorvilleidae). Mar. Biol. Res. 3, 412–420.

HILBIG, B., BLAKE, J.A. 1991. Dorvilleidae (Annelida, Polychaeta) from the U.S. Atlantic slope and rise. Description of two new genera and 14 new species, with a generic revision of *Ophryotrocha*. Zool. Scr. 20, 147–183.

HOISAETER, T., SAMUELSEN, T.J. 2006. Taxonomic and biological notes on a species of Iphitime (Polychaeta, Eunicida) associated with Pagurus prideaux from western Norway. Mar. Biol. Res. 2, 333–354.

HUTH, W. 1933. *Ophryotrocha*-Studien. Zur Cytologie der Ophryotrochen. Z. Zelfforsch. mikrosk Anat. Berlin 20, 309–381.

JOSEFSON, A. 1975. *Ophryotrocha longidentata* sp.n. and *Dorvillea erucaeformis* (Malmgren) (Polychaeta, Dorvilleidae) from west coast of Scandinavia. Zool. Scr. 4, 49–54.

JUMARS, P.A. 1974. A generic revision of the Dorvilleidae (Polychaeta), with six new species from the deep North Pacific. Zool. J. Linn. Soc. 54, 101–135.

LAGRECA, M., BACCI, G. 1962. Una nuova specie di *Ophryotrocha* delle coste tirreniche. Boll. Zool. 29, 13–23.

LEE, H.W., BAILEY-BROCK, J.H., MCGURR, M.M. 2006. Temporal changes in the polychaete infaunal community surrounding a Hawaiian mariculture operation. Mar. Ecol. Prog. Ser. 307, 175–185.

LENIHAN, H.S., OLIVER, J.S. 1995. Anthropogenic and natural disturbances to marine benthic communities in Antarctica. Ecol. Appl. 5, 311–326.

LENIHAN, H.S., OLIVER, J.S., OAKDEN, J.M., STEPHENSON, M.D. 1990. Intense and localized benthic marine pollution around McMurdo Station, Antarctica. Mar. Pollut. Bull. 21, 422–430.

LÉVI, C. 1954. *Ophryotrocha minuta*, nov. sp., nouveau polychète mésopsammique. Bull. Soc. Zool. France 79, 466–469.

LEVIN, L.A., THOMAS, C.L., WISHNER, K. 1991. Control of deep-sea benthic community structure by oxygen and organic-matter gradients in the eastern Pacific Ocean. J. Mar. Res. 49, 763–800.

LEVIN, L.A., ZIEBIS, W., MENDOZA, G.F., GROWNEY, V.A., TRYON, M.D., BROWN, K.M., MAHN, C., GIESKES, J.M., RATHBURN, A.E. 2003. Spatial heterogeneity of macrofauna at northern California methane seeps: influence of sulfide concentration and fluid flow. Mar. Ecol. Prog. Ser. 265, 123–139.

LEVINSEN, G. 1879. Om to nye Slaetger af arctiske Chaetopode Annelider. Vidensk- Meddr dansk naturh. Foren. 1879, 9–18.

LORENZI, M.C., SELLA, G., SCHLEICHEROVA, D., RAMELLA, L. 2005. Outcrossing hermaphroditic polychaete worms adjust their sex allocation to social conditions. J. Evol. Biol. 18, 1341–1347.

LU, H., FAUCHALD, K. 2000. *Ophryotrocha lipscombae*, a new species and a possible connection between ctenognath and labidognath-prionognath eunicean worms (Polychaeta). Proc. Biol. Soc. Wash. 113, 486–492.

MCINTOSH, W. 1885. Report on the Annelida Polychaeta collected by H.M.S. Challenger during the years 1873–1876. Rep. Challenger, (Zool.). 12, 1–554.

MARTIN, D., ABELLO, P., CARTES, J. 1991. A new species of *Ophryotrocha* (Polychaeta, Dorvilleidae) commensal in *Geryon longipes* (Crustacea, Brachyura) from the Western Mediterranean Sea. J. Nat. Hist. 25, 279–292.

MEEK, C.F.U. 1912. A metrical analysis of chromosome complexes, showing correlation of evolutionary development and chromatin thread-width throughout the animal kingdom. Philos. Trans. R. Soc. Lond. B Biol. Sci. 203, 1–74.

MIURA, T. 1997. Two new species of the genus *Ophryotrocha* (Polychaeta, Iphitimiidae) from Kagoshima Bay. Bull. Mar. Sci. 60, 300–305.

MULLINEAUX, L.S., PETERSON, C.H., MICHELI, F., MILLS, S.W. 2003. Successional mechanism varies along a gradient in hydrothermal fluid flux at deep-sea vents. Ecol. Monogr. 73, 523–542.

OCKELMANN, K.W., ÅKESSON, B. 1990. *Ophryotrocha socialis* n. sp., a link between two groups of simultaneous hermaphrodites within the genus (Polychaeta, Dorvilleidae). Ophelia 31, 145–162.

ORENSANZ, J.M. 1990. The eunicemorph polychaete annelids from Antarctic and Subantarctic Seas. Antarct. Res. Ser. 52, 1–183.

OUG, E. 1978. New and lesser known Dorvilleidae (Annelida. Polychaeta) from Scandinavian and Northeast American waters. Sarsia 63, 285–303.

OUG, E. 1990. Morphology, reproduction, and development of a new species of *Ophryotrocha* (Polychaeta, Dorvilleidae) with strong sexual dimorphism. Sarsia 75, 191–201.

PAAVO, B., BAILEY-BROCK, J.H., ÅKESSON, B. 2000. Morphology and life history of *Ophryotrocha adherens* sp. nov. (Polychaeta, Dorvilleidae). Sarsia 85, 251–264.

PARENTI, U. 1961. *Ophryotrocha puerilis siberti, O. hartmanni* ed *O. bocci* nelle acque di Roscoff. Cah. Biol. Mar. 2, 437–445.

PAXTON, H. 2004. Jaw growth and replacement in *Ophryotrocha labronica* (Polychaeta, Dorvilleidae). Zoomorphology 123, 147–154.

PAXTON, H., ÅKESSON, B. 2007. Redescription of *Ophryotrocha puerilis* and *O. labronica* (Annelida, Dorvilleidae). Mar. Biol. Res. 3, 3–19.

PETRECCA, R., GRASSLE, J.F. 1987. Notes on fauna from several deep-sea hydrothermal vent communities. In *Gorda Ridge, a Seafloor Spreading Center in the United States' Exclusive Economic Zone*, edited by G.R. McMurray. New York: Springer-Verlag, pp. 279–283.

PFANNENSTIEL, H.D. 1972. New *Ophryotrocha* species (Polychaeta, Eunicidae) from Japan. Helgoland Wiss. Meer. 23, 117–124.

PFANNENSTIEL, H.D. 1975a. Mutual influence on the sexual differentiation in the protandric polychaete *Ophryotrocha puerilis*. In *Intersexuality in the Animal Kingdom*, edited by R. Reinboth. Heidelberg, Germany: Springer, pp. 48–56.

PFANNENSTIEL, H.D. 1975b. *Ophryotrocha natans* n.sp. (Polychacta, Dorvilleidae): Ein Simultanzwitter mit acht männlichen Segmenten aus dem Golf von Aqaba. Zool. Anz. 195, 1–7.

PFANNENSTIEL, H.D. 1976. Ist der Polychaete *Ophryotrocha labronica* ein proteandrischer Hermaphrodit? Mar. Biol. 38, 169–178.

PLEIJEL, F., EIDE, R. 1996. The phylogeny of *Ophryotrocha* (Dorvilleidae: Eunicida: Polychaeta). J. Nat. Hist. 30, 647–659.

PREMOLI, M.C., SELLA, G. 1995. Sex economy in benthic polychaetes. Ethol. Ecol. Evol. 7, 27–48.

PREMOLI, M.C., SELLA, G., BERRA, G.P. 1996. Heritable variation of sex ratio in a polychaete worm. J. Evol. Biol. 9, 845–854.

PREVEDELLI, D., SIMONINI, R. 2003. Life cycles in brackish habitats: adaptive strategies of some polychaetes from the Venice Lagoon. Oceanol. Acta. 26, 77–84.

PREVEDELLI, D., N'SIALA, G.M., SIMONINI, R. 2005. The seasonal dynamics of six species of Dorvilleidae (Polychaeta) in the harbour of La Spezia (Italy). Mar. Ecol. Evol. Persp. 26, 286–293.

PREVEDELLI, D., N'SIALA, G.M., SIMONINI, R. 2006. Gonochorism vs. hermaphroditism: relationship between life history and fitness in three species of *Ophryotrocha* (Polychaeta: Dorvilleidae) with different forms of sexuality. J. Anim. Ecol. 75, 203–212.

PUURTINEN, M., KAITALA, V. 2002. Mate-search efficiency can determine the evolution of separate sexes and the stablity of hermaphroditism in animals. Am. Nat. 160, 645–660.

RICHARDSON, M.D., HEDGPETH, J.W. 1977. Antactica soft-bottom, macrobenthic community adaptations to a cold, stable, highly productive, glacially affected environment. In *Adaptations within Antarctic Ecosystems. Proceedings of the Third SCAR Symposium on Antarctic Biology*, edited by G.A. Llano. Washington, DC: Smithsonian Institution, pp. 181–196.

ROBINSON, C.A., BERNHARD, J.M., LEVIN, L.A., MENDOZA, G.F., BLANKS, J.K. 2004. Surficial hydrocarbon seep infauna from the Blake Ridge (Atlantic Ocean, 2150 m) and the Gulf of Mexico (690–2240 m). Mar. Ecol. P. S. Z. N. I. 25, 313–336.

ROLANDO, A. 1981. Early courtship and sexual differentiation in *Ophryotrocha labronica* LaGreca and Bacci (Polychaeta, Dorvilleidae). Monit. Zool. Ital. (N.S.) 15, 53–61.

ROLANDO, A. 1982. Sexual condition in a population of *Ophryotrocha robusta* (Annelida, Polychaeta) from Genova. Atti Soc. Tosc. Sci. Nat. (B) 89, 145–152.

SAHLING, H., RICKERT, D., LEE, R.W., LINKE, P., SUESS, E. 2002. Macrofaunal community structure and sulfide flux at gas hydrate deposits from the Cascadia convergent margin, NE Pacific. Mar. Ecol. Prog. Ser. 231, 121–138.

SELLA, G. 1985. Reciprocal egg trading and brood care in a hermaphroditic polychaete worm. Anim. Behav. 33, 938–944.

SELLA, G. 1988. Reciprocation, reproductive success, and safeguards against cheating in a hermaphroditic polychaete worm, *Ophryotrocha diadema* Åkesson, 1976. Biol. Bull. 175, 212–217.

SELLA, G. 1990. Sex allocation in the simultaneously hermaphroditic polychaete worm *Ophryotrocha diadema*. Ecology 71, 27–32.

SELLA, G. 1991. Evolution of biparental care in the hermaphroditic polychaete worm *Ophryotrocha diadema*. Evolution 45, 63–68.

SELLA, G., LORENZI, M.C. 2000. Partner fidelity and egg reciprocation in the simultaneously hermaphroditic polychaete worm *Ophryotrocha diadema*. Behav. Ecol. 11, 260–264.

SELLA, G., RAMELLA, L. 1999. Sexual conflict and mating systems in the dorvilleid genus *Ophryotrocha* and the dinophilid genus *Dinophilus*. Hydrobiologia 402, 203–213.

SELLA, G., ZAMBALDI, M. 1985. Self fertilization effects of some fitness traits in *Ophryotrocha labronica* (Polychaeta, Dorvilleidae). Atti Assoc. Genet. It. 31, 191–192.

SELLA, G., PREMOLI, M.C., TURRI, F. 1997. Egg trading in the simultaneously hermaphroditic polychaete worm *Ophryotrocha gracilis* (Huth). Behav. Ecol. 8, 83–86.

SIMONINI, R. 2002. Distribution and ecology of the genus *Ophryotrocha* (Polychaeta: Dorvilleidae) in Italian harbors and lagoons. Vie Milieu Paris 52, 59–65.

SIMONINI, R., PREVEDELLI, D. 2003. Life history and demography of three populations of *Ophryotrocha japonica* (Polychaeta: Dorvilleidae). Mar. Ecol. Prog. Ser. 258, 171–180.

SMITH, C.R., MAYBAUM, H.L., BACO, A.R., POPE, R.H., CARPENTER, S.D., YAGER, P.L., MACKO, S.A., DEMING, J.W. 1998. Sediment community structure around a whale skeleton in the deep Northeast Pacific: macrofaunal, microbial and bioturbation effects. Deep Sea Res. II 45, 517–567.

STRUCK, T.H., HALANYCH, K.M., PURSCHKE, G. 2005. Dinophilidae (Annelida) is not a progenetic Eunicida; evidence from 18S and 28S rDNA. Mol. Phylogenet. Evol. 37, 619–623.

STRUCK, T.H., PURSCHKE, G., HALANYCH, K.M. 2006. Phylogeny of Eunicida (Annelida) and exploring data congruence using a partition addition bootstrap alteration (PABA) approach. Syst. Biol. 55, 1–20.

STUDER, TH. 1878. Beiträge zur Naturgeschichte wirbelloser Thiere von Kerguelensland. Anatomie von *Brada mamillata* und neue art von *Ophryotrocha* (Polychaeta, pp. 111–121). Arch. Naturgesch. 44, 102–121.

TZETLIN, A.B. 1980a. *Ophryotrocha schubravyi* sp. n. and the problem of evolution of the mouth parts in the Eunicemorpha (Polychaeta). Zool. Zhurnal. 59, 666–676.

TZETLIN, A.B. 1980b. Two new species of the family Dorvilleidae (Polychaeta) from the White and Barents Seas. Zool. Zhurnal. 59, 17–22.

TZETLIN, A.B., MOKIEVSKY, V.O., MELNIKOV, A.N., SAPHONOV, M.V., SIMDYANOV, T.G., IVANOV, I.E. 1997. Fauna associated with detached kelp in different types of subtidal habitats of the White Sea. Hydrobiologia 355, 91–100.

VARELLA-PETTI, M.A., NONATO, E.F., SKOWRONSKI, R.S.P., CORBISIER, T.N. 2006. Bathymetric distribution of the meiofaunal polychaetes in the nearshore zone of Martel Inlet, King George Island, Antarctica. Antarct. Sci. 18, 163–170.

WESTHEIDE, W. 1984. The concept of reproduction in polychaetes with small body size: adaptation in interstitial species. Fortschr. Zool. 29, 265–278.

ZAVARZINA, E.G., TZETLIN, A.B. 1986. Biology of *Ophryotrocha dimorphica* sp.n. (Polychaeta, Eunicida) from the Peter the Great Bay (the Japan Sea). Zool. Zhurnal. 65, 1808–1817.

Chapter 14

Cosmopolitan Earthworms— A Global and Historical Perspective

Robert J. Blakemore

COE Soil Ecology Group, Graduate School of Environment and Information Sciences, Yokohama National University, Tokiwadai, Yokohama, Japan

14.1 INTRODUCTION

Apart from their manifest ecological importance, megadrile oligochaetes—the so-called "common earthworms"—are of particular phylogenetic and biogeographical interest since they are an ancient and diverse group with generally feeble powers of overland travel so that most are confined to their areas of origin. Sims (1980) said the slow-moving Oligochaeta could be regarded as excellent subjects for zoogeographical studies were it not for the paucity of fossil records for these soft-bodied animals. Conversely, due to their antiquity and relative stability in soil, those that persist today, especially in primitive families, are more akin to "living fossils." Thus, the main obstacle to studying their zoogeography is the lack of comprehensive ecological and taxonomic surveys for most regions.

Precursors to earthworms possibly emerged in the late Precambrian some 650–570 million years ago (Valentine 1980), or in the Proterozoic one billion years BP (years before present) (Seilacher et al. 1998). Classified as ecosystem engineers or "bioneers" in the colonization and preparation of soil, these ancestral forms predate the invasion of land by plants, fungi, insects (400–500 million years BP) and occur well before others with similar descent to these early worms, such as reptiles or mammals (200 million years BP), arose (Launer 2006). Earthworms are ubiquitous in all but the driest and coldest regions, and present-day global distributions of the 20 earthworm families have been used to validate Wegener's hypothesis of continental drift and theories of plate tectonics (e.g. Wegener 1915; Michaelsen 1922; Sims and Easton 1972; Sims 1980; Lee 1981, 1985, 1994).

Only recently have some species with common family ancestries, separated over geological time by the drifting continents, been reunited due to unintentional and sometimes deliberate transportation, literally following in the tracks of human endeavors, particularly those involving agriculture, trade and commerce rather than by natural means. Schwert and Dance (1979) netted lumbricid cocoons from Canadian streams, but there are few actual observations of natural dispersal (cf. Schwert 1990); thus, we can

Annelids in Modern Biology, Edited by Daniel H. Shain
Copyright © 2009 John Wiley & Sons, Inc.

only assume these modes of transportation from outcomes such as the relatively wide distribution of native *Heteroporodrilus* in Australia corresponding with (present and past) catchments of the Murray–Darling river system (Blakemore 2000b, 2006b). And while no earthworm genus is known to naturally transcend wide oceanic barriers, some exception may be found in those few genera with euryhaline species (saltwater tolerant as adults and/or cocoons) or in those that obtain assistance by rafting on flotsam (e.g. in soil within tree root boluses) or carriage in mud adhering to feet of birds. Thus, Michaelsen (1905) thought it likely that *Microscolex* originated in South America and its circum-Antarctic distribution was largely due to rafting on prevailing West Wind Drift in the last few thousand years. Examples of euryhaline species are found in genera *Diachaeta, Eukerria, Microscolex, Pontodrilus* and—from Sims (1980) and Blakemore (1999)—*Pontoscolex*, but many of these are also thought to have had their ranges extended considerably by human-mediated transportation. Two of the most widely distributed species are *Pontoscolex corethrurus* (Müller 1856) and *Pontodrilus litoralis* (Grube 1855); the latter, found on tropical beaches around the world and probably originating from the Indo-Australasia region, was likely spread initially as a ships' beach-sand "ballast waif" and secondarily by its cocoons rafting on flotsam (Subba Rao and Ganapati 1974; Blakemore 2007a).

In general, the degree of regional endemism depends on palaeogeological history and the extent of glaciation and/or volcanism, whereas present-day distribution of the relatively few exotics is strongly influenced by recent, historic and prehistoric human trade and migrations (Stephenson 1930; Gates 1972; Lee 1985, 1987; Sims and Gerard 1985; Blakemore 2002, 2005, 2006a). Accordingly, an earthworm population potentially comprises four components (Blakemore 1999): *resident natives* that are often highly endemic; *introduced exotics* that tend to be more widespread; *translocated natives*, i.e. endemic species that have been transported or redistributed outside their natural range within a bioregion; and *neoendemics*, i.e. members of nonnative genera that have sojourned for sufficient time following their introduction to have undergone speciation in a new region (or are currently unknown or extinct in their places of origin).

Examples of neoendemics may be the *Dichogaster* species (Octochaetidae from Africa) that Stephenson (1931) thought diversified on the Pacific Isles, and several *Microscolex* species (Acanthodrilidae from South America) apparently found uniquely on sub-Antarctic islands. In Europe, Sims and Gerard (1985, 1999) said, "Most British species are allochthonous [= exogenous or non-native, cf. autochthonous] of the remaining half-a-dozen we can only speculate how they came to occupy these lands and whether more remain to be discovered"—although to which taxa they refer is unclear as all their lumbricids are common cosmopolitans. Australia's neoendemic species are exemplified by *Rhododrilus* (Acanthodrilidae from New Zealand) and *Begemius* (Megascolecidae from Papua New Guinea), both having several taxa in northern Queensland; also *Microscolex macquariensis* (Beddard 1896), which is confined to Macquarie Island, and *Microscolex kerguelarum* (Grube 1877) on Kerguelen and Heard Island in the Southern Ocean. The long-anticipated "missing link" of native Octochaetidae in Australia, intermediate between those in India and New Zealand, was recently determined, viz *Octochaetus ambrosensis* (Blakemore 1997), yet no claim is made that some other taxa now found in Australia, such as the circumundane *Microscolex dubius* (Fletcher 1886) from South America, nor *Octochaetona beatrix* (Beddard 1902) from India, are neoendemic; rather, these qualify as cosmopolitan species (Table 14.1).

Cosmopolitan species, the main consideration here, are those species both exotic to a region and widespread due to transportation mainly via human activities. Whereas the

Table 14.1 Nonendemic species in Pacific compared to North American and Indian Ocean regions (family classification after Blakemore 2000)

Families (origins) and Species from Regions	Mainland Australia (excluding Tasmania)	Tasmania	N.Z.	Japan (excluding Ryukus)	Ryukyu Islands	Korea (including Cheju Island)	Taiwan	China (including Hainan)	SE Asia	USA and Canada	Mexico	Hawaii	P.I.	India and Sri Lanka	Myanmar (Burma)
Moniligastridae (India and Oriental)															
Desmogaster sinensis Gates 1930	*							+							
Drawida barwelli (Beddard 1886)								*(H)	+		+		+		+
Drawida japonica (Michaelsen 1892)				+		+	+	+?	+						+
Drawida longatria longatria Gates 1925									+						+
Drawida nepalensis Michaelsen 1907								+	+						+
Glossoscolecidae (Neotropical)															
Pontoscolex corethrurus (Müller 1856)	++(Cl)		+	+(B)	*		+	+	+	+	+	+	+	+	+
Almidae (Tropics)															
Glyphidrilus papillatus (Rosa 1890)								+(H) ?						+?	+?
Hormogastridae (Mediterranean)															
Hormogaster redii Rosa 1887										+					
Criodrilidae (S-W Palaearctic)															
Criodrilus lacuum Hoffmeister 1845										+				+	
Sparganophilidae (Nearctic)											+?				
Sparganophilidae tamesis Benham 1892															
Lumbricidae (Holarctic)															
Allolobophora chlorotica (Savigny, 1826)	+	+	+							+	+				
Allolobophoridella eiseni (Levinsen, 1884)		*	+							+		+		+	
Aporrectodea caliginosa (Savigny 1826)	+	+		+		+	+?	+		+	+	+	+	+	+?
Aporrectodea icterica (Savigny 1826)										+					

(Continued)

Table 14.1 *(Continued)*

Families (origins) and Species from Regions	Mainland Australia (excluding Tasmania)	Tasmania	N.Z.	Japan (excluding Ryukus)	Ryukyu Islands	Korea (including Cheju Island)	Taiwan	China (including Hainan)	SE Asia	USA and Canada	Mexico	Hawaii	P.I.	India and Sri Lanka	Myanmar (Burma)
Aporrectodea limicola (Michaelsen 1890)	B+									+					
Aporrectodea longa (Ude 1885)	+	+	+	+						+	+	+			
Aporrectodea rosea (Savigny 1826)	+	+	+			+		+		+	+	+		+	
Aporrectodea trapezoides (Dugès 1828)	+	+	+	+		+	+	+	+	+	+	+		+	
Aporrectodea tuberculata (Eisen 1874)	+		+	+?		+	+?	+?		+	+			+	
Bimastos parvus (Eisen 1874)	+		+?							+					
Bimastos tumidus (Eisen 1874)				+		+	+	+	+	+?	+	+		+	+
Dendrobaena attemsi (Michaelsen 1902)	?	?													
Dendrobaena hortensis (Michaelsen 1890)	B+	*								+				+	
Dendrobaena octaedra (Savigny 1826)				+				+?		+	+	+		+	
Dendrobaena pygmaea (Savigny 1826)				*						+					
Dendrobaena veneta veneta (Rosa 1886)	B+	*	*							+					
Dendrodrilus rubidus rubidus (Savigny 1826)	+	*	+	+		+		+		+	+	+	+	+	
Dendrodrilus rubidus subrubicundus (Eisen 1874)	+	+(MI)								+?					
Dendrodrilus rubidus tenuis (Eisen 1874)	+(HI)			+		+		+?		+?					
Eisenia fetida andrei Bouché 1972	+?		+?	+?		+?				+?	+?	+?			
Eisenia fetida fetida (Savigny 1826)	+	*	+	+		+	*	+	+	+	+	+	+	+	
Eisenia japonica (Michaelsen 1892)				+?		+?		+?						+	
Eisenia nordenskioldi (Eisen 1879) ssp. Emend.						+?		+				+?			

Species																
Eiseniella tetraedra (Savigny 1826)	+	+				+		*			+	+		+		+
Eophila moebii (Michaelsen 1895)		+									+					
Eophila molleri (Rosa 1889: 3)	*			*							+?			+?		+
Lumbricus castaneus (Savigny 1826)		+				+					+	+	+			
Lumbricus festivus (Savigny 1826)	+?											+		+		
Lumbricus friendi Cognetti 1904												+		+		
Lumbricus rubellus Hoffmeister 1845	+	+		+		+??					+	+		+	+	+
Lumbricus terrestris Linnaeus 1758		+	*			+??					+	+		+	+	+
Murchieona minuscula (Rosa 1906)												+?				+?
Murchieona muldali (Omodeo 1956)												+?				+?
Octolasion cyaneum (Savigny 1826)	+	+	+								+	+		+	+	+
Octolasion tyrtaeum lacteum (Örley 1881)	*?		+		+						+			+	+	+
Octolasion tyrtaeum tyrtaeum (Savigny 1826)	+			+?							+	+		+	+	
Satchellius mammalis (Savigny 1826)											+	+				
Ocnerodrilidae (Tropical America and Africa)	*											+				
Gordiodrilus elegans Beddard 1892					+							+		+		+
Nematogenia panamaensis (Eisen 1900)	*		+					+			+	+		+	+	+
Ocnerodrilus occidentalis Eisen 1878	+[CI]		+								+	+		+	+	+
Eukerria kuekenthali (Michaelsen 1908)	+	*?	+	*							+	+		+	+	+
Eukerria saltensis (Beddard 1895)	+[H]										+			+	+	+
Malabaria levis (Chen 1938)												+		?		?
Thatonia exilis Gates 1945												+				+

(Continued)

Table 14.1 (*Continued*)

Families (origins) and Species from Regions	Mainland Australia (excluding Tasmania)	Tasmania	N.Z.	Japan (excluding Ryukus)	Ryukyu Islands	Korea (including Cheju Island)	Taiwan	China (including Hainan)	SE Asia	USA and Canada	Mexico	Hawaii	P.I.	India and Sri Lanka	Myanmar (Burma)
Thatonia gracilis Gates 1942															
Acanthodriidae (Pangean?)															
Microscolex dubius (Fletcher 1886)	+	*							+						+
Microscolex kerguelarum (Grube 1877)	+[HI]									+	+			+	
Microscolex macquariensis (Beddard 1896)		+[MI]	+	+											
Microscolex phosphoreus (Dugès 1837)	+	*	+							+	+			+	
Rhododrilus kermadecensis Benham 1905	B+?	*											+		
Rhododrilus queenslandicus Michaelsen 1916	+														
Rhododrilus sp. north Queensland	+?														
Octochaetidae (Circumtropical, plus Indo-Australasia)															
Dichogaster affinis (Michaelsen 1890)	*							+[H]	+	+	+		+	+	+
Dichogaster annae (Horst 1893)	*								+	+	+			+	
Dichogaster bolaui (Michaelsen 1891)	++[CI]						+	+[H]	+	+	+	+	+	++	+
Dichogaster corticis (Michaelsen 1899)					+				+						
Dichogaster modiglianii (Rosa 1896)	+[CI]								+	+	+	+		+	+
Dichogaster reinckei Michaelsen 1898													+?		
Dichogaster saliens (Beddard 1893)	*+[CI]				+				+	+	+	+		+	+
Dichogaster sp. nov.? in NT	*														
Lennogaster pusillus (Stephenson 1930)									+						
Octochaetona beatrix (Beddard 1902)	*								+						+

Species													
Octochaetona surensis (Michaelsen 1910)												+	
Ramiella bishambari (Stephenson 1914)	+[CI]					+					+	
Megascolecidae (NW American and mostly Indo-Australasian)												
Argilophilus marmoratus Eisen 1893	??					+					+	
Pontodrilus litoralis (Grube 1855)	+	+	+	+	*	+[H]+	+	+	+	+	+	
Pontodrilus primoris Blakemore 2000		*?										+	
Perionyx excavatus Perrier 1872	*	*	+	+	*		+	+	+	+	+?	+
Perionyx macintoshii Beddard 1883												+
Perionyx violaceus Horst 1893				+?		+						
Amynthas agrestis (Goto & Hatai 1899)						+		+				
Amynthas aeruginosus Kinberg 1867						+	+	+	+	+	+	
Amynthas alexandri Beddard 1901											+	
Amynthas asiaticus Michaelsen 1900					+?	+?	+?	+?	+?			
Amynthas aspergillum (Perrier 1872)				+	+	+?	+				
Amynthas carnosus (Goto & Hatai 1899)				+?	+?	+?	+?	+?			
Amynthas corticis (Kinberg 1867)	+	*	+	+	+	+	+	+	+	+	+	+
Amynthas glabrus (Gates 1932)			+	+	+	+?	+?	+?	+	+	+	+?
Amynthas gracilis (Kinberg 1867)	+	+	+	+	+	+?	+	+	+	+	+	+?
Amynthas hupeiensis (Michaelsen 1895)	+[T]?	+?	+	+	+	+	+	+	+	+	+	
Amynthas incongruus (Chen 1933)					+	+?	+?				
Amynthas lautus (Horst 1883)— possibly synonym of *A. robustus*			+?	+?	+?	+?	+?				

(Continued)

Table 14.1 (Continued)

Families (origins) and Species from Regions	Mainland Australia (excluding Tasmania)	Tasmania	N.Z.	Japan (excluding Ryukus)	Ryukyu Islands	Korea (including Cheju Island)	Taiwan	China (including Hainan)	SE Asia	USA and Canada	Mexico	Hawaii	P.I.	India and Sri Lanka	Myanmar (Burma)
Amynthas lavangguanus (Gates 1959)												+?	+		
Amynthas loveridgei (Gates 1968)										+					
Amynthas manicatus (Gates 1931)									+?						
Amynthas minimus (Horst 1893)	+			+	+		+	+?	+	+		+	+	+	+
Amynthas morrisi (Beddard 1892)	+			+	+		+	+?	+	+	+	+	+	+	+
Ann. morrisi group sp. nov.?	*														
Amynthas papilio papilio (Gates 1930)							+								
Amynthas papulosus (Rosa 1896)				+	+		+	+?	+						+
Amynthas perkinsi (Beddard 1896)												+?			
Amynthas rechingeri (Cognetti 1909)												+			
Amynthas robustus (Perrier 1872)	+(CI)			+	+	+	+	+?	+					+	+
Amynthas rodericensis (Grube 1879)								+?	+	+			+		+
Amynthas taipeiensis (Tsai 1964)						+?	+?	+?							
Amynthas tokioensis (Beddard 1892)										+?					
Anisochaeta dorsalis (Fletcher 1886)		+													
Anisochaeta gracilis (Fletcher 1886)		+?													
Anisochaeta sebastiana (Blakemore 1997)		+													
Begemius queenslandicus (Fletcher 1886)	+														
Didymogaster sylvatica Fletcher 1886			+												
Lampito mauritii Kinberg 1867	+(CI)							+	+			+		+	
Metapheretina speiseri (Michaelsen 1913)													+?		+

Taxon	Character markers (as shown)
Metaphire anomala (Michaelsen 1907)	+ +
Metaphire bahli (Gates 1945)	+ + +
Metaphire bipora (Beddard 1900)	+?
Metaphire birmanica (Rosa 1888)	+
Metaphire californica (Kinberg 1867)	+ + + + + + + + + +
Metaphire hilgendorfi (Michaelsen 1892)	+? +?
Metaphire houlleti (Perrier 1872)	+ + + + +? +?
Metaphire javanica (Kinberg 1867)	? +? +?
Metaphire malayana (Beddard 1900)	+? +
Metaphire pajana (Michaelsen 1928)	+
Metaphire peguana (Rosa 1890)	+? + + + + +?
Metaphire planata (Gates 1926)	+ +? +
Metaphire posthuma (Vaillant 1868)	+^(CI) + + + + +?
?*Metaphire sandvicensis* (Beddard 1896) (? = *M. californica*)	+? +?
Metaphire schmardae macrochaeta (Michaelsen 1899)	+ +
Metaphire schmardae schmardae (Horst 1883)	+ + + + + +
Metaphire soulensis (Kobayashi 1938)	+? +? +?
Metaphire virgo (Beddard 1901)	+? +? +?
Notoscolex javanica (Michaelsen 1910)	+
Notoscolex pumila (Stephenson 1931)	+
Pheretima darnleiensis (Fletcher 1886)	+^(T, CI) + +
Pheretima montana Kinberg 1867	+ +

(Continued)

Table 14.1 (*Continued*)

Families (origins) and Species from Regions	Mainland Australia (excluding Tasmania)	Tasmania	N.Z.	Japan (excluding Ryukus)	Ryukyu Islands	Korea (including Cheju Island)	Taiwan	China (including Hainan)	SE Asia	USA and Canada	Mexico	Hawaii	P.I.	India and Sri Lanka	Myanmar (Burma)
Pithemera bicincta (Perrier 1875)	++[CI]			+	+	+	+	+	+	+	+	+	+	+	+
Polypheretima annulata (Horst 1883)									+						
Polypheretima brevis (Rosa 1898)	+[CI]												+		
Polypheretima elongata (Perrier 1872)	+			+	+		+		+	+?	+	+	+	+	+
Polypheretima taprobanae (Beddard 1892)	+										+	+	+	+	+
Polypheretima voeltzkowi (Michaelsen 1907)													+		
Eudrilidae (West African)															
Eudrilus eugeniae (Kinberg 1867)	*									+	+			+	
Total nonendemic exotic spp. (E)	66	27	27	34	18	23	30	47	58+	60	46	33	30	50	50
Endemic natives in region (N)	449	203	172	40	10	70	41	196	?	122	57	0	20	455	135
Approximate total spp. (E + N)	515	230	199	74	28	93	71	243	?	182	103	33	50	505	185
Exotics [E/(E + N)] approximate (%)	12.8	11.7	13.6	46.0	64.3	24.7	42.3	20.0	?	33.0	44.6	100	60.0	9.9	27.0
Region	Mainland Australia	Tasmania	N.Z.	Japan	Ryukyu	Korea	Taiwan	China	SE Asia	USA and Canada	Mexico	Hawaii	P.I.	India	Myanmar
Approximates															
Lumbricidae as % total exotic	35	60	67	38	0	57	20	28	6	53	46	33	10	34	4
Megascolecidae as % total exotic	35	18	22	47	72	39	73	51	63	28	26	51	66	34	56

Abbreviations: + = present as a nonendemic species; * = first records from author's (R.J. Blakemore) studies; ? = indicates some ambiguity of taxonomy, endemicity (e.g. possibility that the taxon is native to a region) or veracity of report; B = J.C. Buckerfield, pers. comm.; [BI] = Bonin Island; [CI] = Christmas Island; [HI] = Hainan; [HII] = Heard and McDonald Islands; [MI] = Macquarie Island; [T] = Torres Straits Islands; N.Z. = New Zealand; P.I. = Pacific Isles (Fiji, Samoa, Tonga, Niue I., Kermadec, Vanuatu, Solomons, Cook Isles, Tahiti, Marianas, Marshalls, Carolines, Marquesas, New Caledonia with Loyalty Islands)—reports from Easton (1984) and Lee (1981).

terms "alien," "exotic" and "invasive" are often used interchangeably, and "peregrine" meaning "wanderer" or "wandering" as first used by Michaelsen (1903) to describe earthworms that are dispersed over a wide range and in geographically remote localities, the definition of "cosmopolitan" is less precise, meaning either widely distributed in many parts of the world or merely present in more than one country, but with the ecological implication of passive or unintentional transportation.

Of some 9,000+ named terrestrial and aquatic oligochaetes (as calculated below), about 6,000 are earthworms with only 120 species (plus synonyms) widely distributed and commonly encountered throughout the world—comprising 47 Megascolecidae (mainly pheretimoids), 33 holarctic Lumbricidae and 40 or so species from some of the 18 other families (Blakemore 2006a). Lee (1985) proposed that only about six lumbricids and the same number of tropical species have life histories adequately revealed, and Reynolds (1998) claimed that detailed ecological studies have been made on fewer than 20 earthworm species. However, Blakemore (1994, 1997a) surveyed and studied 30 earthworm taxa—including several species new to science—in a series of laboratory, glasshouse and field experiments. Since then, Barois et al. (1999) provided basic ecological data for nearly 60 tropical species, and Omodeo et al. (2004) reported ecological and biogeographical traits of some of their 38 Maghreb megadriles.

14.2 NUMBER OF EARTHWORM SPECIES

In the century following Michaelsen's (1900) *Das Tierreich: Vermes* review of all 1,200 oligochaete species then known, taxonomic efforts have overwhelmed and polarized researchers either toward the aquatic and usually smaller microdriles or toward the generally larger terrestrial megadriles (the "true" earthworms). Subsequent totals were based on estimates by Lee (1985, 1987) of ~3,000 megadriles (with just 100 peregrine species), and on a statement by Reynolds (1994) of nearly 8,000 described Oligochaeta, though citing only "Reynolds and Cook (1993)," who tallied just 7,254 species with 3,627 (~50%) terrestrial megadriles (and presumably 3,627 microdriles).

The current, incomplete ZooBank total for Oligochaeta is 7,585 taxa including both megadrile and microdrile species, while other estimates exceed 8,000. A reasonable total, however, is closer to ca. 5,900 named terrestrial megadriles, and, including aquatic microdrile estimates, we reach a conservative total of 9,500–10,200 nominal Oligochaeta taxa with an unknown number of synonyms. Yet probably less than 25–30% of earthworms from around the world have been described. For instance, Tasmanian studies have increased the species total of this allegedly well-studied island fourfold, from 55 to 230 taxa (Blakemore 2000c), and recently added names have nearly doubled the faunas of Australia, Japan, Taiwan and some other Asian countries (see Blakemore 2002, 2003, 2005, 2006a, b), while South American, Middle Eastern and African faunal lists are far from complete.

In comparison, the marine Polychaeta has ~13,000 named taxa (and ZooBank lists just 10,755 spp./genera), although only ~8,000 of these are considered reasonable species. The apparently high diversity among polychaetes is due partly to a greater number (perhaps 10-fold) of full-time researchers for this group. Considering that >99% of the world's food (and fiber) supply is produced on land, whereas only 0.6% comes from oceans and other aquatic ecosystems (FAO 1991), underfunding of soil ecotaxonomic research in favor of marine taxonomic ventures is arguably an unbalanced economic decision.

14.3 CHARACTERISTICS AND ORIGINS OF COSMOPOLITAN EARTHWORMS

Cosmopolitan earthworms have a propensity for transportation and tend to have some or all of the following characteristics: small size, parthenogenic or asexual reproduction such that a single specimen may start a colony (Gates 1972; Sims and Gerard 1999), high fecundity, resistant cocoons, wide environmental or feeding tolerances (eurytopicity), and rapid dispersal rates (Lee 1985, 1987; Blakemore 2002); they may also lack specialized predators in their new lands. Moreover, Gates (1968, 1972) commented that parthenogenesis, often accompanied by polyploidy, permits more rapid accumulation of mutations than if reproduction had remained amphimictic, and that parthenogenetic morphs seem to withstand higher parasitic burdens, possibly because less bodily resources are allocated to reproduction. The ability of an individual or species to survive in a location after its introduction is thereafter influenced by local climate, soil ecology and soil management practices.

Family origins of cosmopolitan earthworms often correspond to the eight independent centers of the world's major cultivated plants (Vavilov 1951; see Table 14.1). Species associations indicate that certain earthworms may have accompanied some plants via agriculture and world trade, but transportation of earthworms into nonindigenous regions, as with crops, is not necessarily via the most direct route. Evidence for such spread is the faunal lists for the Levant and Maghreb (Csuzdi and Pavlicek 1999, 2005; Omodeo et al. 2004; Pavlicek et al. 2004) having several earthworm names familiar to workers in other regions of the world where cultivated plants originally from the Middle East and Mediterranean now grow, and vice versa. The likelihood that routes may be indirect and that the most probable mechanism of introduction is human-mediated transportation are supported by continued interceptions by national quarantine services (e.g. those reported by Gates 1972 and Lee 1987).

Native species may also be translocated within a bioregion by mechanisms similar to those pertaining to exotics: some examples are reported within the Australian genus *Anisochaeta* spp. in Tasmania (Blakemore 1999, 2000a, c; see Table 14.1) and probably for *Bimastos* spp. in North America (Hendrix 1995). Establishing translocation for a native relies on evidence including exact morphological and/or molecular match, a noticeably disjunct distribution and prevalence in the new domain mainly in disturbed habitats such as gardens or fields or other probable mechanisms of transport such as fishing bait.

Apart from determining whether native taxa are new or translocated, one of the challenges in ecological taxonomy is to reliably distinguish exotics from natives and to assess the diversity and distribution of both (e.g. Blakemore 1999; Omodeo et al. 2004; Pavlicek et al. 2004). Faunal comparisons help us appreciate mechanisms of ecosystem functioning and are crucial for understanding regional biodiversity and monitoring the spread of human-mediated species transportations and extinctions.

14.4 OVERVIEW OF RESULTS

This chapter summarizes field and literature surveys, compiled into regional faunal lists allowing for most current taxonomic treatments, based on the author's independent studies as well as on various other sources (as acknowledged in the text). Previous regional species totals from the literature are found, for example, in Michaelsen (1900, 1903), Gates (1972), Sims (1980), Sims and Gerard (1985, 1999), Lee (1985, 1987), Julka (1988), Hendrix

(1995), Fragoso et al. (1995, 1999, 2004), Blakemore (1999, 2000c), Tsai et al. (2000), Fragoso (2001) and Blakemore (2003, 2004a–f, 2005, 2006a, b). Relative proportions of natives to exotics from selected regions are compared in Fig. 14.1, with data extracted from distributional species lists found outside their endemic domains, as presented in Table 14.1. Not all cosmopolitans are listed and, naturally, species are not considered exotic in regions where they are thought to be endemic; nevertheless, they may be subject to translocation there. Moreover, several cosmopolitan species are now so ubiquitous that their origins are obscure or questionable. Blakemore (2002, 2006a) provides detailed descriptions with figures, synonymies, distributions and ecological information for all ~120 known cosmopolitan earthworms, and these will be the main subjects of further discussion. Regions of the globe for which total species checklists are currently compiled and available (e.g. Blakemore 2006b) are shown in Fig. 14.2.

14.5 DISCUSSION

Meaningful biodiversity comparisons require representative estimations from reliable survey (e.g. considering numbers and size of samples, soil depth, various habitats and seasonal differences) and correct identification of community components (specific and ecological categorization). Difficulties include the amount of work required for adequate sampling and hand sorting of soils, and the lack of complete and consistent species guides (Table 14.2).

One day's field sampling often results in at least a fortnight's laboratory / library work for identification, which is why museum shelves around the world are stacked with voucher specimens awaiting identification by specialists. Solutions include implementing standard quantitative and qualitative sampling methods, such as those provided by ISO 23611-1 (2006), and support for the compilation of detailed, comprehensive taxonomic guides that include all known species (as well as those potentially occurring) within a region, rather than using limited "species identifiers" that tend to force choices into one or two species and often lead to false results (Blakemore 1999, 2002, 2006a).

Some species tend to be small in size and, apart from being easily overlooked, are often difficult to identify or are dismissed as immatures. Examples of small worms include several *Drawida* spp., *Eukerria saltensis, Gordiodrilus elegans, Ocnerodrilus occidentalis, Dichogaster* spp., *Dendrobaena* spp., *Helodrilus oculatus, Murchieona minuscula* and some *Amynthas* spp. such as *Amynthas minimus* (see Table 14.1). When such omissions and identification errors are corrected, biodiversity results can be considerably higher. Lavelle (1978), for example, recorded up to 13 species from savannas and gallery forest sites at Lamto, Ivory Coast; Abbott (1985a, b) identified exotics and 62 native "morphospecies" representing an unknown number of taxa from jarrah (*Eucalyptus marginata*) forests in Western Australia; a regional study by Bano and Kale (1991) in southern Karnataka, India recorded 44 species (eight exotic); Omodeo et al. (2004) reported 38 megadrile species in 2,110 specimens from 177 sites of the Maghreb, North Africa; and Omodeo and Rota (1989) described 51 species (14 new) from ca. 2,000 specimens from 50 localities in Turkey collected between 1987 and 1988. In the Levant, 34 species are currently known with about a dozen (40%) cosmopolitan or introduced (Pavlicek et al. 2004). Recent surveys in Taiwan by Chen et al. (2003) found 42 species, and a transect by Tsai et al. (2004) collected 2,163 specimens from 43 samples at 27 localities and recorded 34 earthworm species including 14–15 exotics. A weeklong survey on shorelines at Lake Pedder in western Tasmania revealed 21 species with 16 natives, 5 exotics plus 3 aquatic

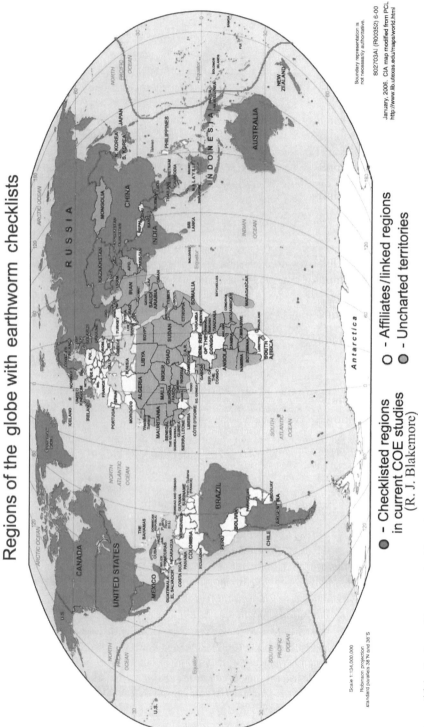

Figure 14.1 Regions with current checklists of earthworm species. (See color insert.)

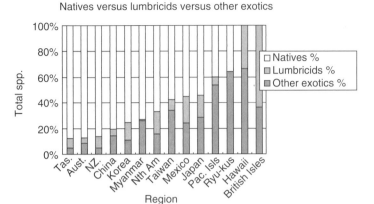

Figure 14.2 Lumbricidae as a proportion of total species from selected regions (only British Isles have >40%). Abbreviations: Tas = Tasmania; Aust = Mainland Australia; N.Z. = New Zealand; Nth Am = North America (Rio Grande to Arctic); Pac Isls = Pacific Isles (not Hawaii); China = China + Hainan. (See color insert.)

Table 14.2 Contingency table of sampling reliability

Case	Ecological Sampling	Taxonomic Treatment	Results
1	+	+	Representative data
2	+	−	Under / overestimate
3	−	+	Under / overestimate
4	−	−	Unrepresentative data

Abbreviations: + = good; − = poor.

microdriles (Blakemore 2000a), while a dozen species were found in a brief survey around Lake Biwa, Japan (Blakemore 2007b). Other notable earthworm biodiversity surveys include 24 species (with 16 exotics) from ca. 100 ha of a mixed farm property systematically sampled monthly for 2 years at Samford, southeast Queensland where a Brisbane regional total exceeded 40 species (Blakemore 1994, 1997a), and 14 species (with 10 exotics) collected one weekend on a 45-ha farm in the Southern Highlands of New South Wales (Blakemore 2001a, b). These high figures exemplify the need for more extensive and, at the same time, more intensive combinations of both ecological survey and taxonomic analysis. Therefore, the statement by Lee (1985) that earthworm species diversity is fairly consistent in different regions and habitats, ranging from 1 to 11 species and most commonly just two to five, should perhaps be revised upwards.

14.6 REGIONAL SPECIES TOTALS AND PROPORTIONS OF EXOTICS

Checklists show that the Oriental pheretimoids (family Megascolecidae) now comprise some 980 valid names from >1,200 nominal taxa, and the holarctic lumbricids (family Lumbricidae) total 670 valid names from ~1,150 nominal taxa (Blakemore 2006b). These two groups contribute two-thirds of the ~120 total cosmopolitan species, while the remaining third has diverse origins in various other families (see Table 14.1). Earlier accounts of exotics have tended to overemphasize the importance of the lumbricids, perpetuating misinformation that they are the major earthworm group, partly because they are common in temperate regions, but mainly because lumbricids are more familiar to workers in Europe and in North America, whereas identification guides for other groups are generally lacking (cf. Gates 1972). Based on current data, interchanging the terms "earthworms" and "Lumbricidae" appears unwarranted as lumbricids are dominant only in temperate regions of northern Europe and the Middle East, and in other regions are often a minor component of the total fauna (see Table 14.1). For instance, when Fragoso et al. (1999) compiled survey data from the humid tropics worldwide, they found 51 common exotics, but just 17 (33%) of these were Lumbricidae. The relative proportions of Lumbricidae including *Eisenia japonica* (Michaelsen 1891), which is supposedly endemic to Korea, Japan and the Siberian Kurils, and the Megascolecidae—mainly Asiatic pheretimoids—are compared to other exotics as summarized in Table 14.1 (cf. Fig. 14.2 derived from this data).

Earthworm diversity in Australia including Tasmania now totals 715 (sub)species, of which 65 are exotics (~9% of total with just ~3% lumbricids); of these, ca. 20 are new records from the author's modest studies including the first Australian report of *Lumbricus terrestris* Linnaeus 1758 from Tasmania (Blakemore 1997b). Continued ecotaxonomic surveys are likely to double the number of Australian natives, and further exotics are expected. Moreover, if all *Begemius* Easton 1984 (Megascolecidae from Papua New Guinea) and *Rhododrilus* Beddard 1889 (Acanthodrilidae from New Zealand) species are included as exotics/neoendemics, then the total of ~80 exotics is considerably higher than a previous calculation of just 27 mainly Lumbricidae species (see Blakemore 1999). Unlike mainland Australia and New Zealand, no native Acanthodrilidae or Octochaetidae are known from Tasmania.

India and adjacent countries including Sri Lanka have 505 described earthworm taxa, only 50 (<10%) of which are exotics and fewer than 18 (3%) Lumbricidae (the native species tend to be within the families Moniligastridae, Octochaetidae and Megascolecidae). In comparison, the tally for China (including Hainan but excluding Taiwan), which is dominated by Megascolecidae (~200 spp.) followed by Moniligastridae (~20 spp.), is now 244 species with 47 (20%) exotics and just 13 (5%) Lumbricidae. Both India and China have long and diverse histories of cultural exchange and are likely sources of several cosmopolitan earthworms. Neither country has confirmation of any native Lumbricidae, Acanthodrilidae nor Glossoscolecidae.

Japan, including the Ryukyus (Okinawan Islands), has 80 earthworm species (50% exotic), and while no Lumbricidae are yet known from tropical Okinawa, a new record from there (Blakemore et al. 2007) is for *P. corethrurus*. The Korean peninsula including volcanic Cheju (Quelpart Island) has 93 species with 23 exotics (24%), and about half of these are Lumbricidae; however, unlike most other oriental countries where both native and exotic faunas are more often dominated by Megascolecidae and Moniligastridae, the cooler Korean climate appears to allow a relatively greater abundance of holarctic Lumbricidae including putative natives *E. japonica* (Michaelsen 1892) and *Eisenia koreana* (Zicsi 1972). The island of Taiwan has 71 known earthworm species, 30 of which are considered exotics (42%), but at least an additional 30 undescribed natives are known (Chen et al. 2003; Blakemore et al. 2006), and just six (20%) of the exotics are Lumbricidae. Only three Lumbricidae are known from Myanmar (formerly called Burma and part of British India), which has 185 species with 50 (<30%) exotics (Gates 1972; Blakemore 2005). Totals for all other Southeast Asian countries are unknown, although 41 exotic species are reported and just three of these are Lumbricidae (see Table 14.1).

Several Asiatic species are now distributed worldwide, for example, *Drawida barwelli* (Beddard 1886), *Amynthas gracilis* (Kinberg 1867) and the *Amynthas corticis* (Kinberg 1867) species complex that Gates (1972) thought might have been transported more widely than the better-known lumbricids. Erstwhile reports from North America of components of the *Metaphire hilgendorfi* (Michaelsen 1892)/*Amynthas tokioensis* Beddard 1892 species complex (e.g. Gates 1958; Hendrix 1995; Hendrix and Bohlen 2002) are thought to have originated from Japan (or possibly China) (see note on Perry's "Black Ships" below).

Continental North America, north of the Rio Grande, has relatively poor faunal diversity due to its division by epieric seaways in the Cretaceous, and subsequently by extensive glaciation as recently as the Pleistocene (~18,000 years ago). Two endemic lumbricid genera are *Bimastos* and *Eisenoides*; thus, the earliest likely period for commencement of the wider distribution of cosmopolitan *Bimastos parvus* (Eisen 1874) outside America follows the European colonization 500–1,000 years ago. A total of 182 species in 12 families, with about 60 (~33%) exotics was compiled by Blakemore (2005), compared to Reynolds and Wetzel (2004) and Wetzel (2003, 2007), who report 161 megadriles in 10 families from the same region, including 45 exotics for which several names are invalid or have long been superceded (e.g. *Aporrectodea turgida, Eisenia fetida, Amynthas hawayanus, Amynthas diffringens*; cf. Sims and Gerard 1985, 1999; Blakemore 1999, 2002).

Mexico has 104 described species, with 46 (~45%) thought to be post-Columbian introductions; of these exotics, less than half are Lumbricidae. Other South American totals include 240–260 oligochaetes in Brazil and 320–350 neotropical species (Fragoso et al. 2004), while Latin America as a whole now has 830 described earthworm species with ~70 (8.4%) exotics and just 35 Lumbricidae (i.e. half of the exotics or just 4.2% of total species) as catalogued by Dr. G. Brown (see Brown and Fragoso 2007).

Volcanic Hawaii harbors 50 nominal earthworm taxa but with just 33 valid names (Blakemore 2006b), and all are introduced exotics, whereas all other Pacific Isles have about 50 species with 30 (~60%) considered exotics, of which just three are Lumbricidae (Blakemore 2005, 2006b; see Table 14.1).

14.7 EARTHWORMS, ARCHAEOLOGY AND HUMAN HISTORY

Regarding the "humble earthworm," Darwin (1881) stated, "It may be doubted whether there are many other animals which have played so important a part in the history of the world, as have these lowly organized creatures ..." and "The vegetable mould [humus] which covers, as with a mantle, the surface of the land, has all passed many times through their bodies."

Although Darwin mainly commented on their intimate ecological relationship with soils and plant growth, his contributions to earthworm study also extended into the realms of anthropology and archaeology; e.g. an experiment on surface casting and the burial of objects was judged as being essential to archaeological excavation (Darwin 1840), and was corroborated by Keith (1942) and Jewell (1958)—see also Hart and Terrell (2002). Remains of invertebrates or other animals often provide archaeological evidence (Steadman 1995; Morrison 1996a, b; Grayson 2001), but unlike small or winged organisms that are carried by the wind, earthworms are less actively mobile, and most (native) species have highly restricted distributions—suitable criteria for indicator taxa to investigate human transportations. Current earthworm populations may be indicative of past human movements, but because their soft bodies rapidly decompose, only the preservation of their chitinous setae (McCobb et al. 2004), fossilized embryos and cocoons (Schwert 1979; Piearce et al. 1990, 1992; Manum and Bose 1991), and calciferous glands (Gates 1972) provide evidence for environmental archaeology.

Recent DNA analyses show that early humans traveled widely from an African origin. Only in relatively recent times have humans had the technical ability to transport substantial amounts of rocks or soils with resident earthworms. Preagricultural people carried few tangibles, and only after the advent of the agricultural era did muscle power from domesticated draft animals (or humans) cultivate and move large quantities of soil. These processes intensified during the mechanization of the Industrial Age and have accelerated under the current globalization of world trade associated with the transition to an information-based economy (Toffler 1980), now tempered with implementations of quarantine barriers, wildlife legislation and import restrictions as applied to plant, animals, soils and soil products.

Agriculture or pastoralism developed independently in several regions including the Middle East, Africa, India, Asia, Sahul, Mesoamerica and the Andes about 10–12,000 years ago (Vavilov 1951). Transportation of earthworms out of the Middle East may have followed the spread of agriculture throughout Europe in the Neolithic era, yet voyages and settlements by Europeans responsible for the global transportation of earthworms into the New World, South Africa and Australasia are thought to have commenced only in the last 1,000 years (Gates 1972; Enckell and Rundgren 1988). *Dendrobaena octaedra* (Savigny 1826) was found in Greenland "in districts inhabited by the Norse ten centuries ago" (Gates 1972), and it also inhabits Hokkaido (Stöp-Bowitz 1969), although just how much of this species' present range is self-acquired is unclear as its dispersal is thought to be restricted only by its freeze tolerance (Berman et al. 2001). Possible examples of prior species redistributions within Western Europe are the small *Satchellius mammalis* (Savigny 1826) found in Germany, Spain, the British Isles and Ireland, an area coincident to the region settled by the Celts, as noted by its synonym *Allolobophora celtica* Rosa 1886, and a similar distribution pertains to *Dendrobaena attemsi* (Michaelsen 1902) (see Rota and Erséus 1997; Fig. 14.1)—both taxa recently recorded as introduced to the USA. Finding the similarly small *H. oculatus* Hoffmeister 1845, with a natural distribution in western and eastern Europe, in a Roman ditch in Verulamium (St. Albans, UK) led to speculation that the species may have been introduced to Britain during the Roman occupation, if not earlier (Dobson and Satchell 1956).

That some exotic Lumbricidae (except perhaps *Bimastos*) originated in the "fertile crescent" of the Middle East and were subsequently transported along with agriculture is supported by presence of ~10 cosmopolitan species in the Levant (Pavlicek et al. 2004; Pavlicek 2006) in common with those found in the British Isles (Sims and Gerard 1999). Reflecting its colonial history, almost all the 30+ Lumbricidae now recorded in Australasia/North America correspond to those known from the British Isles that themselves probably originated from Continental Europe. Australia had remained remote from most world trade until the mid-1800s, and only in the last 500 post-Columbian years can regular exchange account for the distributions of certain European species in the Americas. Fletcher (1886) quotes letters in "*Nature* (1884)" that earthworms did not originally exist in the prairies of the Canadian Northwest nor in parts of the USA (i.e. Kansas, Idaho, and the Indian and Washington Territories).

However, in the mid-15th century (nearly a century before Columbus landed in the Americas), sailors such as Admiral Zheng He established trade routes and bases from China through Southeast Asia as far as India, the Middle East and East Africa. Prior to this,

Indonesian, Indian, Arab, Scandinavian, Japanese and Chinese seafarers, missionaries (e.g. Ibn Battuta), explorers, invaders, refugees, pirates and overland traders using routes such as the Silk Road(s) have exchanged commodities including animals and plants. Ancient trade routes, partially over land and partially by sea, linked Mediterranean Attica and Rome to cultures of the Indus Valley and the east coast of India as long as 2,400 years ago. Also, for at least the last three millennia, Polynesian and Melanesian people (carrying with them food plants such as taro) migrated in the Pacific wherein "coral island taxa," those earthworms dispersed by pre-Columbian voyagers, were distinguished from "tropical tramps," species probably introduced after Europeans entered this region (Lee 1981, 1985, 1987).

Thus, the origins and current distributions of earthworms can likely help track early human migration routes; e.g. dispersal of some cosmopolitan species predate the movement of European explorers and reflect earlier, largely undocumented wanderings and commerce within the Pacific and Indian Oceans. Gates (1972), for instance, found the disjunct distribution of the aquatic Almidae genus *Glyphidrilus* Horst 1889 in Tanzania and in Southeast Asia difficult to interpret by overland transport from an African origin. However, its distribution may be attributed to sea carriage from Asia, possibly via India or Sri Lanka, considering that Madagascar, and possibly the east coast of Africa (where *Glyphidrilus stuhlmanni* Michaelsen 1897 occurs), was colonized from about 1,500 to 2,000 years ago by seafaring people originating from Borneo (where *G. kuekenthali* Michaelsen 1896 occurs), and that these or successive people (see Hules et al. 2005) conceivably transported soil with their plant rootstocks or specimens in their water supplies. Ljungstöm (1972) had considered that six peregrine species of Oriental *Pheretima* found along the eastern coast of South Africa were associated with such early human trade. Conversely, Stephenson (1931) speculated that the often small-sized members of the genus *Dichogaster* "have probably been carried eastwards from Africa in successive stages, perhaps accidentally by early man in his wanderings, in his belongings, or later in merchandise of one sort or another."

For the Asia Pacific region, Sims and Easton (1972) proposed that "man has always been moving among the islands: from prehistoric times onwards successive waves of colonists have spread out from the mainland of Asia while at present some small atolls are occupied only temporarily. In these circumstances, earthworms and their cocoons may unwittingly have been transported in the soil packed around the roots of the crops which these travelers have in all probability taken with them." Dispersal of *Drawida japonica* (Michaelsen 1892) is thought to be from China to Japan, possibly via Taiwan (Blakemore 2003), although Gates (1972) speculated that this species came originally from the Indian Himalayas, and Yunnan and Szechuan in China. Introduction of other species into Japan, e.g. the lumbricid *E. japonica*, may be linked with earlier occupants such as the Ainu people from Siberia. On the other hand, some *Amynthas* and *Metaphire* pheretimoids are either native, neoendemics or are introductions from earlier times, possibly corresponding to the Jomon period when land bridges connected Japan to the Asian mainland. Global transportation of several of these Japanese species probably commenced with reopening of ports after two centuries of isolation following Commodore Perry's "Black Ships" arrival near Yokohama on July 8, 1853.

Other pheretimoids, such as *Pheretima, Begemius* and *Polypheretima*, are indigenous to New Guinea and parts of the Indonesian archipelago—areas of ancient agricultural development—and all three genera have restricted and patchy distributions on adjacent islands and in northern coastal Australia characteristic of imported species (Blakemore 1994, 1999, 2003). Because *Polypheretima brevis* (Rosa 1898) on Christmas Island (Australian territory) and Tonga and *Po. pentacystis* (Rosa 1891) in the Seychelles and

Madagascar both occur outside the normal Oriental "*Pheretima* domain," Easton (1979, 1984) proposed that they were introduced by (early?) human agency and found no evidence to explain transoceanic distributions of some other Pacific *Pheretima* species by natural means. Thus *Polypheretima voeltzkowi* (Michaelsen 1907) is thought to have been introduced to the Comoros, possibly from New Guinea since Lee (1981) called it a "Melanesian species," and was also introduced to Vanuatu. Remote UNESCO World Heritage Easter Island and Henderson Island in the South Pacific, both populated by Polynesian people, have a few pheretimoids [e.g. *Amynthas hendersonianus* (Cognetti 1914)] that have subspecies in New Guinea (Blakemore 2005), as well as *E. saltensis* that originates from South America.

Earthworms may yet help trace undocumented human movements such as those by Macassan seafarers from southwest Sulawesi (formerly Celebes) who are known since the 1300s to have traded a Chinese and Indonesian delicacy, the trepang (dried Holothurian sea slugs) obtained from coastal Arnhem Land in Australia (MacKnight 1976). This could partially account for restriction in Australia of Indonesian *Polypheretima elongata* to a few localities in northern coastal Queensland (Easton 1976; 1982; Blakemore 1994). Distribution of earthworms on the Pacific islands may help explain the enigmatic Lapita culture complex that perhaps originated in the Bismarck Archipelago. The extent of contact and trade that occurred in ancient times is not well-known, but certainly a long period of human occupation in Australia and the Sahul has dates now thought to extend 60,000–120,000 years ago (e.g. Singh et al. 1981; Wright 1986).

In New Zealand, possibly the last habitable place on the planet where permanent human settlement commenced only around 2,000 years ago (Grayson 2001), at least eight kinds of earthworms—including *Rhododrilus edulis* Benham 1904—were formerly used as prized dishes in Maori feasts (Stephenson 1930). Two *Rhododrilus* are known from Australia's coast: *Rhododrilus queenslandicus* Michaelsen 1916 from northern Cape York Peninsula and *Rhododrilus kermadecensis* Benham 1905 from southern Tasmanian tidal mudflats, also reported from Kangaroo Island in South Australia (Blakemore 1999, 2000c). Probably not endemic to Australia, such species may have been introduced and have become naturalized following introduction, or possibly they are relict fauna of ancestral stock shared with other regions and are now isolated (Blakemore 1994). Although prevailing ocean currents are in the opposite direction (from East to West), such worms (along with plants) were possibly carried by boat from New Zealand during unrecorded ephemeral visits by Maori explorers.

Recently, Fragoso et al. (1999) and Gonzalez et al. (2006) attempted to relate Caribbean distributions of exotic earthworms with the African slave trade, noting that the occurrence of some species may be explained by human migration prior to European colonization, possibly due to "island hopping" by indigenous people from South America up to 2,200 years ago.

The process of species transportation continues to date, for example, when neighbors exchange potted plants containing earthworms, or when European cypress (*Cupressus sempervirens*) planted on a dam in Argentina are the likely source of lumbricid *Octodrilus* spp. (Mischis and Brigada 1988; Mischis 2004). Similar accidental imports are also evidenced by the large number of new species described from botanic gardens, although control measures have been stricter since the 1940s such that Gates (1972) cited the interception of 3,430 earthworm specimens by the US Bureau of Plant Quarantine over a 15-year period. Often details of full distributional ranges are speculative and more questions are raised than answers—for several earthworm species even their place of origin is obscured by frequent transportations. Thus, much scope exists for further ecotaxonomic survey to help understand and explain the origins and dispersal mechanisms of cosmopolitan earthworm species, and perhaps the use of molecular analyses of types will help provide answers to fill some of the many gaps in our knowledge.

14.8 BENEFITS AND RISKS OF EARTHWORM TRANSPORTATIONS

Earthworms are reported to sometimes be deliberately transported, e.g. Australian natives from Mt. Kosciusko were exported to the Scottish Cairngorms in an attempt to reduce peat turf mats there (Sims and Gerard 1985, 1999), and a worldwide trade in vermicomposting and fishing bait worms exists. Naturally, a specimen's attendant parasites, pathogens and symbionts accompany it in its travels. Imposition of quarantine barriers and awareness of invasive species problems help regulate such dispersion and, although earthworms are usually considered beneficial or benign, some risk assessment is required as some deleterious effects have emerged (e.g. Hendrix and Bohlen 2002). Examples of environmental risk include exotics *E. saltensis* (Beddard 1895) from South America and recently discovered in Japan (Blakemore et al. 2007), which is considered a pest in Australian rice paddies (Stevens and Warren 2000), and *Dichogaster annae* (Horst 1893), reported as its junior synonym *Dichogaster curgensis* Michaelsen 1921, which is indicated as a serious pest of rice terraces in the Philippine Cordilleras (Barrion and Listinger 1997). Other invasives in lowland tropics are *P. corethrurus* and the *P. elongata* species complex, both of which tend to dominate the native faunas and have some negative agroecological reports (e.g. Stephenson 1930; Gates 1972; Rose and Wood 1980; Blakemore 1994, 1997, 2002, 2006a). Gates (1972) noted that *P. corethrurus*, along with *Po. elongata* (15.6% of the earthworm population), was implicated in rendering a South Indian soil cloddy and unproductive (Puttarudriah and Sastry 1961). Seepages from taro patches on Kauai, Hawaii, from rice paddies in Taiwan and from the famous 2,000-year-old mountain terraces of Ifugao, Philippines were all attributed to morphs identified by Gates as *P. elongata* (see also Joshi et al. 2000). Nevertheless, benefits are also reported (e.g. Spain et al. 1992) and feeding experiments showed Papuan pigs to have a preference for *P. corethrurus* compared with for native species (Sims and Gerard 1999); also, apparent plant growth enhancement has been observed through the activity of this species (Fig. 14.3).

Overwhelmingly, studies show increased plant productivity when earthworms are deliberately added to the soil or when they are encouraged by improved soil management

Figure 14.3 Demonstration of effect of adding earthworms (*Pontoscolex corethrurus*) on soil and litter layers, and plant (*Sorghum bicolor*) growth after 2 weeks: lhs = no worms; rhs + five worms. *Source*: Photo courtesy of Dr. Les Robertson. (See color insert.)

(Lee 1985; Brown et al. 1999; Blakemore 2002, 2006a). Whether exotic species compete with or displace natives is debatable, although Lee (1987) stated, "There is no documented case of direct competition between previously established and newly introduced earthworms." The examples given by Lee (1987) seem to support conclusions by Wood (1974) and Kalisz and Wood (1995) that natives living superficially (but not necessarily subsoil species) mainly decline when natural vegetation is replaced with cultivated vegetation, rather than through direct competition with introduced earthworm species. Moreover, some indigenous earthworms may be more adaptable and persistent under cultivation than is generally realized, as shown by a few reports from ecotaxonomic studies (e.g. Lavelle 1978; Abbott et al. 1985; Blakemore 1994, 1997a; Blakemore and Elton 1994; Blakemore and Paoletti 2007).

14.9 CONCLUSIONS

According to Grayson (2001), "It now appears extremely unlikely that there are any habitable places on earth whose terrestrial biotas were not structured by prehistoric human activities." To this we may add that no region is known to lack introduced earthworms for similar reasons. Earthworms often have remarkably high biodiversity, yet routine agroecological studies tend to overemphasize the importance of just the Lumbricidae components rather than consider contributions of natives and nonlumbricid cosmopolitans. This may be due partly to the difficulty for nonspecialists to identify natives (which require dissection for identification) or the smaller exotics that are easily missed or mistaken for immatures, and partly to studies historically concentrating on just Lumbricidae of European origin even when transplanted to new areas such as North America or Australia (e.g. Edwards and Lofty 1977; Edwards and Bohlen 1996). While the natural spread of earthworms can be slow or negligible, the dispersion of exotics appears to closely follow cultural and technological trade and exchanges: gradually in preagricultural societies, more extensively following the Agricultural Revolution, accelerating during the Industrial Era and subject to globalization in the current period. Thus, earthworm transportation is considered to be mainly human mediated with exotic species' origins often corresponding with provenances of traditional agricultural and horticultural crops.

The compilation and online presentation of earthworm distribution and diversity data is important. Concerted efforts could readily generate a unified checklist of all 6,000 earthworm species, having the most current binomial as the universal identifier under the *International Code of Zoological Nomenclature* (ICZN 1999), with the realization that each species is an ecologically unique entity and that casting light on its origins and peregrinations may help illuminate human history. Ideally, a database matrix should be cross referenced by region (and habitat) to gauge biodiversity and to plot the current or potential spread of exotics and their possible competition with natives, or any unexpected consequences such as acting as intermediate hosts or disease vectors. Earthworms are generally considered beneficial or benign, albeit deleterious environmental effects of some invasive earthworm species have been reported and their environmental risk/benefit needs to be assessed on a case-by-case basis.

ACKNOWLEDGMENTS

The current work was completed under a 21st century Center of Excellence (COE) program sponsored by the Ministry of Education, Culture, Sports, Science and Technology, Japan and entitled "Bio-Ecological Environmental Risk Management" fellowship at Yokohama National University. I thank members of the Soil Ecology Group for help in the preparation of this chapter, especially Drs. M.T. Ito, T. Kamitani and N. Kaneko. Several ideas have been developed following publications or discussion with colleagues as noted. Dr. Dan Shain and Sonia Krutzke made editorial comments that greatly improved the text.

REFERENCES

Taxonomic authorities are not generally presented as these may be found elsewhere.

ABBOTT, I. 1985a. Distribution of introduced earthworms in the northern jarrah forest of Western Australia. Aust. J. Soil Res. 23, 263–270.

ABBOTT, I. 1985b. Influence of some environmental factors on indigenous earthworms in the northern jarrah forest of Western Australia. Aust. J. Soil Res. 23, 271–290.

ABBOTT, I., ROSS, J.S., PARKER, C.A. 1985. Ecology of the large indigenous earthworm *Megascolex imparicystis* in relation to agriculture near Lancelin, Western Australia. J. R. Soc. West. Aust. 68, 13–15.

BANO, K., KALE, R.D. 1991. Earthworm fauna of southern Karnataka, India. In *Advances in Management and Conservation of Soil Fauna*, edited by G.K. Veeresh, D. Rajagopal and C.A. Viraktamath. New Delhi, India: Oxford and IBH, pp. 627–634.

BAROIS, I., LAVELLE, P., BROSSARD, M., TONDOH, J., ANGELES-MARTINEZ, M., ROSSI, J.P., SENAPATI, B.K., ANGELES, A., FRAGOSO, C., JIMENEZ, J.J., DECAËNS, T., LATTAUD, C., KANYONYO, J., BLANCHART, E., CHAPUIS, L., BROWN, G., MORENO, A. 1999. Ecology of earthworms species with large environmental tolerance and/or extended distribution. In *Earthworms Management in Tropical Agroecosystems*, edited by P. Lavelle, L. Brussaard and P. Hendrix. London: CAB International, pp. 57–84.

BARRION, A.T., LITSINGER, J.A. 1997. *Dichogaster nr. curgensis* Michaelsen (Annelida: Octochaetidae): An earthworm pest of terraced rice in the Philippine Cordilleras. Crop Protection 16 (1), 89–93 (5).

BERMAN, D.I., MESHCHERYAKOVA, E.N., ALFIMOV, A.V., LEIRIKH, A.N. 2001. Spread of the earthworm *Dendrobaena octaedra* (Lumbricidae: Oligochaeta) from Europe to Northern Asia is restricted by insufficient freeze tolerance. Dokl. Akad. Nauk 377, 415–418.

BLAKEMORE, R.J. 1994. Earthworms of south-east Queensland and their agronomic potential in brigalow soils. Unpublished PhD Thesis, University of Queensland, p. 613.

BLAKEMORE, R.J. 1997a. Agronomic potential of earthworms in brigalow soils of South-East Queensland. Soil Biol. Biochem. 29, 603–608.

BLAKEMORE, R.J. 1997b. First "common earthworm" found in Tasmania. Invertebrata 9, 1–5.

BLAKEMORE, R.J. 1999. The diversity of exotic earthworms in Australia—a status report. In *The Other 99%*, edited by W. Ponder and D. Lunney. Trans. R. Zoo. Soc. NSW, 182–187.

BLAKEMORE, R.J. 2000a. Taxonomic and conservation status of earthworms from Lake Pedder, Tasmania Wilderness World Heritage Area. Rec. Queen Victoria Mus. 109, 1–36.

BLAKEMORE, R.J. 2000b. Native earthworms (Oligochaeta) from southeastern Australia, with the description of fifteen new species. Rec. Aust. Mus. 52, 187–222.

BLAKEMORE, R.J. 2000c. *Tasmanian Earthworms*. CD-ROM Monograph with Review of World Families. Canberra, Australia: VermEcology, p. 800.

BLAKEMORE, R.J. 2001a. Finding Fletcher's giant worms—from Burrawang to Budderoo. Eucryphia 54, 6–7.

BLAKEMORE, R.J. 2001b. Australian giant worms. Aust. Geog. Mag. 64 (Spring issue).

BLAKEMORE, R.J. 2002. *Cosmopolitan Earthworms—an Eco-Taxonomic Guide to the Peregrine Species of the World*. CD-ROM. Canberra, Australia: VermEcology, p. 500.

BLAKEMORE, R.J. 2003. Japanese earthworms (Annelida: Oligochaeta): a review and checklist of species. Org. Divers. Evol. 3, 241–244.

BLAKEMORE, R.J. 2004a. A provisional list of valid names of Lumbricoidea (Oligochaeta) after Easton, 1983. In *Avances en taxonomia de lombrices de tierra/Advances in earthworm taxonomy (Annelida: Oligochaeta)*, edited by A.G. Moreno and S. Borges. Madrid, Spain: Editorial Complutense, Universidad Complutense, pp. 75–120.

BLAKEMORE, R.J. 2004b. Checklist of the earthworm family Exxidae Blakemore, 2000 (and renaming of *Sebastianus* Blakemore, 1997). In *Avances en taxonomia de lombrices de tierra/Advances in earthworm taxonomy (Annelida: Oligochaeta)*, edited by A.G. Moreno and S. Borges. Madrid, Spain: Editorial Complutense, Universidad Complutense, pp. 121–125.

BLAKEMORE, R.J. 2004c. Checklist of Pheretimoid earthworms after Sims & Easton (1972). In *Avances en taxonomia de lombrices de tierra/Advances in earthworm taxonomy (Annelida: Oligochaeta)*, edited by A.G. Moreno and S. Borges. Madrid, Spain: Editorial Complutense, Universidad Complutense, pp. 126–154.

BLAKEMORE, R.J. 2004d. Checklist of Japanese Earthworms updated from Easton (1981). In *Avances en taxonomia de lombrices de tierra/Advances in earthworm taxonomy (Annelida: Oligochaeta)*, edited by A.G. Moreno and S. Borges. Madrid, Spain: Editorial Complutense, Universidad Complutense, pp. 155–162.

BLAKEMORE, R.J. 2004e. Checklist of Tasmanian Earthworms updated from Spencer (1895). In *Avances en taxonomia de lombrices de tierra/Advances in earthworm taxonomy (Annelida: Oligochaeta)*, edited by A.G. Moreno and S. Borges. Madrid, Spain: Editorial Complutense, Universidad Complutense, pp. 163–173.

BLAKEMORE, R.J. 2004f. Checklist of New Zealand Earthworms updated from Lee (1959). In *Avances en taxonomia de lombrices de tierra/Advances in earthworm taxonomy (Annelida: Oligochaeta)*, edited by A.G. Moreno and S. Borges. Madrid, Spain: Editorial Complutense, Universidad Complutense, pp. 175–185.

BLAKEMORE, R.J. 2005. CD-ROM. In *A Series of Searchable Texts on Earthworm Biodiversity Ecology and Systematics from Various Regions of the World*, edited by M.T. Ito and N. Kaneko. Yokohama, Japan: Soil Ecology Research Group, Graduate School of Environment & Information Sciences, Yokohama National University.

BLAKEMORE, R.J. 2006a. *Cosmopolitan Earthworms—an Eco-Taxonomic Guide to the Peregrine Species of the World*, 2nd edn. Japan: VermEcology, p. 600.

BLAKEMORE, R.J. 2006b. CD-ROM. *A Series of Searchable Texts on Earthworm Biodiversity, Ecology and Systematics from Various Regions of the World*, edited by M.T. Ito and N. Kaneko. Yokohama, Japan: Soil Ecology Research Group, Yokohama National University.

BLAKEMORE, R.J. 2007a. Origin and means of dispersal of cosmopolitan *Pontodrilus litoralis* (Oligochaeta: Megascolecidae). Euro. J. Soil Biol. 43, S3–8.

BLAKEMORE, R.J. 2007b. Review of Criodrilidae (Annelida : Oligochaeta) including *Biwadrilus* from Japan. Opusc. Zool. 37, 1–12.

BLAKEMORE, R.J., ELTON, K.L. 1994. A hundred-year old worm? Aust. Zool. 29, 251–254.

BLAKEMORE, R.J., PAOLETTI, M.G. 2007. Australian earthworms as a natural agroecological resource. Annals of Arid Zone 45, 309–330.

BLAKEMORE, R.J., CHANG, C-H., CHEN, J-H., CHUANG, S-C., ITO, M.T., JAMES, S., WU, S-H. 2006. Biodiversity of Earthworms in Taiwan: a species checklist with the confirmation and new records of the exotic lumbricids *Eisenia fetida* and *Eiseniella tetraedra*. Taiwania 51 (3), 226–236.

BLAKEMORE, R.J., ITO, M.T., KANEKO, N. 2007. Alien earthworms in the Asia/Pacific region with a checklist of species and the first records of *Eukerria saltensis* (Oligochaeta:Ocnerodrilidae) and *Eiseniella tetraedra* (Lumbricidae) from Japan, and *Pontoscolex corethrurus* (Glossoscolecidae) from Okinawa. In *Assessment and Control of Biological Invasion Risks*, edited by F. Koike, M.N. Clout, M. Kawamichi, M. De Poorter and K. Iwatsuki. Gland, Switzerland and Cambridge, UK: IUCN; Kyoto, Japan: Shoukadoh Book Sellers, pp. 173–181.

BROWN, G.G., FRAGOSO, C. 2007. *Minhocas na América Latina: Biodiversidade e Ecologia*. Londrina, Parana, Brazil: Embrapa Soja, pp. 1–539.

BROWN, G.G., PASHANASI, B., VILLENAVE, C., PATRÓN, J.C., SENAPATI, B.K., GIRI, S., BAROIS, I., LAVELLE, P.,

BLANCHART, E., BLAKEMORE, R.J., SPAIN, A.V., BOYER, J. 1999. Effects of earthworms on plant production in the tropics. In *Earthworm Management in Tropical Agroecosystems*, edited by P. Lavelle, L. Brussaard and P.F. Hendrix. Wallingford, Seattle, WA: CAB International, pp 87–147.

CHEN, I., CHANG, C-H., CHEN, J-H. 2003. The species composition and distribution of earthworms in Ilan. Chin. Biosci. 46, 56–65.

COGNETTI DE MARTIIS L. 1914. On a small collection of earthworms from Henderson Island. Ann. Mag. Nat. Hist (ser. 8) 13, 255–257.

CSUZDI, Cs., PAVLICEK, T. 1999. Earthworms from Israel. I. Genera *Dendrobaena* Eisen, 1874 and *Bimastos* Moore, 1891 (Oligochaeta: Lumbricidae). Israel J. Zool. 45, 467–486.

CSUZDI, Cs., PAVLICEK, T. 2005. Earthworms (Oligochaeta) from Jordan. Zool. Mid. East 34, 71–78.

DARWIN, C.R. 1840. On the formation of mould. Trans. Geol. Soc. 5, 505–509.

DARWIN, C.R. 1881. *The Formation of Vegetable Mould through the Action of Worms with Observations on their Habits*. London: Murray, pp. vii, 326.

DOBSON, R.M., SATCHELL, J.E. 1956. *Eophila oculata* at Verulamium: a Roman earthworm population? Nature, Lond. 177, 796–797.

EASTON, E.G. 1976. Taxonomy and distribution of the *Metapheretima elongata* species-complex of Indo-Australasian earthworms (Megascolecidae: Oligochaeta). Bull. Brit. Mus. (Nat. Hist.) Zool. 30, 29–53.

EASTON, E.G. 1979. A revision of the "acaecate" earthworms of the Pheretima group (Megascolecidae: Oligochaeta): *Archipheretima, Metapheretima, Planapheretima, Pleinogaster* and *Polypheretima*. Bull. Brit. Mus. (Nat. Hist.) Zool. 35, 1–128.

EASTON, E.G. 1982. Australian pheretimoid earthworms (Meagscolecidae: Oligochaeta): a synopsis with the description of a new genus and five new species. Aust. J. Zool. 30, 711–735.

EASTON, E.G. 1984. Earthworms (Oligochaeta) from islands of the south-western Pacific, and a note on two species from Papua New Guinea. N.Z. J. Zool. 11, 111–128.

EDWARDS, C.E., BOHLEN, P.J. 1996. *Biology and Ecology of Earthworms*, 2nd edn. London: Chapman & Hall, pp. xii, 426.

EDWARDS, C.A., LOFTY, J.R. 1977. *Biology of Earthworms*, 2nd edn. London: Chapman & Hall, pp. 333.

ENCKELL, P.H., RUNDGREN, S. 1988. Anthropochorous earthworms (Lumbricidae) as indicators of abandoned settlements in the Faroe Islands. J. Archaeol. Sci. 15, 439–451.

FAO. 1991. *Food Balance Sheets*. Rome: Food and Agriculture Organization of the United Nations.

FLETCHER, J.J. 1886. Notes on Australian Earthworms. Part I. Proc. Linn. Soc. N. S. W. 1 (2), 523–576.

FRAGOSO, C. 2001. Las lombrices de tierra de México (Annelida, Oligochaeta). Diversidad, Ecología y manejo. Acta Zool. Mex. (n.s.) 1, 131–171.

FRAGOSO, C., JAMES, S.W., BORGES, S. 1995. Native earthworms of the north. Neotropical region: current status and controversies. In *Earthworm Ecology and Biogeography in North America*, edited by P. Hendrix. Boca Raton, FL: CRC Press, Inc., pp. 67–115.

FRAGOSO, C., KANYONYO, J., MORENO, A., SENAPATI, B.K., BLANCHART, E., RODRIGUEZ, C. 1999. A survey of tropical earthworms: taxonomy, biogeography and environmental plasticity. In *Earthworm Management in Tropical Agroecosystem*, edited by P. Lavelle, L. Brussaard and P. Hendrix. London: CAB Intertional, pp. 1–26.

FRAGOSO, C., BROWN, G., FEIJOO, A. 2004. The influence of Gilberto Righi on tropical earthworm taxonomy: the value of a full-time taxonomist. Pedobiologia 47, 400–404.

GATES, G.E. 1958. On some species of the Oriental earthworm genus *Pheretima* Kinberg, 1867, with key to species reported from the Americas. Am. Mus. Novit. 1888, 1–30.

GATES, G.E. 1968. On a new anthropochorous species of the earthworm genus *Pheretima* (Megascolecidae, Oligochaeta). J. Nat. Hist. 2, 253–261.

GATES, G.E. 1972. Burmese earthworms, an introduction to the systematics and biology of Megadrile oligochaetes with special reference to South-East Asia. Trans. Am. Phil. Soc. 62, 1–326.

GONZALEZ, G., HUANG, C.Y., ZOU, X., RODRIGUEZ, C. 2006. Earthworm invasions in the tropics. Biol. Invasions 8, 1247–1256.

GRAYSON, D.K. 2001. The archaeological record of human impacts on animal populations. J. World Prehist. 15, 1–68.

HART, J.R., TERRELL, J.E. 2002. *Darwin and Archaeology: A Handbook of Key Concepts*. Westport, CT: Bergin & Garvey publishers, p. 259.

HENDRIX, P.F. 1995. *Ecology and Biogeography of Earthworms in North America*. Boca Raton, FL: CRC Publishing, p. 244.

HENDRIX, P.F., BOHLEN, P. 2002. Exotic earthworm invasions in North America: ecological and policy implications. Bioscience 52, 801–811.

HULES, M.E., SYKES, B.C., JOBLING, M.A., FORSTER, P. 2005. The dual origin of the Malagasy in Island Southeast Asia and East Africa: evidence from maternal and paternal lineages. Am. J. Hum. Genet. 76, 894–901.

ICZN. 1999. *International Code of Zoological Nomenclature*, 4th edn. London: International Trust for Zoological Nomenclature, c/o Natural History Museum, p. 306.

ISO 23611-1. 2006. *Soil Quality–Sampling of Soil Invertebrates–Part 1: Hand-Sorting and Formalin Extraction of Earthworms*. Geneva: International Standards Organization.

JEWELL, P.A. 1958. Natural history and experiment in archaeology. Adv. Sci. Brit. Assoc. 59, 165–172.

JOSHI, R.C., MATCHOC, O.R., CABIGAT, J.C., JAMES, S.W. 2000. Survey of the earthworms in the Ifugao rice terraces, Philippines. J. Environ Sci. Manage. 2 (2), 1–12.

JULKA, J.M. 1988. *The Fauna of India and the Adjacent Countries*. Zoological Survey of India. Calcutta: Zoological Survey of India, p. 400.

KALISZ, P.J., WOOD, H.B. 1995. Native and exotic earthworms in wildland ecosystems. In *Earthworm Ecology and Biogeography in North America*, edited by P.F. Hendrix. Boca Raton, FL: Lewis Publishers, pp. 117–126.

KEITH, A. 1942. A postscript to Darwin's "formation of vegetable mould through the action of worms." Nature, Lond. 149, 716.

LAUNER, J. 2006. The descent of man. QJM 99, 275–276.

LAVELLE, P. 1978. Les vers de terre de la savane de Lamto (Cote d'Ivoire). Peuplements, populations et fonctions dans l'ecosysteme. Pub. du Laboratoire de Zool. de l'École Normale Supérieur, Paris 12, 301.

LEE, K.E. 1981. Earthworms (Annelida : Oligochaeta) of Vanua Tu (New Hebrides Islands). Aust. J. Zool. 29, 535–572.

LEE, K.E. 1985. *Earthworms—Their Ecology and Relationships with Soils and Land Use*. Sydney, Australia: Academic Press, p. 411.

LEE, K.E. 1987. Peregrine species of earthworms. In *On Earthworms*, edited by A.M. Bonvicini Pagliai and P. Omodeo. Modena, Italy: U.Z.I., 2, Mucchi, pp. 315–327.

LEE, K.E. 1994. Earthworm classification and biogeography: Michaelsen's contribution, with special reference to southern lands. Mitt. Hamb. Zool. Mus. Inst. 89 (2), 11–21.

LJUNGSTÖM, P-O. 1972. Introduced earthworms of South Africa. On their taxonomy, distribution, history of introduction and on the extermination of endemic earthworms. Zool. Jahrb. Syst. 99, 1–81.

MCCOBB, L.M.E., BRIGGS, D.E.G., HALL, A.R., KENWARD, H.K. 2004. The preservation of invertebrates in 16th-century cesspits at St. Saviourgate, York. Archaeometry 46, 157–169.

MACKNIGHT, C.C. 1976. *The Voyage to Marege: Macassan Trepangers in Northern Australia*. Melbourne, Australia: Melbourne University Press.

MANUM, C.B., BOSE, M.N. 1991. Clitellate cocoons: common but neglected fossils amongst Mesozoic plant litter. In *Fifth Symposium on Mesozoic Terrestrial Ecosystmes*, vol. 364, edited by Z. Kielan-Haworowska, N. Heintz and H.A. Nakrem. Extended Abstracts. Oslo: Contributions from the Paleontological Museum, University of Oslo, pp. 43–44.

MICHAELSEN, W. 1900. *Das Tierreich Vol. 10: Oligochaeta*. Berlin, Germany: Friedländer & Sohn. pp. XXIX, 575.

MICHAELSEN, W. 1903. *Die geographische Verbreitung der Oligochaeten*. Berlin, Germany: Friedländer & Sohn.

MICHAELSEN, W. 1905. Die Oligochaeta der deutschen Sudpolar-expedition. Zoologie 1, 1901–1903.

MICHAELSEN, W. 1922. Die Verbreitung der Oligochaten im Lichte der Wegenerschen Theorie der Kontinentverschiebung und anderer Fragen zur Stammesgeschichte und Verbreitung dieser Tiergruppe. Verh. Naturw. Ver. Hamb. 3, 45–79.

MISCHIS, C.C. 2004. Lombrices de tierra de Argentina (Earthworms in Argentina: faunistic and biogeoraphic aspects). In *Avances en taxonomia de lombrices de tierra/Advances in earthworm taxonomy (Annelida: Oligochaeta)I*, edited by A.G. Moreno and S. Borges. Madrid, Spain: Editorial Complutense, Universidad Complutense, pp. 261–274.

MISCHIS, C.C., BRIGADA, A.M. 1988. *Octodrilus transpadanus* (Rosa, 1884) (Oligochaeta: Lumbricidae), from the province of San Luis (Argentina). Part II. Megadrilogica 4, 139–140.

MORRISON, L.W. 1996a. Community organization in a recently assembled fauna: the case of Polynesian ants. Oecologia 107, 243–256.

MORRISON, L.W. 1996b. The ants (Hymenoptera: Formicidae) of Polynesia revisited: species numbers and the importance of sampling intensity. Ecography 19, 73–84.

OMODEO, P., ROTA, E. 1989. Earthworms of Turkey. Boll. Zool. 56, 167–199.

OMODEO, P., ROTA, E., BAHA, M. 2004. The megadrile fauna (Annelida:Oligochaeta) of Maghreb. Pedobiologia, 47, 458–465.

PAVLICEK, T., CSUZDI, CS., NEVO, E. 2004. Species richness and zoogeographic affinities of earthworms in the Levant. Pedobiologia 47, 452–457.

PIEARCE, T.G., OATES, K., CARRUTHERS, W.J. 1990. Fossil earthworm cocoons from a Bronze Age site in Wiltshire. J. Zool. (Lond.) 220, 537–542.

PIEARCE, T.G., OATES, K., CARRUTHERS, W.J. 1992. Fossil earthworm cocoons from a Bronze Age site in Wiltshire, England. Soil Biol. Biochem. 24, 1255–1258.

PUTTARUDRIAH. M., SASTRY, K.S.S. 1961. A preliminary study of earthworm damage to crop growth. Mysore Agricultural Journal 36, 2–11.

REYNOLDS, J.W. 1994. Earthworms of the world. Glob. Biodivers. 4, 11–16.

REYNOLDS, J.W. 1998. The status of earthworm biogeography, diversity, and taxonomy in North America revisited with glimpses into the future. In *Earthworm Ecology*, edited by C.A. Edwards. Boca Raton, FL: St. Lucie Press, pp. 15–36.

REYNOLDS, J.W., WETZEL, M.J. 2004. Terrestrial Oligochaeta (Annelida: Clitellata) in North America north of Mexico. Megadrilogica 9 (11), 71–98.

ROSE, C.J., WOOD, A.W. 1980. Some environmental factors affecting earthworm populations and sweet potato production in the Tari Basin, Papua New Guinea Highlands. P. N. G. Agric. J. 31, 583–587.

ROTA, E., ERSÉUS, C. 1997. First record of *Dendrobaena attemsi* (Michaelsen) (Oligochaeta, Lumbricidae) in Scandinavia, with a critical review of its morphological variation, taxonomic relationships and geographical range. Ann. Zool. Fenn. 34, 89–104.

SCHWERT, D.P. 1979. Description and significance of a fossil earthworm (Oligochaeta: Lumbricidae) cocoon from postglacial sediments in southern Ontario. Can. J. Zool. 57 (7), 1402–1405.

SCHWERT, D.P. 1990. Active and passive dispersal of lumbricid earthworms. In *Soil Biology as Related to Land Use Practices*, edited by D.L. Dindal. Washington, DC: EPA, pp. 182–189.

SCHWERT, D.P., DANCE, K.W. 1979. Earthworm cocoons as a drift component in a southern Ontario stream. Can. Field-Nat. 93, 180–183.

SEILACHER, A., BOSE, P.K., PFLUGER, F. 1998. Triploblastic animals more than 1 billion years ago: trace fossil evidence from India. Science 282, 80–83.

SIMS, R.W. 1980. A Classification and the distribution of earthworms, suborder Lumbricina (Haplotaxida: Oligochaeta). Bull. Brit. Mus. (Nat. Hist.) Zool. 39, 103–124.

SIMS, R.W., EASTON, E.G. 1972. A numerical revision of the earthworm genus *Pheretima* auct. (Megascolecidae:Oligochaeta) with the recognition of new genera and an appendix on the earthworms collected by the Royal Society North Borneo Expedition. Biol. J. Linn. Soc. Lond. 4, 169–268.

SIMS, R.W., GERARD, B.M. 1985. *Earthworms. Keys and Notes to the Identification and Study of the Species. Synopsis of the British Fauna (New Series)*, vol. 31. Leiden: EJ Brill, p. 171.

SIMS, R.W., GERARD, B.M. 1999. *Earthworms: Notes for the Identification of British Species*, revised edn. Shrewsbury, UK: Field Studies Council, Montford Bridge, pp. 1–169.

SINGH, G., KERSHAW, A.P., CLARK, R. 1981. Fire and the Australian Biota. In *Quaternary Vegetation Fire History of Australia*, edited by A.M. Gill, R.H. Groves and I.R. Noble. Canberra, Australia: Academy of Science, pp. 23–54.

SPAIN, A.V., LAVELLE, P., MARIOTTI, A. 1992. Stimulation of plant growth by tropical earthworms. Soil Biol. Biochem. 23, 1629–1633.

STEADMAN, D.W. 1995. Prehistoric extinctions of Pacific Island birds: biodiversity meets zooarcheology. Science 267, 1123–1131.

STEPHENSON, J. 1930. *The Oligochaeta*. Oxford: Oxford University, Clarendon Press, p. 978.

STEPHENSON, J. 1931. Oligochaeta from Burma, Kenya and other parts of the world. Proc. Zool. Soc. Lond. 1931, 33–92.

STEVENS, M.M., WARREN, G.N. 2000. Laboratory studies on the influence of the earthworm *Eukerria saltensis* (Oligochaeta: Ocnerodrilidae) on overlying water quality and rice plant establishment. Int. J. Pest Manage. 46, 303–310.

STÖP-BOWITZ, C. 1969. A contribution to our knowlege of the systematics and zoogeography of Norwegian earthworms (Annelida, Oligochaeta: Lumbricidae). Nytt Mag. Zool. 17, 169–280.

SUBBA RAO, B.V.S.S.R., GANAPATI, P.N. 1974. On the breeding and cocoons of a littoral oligochaete *Pontodrilus bermudensis* Beddard. Proc. Ind. Acad. Sci. 1, 18.

TOFFLER, A. 1980. *The Third Wave*. New York: Bantam Books, p. 537.

TSAI, C-F., SHEN, H-P., TSAI, S-C., 2000. Native and exotic species of terrestrial earthworms (Oligochaeta) in Taiwan with reference to Northeast Asia. Zool. Stud. 39, 285–294.

TSAI, C-F., SHEN, H-P., TSAI, S-C. 2004. Endemicity and altitudinal stratification in distribution of megascolecid earthworms in the centro-western Taiwan. Endemic Spp. Res. 6, 1–18.

VALENTINE, J.W. 1980. L'origine des grands groupes d'animaux. Recherche 112, 666–674.

VAVILOV, N.I. 1951. The origin, variation, immunity and breeding of cultivated plants. In *Chronica Botanica 13, 1–366 (1949–50)*, edited by K.R. Chester. New York: Ronald Press, p. 366.

WEGENER, A. 1915. *Die Entstehung der Kontinenten und Ozeane*. Brunswick: Friederich Vieweg & Sohn, p. 94.

WOOD, T.G. 1974. The distribution of earthworms (Megascolecidae) in relation to soils, vegetation and altitude on the slopes of Mt Kosciusko, Australia. Aust. J. Anim. Ecol. 43, 87–106.

WRIGHT, R.V.S. 1986. How old is zone F at Lake George? Arch. Oceania 21, 138–139.

Part V

Extreme Environments and Biological Novelties

Chapter 15

Hydrothermal Vent Annelids

Florence Pradillon and Françoise Gaill

UMR CNRS 7138 Systématique, Adaptation et Evolution, Université Pierre et Marie Curie, Paris, France

15.1 INTRODUCTION

Since their discovery ~30 years ago (Lonsdale 1977; Corliss et al. 1979), hydrothermal vent communities have been investigated as models of adaptation to the extreme conditions prevailing in this environment (Gaill and Hunt 1991; Van Dover and Lutz 2004). Frequently located below 1,000-m depth on oceanic ridges, deep-sea hydrothermal vents are considered extreme environments. Hot fluids with temperatures up to 400 °C exit from "black smokers" (Von Damm et al. 1995), thus creating sharp thermal gradients with the surrounding cold abyssal seawater.

Habitats associated with deep-sea hydrothermal vents have been described as some of the most challenging for animals on earth. The high instability of their geological setting, subject to frequent volcanic and tectonic events, makes these habitats particularly ephemeral. The chaotic change in fluid emission intensity induces dramatic changes in community composition over time. Extreme physicochemical characteristics, as compared to most oceanic environments, include temperatures exceeding 100 °C (Desbruyères et al. 1985; Chevaldonné et al. 1992), acidic pH and hundreds of micromolar sulfide concentrations (e.g. H_2S; Le Bris et al. 2003, 2005). To what extent organisms are exposed to these physicochemical extremes is unknown, but new tools and methods have improved our knowledge of environmental variability at decimeter scales, enabling us to estimate the conditions experienced by vent organisms (Le Bris and Gaill 2007).

Annelids are one of the most represented animal phyla at vents, comprising ~130 of the ~700 total described species (Wolf 2005); polychaetous annelids such as the giant tube worm *Riftia pachyptila* and the highly thermotolerant *Alvinella pompejana* (Fig. 15.1) are considered emblematic vent species (Gaill 1993). Indeed, *A. pompejana* is thought to be one of the most thermotolerant metazoans on earth (Chevaldonné et al. 1992; Cary et al. 1998). Recently, its northern Pacific relative, *Paralvinella sulfincola*, has been confirmed to tolerate 50–55 °C (Girguis and Lee 2006), the highest temperature known for a marine metazoan. Even more surprising, *P. sulfincola* has been shown to prefer temperatures between 40 and 50 °C.

Annelids in Modern Biology, Edited by Daniel H. Shain

Figure 15.1 *Alvinella pompejana* specimen out of its tube (a) and on a smoker wall after an anthropogenic disturbance (b) (worm sampling by Nautile). (See color insert.)

The vent habitat is also characterized by its patchiness and high instability. Depending on the geological context, vents may persist over 10–100 years. Dispersal must then be a key process that allows vent organisms to colonize new sites over time scales in the range of their life cycle and to survive over geological time (Lutz et al. 1984; Mullineaux and France 1995). For this reason, many studies focus on the dispersal abilities of vent organisms to understand how they colonize new sites that may be hundreds of kilometers away.

In this chapter, we will focus on thermal adaptations of vent organisms since high and fluctuating thermal environments appear to challenge all metazoan life at vent sites. But first, we will briefly present the life history characteristics of vent annelids; both adult and larval stages will be considered, and we will see that different strategies of thermal adaptation are used at different periods of their life cycle. Additionally, studies on their early stages will provide hypotheses concerning the mechanisms of dispersal between vent sites.

15.2 *ALVINELLA POMPEJANA*: A SYMBIOTIC SYSTEM

Alvinella pompejana (Pompeii worm) is found in association with high-temperature venting at the surface of hydrothermal chimneys of the East Pacific Rise (Fig. 15.1). These tubicolous worms (~10 cm in length; Fig. 15.1) assemble in colonies, forming patches of various sizes on smoker walls (Desbruyères et al. 1985; Gaill and Hunt 1991). Unlike the giant tube worm *R. pachyptila, A. pompejana* does not harbor endosymbionts, but rather abundant microbial communities have been described that cover the inner part of its tube and attach to posterior appendages of the animal (Gaill et al. 1984, 1988; Gulik and Gaill 1988). The contribution of the associated microflora to the mechanisms that protect these worms from environmental threats remains putative; for example, its role in buffering temperature at the surface of the animal has been suggested (Di Meo-Savoie et al. 2004), and although the role of these microbes in forming dense sulfur filamentous mats is unknown, the worm's nutrition is thought to derive from grazing on surrounding free-living microflora (Gaill and Hunt 1991).

Hydrothermal vent animal communities are supported solely on carbon originating from microbial chemoautotrophy instead of phototrophy. Microorganisms use reduced chemicals such as hydrogen sulfide, methane or hydrogen found in vent fluids as their energy source for fixing carbon (Karl 1995). Many are free living, but some exist in close association with vent metazoans (Bright and Giere 2005). One of the most studied chemoautotrophic systems is that used by gamma proteobacterial endosymbionts of the giant siboglinid tube worm *R. pachyptila*. These sulfide-oxidizing symbiotic bacteria live in the cells of a specialized tissue called a trophosome (Felbeck et al. 1981; Cavanaugh 1983). The endosymbionts function via the Calvin cycle to fix carbon and to transfer it to their host, deriving energy from the oxidation of hydrogen sulfide.

The alvinellid polychaetes *A. pompejana* and *Alvinella caudata* support epibiotic bacteria on their outer surfaces (Gaill et al. 1984). High densities of microbial sulfur filaments partly cover the inner face of the tube or attach to appendages on the dorsal part of the worm's body (Gaill et al. 1984; Gaill and Hunt 1986, 1991). Bacteria associated with the dorsal epidermis of *A. pompejana* are morphologically and metabolically diverse (Gaill et al. 1988; Prieur et al. 1990). Approximately 80% of this epibiotic community are epsilon proteobacteria, dominated by two phylotypes that are also found on rocks and tubes in the vent ecosystem (Cary et al. 1997). Gaill et al. (1984) have suggested that these epibiotic bacteria use inorganic compounds from vent fluids and the worm's metabolism, as well as organic compounds secreted by the worm. In return, the worm presumably uses dissolved organic compounds produced by the bacteria. Other relationships can be considered, such as detoxification of the worm's immediate environment (Alayse-Danet et al. 1987). Campbell et al. (2003) showed that two genes (adenosine 5'-triphosphate [ATP] citrate lyase and 2-oxoglutarate:acceptor oxidoreductase) involved in the reverse tricarboxylic acid (rTCA) cycle, a CO_2 fixation pathway, were expressed within the *A. pompejana* episymbiont community. They also found that phylogenetically distinct forms of ATP citrate lyase were associated with and expressed by bacteria extracted from *A. pompejana* tubes. Interestingly, a recent proteomic approach based on the metagenome sequence of *R. pachyptila* endosymbionts has revealed that they also use the rTCA cycle in addition to the Calvin cycle for CO_2 fixation (Markert et al. 2007). The rTCA cycle within the bacterial communities may thus be a significant CO_2 fixation pathway for sustaining vent metazoans.

15.3 TEMPERATURE ADAPTATION

At hydrothermal vents, undiluted fluids enriched in hydrogen sulfide escape chimneys at temperatures that may reach 400 °C. Metazoans living around and on the chimney walls are bathed by fluids diluted with seawater to different extents. The polychaete *A. pompejana* is described as one of the most thermotolerant marine invertebrates because temperatures above 80 °C have been recorded on chimney walls where they build their tubes (Le Bris et al. 2005; Table 15.1); therefore, the species *A. pompejana* has become a model for investigations on high-temperature adaptation.

Due to the great variability in both spatial and time scales, assessing *in situ* temperature precisely in the habitat of *Alvinella* species has been difficult, and the extent to which these organisms might be thermotolerant has been the subject of much controversy. Nonetheless, adaptations to high-temperature environments have been investigated at various levels in *A. pompejana*, including the molecular characteristics of extracellular matrices (ECMs), the cellular characteristics of developing embryos and the behavioral characteristics of adult worms.

15.4 TEMPERATURE ADAPTATION AT A MOLECULAR LEVEL

At hydrothermal vents, animals must protect themselves from chemical and physical fluctuations in the environment. ECMs are a first protection barrier from external conditions (Fig. 15.2). Their composition and resistance properties have been examined to assess

Table 15.1 Temperatures reported in close surroundings of *Alvinella pompejana* colonies

Origin of Data	Among Tubes (Mean)	Inside Tubes (Mean)
Desbruyères et al. (1985)	~100 °C	nd
Chevaldonné et al. (1992)	40–80 °C	nd
Cary et al. (1998)	nd	68 ± 6 °C
Le Bris et al. (2003)	60 ± 10 °C	nd
Di Meo-Savoie et al. (2004)	nd	29–84 °C
Le Bris et al. (2005)	61 ± 38 °C	59 ± 6 °C

Abbreviations: nd = nondetermined.
Source: Adapted from Le Bris and Gaill (2007), with permission from Springer.

Figure 15.2 Experimentally obtained Alvinellids. These tubicolous animals have colonized a titanium ring for alvinellid colonization (TRAC, Hot '96 expedition).

the degree of thermal adaptation by *Alvinella* sp., which is thought to reflect the *in situ* temperatures at which these organisms are exposed.

Alvinella sp. are protected by two types of integument: a secreted tube that adheres to the chimney wall, and an ECM that contains an epidermis and cuticle (Gaill 1993). One main component of an organism's cuticle is collagen, and its characteristics may provide a relevant set of information relative to the environment. Collagen is one of the most well-known extracellular proteins in the animal kingdom and is a marker of thermal adaptation that has been well characterized in *A. pompejana* (Gaill et al. 1991, 1995) and in other vent species (Mann et al. 1996). Molecular mechanisms underlying the thermostability of *A. pompejana* collagen have been described (Sicot et al. 2000; see below).

15.5 *ALVINELLA* TUBES

Tubes secreted by *Alvinella* spp. mainly consist of proteins and sulfated glycosaminoglycans. Less than 10% of the organic content of the tube is sugar, presumably as poly- or oligosaccharide material bound to the proteins. The inner surface is covered by filamentous bacteria that—as tubes are concentrically deposited—become trapped under consecutive layers of material. Within the layers of secreted protein, a novel distribution of fibrils is displayed, reminiscent of the arrangement of polymeric units in a cholesteric liquid crystal (Gaill and Bouligand 1987). An observed twist variation in the arrangement of secreted proteins may result from the worm's movements during new layer deposition, as described in vestimentiferan tubes (Gaill et al. 1997). Mucous secretion and its associated bacteria may provide the initial structure-dictating constraint, later modified by the fibrous structure itself as it is deposited, or by pulses of additional mucus production and bacterial growth. Such biopolymeric organization likely provides a specific thermal resistance for the worm (Gaill and Hunt 1991).

Unlike siboglinid tubes (Gaill 1993), *Alvinella* tubes do not contain chitin. X-ray diffraction studies do not indicate an ordered protein secondary structure, although their amino acid composition, with high glycine, alanine and serine levels, is typical of beta-pleated fibrous proteins such as silk fibroins (Gaill and Hunt 1986). Polychaete worms exhibit a peculiar versatility with regard to their tube compositions and, therefore, the amino acid content of the alvinellid tube is probably of little use in establishing comparisons with other polychaetes. However, one characteristic must be retained for its thermal stability, namely, its highly hydrophobic nature.

The material of the *Alvinella* tube has considerable chemical stability. While invertebrate structural materials are often chemically stable, usually as a consequence of extensive cross-linking, most will swell and eventually disrupt at room temperature in strongly acidic or alkaline solutions, or in chaotropic agents such as anhydrous formic and haloacetic acids or lithium thiocyanate. The *Alvinella* tube shows little response to these or to disulfide bond-breaking agents, although a cycle of concentrated hydrochloric acid and potassium hydroxide treatments will cause delamination, swelling and some solubilization (Gaill and Hunt 1986). The tube's thermal stability is also great, with little swelling or shrinkage taking place between 0 and 100 °C. This, too, probably reflects a high degree of cross-linking.

Tubes contain an inorganic fraction with up to 29% minerals including phosphorus, calcium, iron, magnesium and zinc. Tubes also contain between 12 and 25% of elemental free sulfur, the amount depending upon the tube's age and the area of origin (Gaill and Hunt 1986). Specific patterns of mineral deposition within or at the surface of tubes suggest that the microenvironment inside tubes is much more stable than the outside. Zinc–iron sulfide nanocrystals grouped in submicrometer-sized clusters have been described between proteinaceous tube layers (Zbinden et al. 2001). These minerals show a specific zinc–iron signature and have a conserved size, contrary to mineral precipitations found on the outside of tubes. The nanometer size of individual minerals and their constant composition within tubes suggest that these crystals most probably originate from tube-associated bacteria (Zbinden et al. 2001). Mineral particles are also useful markers for evaluating the chemical characteristics of the microenvironment (Zbinden et al. 2003). Gradients in mineral crystal size and composition may reflect gradients in chemical characteristics between the inside and outside of tubes, as well as decimeter-scale gradients between tubes located within or at the surface of *Alvinella* colonies. Collectively, these observations suggest that the tube acts as an efficient protective barrier with respect to the external environment.

15.6 COLLAGENS

Collagens belong to a family of extracellular proteins characterized by a triple helical domain formed by the association of three similar peptides, called α chains, that comprise a succession of Gly-X-Y amino acid triplets, with the Y position usually occupied by a hydroxylated proline residue. Collagen is synthesized inside the cell and secreted into the extracellular compartment. Intracellular collagen comprises a central triple helical domain, which is maintained during the life of the molecule. The carboxy- and amino-propeptides (C-pro and N-pro, respectively), which end the central triple helical component, are cleaved when the molecule is secreted into the ECM.

Vent annelid species possess two abundant collagen types that differ in composition, size, domain structure (Gaill et al. 1994) and immunological properties (Gaill et al. 1991; Table 15.2). Whereas interstitial collagen (found in worm tissues) is similar in morphology to the fibrillar collagen of vertebrates, cuticular collagen (covering the epidermis) is rather unusual. With lengths up to 1.5 and 2.5 μm in *R. pachyptila* and *A. pompejana*, respectively, these collagens are the longest collagenous proteins known (Gaill et al. 1991, 1995). Curiously, they have a terminal globular domain, like that of nonfibrillar collagen molecules, and no comparable structure has thus far been identified in other animals. The cuticular collagen of tube worms from various chemosynthetic environments, including cold seeps, is similar in length, suggesting that this characteristic is phylogenetically related.

Alvinella pompejana has the most thermostable collagen known (Gaill et al. 1995), a property that appears unrelated to the high-pressure characteristics of vent environments (Auerbach et al. 1995). The temperature at which these collagens are denatured (Tm) is 46 °C for cuticular and 45 °C for interstitial collagen (Gaill et al. 1995). Among the fibrillar collagens of 40 other vertebrates and invertebrates, *A. pompejana* collagen is positioned at the upper limit for melting temperature, only behind that of thermostable synthetic collagens (Le Bris and Gaill 2007). The thermal stability of *A. pompejana* cuticular and interstitial collagen is significantly higher than that of other vent annelids. In comparison, *Paralvinella grasslei* collagen has a denaturation temperature of 35 °C, and *R. pachyptila* collagen stability only reaches 29 °C (Gaill et al. 1991; Mann et al. 1996).

Except for the thermostability of the molecule, cuticular collagens from coastal and vent species share similar structural characteristics (Gaill et al. 1991). This is also true for the interstitial collagen of annelids from various habitats. These characteristics, including a substantial and relatively constant level of 4-hydroxyproline in the Y position

Table 15.2 Amino acid composition of interstitial collagen from the three annelid worms

Residues		*Alvinella pompejana* Tm = 46 °C		*Arenicola marina* Tm = 28 °C		*Riftia pachyptila* Tm = 29 °C		
		BC	cDNA	BC	cDNA	BC	BS	cDNA
A	Ala	6.1	5.7	4.9	4.1	8.2	7.7	7.8
C	Cys	0.0	0.0	0.0	0.0	0.0	0.0	0.0
D	Asp	5.5	5.8	5.6	4.7	5.0	4.4	3.9
N	Asn							
E	Glu	10.2	9.4	10.6	10.4	10.2	9.5	10.0
Q	Gln							
F	Phe	1.2	0.4	0.8	0.2	0.6	0.5	0.0
G	Gly	29.6	33.5	32.3	35.7	33.5	34.2	35.7
H	His	1.1	0.4	0.5	0.0	1.4	0.5	0.4
I	Ile	1.7	1.0	0.8	1.0	1.4	1.6	0.4
K	Lys	4.2	3.3	3.1	2.4	4.3	3.8	3.0
L	Leu	4.8	3.9	3.2	3.1	3.0	3.1	3.9
M	Met	0.8	1.4	0.8	1.6	0.3	1.8	3.0
P	Pro	19.7	25.3	18.8	19.4	15.7	16.1	17.0
R	Arg	7.8	7.1	7.1	6.9	6.9	6.8	6.5
S	Ser	3.2	2.2	7.9	6.5	5.0	5.1	4.3
T	Thr	2.2	1.6	2.1	2.4	2.8	2.6	1.7
V	Val	1.5	0.8	1.3	1.4	1.6	1.9	1.7
W	Trp	0.0	0.2	0.0	0.2	0.0	0.1	0.4
Y	Tyr	0.4	0.0	0.2	0.2	0.1	0.3	0.0

Source: Adapted from Sicot et al. (2000), with permission from Elsevier Limited.
BC, biochemical composition; BS, data obtained from the peptidic sequence.

of the Gly-X-Y sequence triplets, have apparently been conserved in various annelid families.

The origin of collagen stability is not well understood, but the degree of proline hydroxylation is likely an important factor. In collagen-like peptides that form triple helices, the substitution of hydroxyproline (Hyp) for proline in the Y position of the repeating Gly-X-Y triplets provides additional sites for hydrogen bonding of water molecules in peptide crystals (reviewed in Sicot et al. 2000). The water bridges, however, only partially contribute to the molecule's thermostability. Different explanations have been proposed to explain this property, involving the entropy state of the peptide chain and an electron-withdrawing inductive effect of the hydroxyl group, but the exact correlation between the thermal stability score and the different processes involved in its molecular stability remain unclear.

In *A. pompejana*, almost all collagen proline residues in the Y position of the Gly-X-Y triplets are fully hydroxylated (Gaill et al. 1995), suggesting that the mechanism of collagen stability is comparable to vertebrate fibrillar collagen. Sicot et al. (2000) have demonstrated a clear correlation between thermal stability and proline content in the Y position, indicating that proline in the Y position of the Gly-X-Y triplet is a decisive factor involved in collagen thermal stability.

The relative percentage of proline in the Y position of Gly-X-Y triplets is three times higher in *A. pompejana* than in *R. pachyptila*, the latter of which has the lowest relative percentage. The frequency of double-P triplets (GPP) relies only on the frequency of the overall proline content among the various amino acids present at the second and third positions of the triplets. Hence, an increase in proline content would automatically lead to an increase in GPP triplets. Since GPP triplets are among the most thermally stabilizing triplets (reviewed in Gaill et al. 1991), increases in proline content subsequently result in an increase in the thermal stability of the triple helix.

Bächinger and Davis (1991) have proposed the concept of sequence-specific relative stability of the collagen triple helix to quantify thermal stability. Sicot et al. (2000) have shown that *A. pompejana* has the highest score among the fibrillar collagens analyzed to date, with the highest number of stabilizing triplets and the highest GPP frequency. Moreover, all the stabilizing factors currently known are amplified in *A. pompejana* collagen, including the percentage of stabilizing triplets, proline content and frequency of hydroxyproline in the Y position of Gly-X-Y triplets.

Results obtained on cuticular collagen, observed in alvinellid and siboglinid worms (Gaill et al. 1994; Mann et al. 1993, 1996), indicate that the same type of protein family exhibits different strategies of thermal stability. One strategy with great evolutionary success is found in alvinellid interstitial collagen, relying on proline hydroxylation of residues in the Y position of Gly-X-Y triplets. Another that seems to be novel relies on the glycosylation of threonine residues in the same position of the triplet (Mann et al. 1996) and is found in the cuticular collagen of *R. pachyptila*. Both strategies highlight the importance of posttranslational modifications in thermal stability. Whether these differences are phylogenetically related or collagen type-specific remains unknown and await further investigation.

Another potential stabilizing property is the number and distribution of Gly-X-Y triplets in collagen α chains. In principle, more than 400 Gly-X-Y triplets are possible, but analysis of sequences from fibrillar and nonfibrillar collagens shows that only a limited set of triplets is found in significant numbers, and many are never observed (Sicot et al. 2000). The frequency distribution of triplets throughout the chain length of these collagens has not yet been analyzed precisely, but may be revealing.

Analyses of alvinellid and *R. pachyptila* collagen sequences has revealed the evolutionary mechanisms involved in their thermal stability (Sicot et al. 1997). Depending on which collagen domain is considered (i.e. the triple helix or C-propeptide) results in different phylogenetic trees. If the phylogenetic analysis is conducted solely on the C-propeptide, the obtained tree reflects the species classification published in Sicot et al. (1997), namely, collagen domains of *A. pompejana* and a coastal annelid, *Arenicola marina*, are grouped, whereas *R. pachyptila* has a distinct position. In contrast, *R. pachyptila* collagen groups with that of *A. marina* when the triple helical domain is analyzed, suggesting that different selective constraints have been applied on the two domains during evolution. This difference has been correlated with the habitat temperatures of the worms, since *A. marina* and *R. pachyptila* live in colder habitats than the thermophilic worm *A. pompejana*.

Collagen is a modular protein wherein the triple helical domain has a longer life than the C-propeptide (which is removed once the collagen is secreted). Therefore, the triple helix and the C-Pro domains would have been subjected to different selective pressures and would have evolved independently. Selective pressure would be stronger on the triple helical domain, which is exposed to high temperature in the *A. pompejana* ECM. With collagen domain evolution not being neutral, the conclusion of Sicot et al. (2000) leads to the hypothesis that the collagen triple helix region evolves in a different direction according to the habitat temperature of the animal, with the evolution of the C-propeptide remaining constant.

With a thermal stability of 45 °C, synthesis of *A. pompejana* collagen can be assumed to stop at higher temperatures, suggesting that its maximum body temperature would be less than 45 °C. However, collagen is not found in tissues in a molecular form, but rather in a supramolecular state. Once synthesized, collagen molecules assemble into fibrils, and such a polymeric organization has a thermal stability that can exceed typical collagen molecules by 20 °C (Gaill et al. 1988). *A. pompejana* fibrillar collagen assemblages may thus be stable up to 65 °C, which is consistent with experimental data (reviewed in Le Bris and Gaill 2007). If we assume that collagen cannot be synthesized by epidermal cells above 45 °C, the worm could still sustain such a high temperature without any damage to its collagen assemblages, meaning that it could survive thermal fluctuations between 40 and 60 °C. This would also support the idea of a thermal gradient along the animal's body length (Cary and Stein 1998), which is consistent with the observation that the cuticle of the oldest worms disappears on their posterior regions (Gaill et al. 1984). Data on *A. pompejana* prolylhydroxylase (Kaule et al. 1998), the enzyme responsible for proline hydroxylation, indicate that the worm not only survives in the highest temperature known for marine invertebrates but also has metabolic machinery adapted for working in low oxygen environments (Le Bris and Gaill 2007).

15.7 TEMPERATURE ADAPTATION AT A CELLULAR LEVEL: THE CASE OF DEVELOPING EMBRYOS

Adult organisms (e.g. *A. pompejana*) living on chimney walls display several thermal adaptations that allow them to survive in extreme and fluctuating conditions. Embryonic forms, however, do not exhibit collagen cuticles or tubes, and appear to be more sensitive than adults to thermal conditions. While dispersing, embryonic forms are exposed to mostly abyssal conditions (i.e. low and stable temperatures), and thus their lack of adaptation to high temperature might not be too surprising. Therefore, a dual life cycle including a nonthermotolerant phase (embryos) and a thermotolerant phase (adult) could have evolved as an adaptative strategy to colonize hot but patchy habitats.

We conducted developmental studies to examine the embryos' tolerance to physicochemical parameters to determine favorable developmental conditions *in situ*, which could be used to assess possible development areas and dispersal capabilities. *A. pompejana* is a thermotolerant species as an adult, but for embryos, two main developmental strategies were considered. First, embryos may develop in abyssal seawater with temperatures typically ~2 °C, or secondly, embryos may be thermotolerant and may develop on vent chimney walls within adult colonies (>20 °C). In the first scenario, dispersal could occur through transportation with marine currents over tens to hundreds of kilometers, allowing colonization of new distant sites. In the second scenario, embryos would develop without dispersal.

In vitro, embryos of *A. pompejana* exhibit low temperature tolerance, being unable to survive >20 °C. At optimal temperatures (~10 °C), cleavage rates are very slow with ~1 division every 24 h. By comparison, many polychaete species inhabiting coastal environments develop to larval stage after only 24 h (Fischer and Dorresteijn 2004). Additionally, the developmental process in *A. pompejana* was shown to be arrested at 2 °C, but a transient temperature increase could trigger development of arrested embryos (Pradillon et al. 2001).

The thermal tolerance window determined for early embryos of *A. pompejana* is therefore restricted to temperatures lower than those encountered in adult colonies, suggesting that embryos cannot develop near the vents (Fig. 15.3). However, recent studies

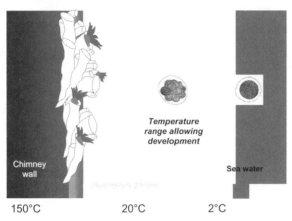

Figure 15.3 Diagram of thermal conditions *in situ* allowing *Alvinella pompejana* embryonic development. Above 20 °C, near adult tubes, development is prevented by high temperatures. At the abyssal temperature (2 °C), development is arrested. (See color insert.)

identify a diversity of habitats that are influenced by hydrothermal vents (Le Bris et al. 2005). Diffuse flow areas with mild temperatures would be compatible with *A. pompejana* embryonic development, and some embryos could be entrained with bottom currents where they would be exposed to very low temperatures that would arrest development. This kind of dormancy would then be stopped by temperature increase if the embryos arrived close to a vent. In principle, this mechanism could allow wide dispersal capabilities, though the potential duration of this dormancy is still in question.

To examine these hypotheses, incubators containing embryos of *A. pompejana* were deployed in different habitats of a single edifice along a gradient of hydrothermal influence (Pradillon et al. 2005). Only 10% of embryos incubated above an adult *Alvinella* colony survived after 5 days; among the surviving embryos, none of the surviving embryos developed. In contrast, 70% of embryos incubated in a milder area (a *R. pachyptila* clump, 1 m below the *Alvinella* colony and on the bare mineral seafloor) developed normally (Pradillon et al. 2005). These results support the idea that embryonic development outside of an adult *Alvinella* colony is possible, while embryos would not be viable in the colony itself. Temperature measurements close to the incubation device in the *Alvinella* colony indicated an average of 13 °C during the 5 days of the experiment, but frequent bursts above 20 °C were also recorded. Additionally, sulfide levels were up to 10-fold higher in the *Alvinella* colony than in the two other habitats. These experiments, however, did not allow us to decipher which parameters predominantly affect embryo survival and development.

Both *in vitro* and *in situ* studies indicate that *A. pompejana* embryos can disperse through abyssal seawater and develop when they find conditions around 10 °C. Development in the shallow part of the ocean can be excluded since embryos cannot develop at atmospheric pressure (Pradillon et al. 2005). Since pressure tolerance has not been precisely determined, the range of possible vertical movement is still a matter of speculation: embryos might be entrained far enough above the sea bottom by the hydrothermal plume to be further entrained by upper layers of currents, with possibly different regimes than currents running on the bottom (Mullineaux and France 1995). Recent studies show that relatively low temperatures may be found at the surface of adult colonies because the tubes built by *A. pompejana* may isolate the surface of the colony from the hot chimney wall (Le Bris et al. 2005). This could provide a suitable habitat for embryonic development and could lead to reconsideration of the assumption that early development is excluded from the adult environment.

15.8 BEHAVIORAL ADAPTATION TO A HIGH-TEMPERATURE ENVIRONMENT

Life on chimney walls does not always require specific adaptations of molecular or cellular processes to high temperature. In fact, most metazoans at vents live below room temperature (20 °C). The recent development of aquaria that can simulate both pressure and temperature of the vent environment (Shillito et al. 2001, 2004; Pradillon et al. 2004; Girguis and Lee 2006) now allows the question of temperature resistance of vent animals to be addressed by submitting them to various temperature regimes. Using a pressurized incubator equipped with video facilities (IPOCAMP™), Shillito et al. (2001) found that the "carterpillar worm" *Hesiolyra bergi*, which is known to forage around and in tubes of the thermophilic *Alvinella* sp., become hyperactive at temperatures above 35 °C. In the 41–46 °C temperature interval, animals lose coordination and finally die. Transmission electronic microscopy analysis of the worm cuticle revealed important damages after exposure to 39 °C. These experiments suggest that the critical thermal maximum (CTMax), defined as a "thermal trap," i.e. the temperature at which the worm is no longer capable of proper locomotion, is between 41 and 46 °C, and 35 °C represents the thermal threshold that triggers the worms' escape response. *In situ*, behavioral response (i.e. escape) rather than biochemical adaptations may allow *H. bergi* survival in *Alvinella* spp. colonies.

Although alvinellids have been suggested to thrive at temperatures up to 80 °C (Cary et al. 1998; Le Bris et al. 2005), the rapid mixing of vent fluids and cold seawater produces a dynamic thermal regime that is difficult to characterize. Therefore acertaining alvinellid thermal tolerance via *in situ* measurements is difficult. Recently, however, the thermal preferences of two alvinellids from Juan de Fuca vent sites were directly assessed by experimentally producing a thermal gradient in a pressure vessel (Girguis and Lee 2006). *P. sulfincola*, which occupies an analogous niche with *A. pompejana*, i.e. the hottest part of the habitat, was shown to prefer temperatures ~50 °C, whereas *Paralvinella palmiformis* consistently avoided temperatures >35 °C (Girguis and Lee 2006).

Alvinella pompejana has, until recently, been a delicate species that does not easily survive transport to shipboard laboratories and recompression in pressure vessels. However, during the Phare cruise in 2002, ~20 *A. pompejana* were successfully recompressed and two individuals remained fully active for 20 h after recompression, at which time the experiment was stopped (Shillito et al. 2004). Although these experiments need further validation to clarify the basis of survival, *A. pompejana* is now established as a model species for physiological analyses of thermal resistance. Additionally, the use of isobaric collection devices may improve the recovery of live *A. pompejana* by shortening or eliminating the traumatic decompression process.

15.9 FUTURE DEVELOPMENT OF THERMAL ADAPTATION STUDIES

With the development of pressure systems allowing the maintenance of vent organisms in good physiological condition for several hours or days, assessing the physiological response of organisms exposed to controlled stress situations is now possible. At the cellular level, the induction of a highly conserved set of proteins called heat shock proteins (HSPs) is one of the best characterized heat shock responses. More generally, HSPs are part of cellular mechanisms that protect endogenous proteins and membranes from environmental challenges. Under stress conditions, but also in unstressed cells, HSPs prevent aggregation and assist in protein folding. In these two situations, different sets of HSP are involved, namely, constitutively expressed and stress-inducible HSPs. In vent organisms that are constantly exposed to fluctuations in their thermal environment, specific HSP expression might be expected. For example, studies on vent shrimps have shown that exposure to temperatures >25 °C triggers an HSP response (Ravaux et al. 2003, 2007). HSP synthesis is now being used as a marker to identify the thermal resistance of alvinellids and vent organisms.

15.10 PERSPECTIVES

Vent annelids may be viewed as representative taxa for understanding adaptation mechanisms and survival strategies in their extreme environment. They exhibit a wide range of characteristics, including the originality of molecular mechanisms, specificity of some cell metabolism, various physiological responses, links between different stress responses and the regulation of genomic activity. Using such models at different stages of their life cycle will reveal considerable knowledge about the mechanisms and strategies developed to cope with extreme environments.

Molecular mechanisms involved in the increased thermostability of vent worm collagens have been identified and reveal the role of environmental stress in the evolution of the molecule. Not only is *A. pompejana* adapted to an extremely high-temperature environment via molecular mechanisms, but adaptation also occurs at the scale of the whole organism through a dual strategy during its life cycle. Embryos do not tolerate temperatures above 14 °C, whereas adults are able to cope with temperatures exceeding 45 °C. *A. pompejana* thus appears to have developed adaptative strategies at a variety of biological levels.

Environmental stresses have played, and still play, an important role in the evolution of biological systems. Genomic approaches are now being developed in *A. pompejana* that may reveal novel molecules, metabolic pathways that stabilize cellular macromolecules, novel protection systems and nonconventional pathways. Genome data on models from extreme environments such as the thermophilic *A. pompejana* would therefore provide insight not only in general evolutionary processes such as the evolution of body plans, but also in genome plasticity (e.g. correlations between levels of environmental stress and the proportion of active transposable elements), and in deciphering molecular and structural mechanisms of protein thermostability. Mechanisms allowing worms to cope with other sources of stress such as pressure or high sulfide concentration at different stages of their life cycle might also be addressed through the use microarray technology. Express sequence tag (EST) data will soon allow the development of microarray approaches to characterize the global stress response pathways to environmental injuries.

REFERENCES

ALAYSE-DANET, A.M., DESBRUYÈRES, D., GAILL, F. 1987. The possible nutritional or detoxification role of the epibiotic bacteria of Alvinellid polychaetes: review of current data. Symbiosis 4, 51–62.

AUERBACH, G., GAILL, F., JAENICKE, R., SCHULTHESS, T., TIMPL, R., ENGEL, J. 1995. Pressure dependence of collagen melting. Matrix Biol. 14, 589–592.

BÄCHINGER, H.P., DAVIS, J.M. 1991. Sequence specific thermal stability of the collagen triple helix. Int. J. Biol. Macromol. 13, 331–338.

BRIGHT, M., GIERE, O. 2005. Microbial symbiosis in Annelida. Symbiosis 38, 1–45.

CAMPBELL, B.J., STEIN, J.L., CARY, S.C. 2003. Evidence of chemolithoautotrophy in the bacterial community associated with *Alvinella pompejana*, a hydrothermal vent polychaete. Appl. Environ. Microbiol. 69, 5070–5078.

CARY, C.S., STEIN, J.L. 1998. Spanning the thermal limits: an extreme eurythermal symbiosis. Cah. Biol. Mar. 39, 275–278.

CARY, C.S., COTTRELL, M.T., STEIN, J.L., CAMACHO, F., DESBRUYÈRES, D. 1997. Molecular identification and localization of filamentous symbiotic bacteria associated with the hydrothermal vent annelid *Alvinella pompejana*. Appl. Environ. Microbiol. 63, 1124–1130.

CARY, C.S., SHANK, T., STEIN, J.L. 1998. Worms bask in extreme temperatures. Nature 391, 545–546.

CAVANAUGH, C.M. 1983. Symbiotic chemoautotrophic bacteria in marine invertebrates from sulphide-rich habitats. Nature 302, 58–61.

CHEVALDONNÉ, P., DESBRUYÈRES, D., CHILDRESS, J.J. 1992. Some like it hot… and some even hotter. Nature 359, 593–594.

CORLISS, J.B., DYMOND, J., GORDON, L.I., EDMOND, J.M., VON HERZEN, R.P., BALLARD, R.D., GREEN, K., WILLIAMS, D., BAINBRIDGE, A., CRANE, K., VON ANDEL, T.H. 1979. Submarine thermal springs on the Galapagos Rift. Science 203, 1073–1083.

DESBRUYÈRES, D., GAILL, F., LAUBIER, L., FOUQUET, Y. 1985. Polychaetous annelids from hydrothermal vent ecosystems: an ecological overview. Biol. Bull. 6, 103–116.

DI MEO-SAVOIE, C.A., LUTHER, G.W. III, CARY, S.C. 2004. Physico-chemical characterization of the microhabitat of the epibionts associated with *Alvinella pompejana*, a hydrothermal vent annelid. Geochim. Cosmochim. Acta 68, 2055–2066.

FELBECK, H., CHILDRESS, J.J., SOMERO, G.N. 1981. Calvin-Benson cycle and sulphide oxidation enzymes in animals from sulphide-rich habitats. Nature 293, 291–293.

FISCHER, A., DORRESTEIJN, A. 2004. The polychaete *Platynereis dumerilii* (Annelida): a laboratory animal with spiralian cleavage, lifelong segment proliferation and a mixed benthic/pelagic life cycle. Bioessays 26, 314–325.

GAILL, F. 1993. Aspects of life development at deep sea hydrothermal vents. FASEB J. 7, 558–565.

GAILL, F., BOULIGAND, Y. 1987. Alternating positive and negative twist of polymers in an invertebrate integument. Mol. Liq. Cryst. 153, 31–41.

GAILL, F., HUNT, S. 1986. Tubes of deep sea hydrothermal vent worms *Riftia pachyptila* (Vestimentifera) and *Alvinella pompejana* (Annelida). Mar. Ecol. Prog. Ser. 34, 267–274.

GAILL, F., HUNT, S. 1991. The biology of annelid worms from high temperature hydrothermal vent regions. Reviews in Aquatic Sciences 4, 107–137.

GAILL, F., DESBRUYÈRES, D., PRIEUR, D., GOURRET, J. 1984. Mise en évidence de communautés bactériennes épibiontes du "Ver de Pompéï" (*Alvinella pompejana*). C. R. Acad. Sci. III 298, 553–558.

GAILL, F., DESBRUYÈRES, D., LAUBIER, L. 1988. Relationships between the "Pompeii worms" and their epibiotic bacteria. Oceanol. Acta n spécial. 8, 147–154.

GAILL, F., WIEDEMANN, H., MANN, K., KÜHN, K., TIMPL, R., ENGEL, J. 1991. Molecular characterization of cuticular and interstitial collagens from worms collected at deep-sea hydrothermal vents. J. Mol. Biol. 221, 157–163.

GAILL, F., HAMRAOUI, L., SICOT, F-X., TIMPL, R. 1994. Immunological properties and tissue localization of two different collagen types in annelid and vestimentifera species. Eur. J. Cell Biol. 65, 392–401.

GAILL, F., MANN, K., WIEDEMANN, H., ENGEL, J., TIMPL, R. 1995. Structural comparison of cuticle and interstitial collagens from annelids living in shallow sea-water and at deep-sea hydrothermal vents. J. Mol. Biol. 246, 284–294.

GAILL, F., SHILLITO, B., MÉNARD, F., GOFFINET, G., CHILDRESS, J.J. 1997. The rate and process of tube production by the deep-sea hydrothermal vent tubeworm *Riftia pachyptila*. Mar. Ecol. Prog. Ser. 148, 135–143.

GIRGUIS, P.R., LEE, R.W. 2006. Thermal preference and tolerance of Alvinellids. Science 312, 231.

GULIK, A., GAILL, F. 1988. Ultrastructural characteristics of *Alvinella pompejana* associated bacteria. Oceanol. Acta No. Sp. 8, 161–164.

KARL, D.M. 1995. Ecology of free-living, hydrothermal vent microbial communities. In *The Microbiology of Deep-Sea Hydrothermal Vents*, edited by D.M. Karl. Boca Raton, FL: CRC Press, 35–124.

KAULE, G., TIMPL, R., GAILL, F., GUNZLER, V. 1998. Prolyl hydroxylase activity in tissue homogenates of annelids from deep sea hydrothermal vents. Matrix Biol. 17, 205–212.

LE BRIS, N., GAILL, F. 2007. How does the annelid *Alvinella pompejana* deal with an extreme hydrothermal environment? Rev. Environ. Sci. Biotechnol. 6, 197–221.

LE BRIS, N., SARRADIN, P.M., CAPRAIS, J.C. 2003. Contrasted sulphide chemistries in the environment of 13°N EPR vent fauna. Deep Sea Res. Part I Oceanogr. Res. Pap. 50, 737–747.

LE BRIS, N., ZBINDEN, M., GAILL, F. 2005. Processes controlling the physico-chemical micro-environments associated with Pompeii worms. Deep Sea Res. Part I Oceanogr. Res. Pap. 52, 1071–1083.

LONSDALE, P.F. 1977. Clustering of suspension-feeding macrobenthos near abyssal hydrothermal vents at oceanic spreading centers. Deep Sea Res. 24, 857–863.

LUTZ, R.A., JABLONSKI, D., TURNER, R.D. 1984. Larval development and dispersal at deep-sea hydrothermal vents. Science 226, 1451–1453.

MANN, K., GAILL, F., TIMPL, R. 1993. Amino-acid sequence and cell-adhesion activity of a fibril-forming collagen from the tube worm *Riftia pachyptila* living at deep sea hydrothermal vents. Eur. J. Biochem. 210, 839–847.

MANN, K., MECHLING, D., BACHINGER, H., ECKERSON, C., GAILL, F., TIMPL, R. 1996. Glycosylated threonine but not 4-hydroxyproline dominates the triple helix stabilizing positions in the sequence of a hydrothermal vent worm cuticle collagen. J. Mol. Biol. 261, 255–266.

MARKERT, S., ARNDT, C., FELBECK, H., BECHER, D., SIEVERT, S.M., HÜGLER, M., ALBRECHT, D., ROBIDART, J., BENCH, S., FELDMAN, R.A., HECKER, M., SCHWEDER, T. 2007. Physiological proteomics of the uncultured endosymbiont of *Riftia pachyptila*. Science 315, 247–250.

MULLINEAUX, L.S., FRANCE, S.C. 1995. Dispersal mechanisms of deep-sea hydrothermal vent fauna. In *Seafloor Hydrothermal Systems: Physical, Chemical, Biological and Geological Interactions. Geological Monograph 91*, edited by S.E. Humphris, R.A. Zierenberg, L.S. Mullineaux and R.E. Thomson. Washington, DC: American Geophysical Union, pp. 408–424.

PRADILLON, F., SHILLITO, B., YOUNG, C.M., GAILL, F. 2001. Developmental arrest in vent worm embryos. Nature 413, 698–699.

PRADILLON, F., SHILLITO, B., CHERVIN, J-C., HAMEL, G., GAILL, F. 2004. Pressure vessels for in vitro studies of deep-sea fauna. High Pressure Res. 24, 237–246.

PRADILLON, F., LE BRIS, N., SHILLITO, B., YOUNG, C.M., GAILL, F. 2005. Influence of environmental conditions on early development of the hydrothermal vent polychaete *Alvinella pompejana*. J. Exp. Biol. 208, 1551–1561.

PRIEUR, D., CHAMROUX, S., DURAND, P., ERAUSO, G., FERA, P., JEANTHON, C., LE BORGNE, L., MÉVEL, G., VINCENT, P. 1990. Metabolic diversity in epibiotic microflora associated with the Pompeii worms *Alvinella pompejana* and *A. caudata* (Polychaete: Annelida) from deep-sea hydrothermal vents. Mar. Biol. 106, 361–367.

RAVAUX, J., GAILL, F., LE BRIS, N., SARRADIN, P-M., SHILLITO, B. 2003. Heat-shock response and temperature resistance in the deep-sea vent shrimp *Rimicaris exoculata*. J. Exp. Biol. 206, 2345–2354.

RAVAUX, J., TOULLEC, J-Y., LÉGER, N., LOPEZ, P., GAILL, F., SHILLITO, B. 2007. First hsp70 from two hydrothermal vent shrimps, Mirocaris fortunata and Rimicaris exoculata: characterization and sequence analysis. Gene 386, 162–172.

SHILLITO, B., JOLLIVET, D., SARRADIN, P-M., RODIER, P., LALLIER, F.H., DESBRUYÈRES, D., GAILL, F. 2001. Temperature resistance of *Hesiolyra bergi*, a polychaetous annelid living on deep-sea vent smoker walls. Mar. Ecol. Prog. Ser. 216, 141–149.

SHILLITO, B., LE BRIS, N., GAILL, F., REES, J-F., ZAL, F. 2004. First access to live Alvinellas. High Pres. Res. 24, 169–172.

SICOT, F-X., EXPOSITO, J-Y., MASSELOT, M., GARRONE, R., DEUTSCH, J., GAILL, F. 1997. Cloning of an annelid fibrillar collagen gene and phylogenetic analysis of vertebrate and invertebrate collagens in nucleic acids, protein synthesis and molecular genetics. Eur. J. Biochem. 590, 1–9.

SICOT, F-X., MESNAGE, M., MASSELOT, M., EXPOSITO, J., GARRONE, R., DEUTSCH, J., GAILL, F. 2000. Molecular adaptation to an extreme environment: origin of the thermal stability of the Pompeii worm collagen. J. Mol. Biol. 302, 811–820.

VAN DOVER, C.L., LUTZ, R.A. 2004. Experimental ecology at deep-sea hydrothermal vents: a perspective. J. Exp. Mar. Biol. Ecol. 300, 273–307.

VON DAMM, K., OOSTING, S., KOZLOWSKI, R., BUTTERMORE, L., COLODNER, D., EDMONDS, H., EDMOND, J., GREBMEIER, J. 1995. Evolution of East Pacific Rise hydrothermal vent fluids following a volcanic eruption. Nature 375, 47–50.

WOLF, T. 2005. Composition and endemism of the deep-sea hydrothermal vent fauna. Cah. Biol. Mar. 46, 97–104.

ZBINDEN, M., MARTINEZ, I., GUYOT, F., CAMBON-BONAVITA, M.A., GAILL, F. 2001. Zinc-iron sulphide mineralization in tubes of hydrothermal vent worms. Eur. J. Mineral 13, 653–658.

ZBINDEN, M., LE BRIS, N., COMPERE, P., MARTINEZ, I., GUYOT, F., GAILL, F. 2003. Mineralogical gradients induced by Alvinellids at hydrothermal vents (9°N, EPR). Deep Sea Res. Part I Oceanogr. Res. Pap. 50, 269–328.

Chapter 16

Glacier Ice Worms

Paula L. Hartzell* and Daniel H. Shain†

*Mountain Research Institute, Kealakekua, HI
†Biology Department, Rutgers, The State University of New Jersey, Camden, NJ

O ice worms of Northern mythology,
Mysterious as the Borealis,
Dark secrets of your biology
Thrill me to each vena digitalis.

Erupting at sunset from the glacial plains,
Diminutive strands of algid lava,
You return at dawn through the land's iced veins
As do blood cells through a vena cava.

Mesenchytraeus solifugus threads,
Streamlined black beauties, Nature's selection,
You cruise the ice on invisible sleds,
In cool hyperborean perfection.

Alas! Winter depletes my energy;
I long for your levels of ATP.

—*Sonia Krutzke*

16.1 INTRODUCTION

Glacier ice worms (Enchytraeidae: *Mesenchytraeus solifugus*, ssp. *Mesenchytraeus solifugus rainierensis* and *Sinenchytraeus glacialis*) are the only known annelids—and among only a few metazoans—adapted to life in glacial ice and snow. Other members of their family, Enchytraeidae, are abundant in Arctic regions, and many *Mesenchytraeus* species (e.g. *Mesenchytraeus gelidus, Mesenchytraeus altus, Mesenchytraeus hydrius*) endure permafrost and temporary snow (Welch 1916, 1917a, b, 1919; Sømme and Birkemoe 1997), but these latter species all require that a portion of their life cycle occurs at seasonal (i.e. ice-free) conditions. Thus, glacier ice worms appear to be the only annelid species to take the evolutionary step(s) necessary to reside permanently in glacier ice.

North American ice worms are geographically restricted to temperate glaciers of the Pacific Northwest (PNW), from Alaska to Oregon (Tynen 1970; Hartzell et al. 2005)

Annelids in Modern Biology, Edited by Daniel H. Shain
Copyright © 2009 John Wiley & Sons, Inc.

(Fig. 16.1). With the exception of an ice worm population reported on the Tibetan Plateau (i.e. *S. glacialis*; Liang et al. 1979), other credible ice worm reports are nonexistent despite suitable habitat and extensive scientific explorations of suitable temperate habitats throughout the world (e.g. Patagonia, Greenland). Temperate glaciers are distinguished from harsher environments (e.g. inland glaciers, polar ice sheets) by their internal temperatures, which remain near 0 °C throughout the year; thus, organisms inhabiting temperate ice do not have to cope with temperature extremes or dehydration due to lack of liquid water. They are exposed, however, to high levels of ultraviolet light, and most organisms inhabiting these environments have developed protective pigmentation.

The unusual habitat of glacier ice worms makes them a species of evolutionary and ecological interest because of their confinement to fragmented, shrinking glaciers, with a limited capacity for dispersal. Their adaptations to life in ice make them a species of physiological and cryogenic interest, with possible industrial applications (e.g. biomedical, food storage). Because they are limited to glacier environments, these worms are threatened by habitat reduction in our changing global climate, particularly in the southern portion of their range.

Glacier ice worms present an opportunity to examine the history and dynamics of isolated glacier populations. Unlike most glacial taxa, they are easily collected from large populations, often occurring at densities of several hundred individuals per square meter on glacial surfaces. Based on the lack of gene flow between populations, they do not have an efficient mechanism for traveling between glaciers, and thus appear to rely on active dispersal (i.e. crawling) to expand their geographic range (Hartzell et al. 2005). Restriction

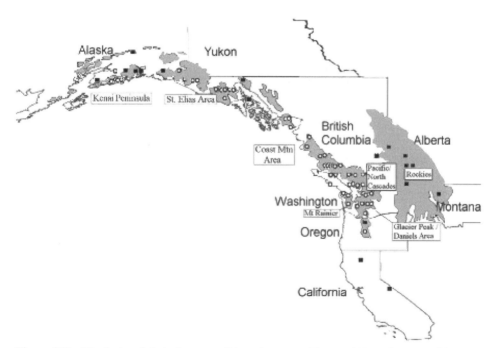

Figure 16.1 Distribution of glacier ice worms (*Mesenchytraeus solifugus* and *Mesenchytraeus solifugus rainierensis*) in North America. Open circles indicate sites where ice worms are confirmed; solid squares indicate sites visited, but where no ice worms were found. Shaded areas identify regions with current glaciation.

Source: Reproduced from Hartzell et al. (2005), with permission from NRC Research Press.

to glaciers makes this taxa a classic "island" population, where processes of hierarchical fragmentation can be observed in a relatively easily documented temporal framework, through glaciological, geomorphological, pollen and tree-ring dating. The evolutionary history and population dynamics of ice worms may be useful in postulating the effects of future climate change on glacier communities, as well as providing a model for the effects of fragmentation on populations.

16.2 NATURAL HISTORY

Glacier ice worms complete their life cycle in and on glaciers, at relatively constant temperatures (~0 °C on temperate glaciers). They are found on various glacier substrates including snow, firn (snow in transition into ice), glacial pools, streams and hard glacier ice. Ice worm populations have been confirmed on >100 independent glaciers in the PNW region of North America (Fig. 16.1; Hartzell et al. 2005), ranging from south-central Alaska to Oregon (Russell 1892; Moore 1899; Emery 1898a–c, 1900; Eisen 1904; Welch 1916; Gudger 1923; Odell 1949; Tynen 1970; Shain et al. 2001; Hartzell 2005; Hartzell et al. 2005); they are also documented on Glaceir Jaoguo (4,600 m) of the Tibetan Plateau (Liang et al. 1979). On Byron Glacier, AK, ice worms were distributed along eastern ridge avalanche cones, and totaled more than 20 million in 2000 (Fig. 16.2; Shain et al. 2001), though that number is likely considerably lower now. Glaciers in central and northern

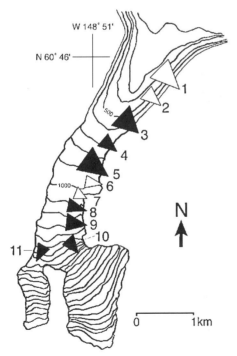

Figure 16.2 Distribution of glacier ice worms on Byron Glacier, Alaska. The lower valley of Byron Glacier (<2,000 ft) contains 11 distinct avalanche cones denoted by triangles. Cones 3, 4, 5, 8, 9, 10 and 11 contained ice worm populations (filled triangles); cones 1, 2, 6 and 7 (open triangles) lacked ice worms. In total, >20 million ice worms inhabit Byron Glacier.

Source: Reproduced from Shain et al. (2001), with permission from NRC Research Press.

British Columbia, the Yukon and noncoastal glaciers in Alaska including those in Denali National Park are reported to support glacier ice worm populations, but none have been confirmed at these locations.

Glacier ice worms and springtails (i.e. Collembola, or "snowfleas") comprise the bulk of macroinvertebrate populations on PNW glaciers, constituting ~29 and ~69% of estimated abundance, respectively; other macroinvertebrates (e.g. Diptera) comprise the remaining ~2% (Hartzell 2005). Because of their relatively large size, ice worms typically constitute >90% of the total macroinvertebrate biomass on glaciers where they are present—Collembola dominate macroinvertebrate biomass on glaciers where ice worms are absent.

Ice worm densities vary by time of day, cloud cover, elevation and latitude, but do not appear affected by wind speed, air temperature (during summer months) or barometric pressure. Densities are typically highest in water pools and snow, moderate in ice and lowest in glacial streams. Mean densities vary diurnally in glacial snowpack, which is exposed to high levels of ultraviolet light. Ice worms first appear each afternoon in the shade created by snow cups or runnels (i.e. a rivulet or water course, although water may not be visibly present) as the sun lowers, and continue to appear along the forefront of shade, until they may cover an entire accumulation zone by evening. Hartzell (2005) observed a peak density of ~2,300 ice worms/m^2 in the shade of a runnel one late afternoon, while only 2 ice worms/m^2 were observed in the sunny portion of the same quadrant. Two hours later, densities greater than 20 individuals/m^2 were observed at every station along the transect. Similar observations have been reported by Goodman (1970) and Shain et al. (2001). Peak ice worm densities on a local scale thus depends on a glacier's topography and surrounding landscape.

Goodman (1970) documented an inverse relationship between peak ice worm densities at the surface of Casement Glacier, AK versus net radiation; indeed, the species name *solifugus* (Latin for sun avoiding) reflects their aversion to sunlight. Ice worm densities, however, do not vary diurnally in other habitats, including glacial pools, streams and shaded crevasses where they can be found even at midday, emphasizing the importance of their light avoidance behavior (Goodman 1970; Shain et al. 2001). Ice worm densities vary directly with percent dirt cover (i.e. a visual estimate of ground covered by dirt; $r^2 = 0.25$, $n = 19$; Hartzell 2005), which may provide protection from light, temperature extremes and/or predation from birds. Curiously, ice worms appear to be attracted to heat (Shain et al. 2001), which may help to explain their putative overwintering behavior; specifically, ice worms likely burrow within insulating snowfall (possibly at the nutrient-rich boundary defining that year's algal growth, and also a logical place and time to deposit eggs) where temperatures are relatively warm in comparison to the surface. Note that ice worms have not been observed on glacial surfaces during winter months, suggesting that their typical diurnal movements are halted and they may enter a state of reduced metabolism or diapause.

In North America, ice worms are more abundant in the southern half of their geographic range (i.e. British Columbia, Washington; Hartzell et al. 2005). They have not been observed east of the Pacific/Cascade Range, likely due to climatic severity on inland glaciers. Ice worms are typically observed during the snowmelt period (i.e. ablation; May to September in the south, July and August in Alaska). Ice worms have not been observed on glacier surfaces during snow accumulation months, and likely require some time before appearing on the surface after the ablation period begins. Due to the PNW regional trend of warmer, wetter winters, their foraging season (and body length) generally increases in length as a function of more southerly latitudes.

Appearance of glacier ice worms during the ablation period corresponds to the algae growing season on PNW glaciers. Small-sized, red algae (e.g. *Chlamydomonas nivalis*) likely play an important role in their diet, although ice worm ingestion appears to be non-specific (Goodman 1970; Hartzell 2005). The abundance of relatively small microbes (10–40 µm) is reduced when ice worms are abundant, apparently the result of ice worm predation. The mean width of an adult ice worm mouth is 30–36 µm, making this microbial size class an appropriate food source. Note that specific microbial types do not vary in abundance in the presence of glacier ice worms (Hartzell 2005), as expected by nonselective grazing. Ice worm gut contents reveal "just about everything else that may be found in the ice or snow" (Goodman 1970), including windblown pollen and inorganic debris (Goodman 1970; Goodman and Parrish 1971; Hartzell 2005).

Tynen (1970) and Hartzell et al. (2005) searched but found no evidence of glacier ice worms in the Rocky Mountains of Canada or the USA. Shain et al. (2001) suggest this void is due to a more severe climate regime in the Rockies, while Tynen (1970) explained this as being part of a different ice sheet during the Pleistocene. Because ice worms appear capable of dispersal over some distance, at least in rare events, and some ice worms appear cold hardy (i.e. putative ice worms in Denali National Park, AK), an ecological barrier of some sort appears to keep them from colonizing glaciers in the Rocky Mountain region.

16.3 CLASSIFICATION AND PHYLOGENETIC RELATIONSHIPS

Glacier ice worms are members of superclass Clitellata, and are in the family Enchytraeidae. Most enchytraeids are small (<0.5 cm), white or translucent, and include marine, terrestrial and freshwater species. The most described *Mesenchytraeus* species are riverine, and many are found along the western coast of North America. Glacier ice worms (*Mesenchytraeus solifugus*) and Yosemite snow worms (*M. gelidus*) (Welch 1916), however, are relatively large (0.5–2.0 cm in length) and are highly pigmented.

The species *Melanenchytraeus solifugus* was first reported by Wright (1887) and described by Emery (1898a–c, 1900) based on specimens collected from Malaspina Glacier, Mount St. Elias, AK. Aside from their dark pigmentation, ice worms resemble other members of their family (Emery 1898a–c, 1900; Moore 1899; Welch 1916, 1917a, b; Goodman and Parrish 1971). Moore (1899) corrected Emery's designation from *Melanenchytraeus* to the preexisting *Mesenchytraeus* genus and also identified a second species of glacier ice worm that he dubbed *M. niveus*. Welch (1917a, b) later argued that Moore was examining immature *M. solifugus* specimens and not a separate species, and this opinion has remained unchallenged. Upon examining specimens from Mount Rainier, WA and comparing them to specimens from the St. Elias, AK region, Welch (1916) proposed the variant *M. solifugus rainierensis* based on morphological criteria (e.g. smaller number of setae per bundle, two instead of three diverticula on the spermathecae, enclosure of the sperm sacs and ducts by the ovisac).

Holotype locations for *Mesenchytraeus solifugus* and *M. solifugus rainierensis* are unknown, but paratypes are stored at the Smithsonian National Museum of Natural History, Washington, DC [accession numbers for *M. solifugus* (Emery): 16366, 21111-21113, 32917; *M. solifugus rainierensis* (Welch): 16822, 25496]. Paratypes are also stored with Mount Rainier National Park (catalog numbers MORA 15404-15406). More recently, Goodman and Parrish (1971) described and illustrated the cuticle, epithelium, intercellular junctions and basement membrane of *M. solifugus*, while Shain et al. (2000) provided further details on their head pore morphology and setae.

The North American Yosemite snow worm, *M. gelidus*, appears most closely related to glacier ice worms. *M. gelidus* has been reported from Mt. Rainier National Park in Washington (Welch 1916), and Yosemite National Park in California (Aitchison 1977; K. Coates, pers. comm.), and are thus partially allopatric with ice worms. The divergence of glacier ice worms and snow worms was preceded by an earlier split from *Mesenchytraeus armatus* and *Mesenchytraeus flavus*, both European species, supporting the monophyly of the genus *Mesenchytraeus* based on 28S rDNA comparisons (Hartzell et al. 2005). While glacier ice worms clearly form two distinct clades (northern and southern) based on multilocus phylogenetic evidence, they are monophyletic with regard to other described mesenchytraeids (Hartzell 2005; Hartzell et al. 2005).

16.4 ORIGINS

Tynen (1970) hypothesized that all extant North American glacier ice worm populations are fragments of a pan-Cordilleran ancestral population. The Cordilleran ice sheet covered a large part of the PNW, west of the Continental Divide, at its maximum ~40,000 years ago (Clague 1976, 1991; Waitt and Thorson 1983; Sirkin and Tuthill 1987; Lian and Hickin 1993; Ager 1994; Wiles and Calkin 1994; Riedel 2003). Unlike the Laurentide ice sheet, however, the Cordilleran is believed to have melted over a relatively short time period (<4,000 years before present), leaving fragmented glaciers behind that reflect the contemporary landscape. An ancestral Cordilleran ice worm population seems plausible since most current glaciers supporting ice worm populations were at one time part of the Cordilleran ice sheet, and a mid-Pleistocene separation of Alaskan from southern populations is supported by genetic evidence (Hartzell 2005). Thus, a putative panmictic ancestral population may have existed based on a distribution of genetic polymorphisms, with extant populations characterized by increasingly fragmented, genetically isolated units (Hartzell 2005).

However, the most extreme putative northern and southern populations—Denali, AK in the north and the Sisters Range (including Mt. Jefferson) in central Oregon—were not part of the Cordilleran ice sheet during the Pleistocene. Rather, the northern populations may have been part of a larger ice sheet during the Kansan Glaciation (~410–380,000 years before present), suggesting an active means of dispersal; alternatively, ice worms may disperse passively at least on rare occasions (e.g. via birds, see below). Genetic evidence based on the COI locus suggests a tentative divergence from *Mesenchytraeus pedatus* ~350,000 years before present (Hartzell 2005), which in principle could corroborate a Kansan origin. Note, however, that 28S rDNA and CO1 sequence comparisons suggest a more ancient North American ice worm origin (Hartzell et al. 2005). The origins of Tibetan ice worm populations remain unexplored.

The presence of ice worms in the Sisters Range in central Oregon is inconsistent with the primarily active mode of dispersal apparent throughout the remainder of their geographic range (Hartzell et al. 2005). This single example of passive ice worm dispersal may have been facilitated by a bird, such as the gray-crowned rosy finch (*Leucosticte tephrocotis*), which is frequently observed feeding near glaciers in central Oregon during the spring and summer. These finches are normally short-distance migrants, so in addition to the unlikely scenario of successfully transporting live glacier ice worms (which begin to autolyze at temperatures >5 °C; Edwards 1986) or egg-filled cocoons, they would also be unlikely to transport them any significant distance; however, this mechanism need only to have occurred once, from roughly Mt. Rainier or Mt. St. Helens to central Oregon, a distance of less than 500 km. Clearly an unusual bird-borne dispersal to the Sisters Range is more parsimonious than an active trek across nonglaciated terrain (e.g. from the glaciers of Mt. Adams and Mt. Hood). This does not preclude the possibility of putative Denali ice worm populations deriving from a Kansan Cordilleran population, but Denali populations have not yet been subject to scrutiny.

The ancestral glacier ice worm population is divided early into two distinct geographic clades, one northern and one southern (Hartzell et al. 2005). Two scenarios could explain this split. In the first, a common ancestor was derived from a snow worm ancestor, perhaps somewhere in British Columbia. This large, monophyletic population may not have covered the Cordilleran ice sheet, but does appear to have given rise to populations that occupied Alaska, British Columbia and Washington State (presumably by active dispersal). As current Alaskan glacial ice worm populations appear as discrete mobile colonies (Goodman 1970), this pattern may have characterized the habits of the common ancestor as well. In the second scenario, the northern and southern clades may have arisen from a shared, widespread snow worm ancestor, developing glacier adaptations independently (i.e. convergent evolution). The close proximity of *M. gelidus* and *M. pedatus* on phylogenetic trees (Hartzell et al. 2005) and the widespread distribution of snow worms in western North America support a less specialized ancestral state. While the second hypothesis remains a possibility, we judge that it is less likely since the shared biochemical, physiological and reproductive traits of both glacier ice worm clades favor a single, ice-adapted ancestor.

16.5 CLADES

Morphological, biogeographic and genetic evidence supports the division of the distinct northern and southern glacier ice worm clades in North America (Fig. 16.3). Comparing specimens from Mount Rainier, WA with descriptions of St. Elias, AK region specimens, Eisen (1904), Emery (1898a–c, 1900) Moore (1899) and Welch (1916) proposed the subspecies *M. solifugus rainierensis*. According to Welch (1916), differences between the two groups were apparent in the reproductive organs, specifically in the number of diverticula on spermathecae, linearity of the speriducal funnel and the enclosure of the sperm sacs by the oviduct. The two morphotypes occur as geographically discrete and isolated units: one along the southern and southeastern coasts of Alaska (i.e. *M. solifugus*),

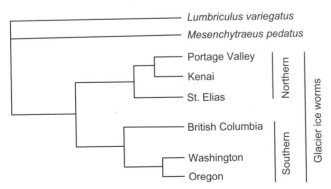

Figure 16.3 Regional phylogeny of glacier ice worms based on 28S ribosomal and cytochrome c oxidase subunit 1 (CO1) DNA sequences. Northern and southern clades are indicated.
Source: Based on Hartzell et al. (2005).

and the other including the Cascade and Olympic Ranges of central and southern British Columbia, Washington and Oregon. Genetic divergence between the two clades is >10% at the COI locus (Hartzell 2005; Hartzell et al. 2005), comparable with that found between other annelid species (e.g. Bely and Wray 2004). Morphological differences between ice worms in each respective clade include their mean and maximum size (i.e. length, width; Hartzell 2005), which may be genetically based, but may also reflect differences in the length of the foraging season (longer in southern latitudes) and hence ice worm growth period.

Because a traditional taxonomist has not reexamined specimens from these two groups to provide a more complete morphological comparison required for reclassification, we will continue to refer to these two groups as clades. Based on known morphological differences, geographic and genetic isolation, however, the northern and southern clades likely represent two different species. Additional sampling (e.g. Denali or the Wrangell-St. Elias regions of Alaska, or the central interior of British Columbia) may result in the identification of new glacier ice worm species.

16.5.1 Northern Clade

Glacier ice worms inhabiting regions north of British Columbia often appear in isolated colonies on the same glacier (e.g. Byron Glacier, AK; see Fig. 16.2), and usually appear on the glacial surface only during the months of June–August. Ice worm distribution in this region is disjunct: they are found on many glaciers in the Kenai Peninsula and along Prince William Sound, AK, but not ~50 km inland on the Matanuska Glacier. Further north, accounts suggest that they appear in patches within Denali National Park, AK (P. Roderick, pers. comm.), though these populations await corroboration and probably represent a new species. Ice worms are relatively abundant in the St. Elias, AK region, coastal glaciers of the Alsek Range (e.g. Grand Pacific, Grand Plateau and Ferris glaciers) and the Glacier Bay region (Goodman 1970; Tynen 1970). Surprisingly, ice worms have not been found on the Juneau Icefield (northern Coastal Range) even after years of intensive observation by members of the Juneau Icefield Research Program.

The northern clade includes at least two genetically and geographically distinct groups, one located on the Kenai Peninsula, and another in the St. Elias area (putative Denali populations may form a third group). Among these *Mesenchytraeus solifugus* populations, the relationship between genetic divergence, geographic distance ($r^2 = 0.76$) and temporal separation ($r^2 = 0.77$, $p < 0.0001$) is strong (Hartzell 2005). This holds true both for the northern clade as a whole, where the average between-population distance is ~800 km, as well as within the St. Elias group, where interglacial distance averages ~680 km. Among Kenai populations, however, where mean geographic distance is only ~80 km, temporal distance is a better predictor of divergence than geographic distance ($r^2 = 0.46$ and 0.23, respectively). This disparity reflects a difference in glacial dynamics over varying spatial scales: temporal distance varies directly with geographic distance over long distances, while over short distances it does not, due to the greater role of local topography on the fragmentation of glaciers, and hence their populations.

Ice worm populations in the St. Elias, AK region exhibit the highest level of genetic variation among North American populations (Hartzell 2005), consistent with the much larger potential population sizes inferred from glacier size (i.e. St. Elias has the most, and largest, glaciers in North America). In contrast, Kenai Peninsula ice worm populations display less genetic variation; they have fewer exclusive polymorphisms than expected if these two groups (i.e. St. Elias and Kenai Peninsula) were part of the same population. Limited geographic distribution, glacial size and theta values all suggest that the subclade comprising Kenai Peninsula ice worm populations is much smaller than either the ancestral population or extant St. Elias populations (Hartzell 2005), and is consistent with the Kenai region as being the leading edge of a northward expansion (Ibrahim et al. 1996).

16.5.2 Southern Clade

The southern clade includes all glacier ice worm populations in Oregon, Washington and southern British Columbia. Populations from outside the Cordilleran boundaries (e.g. Oregonian) form distinct groups within the larger clade (Hartzell 2005; Lee 2005; Hartzell et al. 2005). Welch (1916) collectively described these worms as subspecies *M. solifugus rainierensis* based on specimens collected from Mt. Rainier, WA, although subsequent geographic and genetic evidence suggests that they may represent a separate species (see above; Hartzell et al. 2005). Southern clade ice worms are significantly larger than those of the northern clade, perhaps reflecting the longer period of primary productivity (growth season) in more southern regions. British Columbia ice worms, however, are significantly larger relative to Washington ice worms, indicating a role for other environmental factors (e.g. larger glaciers, more varied habitat) or perhaps genetic drift.

Ice worms are ubiquitous on glaciers that comprise the southern clade. Indeed, in the Pacific/Cascades Ranges, all 51 active glaciers surveyed had moderate to high ice worm densities; mean density after 17:00 was 460 individuals/m^2, with a maximum of 6,400 individuals/m^2 ($n = 75$) observed during late evening hours in summer (Hartzell 2005). Three nonactive ice remnants surveyed in this subregion exhibited an absence of ice worms or low density counts ($n = 24$). Temporal separation was a better predictor of genetic distance between ice worm populations ($r^2 = 0.20$) relative to geographic distance ($r^2 = 0.04$), reflecting a relatively complex glaciological history in this region (e.g. changes in elevation have a greater influence on the distribution and fragmentation of glaciers at these latitudes, relative to those farther north).

M. solifugus rainierensis populations exhibit genetic evidence of decreased population size and increased genetic divergence relative to their putative ancestral population (Hartzell 2005). Ancestral ice worm populations were probably large, with extant populations formed by fragmentation of parent glaciers. Those populations from within the maximum extents of the Cordilleran ice sheet—the Pacific Range/North Cascades of British Columbia and northern Washington—display the greatest genetic diversity within this clade. In contrast, molecular evidence for Sisters Range (Oregon) populations suggests a relatively recent expansion from a small founder population, consistent with leading-edge expansion models (Ibrahim et al. 1996).

16.6 PHYSIOLOGY

Aside from their heavy pigmentation, glacier ice worms have no obvious morphological adaptations to their extreme environment (Goodman and Parrish 1971; Shain et al. 2000). And while other enchytraeids endure Arctic permafrost by dehydration (Holmstrup and Westh 1994; Sømme and Birkemoe 1997; Pederson and Holmstrup 2003) or by accumulating sugars (Holmstrup and Sjursen 2001; Holmstrup et al. 2002), these mechanisms do not appear useful to ice worms, which survive in a hydrologic environment that vacillates only a few degrees from 0 °C. Ice worm protoplasm can be supercooled to −6.8 °C, which is sufficient for long-term survival on hydrated temperate glaciers; they begin to autolyze upon continuous exposure to temperatures >5 °C (Edwards 1986).

Napolitano et al. (2004) determined that *M. solifugus* specimens contain unusually high 5′ adenosine triphosphate (ATP) levels, which paradoxically increase as temperatures fall well below 0 °C (Fig. 16.4). This has been interpreted as a compensatory mechanism by which temperature-dependent reductions in molecular motion and diffusion (e.g. number of molecular collisions), and reduction in Gibb's free energy of ATP hydrolysis ($\Delta G' = \Delta G° + RT$ ln [ADP][Pi]/[ATP]) are offset by gains in [ATP], which drive most biochemical reactions (Napolitano and Shain 2004, 2005; Napolitano et al. 2004). In principle, these energetic differences help to explain the relatively robust activity levels (e.g. motor, neural) of glacier ice worms, which function comparably with their mesophilic counterparts (e.g. earthworms) who live between 10 and 20 °C. Note that disparate micro-biota (e.g. bacteria, algae, fungi) that co-inhabit Alaskan glaciers, and also Antarctic ice, appear to have made similar energetic adaptations to those observed in ice worms (Napolitano and Shain 2004).

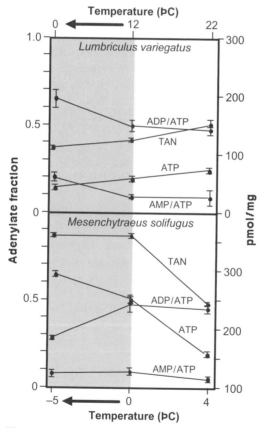

Figure 16.4 Adenylate profiles of North American ice worms versus a mesophilic worm (*Lumbriculus variegatus*). Acclimated temperature points are shown in the center and at the right side of each panel; shaded regions represent a 20-min cold shock at indicated temperatures. Ice worms (bottom) elevated ATP and total adenylate nucleotide (TAN) levels as temperatures fell (right side, white), and upon cold shock (left side, shaded) ($p < 0.05$); the opposite trend was observed in *L. variegatus* (top) ($p < 0.05$).

The underlying mechanism(s) of the ice worm's unusual energy metabolism remains unclear, but may involve a few key enzymes associated with ATP synthesis/degradation. Specifically, ATP synthetic enzymes (e.g. F_1F_0 ATP synthase) seem relatively well adapted to cold temperatures in comparison to energy consumption reactions (Farrell et al. 2004; Napolitano and Shain 2004; Hohenstein and Shain 2006). Additionally, the primary AMP degrading enzymes, AMP phosphatase and AMP deaminase, which control the adenylate pool size by removing AMP in response to excess ATP (Ataullakhanov and Vitvitsky 2002), may be disproportionately downregulated in ice worms (Morrison 2007; Marotta et al. submitted). Consequently, the accumulation of AMP—a potent stimulator of ATP synthetic processes (Atkinson 1977; Hardie and Hawley 2001)—likely contributes to the relatively high steady-state ATP levels observed in glacier ice worms and other cold-adapted taxa (Morrison and Shain 2008).

16.7 CONSERVATION STATUS

Glacier ice worms thrive on relatively large, complex glaciers, i.e. a direct, positive relationship exists between glacier size and ice worm density (Hartzell 2005). The number of habitat types (e.g. firn, pools, crevasses) on a given glacier is directly related to microbial abundance (primary production) as well as macroinvertebrate abundance (secondary production, including ice worms). The number of habitat types is directly related to glacier size; most glaciers $<0.25\,km^2$ have only one habitat type (e.g. firn), while multihabitat glaciers tend to be larger and less isolated (most North American glaciers $>0.25\,km^2$ support multiple habitat types). Relatively small glaciers typically have proportionately smaller ice worm populations, resulting in greater vulnerability to population extinction by stochastic processes. Isolated populations exhibit reduced genetic variability (e.g. Sisters Range ice worm populations), restricting their ability to adapt to changing conditions. Unfortunately, as glaciers retreat with our changing global climate, ice worm populations diminish with them.

Glacier mass balance in the PNW has been decreasing rapidly since the 1970s following a slow decline that began in the 1880s, and a concomitant decrease in summer stream flow has occurred due to decreased glacial contribution (Bach 2002; Molnia 2003, 2007; Chennault 2004). Changes in glacial mass balance are largely accounted for by changes in climate, and can be predicted in the PNW by current climate indices, particularly the El Niño-Southern Oscillation (ENSO) index (Bitz and Battisti 1999; Chennault 2004). PNW glaciers are experiencing a mean terminus retreat of ~28 m/year and thinning of ~12 m/year (Bitz and Battisti 1999; Riedel 2003), resulting in the loss of numerous glacier ranges, particularly in the south (e.g. Cascades, Olympics and Sisters Ranges). These glaciers support many unique and irreplaceable populations, e.g. the Sisters Range in Oregon representing the southernmost glacier ice worm population—yet those glaciers will probably be gone by 2050.

Future studies of glacier ice worms should focus on documentating "gap" populations (e.g. Denali, the Wrangel-St Elias-Kluane region, and central and northern British Columbia) as well as those populations subject to near-term extinction (e.g. the Sisters, Olympic and Cascade Ranges, including Vancouver Island). Preservation of these populations through protection of reserves, and/or through translocation, should be considered. Conservation efforts should focus on documenting and preserving those sources of glacier ice worm diversity.

REFERENCES

AGER, T.A. 1994. Terrestrial Palynological and Paleobotanical Records of Pliocene Age from Alaska and Yukon Territory. USGS file report 94-23.

AITCHISON, S.W. 1977. Ice worms. Summit, August–September, 6–9.

ATAULLAKHANOV, F.I., VITVITSKY, V.M. 2002. What determines the intracellular ATP concentration? Biosci. Rep. 22, 501–511.

ATKINSON, D.E. 1977. *Cellular Energy Metabolism and Its Regulation*. New York: Academic Press.

BACH, A. 2002. Snowshed contributions to the Nooksack River watershed, North Cascades Range, Washington. Geogr. Rev. 92, 192–212.

BELY, A.E., WRAY, G.A. 2004. Molecular phylogeny of naidid worms (Annelida: Clitellata) based on cytochrome oxidase I. Mol. Phylogenet. Evol. 30, 50–63.

BITZ, C.M., BATTISTI, D.S. 1999. Interannual to decadal variability in climate and the glacier mass balance in Washington, Western Canada and Alaska. J. Clim. 12, 3181–3196.

CHENNAULT, J.W. 2004. Modeling the contributions of glacial meltwater to streamflow in Thunder Creek, North Cascades National Park, Washington. MS Thesis, Western Washington University.

CLAGUE, J.J. 1976. Quadra sand and its relation to the late Wisconsin glaciation of southwest British Columbia. Canadian Journal of Earth Science 13, 803–815.

CLAGUE, J.J. 1991. Quaternary glaciation and sedimentation. In *Geology of the Cordilleran Orogen in Canada*, edited by H. Gabrielse and C.J. Yorath. Geological Survey of Canada.

EDWARDS, J.S. 1986. How small ectotherms thrive in the cold without really trying. Cryo Letters 6, 388–390.

EISEN, G. 1904. Enchytraeidae of the west coast of North America. In *Harriman Alaska Expedition, Alaska, XII, The Annelids*, edited by C. Hart Merriam. Washington, DC: Smithsonian Institution, pp. 59–61.

EMERY, C. 1898a. Diagnosi di un nuovi genere e nuova specie di annelidi della famiglia degle Enchytraeidae. Atti R. Accad. Lincei 5, 100–111.

EMERY, C. 1898b. Über einen schwarzen oligochäten von den Alaska gletschern. Verh. Schweiz. Naturforsch. Gesellschaft, Sektion fur Zool. 89.

EMERY, C. 1898c. Sur un oligochete noir des glacier delí Alaska. Rev. Suisse Zool. 5 (Suppl.), 21–22.

EMERY, C. 1900. On *Melanenchytraeus solifugus*. In *The Ascent of Mount St. Elias [Alaska]*, Appendix D, edited by Filippo de Filippi. London, pp. 224–231.

FARRELL, A.H., HOHENSTEIN, K.A., SHAIN, D.H. 2004. Molecular adaptation of the ice worm, *Mesenchytraeus solifugus*: divergence of energetic associated genes. J. Mol. Evol. 56, 666–673.

GOODMAN, D. 1970. *Ecological Investigations of Ice Worms on Casement Glacier, Southeast Alaska*. Report 39. Ohio State University Research Foundation, Institute of Polar Studies, pp. 1–59.

GOODMAN, D., PARRISH, W.B. 1971. Ultrastructure of the epidermis in the ice worm, *Mesenchytraeus solifugus*. J. Morphol. 135, 71–86.

GUDGER, E.W. 1923. Snow worms; enchytraeid worms found in the snow and on the glaciers of high mountains. Nat. Hist. 23, 450–456.

HARDIE, D.G., HAWLEY, S.A. 2001. AMP-activated protein kinase: the energy charge hypothesis revisted. Bioessays 23, 1112–1119.

HARTZELL, P.L. 2005. Pacific northwest glacial ecosystems. PhD Thesis, Clark University, Worcester, MA.

HARTZELL, P.L., NGHIEM, J.V., RICHIO, K.J., SHAIN, D.H. 2005. Distribution and phylogeny of glacier ice worms (*Mesenchytraeus solifugus* and *Mesenchytraeus solifugus rainierensis*). Can. J. Zool. 83, 1206–1213.

HOHENSTEIN, K.A., SHAIN, D.H. 2006. Divergence of F_1 ATP synthase subunits in the ice worm, *Mesenchytraeus solifugus*. Hydrobiologia 564, 51–58.

HOLMSTRUP, M., SJURSEN, H. 2001. Freeze induced glucose accumulation in the enchytraeid, *Fredericia ratzeli*, from Greenland. Cryo Letters 22, 272–276.

HOLMSTRUP, M., WESTH, P. 1994. Dehydration of earthworm cocoons exposed to cold: a novel cold hardiness mechanism. J. Comp. Physiol. [B] 164, 312–315.

HOLMSTRUP, M., BAYLEY, M., RAMLOW, H. 2002. Supercool or dehydrate? An experimental analysis of overwintering strategies in small permeable arctic invertebrates. Proc. Natl. Acad. Sci. U.S.A. 99, 5716–5720.

IBRAHIM, K., NICHOLS, R.A., HEWITT, G.M. 1996. Spatial patterns of genetic variation generated by different forms of dispersal during range expansion. Heredity 77, 282–291.

LEE, B. 2005. A phylogeographic study of ice worm populations, *Mesenchytraeus solifugus*, in the Olympic mountains. Senior Research Thesis, University of Puget Sound, Tacoma, WA.

LIAN, O.B., HICKIN, E.J. 1993. Late Pleistocene stratigraphy and chronology of lower Seymour Valley, southwestern British Columbia. Can. J. Earth Sci. 30, 841–850.

LIANG, Y., HSU, C., CHANG, T. 1979. A new genus and species of Enchytraeidae from Tibet. Acta Zootax. Sinica 4, 312–317.

MAROTTA, R., PARRY, B.R., SHAIN, D.H. Divergence of the AMP deamiase enzyme in the ice worm, *Mesenchytraeus solifugus* (Annelida, Clitellata, Enchytraeidae) (submitted).

MOLNIA, B.F. 2003. *Glaciers of Alaska, with sections on Columbia and Hubbard Tidewater Glaciers, by Krimmel, RM.* Satellite image atlas of glaciers of the world: U.S. Geological Survey Professional Paper 1386-K (Glaciers of Alaska).

MOLNIA, B.F. 2007. Late nineteenth to early twenty-first century behavior of Alaskan glaciers as indicators of changing regional climate. Glob. Planet. Change 56, 23–56.

MOORE, J.P. 1899. A snow inhabiting Enchytraeid (*Mesenchytraeus solifugus* Emery) collected by Mr. Henry G. Bryant on the Malaspina Glacier. Alaska. Proc. Acad. Nat. Sci. Philadelphia 1899,125–144.

MORRISON, B.A. 2007. Elevated ATP levels increase cold tolerance in Escherichia coli. MS Thesis, Rutgers, The State University of New Jersey.

MORRISON, B.A., SHAIN, D.H. 2008. An AMP nucleosidase gene knockout in *Escherichia coli* elevates intracellular ATP levels and increases cold tolerance. Biol. Lett. 4, 53–56.

NAPOLITANO, M., SHAIN, D.H. 2004. Four kingdoms on glacier ice: convergent energetic processes boost energy levels as temperatures fall. Proc. R. Soc. Lond. 271 (Suppl. 5), 276–280.

NAPOLITANO, M., SHAIN, D.H. 2005. Distinctions in adenylate metabolism among organisms inhabiting temperature extremes. Extremophiles 9, 93–98.

NAPOLITANO, M., NAGELE, R.O., SHAIN, D.H. 2004. Ice worms boost energy levels at low physiological temperatures. Comp. Biochem. Physiol. A Physiol. 137, 227–235.

ODELL, N.E. 1949. Ice-worms in Yukon and Alaska. Nature 164, 1098.

PEDERSON, P.G., HOLMSTRUP, M. 2003. Freeze or dehydrate: only two options for the survival of subzero temperatures in the arctic enchytraeid *Fridericia ratzeli*. J. Comp. Physiol. [B] 173, 601–609.

RIEDEL, J.L. 2003. Late Fraser Alpine Glacier Activity in the North Cascades, Washington. Paper No. 86-6. Geological Society of America.

RUSSELL, I.C. 1892. *Second Expedition to Mt. St. Elias in 1891, 13th Annual Report of the US Geological Survey, 1891–1892, Pt. II*. Geology, p. 33.

SHAIN, D.H., CARTER, M.R., MURRAY, K.P., MALESKI, K.A., SMITH, N.R., MCBRIDE, T.R., MICHALEWICZ, L.A., SAIDEL, W.M. 2000. Morphologic characterization of the ice worm, *Mesenchytraeus solifugus*. J. Morphol. 246, 192–197.

SHAIN, D.H., FARRELL, A., MASON, T., MICHALEWICZ, L.A. 2001. Distribution and behavior of ice worms (*Mesenchytraeus solifugus*) in south-central Alaska. Can. J. Zool. 79, 1813–1821.

SIRKIN, L., TUTHILL, S.J. 1987. Late Pleistocene and Holocene deglaciation and environments of the southern Chugach Mountains, Alaska. Geol. Soc. Am. Bull. 99, 376–384.

SØMME, L., BIRKEMOE, T. 1997. Cold tolerance and dehydration in Enchytraeidae from Svalbard. J. Comp. Physiol. [B] 167, 264–269.

TYNEN, M.J. 1970. The geographic distribution of ice worms (Oligochaeta: Enchytraeidae). Can. J. Zool. 48,1363–1367.

WAITT, R.B., THORSON, R.M. 1983. The Cordilleran Ice Sheet in Washington, Idaho and Montana. In *Late-Quaternary Environments of the United States*. Vol. 1. The late Pleistocene, edited by S.C. Porter. St. Paul, MN: University of Minnesota Press, pp. 53–70.

WELCH, P.S. 1916. Snow field and glacier Oligochaeta from Mt. Rainier, Washington. Trans. Am. Microsc. Soc. 35, 85–124.

WELCH, P.S. 1917a. *Alaskan glacier worms (Clitellata)*. Bionomical Leaflets, McGill University, Montreal, No. 2.

WELCH, P.S. 1917b. Enchytraeidae (Oligochaeta) from the Rocky Mountain Region. Trans. Am. Microsc. Soc. 36, 67–81.

WELCH, P.S. 1919. Further studies on North American *Mesenchytraeids* (Oligochaeta). Trans. Am. Microsc. Soc. 38, 175–188.

WILES, G.C., CALKIN, P.E. 1994. Late Holocene, high-resolution glacial chronologies and climate, Kenai Mountains, Alaska. Quat. Sci. Rev. 20, 449–461.

WRIGHT, G.F. 1887. The Muir Glacier. Am. J. Sci. 33, 5.

Chapter 17

Sperm Ultrastructure in Assessing Phylogenetic Relationships among Clitellate Annelids

Roberto Marotta and Marco Ferraguti

Dipartimento di Biologia, Università di Milano, Italy

17.1 INTRODUCTION

Clitellata is a taxon of annelids characterized by its hermaphroditic modality of reproduction and specialized reproductive structure (i.e. clitellum); as a group, Clitellata contains ~5,000 of the >15,000 species of segmented worms described (Erséus 2005). The monophyly of clitellates and several of their constituent taxa is supported by a wide set of data including general morphology (Purschke et al. 1993; Nielson 1995), sperm ultrastructure (Ferraguti 1999) and DNA sequence similarity (McHugh 1997; Siddall et al. 2001; Erséus and Källersjö 2004).

Clitellata have filiform spermatozoa that, following the terminology of Rouse and Jamieson (1987), are classified as introsperm, i.e. sperm that have no contact with water when passed from male to female. A set of autapomorphies characterize the filiform spermatozoa of the Clitellata (Fig. 17.1). Specifically, the presence of an acrosome tube often containing a withdrawn acrosome vesicle, interpolation of mitochondria between the nucleus and the flagellum, peculiar modifications of the axoneme central apparatus, and the presence of glycogen granules external to the axoneme (Jamieson 2006).

As an example of a sperm model in oligochaetous clietellates, we describe the spermatozoon of the Australian phreodrilid *Insulodrilus bifidus* (Phreodrilidae), as observed at the sperm funnels, i.e. the beginning of the male ducts (Fig. 17.2). The spermatozoa of *I. bifidus* are long, filiform cells, containing an acrosome, nucleus, mitochondria and tail (for a better understanding of the following description, refer to Fig. 17.1). The acrosome is formed by an acrosome tube (basally thickened to form a limen) filled almost completely by a vesicle (Fig. 17.2B) that contacts the sperm plasma membrane through the anterior opening of the tube, and is indented at its distal extremity to host an acrosome rod (perforatorium). An apical corona surrounds the anterior extremity of the acrosome. The nucleus (Fig. 17.2A) is gently twisted along its whole length. Six mitochondria

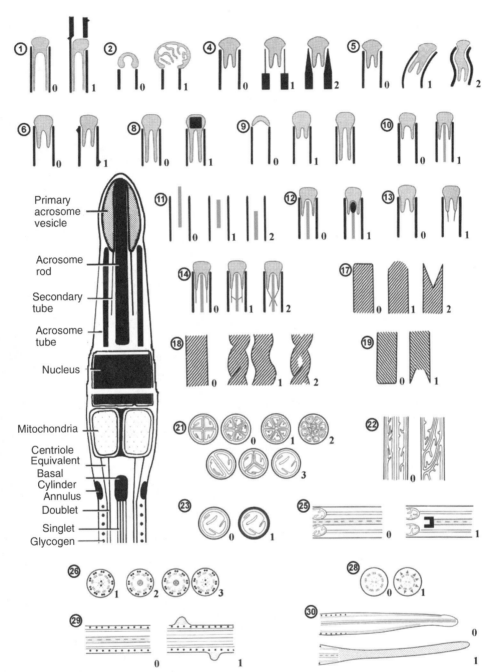

Figure 17.1   Schematic of clitellate spermatozoal characters in different states. Inset, hypothetical plesiomorphic spermatozoon (modified from Jamieson et al. 1987) for the Clitellata, as inferred from ancestral state reconstruction analysis.

Source: Reprinted from Marotta et al. (2008), with permission from Blackwell Synergy.

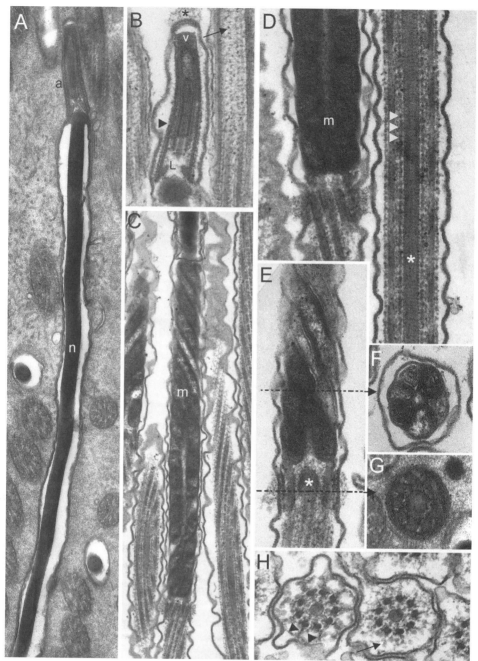

Figure 17.2 Spermatozoa of *Insulodrilus bifidus*. (A) Nucleus (n) and acrosome (a) (×35,000). (B) An acrosome at higher magnification, close to a flagellum. Note the apical corona (asterisk), the acrosome tube (arrowhead) with limen (L) and the acrosome vesicle (v) (×60,000). The flagellum to the right was cut tangentially through the link heads (arrow) connecting the doublets to the central sheath. (C) Long midpiece formed by six twisted mitochondria (m). Flagella at the lower right is cut tangentially, showing the twisting of doublets (×30,000). (D) Base of a mitochondrial midpiece (m) and flagellum cut tangentially (left); a flagellum cut sagittaly shows the prominent central sheath (asterisk) and link heads (arrowheads; compare to B) (×60,000). (E) Basal portion of the midpiece and basal body region showing the basal cylinder (asterisk) (×60,000). (F) Cross-cut midpiece (×65,000). (G) Cross-cut basal body at the center of which a basal cylinder is visible (×65,000). (H) Two cross-cut flagella (compare to longitudinal sections in B, D). Faint connections link the doublets to the plasma membrane (arrowheads). Some of the 18 glycogen granules surrounding the axoneme are visible (arrow) (×65,000).

(Fig. 17.2C–F) interpolate between the nucleus and the flagellum to form a cylindrical midpiece. Each mitochondrion has the shape of a cylindrical sector (Fig. 17.2F) and the six mitochondria are twisted around a small, empty central axis (Fig. 17.2D–F). The tail contains a $9 \times 2 + 2$ axoneme starting from a deeply modified basal body penetrated by a basal cylinder from which the two central tubules of the axoneme start (Fig. 17.2E, G). The axoneme has its peripheral doublets twisted around the longitudinal axis (Fig. 17.2C, D) and the central apparatus is surrounded by a prominent central sheath (Fig. 17.2D, H). Nine couples of glycogen granules external to the axoneme are visible in some fixations.

Within Clitellata, the ultrastructure of spermatozoa has proven useful for phylogenetic assessments, not only at high (Jamieson et al. 1987; Ferraguti and Erséus 1999; Marotta et al. 2008) but also at lower taxonomic levels among branchiobdellids (Cardini and Ferraguti 2004), enchytraeids (Westheide et al. 1991) and tubificids (Marotta et al. 2003). Nevertheless, in a study of tubificid relationships based on sperm morphology, Erséus and Ferraguti (1995) point out that at low taxonomic levels, "the spermatozoal character patterns are complex and contain elements of convergence and probably also reversal, and therefore they should be used ... [in tubificids] ... only in combination with other information." Following this suggestion, to assess the contribution of spermatozoon ultrastructure to clitellate phylogeny, 34 spermatozoal characters—concerning different regions of the sperm cell (Fig. 17.1)—were combined with 18S rDNA gene sequences and general somatic characters for 39 species belonging to 13 different clitellate families in a combined data phylogenetic study of broadly sampled clitellates (Marotta et al. 2008). Although the Bayesian phylogenetic analyses of the spermatological dataset alone were poorly resolved (Fig. 17.3), with only 13 spermatozoal nodes with posterior probability ≥ 0.85, the spermatological dataset strongly supported the monophyly of clitellates (posterior probability 1.00), and inside them that of Hirudinida (posterior probability 1.00), Lumbricidae (posterior probability 0.91), Lumbriculidae (posterior probability 0.90–0.91), Naidinae (posterior probability 0.86–0.87) and Enchytraeidae (posterior probability 0.88). The spermatozoal characters also supported the grouping of branchiobdellids with acanthobdellids and leeches (posterior probability 0.97–0.98), and inside it the sister group relationship between Hirudinida and *Acanthobdella* (posterior probability 0.92–0.93). Spermatozoal characters, together with traditional somatic characters, thus contributed to the 18S rDNA phylogeny not only in supporting some of the groupings obtained by 18S rDNA analyses, but also in suggesting new relationships: specifically, the sister group relationship between *Propappus* and Enchytraeidae, and between *Acanthobdella* and Hirudinida, and inside Tubificidae, the close relationship between Tubificinae and Limnodriloidinae (Marotta et al. 2008) (Fig. 17.4). The contribution of sperm ultrastructure in sustaining these relationships is considered below.

17.2 THE SPERMATOZOON OF *PROPAPPUS VOLKI* (MICHAELSEN 1916)

Propappus volki was first described by Michaelsen in 1916 from Elbe River, near Hamburg, Germany. A new genus was established for this species, which was placed inside Enchytraeidae together with a second species, *Propappus glandulosus*, as an early branch of the family. Based on several morphological characters, the placement of *Propappus* within Enchytraeidae has been questioned, first by Nielsen and Christensen (1959) and later by Coates (1986), who erected a new family, Propappidae, closely related to Enchytraeidae

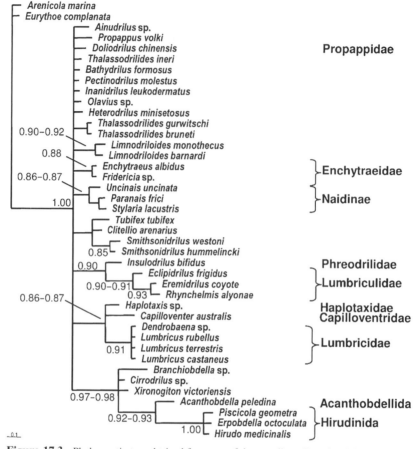

Figure 17.3 Phylogenetic tree obtained from one of three replicate Bayesian inference runs of the spermatozoal dataset. Posterior probability ≥0.85 is indicated on nodes. Parentheses indicate monophyletic groupings.

Source: Modified from Marotta et al. (2008), with permission from Blackwell Synergy.

and Haplotaxidae. In recent years, *P. volki* has been included in a phylogenetic analysis of Clitellata based on 18S rRNA (Erséus and Källersjö 2004), which grouped *Propappus* with Phreodrilidae, Tubificidae and Haplotaxidae, while enchytraeids formed a monophyletic group with Crassiclitellata (*sensu* Jamieson 1988). In their DNA-based phylogeny, Erséus and Källersjö (2004) supported Coates' (1986) removal of Propappidae from Enchytraeidae.

The spermatozoon of *P. volki* (Gustavsson et al. 2008) is a filiform cell characterized by an elongated nucleus and a modified midpiece. Compared to all previously studied clitellate species, the spermatozoa of *P. volki* share many characters with those of Enchytraeidae, thus supporting a relationship between these two taxa (Gustavsson et al. 2008). Concerning the acrosome, both enchytraeids and *P. volki* have a short and straight acrosome tube, with the acrosome vesicle external to the tube, and the acrosome rod reaching the apical portion of the vesicle. The secondary tube is in contact with the basal portion of the acrosome rod in a wider cylindrical body. This structure resembles the distinctive

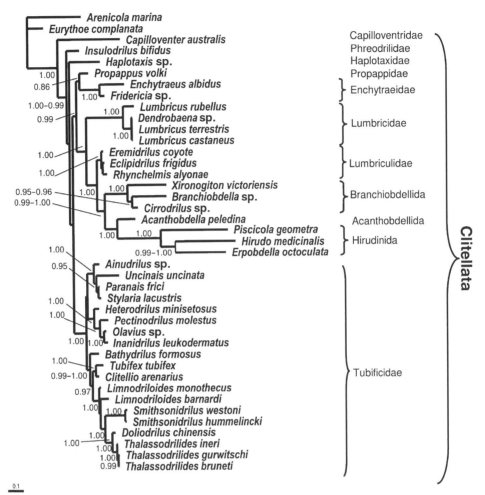

Figure 17.4 Phylogenetic tree obtained from one of three replicate Bayesian inference runs of combined data (18S rDNA, somatic and spermatozoal). Posterior probabilities ≥0.85 are indicated on nodes. Parentheses indicate monophyletic groupings.

Source: Modified from Marotta et al. (2008), with permission from Blackwell Synergy.

basal node as described in spermatozoa of megascolecid earthworms, tentatively considered homologous to the node sheath present in lumbricid spermatozoa (Jamieson 1978). Its presence inside the acrosome of *P. volki* may thus support a relationship between *Propappus* and earthworms. The nucleus, apically corkscrew shaped and basally straight, corroborates the pattern described for the nucleus in the majority of enchytraeid spermatozoa (Jamieson 2006) and, as in enchytraeid spermatozoa, the midpiece in *P. volki* is elongated and twisted. Thus, the grouping of *P. volki* with tubificids, haplotaxids and phreodrilids, as proposed by 18S rRNA parsimony analysis (Erséus and Källersjö 2004), is not supported by spermatozoal data, that instead corroborate the morphology-based traditional view of a close relationship to enchytraeids, in agreement with recent phylogenetic analyses (Rousset et al. 2007; Marotta et al. 2008).

17.3 SPERM ULTRASTRUCTURE IN BRANCHIOBDELLIDS, *ACANTHOBDELLA PELEDINA*, AND HIRUDINEANS

Branchiobdellids are small clitellates, a few millimeters long, without chaetae and with a fixed number of body somites, that live attached with their posterior sucker on freshwater crustaceans of the northern hemisphere. Similar to branchiobdellids, *A. peledina* lives parasitically on arctic salmons; it is the only species of Acanthobdellida, a taxon of leech-like worms, with oligochaete-like chaetae and a constant number of somites. Hirudinida (leeches) are predatory or ectoparasitic clitellates, lacking chaetae and with reduced and specialized coelomic cavities (Siddall et al. 2006). Although in agreement with the clitellate spermatozoon pattern outlined above (see Fig. 17.1), the spermatozoa of branchiobdellids, acanthobdellids and hirudinids represent the most aberrant and complex clitellate spermatozoa (Fig. 17.5), and constitute an interesting field of study to test the power of fertilization biology in shaping the sperm cell (Ferraguti 1999).

The combined phylogenetic analysis of Marotta et al. (2008) strongly supports the grouping of leeches and their allies with Lumbriculidae, in agreement with the phylogenetic analyses of Siddall et al. (2001) and Erséus and Källersjö (2004), which also support the sister group relationship between Hirudinida and Branchiobdellida and, consequently, for *Acanthobdella* as their plesiomorphic sister group. On the other hand, the phylogenetic analysis of the combined dataset strongly suggests a sister group relationship between *A. peledina* and Hirudinida, with Branchiobdellida as their plesiomorphic sister group (Fig. 17.4). This view, despite several hypotheses proposing that similarities between leeches and *A. peledina* have been convergently acquired in relation to commensalism (summarized in Siddall and Burreson 1996), has been supported independently by traditional morphology (Purschke et al. 1993) and sperm ultrastructure (Westheide and Purschke 1996; Ferraguti and Erséus 1999).

Several spermatozoal autapomorphies support the sister group relationship between *A. peledina* and Hirudinida, including the presence of coiled fibers around the nucleus, and a midpiece formed by a single mitochondrion, surrounded by an electron-dense sheath (Fig. 17.5B, C). The secondary loss of the axonemal basal cylinder (*sensu* Ferraguti 1984a) and the presence of a flagellum endpiece filled with dense material are instead autapomorphies for the Branchiobdellida + *Acanthobdella* + Hirudinida assemblage (Marotta et al. 2008).

17.4 SPERM ULTRASTRUCTURE INSIDE TUBIFICIDAE

Tubificidae is a large and cosmopolitan clade of aquatic oligochaete worms living in freshwater, brackish water and marine habitats, comprising ~1,000 species and divided into several subfamilies. In accord with the great diversity in tubificid sperm morphologies (Erséus and Ferraguti 1995), no autapomorphies are present in the family, but several autapomorphies characterize tubificid subfamilies (Marotta et al. 2008). In particular, a double sperm line, i.e. the production of two types of spermatozoa and the presence of complex sperm aggregates called spermatozeugmata *sensu* Ferraguti et al. (1989), has been proposed as synapomorphies for Tubificinae and Limnodriloidinae in various phylogenetic analyses based on somatic and spermatological characters (Erséus 1990; Marotta et al. 2003).

The two sperm types, eusperm and parasperm (Healy and Jamieson 1981), present in Tubificinae and Limnodriloidinae, differ in their structure and function. In both

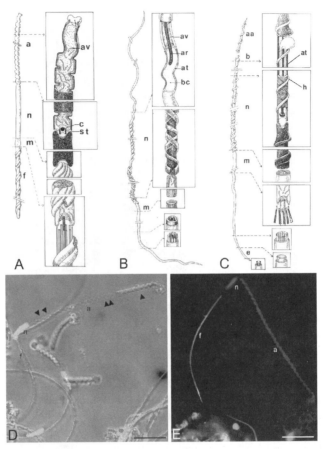

Figure 17.5   Spermatozoa of branchiobdellids, acanthobdellids and hirudineans. (A) Three-dimensional reconstruction of *Branchiobdella italica* spermatozoon showing a large acrosome with a highly introflected inner membrane forming tubules almost filling the subacrosomal space, a straight short nucleus, seven twisted mitochondria and a flagellum with a helical marginal fiber. (B) Three-dimensional reconstruction of *Acanthobdella peledina* spermatozoon. The acrosome vesicle is external to the tube; the rod delimits a basal chamber; the nucleus is a complex spiral; a single mitochondrion is present and is surrounded by a sheath that continues around the first tract of the flagellum. (C) Three-dimensional reconstruction of *Erpobdella octoculata* spermatozoon. The acrosome is long and corkscrew shaped, and presents an anterior extension of the tube, the anterior acrosome. The acrosome vesicle forms a lateral button, and the acrosome tube is surrounded by a helical ridge. The nucleus is a complex spiral; a single mitochondrion is present, as well as a basal body with remnants of the anchoring apparatus. (D, E) Immunofluorescence staining of *B. italica* sperm cells: DNA is blue (DAPI); F-actin is red (TRITC-phalloidin), and alpha tubulin is green (Alexa Fluor 430 anti-alpha tubulin). Panel D shows late spermatids. A large portion of the twisted acrosome tube is empty (double arrowhead). The acrosome vesicle penetrates the apical portion of the acrosome tube (arrowhead). Panel E shows a single spermatozoon. Note the flagellum, the short nucleus, and the elongated and twisted acrosome. Abbreviations: a = acrosome; aa = anterior acrosome; ar = acrosome rod; at = Acrosome Tube; av = acrosome vesicle; b = lateral button; bc = basal chamber; c = cork; e = flagellar end-piece; h = helical ridge; m = mitochondria; n = nucleus; st = secondary tube; f = flagellum.
Source: Modified from Ferraguti (1999), with permission from John Wiley & Sons, Inc. (See color insert.)

subfamilies, the eusperm (i.e. the fertilizing spermatozoa) are filamentous cells with a tapered, tubular, terminal acrosome, an extremely elongate nucleus, a cylindrical sometimes spiral midpiece formed by two to six radially adpressed mitochondria, and a long flagellum containing a characteristic basal cylinder (Ferraguti 1984a), and two modifications of the central axonemal apparatus referred to as "tetragon fibers" and as "prominent central sheath" (Ferraguti 1984b) (Fig. 17.6A–D). Parasperm have been observed in all tubificine investigated, but among Limnodriloidinae, only in *Smithsonidrilus* and *Limnodriloides* species (Marotta et al. 2003). The differences between parasperm and eusperm concerns, in both subfamilies, the same parts of the cell. In parasperm, the various acrosome structures are reduced; the nuclei are smaller and simpler; the number of mitochondria is lower, and differences occur in the axoneme and in the plasma membrane around it (Fig. 17.6E–G). The combined clitellate phylogenetic analysis of Marotta et al. (2008) strongly supports the idea that in tubificids the double sperm line originated only once, in a common ancestor of Tubificinae and Limnodriloidinae. Thus, although it is difficult to envisage a reversal to a single sperm line from such a complex model of dichotomous spermatogenesis, the double sperm line appears to be secondarily lost in those limnodriloidines species that possess only eusperm (*Thalassodrilides* and *Doliodrilus*; Fig. 17.6).

In all species of Tubificinae, with the exception of *Aulodrilus* and related forms (Brinkhurst 1990), the spermatozeugmata are formed by two sperm types (Ferraguti 1999). Within the Limnodriloidinae, sperm aggregates similar to those typical of tubificines are present only in *Limnodriloides* species, whereas all examined species of the genus *Smithsonidrilus* produce separate spermatozeugmata, each formed by parasperm or eusperm (Fig. 17.6H–J; Marotta et al. 2003). Spermatozoal characters, together with traditional morphological and 18S rDNA comparisons, corroborate the idea that the spermatozeugmata composed of both eusperm and parasperm are homologous in the Tubificinae and Limnodriloidinae, in spite of the differences in their morphological details (Marotta et al. 2003).

Figure 17.6   Eusperm, parasperm and sperm aggregates in Limnodriloidinae. Panels A–D show the eusperm of *Thalassodrilides gurwitschi* as an example of Limnodriloidinae eusperm. (A) Acrosome with a long acrosome tube basally thick but then (arrow) abruptly thinner (×60,000). (B) Longitudinal, apical section of the nucleus showing its coiled thread shape (×40,000). (C) The nucleus is basally straight (×40,000). (D) Cross section of the tail showing tetragon fibers (×67,500). Panels E–G show the parasperm of *Isochaetides arenarius* as an example of tubificine parasperm. (E) The short and straight nucleus has, at the top, a simple acrosome (arrow) (×22,500). (F) A longitudinal section of the midpiece and basal portion of the tail shows the reduced basal cylinder (arrow) at the base of the axoneme and the plasma membrane separated from the axoneme for a long tract (×26,000). (G) The midpiece has three mitochondria; cross-sectioned nuclei show an oval outline. An arrow identifies a cross section of parasperm acrosome (×40,000). Panels H–J show Limnodriloidinae spermatozeugmata. (H) Cross-sectioned spermatozeugma of *Limnodriloides monothecus* formed by both eusperm and parasperm. A large, axial cylinder is formed by parallel eusperm (a) and a narrow cortex (c), which is formed by a few rows of parasperm tails (×5,500). Arrow indicates round cross sections of parasperm nuclei. (I) Cross section of part of the large and poorly organized eusperm aggregate of *Smithsonidrilus luteolus* with its characteristic whirlpool-like arrangement (×3,000). (J) Cross-sectioned spermatozeugma of *S. luteolus* formed by parasperm only (×7,000).

Source: Modified from Marotta et al. (2003), with permission from Blackwell Publishing.

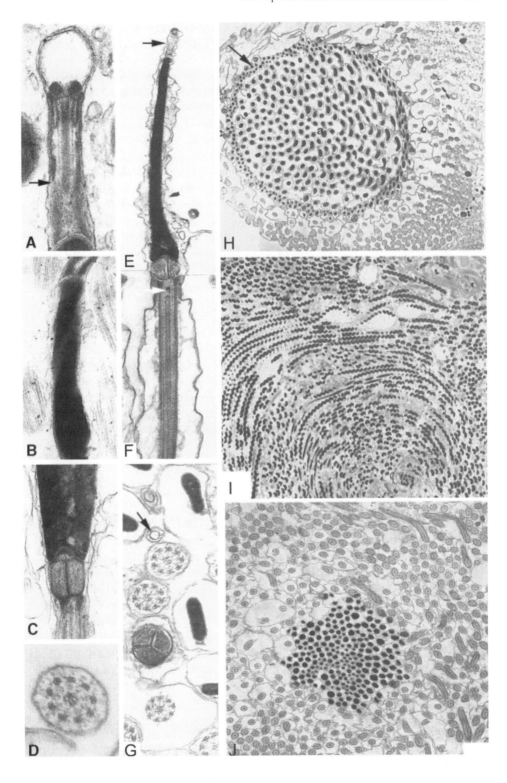

17.5 PLESIOMORPHIC SPERMATOZOON OF CLITELLATES AND SPERMATOLOGICAL APOMORPHIC TRENDS

Jamieson et al. (1987) proposed an intuitive hypothetical plesiomorphic model of the spermatozoon for oligochaetous clitellates. The ancestral state reconstruction on our combined phylogenetic analysis (Marotta et al. 2008) corroborated the hypothetical model of Jamieson et al. (1987), but with some differences. In agreement, the putative plesiomorphic spermatozoon we suggest has a short acrosome tube, with an acrosome vesicle and an acrosomal rod strongly protuberant beyond the tube, a straight and elongated nucleus, and a short unspiralized midpiece formed by four or five mitochondria. But it departs from the Jamieson model in various aspects: first, the thickness of the acrosome tube wall is constant for its whole length; second, the diameter of the nucleus is uniform throughout its length, not tapering at the apex as suggested; third, the nuclear apex is flat and not domed; and fourth, the modification of the central axonemal apparatus is the prominent central sheath instead of the tetragon fibers (Fig. 17.1).

Starting from the hypothetical plesiomorphic spermatozoon, Jamieson et al. (1987) first proposed spermatological apomorphic trends in the evolution of oligochaetous clitellates. The progressive withdrawal of the acrosome rod into the acrosomal tube, the appearance of ornamentations on the axial rod making contact with the secondary tube with connectives and the progressive spiralization of the nucleus are all confirmed in the combined clitellate phylogeny of Marotta et al. (2008), but with less linearity, showing many reversions and convergences (Fig. 17.7A). Some other proposed character changes, including a progressive withdrawal of the acrosome vesicle inside the acrosome tube, and a progressive increase in the number of mitochondria (Jamieson et al. 1987) did not show a precise trend in our analysis (Fig. 17.7B). The proposed evolutionary pattern of clitellate sperm tail (Ferraguti 1984a) was confirmed by the combined dataset; specifically, the prominent central sheath was the plesiomorphic state, and its loss was a secondary event, leading to a model of mature sperm axoneme with tetragonal fibers only (Fig. 17.7C; Marotta et al. 2008).

17.6 PATTERNS OF SPERMATOLOGICAL CHARACTERS AMONG CLITELLATES

The pattern of spermatological characters on the combined phylogenetic tree shows that, among clitellates, the spermatozoal character pattern is complex and contains elements of convergence and reversals (Marotta et al. 2008), thus extending the conclusions of Erséus and Ferraguti (1995) to all clitellates. Indeed, among clitellates, most spermatozoal characters result in homoplasic characters that give support to tree topology, as revealed by the wide gap between retention index (RI) and consistency index (CI) obtained when constraining the sperm characters to fit the combined phylogenetic tree (Marotta et al. 2008). Such a pattern can be explained assuming that the clitellate spermatozoal morphology, compared for example to that of polychaetes (Jamieson and Rouse 1989), is morphologically constrained due to the fact that nearly all clitellates, with the exclusion of hirudineans, share a similar reproductive modality (Ferraguti 1999). Thus, we suggest that in their long evolutionary history, similar spermatozoal traits may have originated several times independently at the base of large clitellate groups, with their spermatozoa having only a small set of permitted morphological changes.

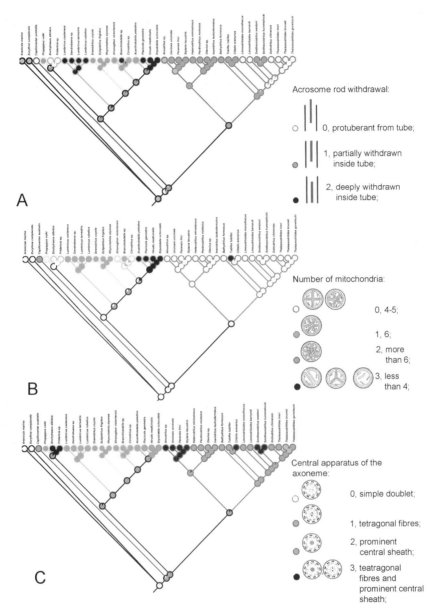

Figure 17.7   Maximum likelihood spermatozoal character state reconstructions on a corrected combined data Bayesian topology using the ancestral state reconstruction packages as implemented in Mesquite version 1.12 (Maddison and Maddison 2006). Colored spots at nodes indicate different character states. Gray spots represent missing characters. (A) Acrosome rod withdrawal character state reconstruction. Note the two reversions from an acrosome rod partially withdrawn inside the tube (green) to an acrosome rod protuberant from the tube, occurring in Enchytraeidae and Tubificidae. (B) Mitochondria character state reconstruction. Note the reduction from 4 to 5 mitochondria (white) to a single mitochondrion (black) in *Acanthobdella* and hirudineans. (C) Central apparatus of the axoneme character state reconstruction. Tubificidae (red), Hirudinida (green), *Acanthobdella* (gold), Branchiobdella (orange), Lumbriculidae (yellow), Lumbricidae (light blue), Enchytraeidae (pink), *Propappus* (goldenrod), Capilloventridae (blue).

Source: Modified from Marotta et al. (2008), with permission from Blackwell Synergy. (See color insert.)

REFERENCES

BRINKHURST, R.O. 1990. A phylogenetic analysis of Tubificinae. Can. J. Zool. 69, 392–397.

CARDINI, A., FERRAGUTI, M. 2004. The phylogeny of Branchiobdellida (Annelida, Clitellata) assessed by sperm characters. Zool. Anz. 243, 37–46.

COATES, K.A. 1986. Redescription of the oligochaete genus *Propappus*, and diagnosis of the new family Propappidae (Annelida, Oligochaeta). Proc. Biol. Soc. Wash. 99, 417–428.

ERSÉUS, C. 1990. Cladistic analysis of the subfamilies within the Tubificidae (Oligochaeta). Zool. Scr. 19, 57–63.

ERSÉUS, C. 2005. Phylogeny of oligochaetous Clitellata. Hydrobiologia 535/536, 357–372.

ERSÉUS, C., FERRAGUTI, M. 1995. The use of spermatozoal ultrastructure in phylogenetic studies of Tubificidae (Oligochaeta). In *Advances in Spermatozoal Phylogeny and Taxonomy*, edited by B.G.M. Jamieson, J. Ausio and J.L. Justine. Paris: Mémoires du Muséum d'Histoire Naturelle, pp. 189–201.

ERSÉUS, C., KÄLLERSJÖ, M. 2004. 18S rDNA phylogeny of Clitellata (Annelida). Zool. Scr. 33, 187–196.

FERRAGUTI, M. 1984a. Slanted centriole and transient anchoring apparatus during spermiogenesis of an Oligochaete (Annelida). Biol. Cell. 52, 175–180.

FERRAGUTI, M. 1984b. The comparative ultrastructure of sperm flagella central sheath in Clitellata reveals a new autapomorphy of the group. Zool. Scr. 13, 201–207.

FERRAGUTI, M. 1999. Euclitellata. In *Reproductive Biology of Invertebrates*, edited by B.G.M. Jamieson, K.G. Adyiodi and R.G. Adyiodi. Chichester: John Wiley & Sons, pp. 125–182.

FERRAGUTI, M., ERSÉUS, C. 1999. Sperm types and their use for a phylogenetic analysis of aquatic clitellates. Hydrobiologia 402, 225–237.

FERRAGUTI, M., GRASSI, M., ERSÉUS, C. 1989. Different models of tubificid spermatozeugmata. Hydrobiologia 278, 165–178.

GUSTAVSSON, L.M., FERRAGUTI, M., MAROTTA, R. 2008. Comparative ultrastructural study of the cuticle and spermatozoa in *Propappus volki* Michaelsen, 1916 (Annelida: Clitellata). Zool. Anz. 247, 123–132.

HEALY, J.M., JAMIESON, B.G.M. 1981. An ultrastructural examination of developing and mature parasperm in *Pyrazus ebeninus* (Mollusca, Gastropoda, Potamididae). Zoomorphology 112, 101–119.

JAMIESON, B.G.M. 1978. A comparison of spermiogenesis and spermatozoal ultrastructure in megascolecid and lumbricid earthworms (Oligochaeta: Annelida). Aust. J. Zool. 26, 225–240.

JAMIESON, B.G.M. 1988. On the phylogeny and higher classification of the Oligochaeta. Cladistics 4, 367–410.

JAMIESON, B.G.M. 2006. Non-leech Clitellata. With contributions by M. Ferraguti. In *Reproductive Biology and Phylogeny of Annelida*, edited by G. Rouse and F. Pleijel. Enfield, NH: Science Publisher Inc., pp. 235–392.

JAMIESON, B.G.M., ROUSE, G.W. 1989. The spermatozoa of the Polychaeta (Annelida): an ultrastructural review. Biol. Rev. Camb. Philos. Soc. 64, 93–157.

JAMIESON, B.G.M., ERSÉUS, C., FERRAGUTI, M. 1987. Parsimony analysis of the phylogeny of some Oligochaeta (Annelida) using spermatozoal ultrastructure. Cladistics 3, 145–155.

MADDISON, W.P., MADDISON, D.R. 2006. Mesquite: a modular system for evolutionary analysis. Version 1.12. http://mesquiteproject.org. Accessed on April 26, 2007.

MCHUGH, D. 1997. Molecular evidence that echiurans and pogonophorans are derived annelids. Proc. Natl. Acad. Sci. U.S.A. 94, 8006–8009.

MAROTTA, R., FERRAGUTI, M., ERSÉUS, C. 2003. A phylogenetic analysis of Tubificinae and Limnodriloidinae (Annelida, Clitellata, Tubificidae) using sperm and somatic characters. Zool. Scr. 32, 255–278.

MAROTTA, R., FERRAGUTI, M., ERSÉUS, C., GUSTAVSSON, L.M. 2008. Combined-data phylogenetics and character evolution of Clitellata (Annelida) using 18S rDNA and morphology. Zool. J. Linn. Soc. 154, 1–26.

NIELSON, C.O. 1995. *Animal Evolution: Interrelationships of the Living Phyla.* Oxford: Oxford University Press.

NIELSEN, C.O., CHRISTENSEN, B. 1959. The Enchytraeidae, critical revision and taxonomy of European species. Nat. Jutl. 8–9, 1–160.

PURSCHKE, G., WESTHEIDE, W., ROHDE, D., BRINKHURST, R.O. 1993. Morphological reinvestigation and phylogenetic relationship of *Acanthobdella peledina* (Annelida, Clitellata). Zoomorphology 113, 91–101.

ROUSE, G.W., JAMIESON, B.G.M. 1987. An ultrastructural study of the spermatozoa of the polychaetes *Eurythoe complanata* (Amphinomidae), *Clymenella* sp. and *Micromaldane* sp. (Maldanidae), with definition of sperm types in relation to reproductive biology. J. Submicrosc. Cytol. 19, 573–584.

ROUSSET, V., PLEIJEL, F., ROUSE, G.W., ERSÉUS, C., SIDDALL, M.E. 2007. A molecular phylogeny of annelids. Cladistics 23, 41–63.

SIDDALL, M.E., BURRESON, E.M. 1996. Leeches (Oligochaeta?: Euhirudinea), their phylogeny and the evolution of life-history strategies. Hydrobiologia 334, 277–285.

SIDDALL, M.E., APAKUPAKUL, K., BURRESON, E.M., COATES, K., ERSÉUS, C., GELDER, S.R., KÄLLERSJÖ, M., TRAPIDO-ROSENTHAL, H. 2001. Validating Livanow: molecular data agree that leeches, branchiobdellidans and *Acanthobdella peledina* form a monophyletic group of oligochaetes. Mol. Phylogenet. Evol. 21, 346–351.

SIDDALL, M.E., BELY, A.E., BORDA, E. 2006. Hirudinida. In *Reproductive Biology and Phylogeny of Annelida*, edited by G. Rouse and F. Pleijel. Enfield, NH: Science Publisher Inc., pp. 235–392.

WESTHEIDE, W., PURSCHKE, G. 1996. Proacrosome and acrosome of the spermatozoon in *Acanthobdella peledina* (Annelida: Clitellata). Invertebr. Reprod. Dev. 29, 223–230.

WESTHEIDE, W., PURSCHKE, G., MIDDENDORF, K. 1991. Spermatozoal ultrastructure of the taxon *Enchytraeus* (Annelida, Oligochaeta) and its significance for species discrimination and identification. Z. zool. Syst. Evolutionsforsch. 29, 323–342.

Chapter 18

Clitellate Cocoons and Their Secretion

Jon'elle Coleman and Daniel H. Shain

Biology Department, Rutgers, The State University of New Jersey, Camden, NJ

Oft disregarded leech, ignored at best;
At worst, in human ignorance reviled;
Lessons to be gained from your nature mild
Are stained instead by blood that you ingest.

Knowledge would alleviate this unrest
And profess the grace in which you are styled;
Perhaps then would others be as beguiled
As I, and their minds be likewise impressed.

One need contemplate your cocoon alone,
Which you secrete with skill and fill with life,
Then slip from yourself, opercula sealed,
Congealed and affixed beneath log or stone,
To see—through science's study and strife—
The delicate handwork of God revealed.

—*Sonia Krutzke*

18.1 INTRODUCTION

Clitellate annelids are a group of metazoan invertebrates comprising more than one-third of ~15,000 described species of segmented worms (Reynolds and Cook 1993; Rouse and Pleijel 2001; Erséus 2005); one feature distinguishing them from polychaetes is a glandular clitellum from which cocoon(s) are secreted. Clitellates are common in both terrestrial (mainly moist soil or vegetation) and aquatic (primarily freshwater) habitats worldwide. As a result of their ubiquity, they encounter diverse environments (e.g. drought, heat, cold) and thus have developed several reproductive strategies, one of which is sexual reproduction involving the formation of protective egg capsules or cocoons (Mehra 1920; Dales 1970; Learner et al. 1978; Loden 1981; Kutschera and Wirtz

Annelids in Modern Biology, Edited by Daniel H. Shain
Copyright © 2009 John Wiley & Sons, Inc.

2001). Cocoons offer developing embryos a homeostatic microenvironment in which they can endure unfavorable conditions (e.g. environmental stress, predation, microbial degradation).

The materials required for cocoon production are synthesized and secreted from the clitellum, which becomes externally prominent when worms are sexually mature as a direct result of the increased size and secretory content of clitellar gland cells (CGCs). A combination of CGC secretions produces the complete cocoon. The location of CGCs within the clitellum dictates the formation of the cocoon, more typically as a sheath surrounding the clitellum, but some species (e.g. *Helobdella papillornata*, *Theromyzon trizonare*) secrete a sac from a ring of CGCs that surround the female gonopore.

The cocoon initially forms as a soft, transparent structure that gradually hardens to form a protective covering. Immediately after cocoon formation, eggs are deposited into the cocoon along with a proteinaceous fluid that nourishes the young and/or maintains a microenvironment suitable for development. The worm typically removes the cocoon by passing it over its head and sealing it with plug-like structures known as opercula (Latin for "little lid"). Once the cocoon is deposited onto a substrate, the parent may shape the cocoon with its mouth and/or anterior sucker to maximize the area contacting the substrate, thus ensuring its secure attachment. Juveniles typically exit the cocoon through the opercula, and then lead an independent life. Some leech species (e.g. *Theromyzon* sp.) are not fully developed upon hatching and attach directly to the venter of the parent, and are often carried to their first meal.

Clitellate annelids secrete three distinct cocoon types—hard-shelled, membranous and gelatinous (Siddall and Burreson 1995)—with diverse morphologies and unique physical properties (e.g. thermal, chemical and proteolytic resiliency; Mason et al. 2004). Differences in the size and shape of cocoons, the manner in which they are secreted and deposited, the time required to produce a single cocoon, and the number of cocoons produced by an individual can be attributed, at least in part, to the physical demands of different habitats. For example, earthworms and some terrestrial leeches have developed a spongy outer covering that aids in moisture preservation (Matthai 1921; Herter 1937; Bhatia 1939; Mather 1954; Maitland et al. 2000). By comparison, cocoons produced in an aquatic environment typically comprise a thin, transparent membrane, which may have developed concurrently with a protective brooding behavior (Apakupakul et al. 1999; see below).

Cocoon production may have evolved as a means to ensure reproductive success among species that live in unpredictable environments. While cocoons are an integral part of the clitellate life cycle, their origin remains uncertain. Several theories suggest that cocoon production in Clitellata has developed from marine polychaetes in response to terrestrial invasions (McHugh 1997; Westheide 1997; Kojima 1998; Purschke 1999; Westheide et al. 1999). Polychaetes, which typically disperse their gametes into the open ocean, may have undergone dramatic changes in reproductive strategies upon invading swamps and lake margins. Due to low light levels and scarce food supplies in these habitats, population densities may have been dramatically decreased, leaving potential partners in limited supply. This may have led to the hermaphroditic nature (i.e. the same animal functions as both male and female) of clitellates, providing them with several reproductive advantages over polychaetes, including the ability to mate with any individual they encounter, and in the case of protandrous hermaphrodites (i.e. those species that exist first as males, then as females), all individuals are guaranteed a partner and are also capable of egg laying themselves.

18.2 REPRODUCTIVE BIOLOGY

Clitellate annelids share a variety of reproductive behaviors. For example, they are her-maphroditic with the sexual reproductive organs located in separate segments (Mann 1962). Some clitellates are capable of both sexual and asexual reproduction (e.g. fission, fragmentation) and may even alternate between modes (Loden 1981; Davies and Singhal 1988). An example of the latter is demonstrated by some Naididae species (Oligochaeta) that usually reproduce asexually, but in times of environmental stress will secrete cocoons and reproduce sexually (Loden 1981).

Though some clitellates are capable of self-fertilization (e.g. the freshwater oligo-chaete *Tubifex tubifex*) or parthenogenesis (e.g. Lee 1985; Fragoso et al. 1999), the majority cross-fertilize by copulation. In leeches, cross-fertilization is accomplished either with an intromittent organ or a sperm-filled sac called a spermatophore (Michaelsen 1926). Spermatophores are found in members of the families Erpobdellidae, Glossiphoni-idae and Piscicolidae and involve hypodermic insemination (Brandes 1901), i.e. the direct transfer of sperm to the ventral body surface of the partner, typically in the clitellar region (Mann 1962). Sperm exit the spermatophore and find their way via the coelomic sinuses to ovaries, where eggs are fertilized (Brandes 1901; Davies and Govedich 2001). The time it takes the sperm to reach the ovaries often results in a delay up to several days between copulation and cocoon production in these species (Davies and Govedich 2001).

In contrast, fertilization in earthworms occurs externally within the cocoon (e.g. *Eisenia fetida*; Grove 1928). Sperm are first stored in chambers called spermathecae, and after cocoon formation, eggs and sperm are simultaneously released into the cocoon. Some leech species (e.g. *Hirudo medicinalis, Helobdella stagnalis*) store sperm in spermathecae during adverse environmental conditions; sperm can be stored, sometimes for months, until favorable conditions arise (Davies and Reynoldson 1976).

18.3 CLITELLUM AND CGCs

The clitellum is a transient structure that becomes externally visible in sexually mature clitellates, forming a vesicular swelling that extends over a variable number of anterior segments. The prominence of the clitellum is variable across clitellate species, ranging from single to multiple cell layers. Mature megadriles (e.g. earthworms) are characterized by a multilayered clitellum that is externally obvious as a dark band. In contrast, the single-layered clitellum of microdriles (e.g. Tubificidae, Enchytraeidae) and nearly all Hirudinea (Erséus 2005) is barely distinguishable from the rest of the tegument. Note that additional clitellar layers correlate with the process of albumenotrophy, in which yolk is secreted into the cocoon fluid instead of being internalized within the egg (e.g. Crassiclitellata; Jamieson 1988).

The morphological changes observed in the clitellum correlate with dynamic, clitel-lum-specific cellular events, specifically the increased size and secretory content of CGCs, which are thought to produce distinct components of the cocoon (Hirao 1965; Richards 1977; Sawyer 1986; Sayers et al. submitted). Authors have identified between two to six different CGC types depending on the species and the histological stains employed (e.g. Grove and Cowley 1927; Hess and Vena 1974; Malecha and Prensier 1974; Richards 1977; Suzutani 1977; Suzutani-Shiota 1980; Fleming and Baron 1982). In *Theromyzon tessula-tum*, four CGC types have been identified (types II–IV), each of which is distinguished by its morphology and secretory content (Sayers et al. submitted). Cocoon-associated

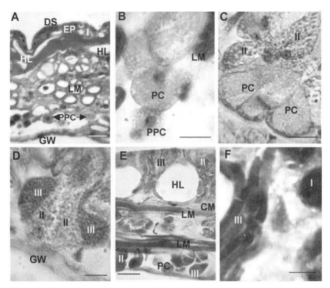

Figure 18.1 Temporal development and differentiation of type II and type III cells. (A) PPCs in deep muscle tissue. (B) PPCs hypertrophied into clusters of precursor cells (PCs) that were initially stained blue in Masson's trichrome. (C) PC differentiated into type II (blue granules) or type III (red granules) cells depending upon their position within the clitellum. (D–G) Type II/III cells elongate and form tubular processes that extend to the epithelial surface and fill with granules in the days leading up to cocoon secretion. Cell types I–III are indicated (I, II, III). Scale bars: A, E = 20 mm; B–D = 10 mm; E = 5 mm. Abbreviations: CM = circular muscle; DS = dorsal surface; EP = epithelium; GW = gut wall; HL = hypodermal coelomic lacunae; LM = longitudinal muscle. (See color insert.)

cellular events in *T. tessulatum* begin ~1 week prior to cocoon secretion with the hypertrophy of clitellum-specific pro-precursor cells (PPCs) and their subsequent proliferation, which forms clusters of precursor cells (PCs) throughout the clitellum (Fig. 18.1). PCs differentiate into either type II or type III cells depending upon their position within the clitellum; the former are highly concentrated along the ventral and dorsolateral edges, while the latter are localized near the dorsal midline. These two cells types contribute material either to the opercula (type II) or the cocoon membrane (type III), while other cell types appear to make only minor contributions to the cocoon.

Piscicola geometra has been reported to have five CGC types: type 1 produces an acid mucopolysaccharide; type 2a produces the cocoon's fibrillar component and type 2b produces operculum material; type 3 produces the dense mass of the cocoon wall, and Type 4 produces an albuminous food source to feed developing embryos (Malecha and Prensier 1974; Malecha and Vinckier 1983). Differences reported in the number and distribution of CGCs among clitellate species seem due in part to the cocoon type, developmental mode (i.e. albumenotrophy versus internalized yolk) and, in some instances, the stage of the life cycle in which specimens have been examined.

18.4 COCOON PRODUCTION

Most studies of clitellate cocoon synthesis have focused on the hardened cocoons of earthworms and erpobdellid leeches (Brumpt 1900; Grove and Cowley 1927; Oishi 1930; Pawlowski 1955; Nagao 1957; Hirao 1965; Sawyer 1970; Albert 1975; Westheide and

Muller 1996). Though this process is superficially similar among terrestrial and aquatic clitellates, different habitat conditions likely contribute to variations in the size and shape of cocoons, the manner in which cocoons are produced and deposited, the number of cocoons produced by an individual and the time it takes to produce a single cocoon (Lee 1985; Edwards and Bohlen 1996).

Size variation among cocoons exists both within and between species, and is largely correlated with the age and / or size of the parent. For example, the giant earthworm *Megascolides australis* can reach 3 m in length and produces a cocoon ~9 cm in length (Van Praagh and Hinkley 2002). Morphological variation in the shape of clitellate cocoons often correlates with the shape of the worm that secretes it. For example, glossiphoniid leeches have dorsoventrally flattened and dorsally convex bodies, and some (e.g. *T. tessulatum*, *Placobdella papillifera*) produce cocoons with a flattened base and a curved upper portion (Davies and Wilkialis 1982; Dimitriu and Shain 2004); species with cylindrical bodies tend to produce spherical cocoons (e.g. piscicolids; Sawyer 1986).

Different habitats may have important selective pressures for generating different cocoon morphologies. Terrestrial-based cocoon production requires special modifications to keep cocoons hydrated and in locations that allow optimal growth (e.g. moist soil and out of direct sunlight). In aquatic environments, the cocoon is usually cemented to a submerged substrate at the water's edge. Differences in the number of cocoons produced by individuals may be attributed to population density (i.e. a decrease in population density typically results in higher number of cocoons per individual) or may be a function of the organism's life cycle. Most oligochaetes and predaceous leeches are semelparous, reproducing only once in their annual or biannual life cycles, while most sanguivorous leeches are iteroparous, requiring several blood meals before reaching maturity, and reproducing multiple times during their life cycle. The former lifestyle (semelparous) requires a higher energy investment in reproduction than the latter (Mann 1957, 1962; Dash and Senapati 1980; Calow and Riley 1982; Sawyer 1986; Bhattacharjee and Chaudhuri 2002), and clutch sizes tend to be larger.

18.4.1 Cocoon Formation and Egg Laying

At the onset of cocoon formation, a mucous secretion [i.e. external mucous layer (EML), slime layer] encompasses the surface of the clitellar epidermis, as demonstrated by the adhesion of sand particles to this region (Hirao 1965). The EML represents the outer cocoon wall and the structural scaffold into which the remainder of the cocoon is deposited (Wilkialis and Davies 1980; Westheide and Muller 1996). Immediately prior to cocoon production, the parent prepares the substrate (e.g. rock, log) with a secretion of cementitous, glue-like material that secures the cocoon and / or may provide a source of resistance, allowing the parent to withdraw itself from the cocoon (Ditlevsen 1904). The process of preparing a substrate with a glue-like material is not restricted to clitellates; some marine arthropods (e.g. barnacles) use a proteinaceous cement to permanently attach to a substrate (Kamino et al. 2000).

In most clitellates, a tubular sheath forms around the clitellum with its ends tightly compressed against the worm's body. The time frame of cocoon production varies from ~20 min to several hours depending upon the cocoon's size. Those species that produce multiple cocoons in a clutch (i.e. glossiphoniid leeches) do so in rapid succession, depositing each cocoon onto the substrate as it is produced (e.g. *T. tessulatum*—Wilkialis and Davies 1980; *P. papillifera*—Davies and Wilkialis 1982).

Following copulation, cocoon secretion and egg laying occur within days (e.g. for *Erpobdella* and most earthworms), weeks or sometimes months (e.g. for *Hirudo* sp.) (Dales 1970; Wilkialis and Davies 1980; Kutschera and Wirtz 2001). Descriptions of egg laying have focused on aquatic leeches (e.g. *Erpobdella punctata*, Sawyer 1970; *T. tessulatum*, Wilkialis and Davies 1980), but this process is likely different in terrestrial clitellates (e.g. megadriles) and in those without suckers (megadriles, microdriles). Egg laying by the aquatic leech *T. tessulatum* involves the attachment of the leech's posterior sucker to the substrate (e.g. rock, log), and the simultaneous raising and twisting of its anterior end in an almost complete 360° rotation, a process that is repeated several times (Wilkialis and Davies 1980). This twisting appears to facilitate the movement of eggs from the female gonopore and the subsequent deposition of eggs (and possibly cocoon fluid) into the cocoon, and may be a mechanism to loosen the cocoon wall from the leech's body to allow cocoon removal (Sawyer 1970).

The number of eggs within a cocoon varies between species and among individuals of the same species, correlating with the age/size of the parent (Sawyer 1986). Most earthworm species (Bhattacharjee and Chaudhuri 2002) and piscicolid leeches (Sawyer 1986) typically have only one hatchling emerge from a cocoon, while other species have >100 embryos per cocoon (e.g. glossiphoniid leeches; Wilkialis and Davies 1980; Mason et al. 2004).

Several factors govern the development time of the young, including temperature (e.g. higher temperatures tend to shorten development time; Kaster 1980), food availability, and the amounts of ventilation and protection. Embryos that internalize yolk within their cells (e.g. primitive oligochaetes and glossiphoniid leeches) are guaranteed a constant food supply even if they develop independently from their cocoon (Shankland and Savage 1997). Such species develop rapidly and hatch at an advanced developmental stage, whereupon the likelihood of survival is significantly increased (Govedich and Davies 1998). Embryos that do not contain internalized yolk (albumenotrophy) receive their nutrition from an albuminous fluid inside the cocoon (Anderson 1973). This proteinaceous fluid is contained within the lumen of all clitellate cocoons; for those species that internalize yolk, cocoon fluid may function as a buffering and/or antimicrobial agent.

18.4.2 Cocoon Deposition

The most common form of clitellate cocoon deposition is a quick withdrawal of the worm's anterior end through the cocoon (Hirao 1965), a process that is briefly halted in clitellates requiring external fertilization (e.g. earthworms; to allow anteriorly positioned spermathecae to release sperm into the cocoon). Alternation between swelling of the worm's posterior and constriction of the preclitellar region likely facilitates movement of the cocoon toward the worm's anterior end (Nagao 1957). As the head is retracted through the cocoon, first the anterior, then the posterior cocoon ends are sealed with glue-like plugs (i.e. opercula). Cocoons secreted from the ventral surface (e.g. *Haementeria ghilianii*), however, form a sac originating from the periphery of the female gonopore and contain only a single operculum.

Following deposition, cocoons may attach either to a substrate or directly to the parent's venter. The site of cocoon attachment is often determined by the type of substrates available (Sawyer 1970). For example, amphibious oligochaetes (e.g. the medicinal leech *H. medicinalis*) deposit their cocoons in moist habitats, such as damp soil or under structures proximal to the water's edge (e.g. rotten leaves), while cocoons deposited by aquatic leeches may be attached to a submerged substrate (e.g. underside of a rock or wood) or to the body of a host organism (e.g. crayfish in the case of *Branchiobdella pentodonta*; Farnesi and Tei 1975). Cocoons may also attach directly to the parent's venter (e.g. *T. trizonare, Helobdella* sp.; Davies and Oosthuizen 1992; Siddall et al. 2005) with the advantage that the parent is mobile following egg laying.

The shape of cocoons likely contributes to how securely they adhere to a substrate (Mann 1962). Spherical cocoons secreted by members of the family Hirudinidae are loosely attached (contacting at only a single point), while cocoons with a flattened base and therefore larger surface area (e.g. Glossiphoniidae, Erpobdellidae) adhere more tightly (Mann 1962). Following deposition, the parent may shape the cocoon using its mouth and/or anterior sucker, stretching the cocoon to maximize the area contacting the substrate (Brumpt 1900; Sawyer 1970). This process likely enables the spherical cocoons of piscicolid leeches to securely adhere to the surface of their fish hosts.

Freshly laid cocoons are soft and vulnerable to predation but later harden to form a protective covering, and often appear dark brown (Sawyer 1970). The time required for hardening ranges from a few hours (e.g. *E. punctata*; Sawyer 1970) to a few days (e.g. *Erpobdella lineata*; Nagao 1957). During the hardening process, the parent may remain over the cocoon and ventilate it, a process that may expedite hardening of the membrane in addition to protecting the embryos (Pawlowski 1955; Sawyer 1970). Many oligochaetes and arhynchobdellid leeches (including *E. lineata*), however, abandon their cocoons immediately after shaping them, and many of these cocoons are lost to predation (Pawlowski 1955; Kutschera 1984).

According to Sawyer (1970), an increased duration of ventilation may have evolved into a protective brooding behavior that is exhibited by members of the family Glossiphoniidae. Many glossiphoniids secrete membranous or gelatinous cocoons that do not have the protective covering typical of hard-shelled cocoons (Apakupakul et al. 1999). In an apparent effort to prevent predation of the offspring, these species brood their eggs until they are at an advanced developmental stage (Sawyer 1971; Davies and Govedich 2001; Kutschera and Wirtz 2001).

Juveniles exit the cocoon through holes formerly occupied by opercula, an event that has been observed in several species (Reinecke and Viljoen 1988; Westheide and Muller 1996; Mason et al. 2005). Opercula are typically located at the terminal ends of the cocoon; however, some leech species (e.g. *T. tessulatum*) produce cocoons in which opercula are positioned asymmetrically on the upper aspect (noncemented side) of the cocoon membrane, which allows direct contact between the brooding parent and its emerging young as they hatch (Mason et al. 2005).

Several ideas have been proposed regarding the mechanism(s) utilized for hatching from the cocoon, all of which portray a weakening of the operculum/membrane boundary as the likely cause, for example, the secretion of chemicals that soften this boundary or mechanical stress resulting from juveniles colliding with inwardly protruding opercula (Westheide and Muller 1996). We have proposed that opercula contain glycosylated proteins, in which case microbial degradation of the "sugar coat" may weaken the opercula and facilitate hatching (Coleman et al. 2008).

18.5   BROODING BEHAVIOR WITHIN GLOSSIPHONIIDAE

Parental care, characterized by a protective brooding behavior, is known only within the family Glossiphoniidae (e.g. *Haementeria, Helobdella* and *Theromyzon*), suggesting that this behavior arose de novo within this family (Stephenson 1930; Herter 1937; Autrum

Figure 18.2 Terminal stages of the *Theromyzon tessulatum* life cycle. Four juveniles attached to the venter of the parent with the posterior sucker. Anterior is up. Scale bar = 2 mm.

1939; Grasse 1959; Sawyer 1971, 1986; Siddall and Burreson 1996). Parental care is unusual among invertebrates and likely evolved in glossiphoniid leeches as a mechanism for protecting the young against predation (Sawyer 1971).

According to Sawyer (1971), parental care can be arranged in a series of increasing complexity, generating an evolutionary time line of this behavior. Beginning with the most primitive mode of parental care, the parent leech deposits its cocoon(s) onto a substrate (e.g. rock, log), covers the cocoon(s) with its flattened body and broods its eggs (Sawyer 1971; Kutschera and Wirtz 2001; Govedich 2004). The next level is demonstrated by *H. stagnalis*, which affixes cocoons directly to its ventral surface rather than to an inanimate object—when the parent is confronted by predators, it forms a temporary brood pouch using its lateral margins and/or a midventral invagination (Moore 1959; Sawyer 1971, 1972; Van der Lande and Tinsley 1979; Kutschera and Wirtz 1986, 2001; Siddall et al. 2005). The South African glossiphoniid leech *Marsupiobdella africana* carries its young on its ventral surface in a specialized internal brood pouch (Sawyer 1971; Siddall et al. 2005), while other glossiphoniids have been observed to feed their young (e.g. some species of *Helobdella* and *Glossiphonia*) (Milne and Calow 1990; Davies and Govedich 2001; Kutschera and Wirtz 2001), often offering up the meal they would otherwise have for themselves.

Upon hatching through the opercula, juveniles often attach to the ventral surface of the parent leech via an embryonic attachment organ (Nagao 1958) and are carried for varying lengths of time before their first meal, at which time the young are able to lead an independent life (Fig. 18.2; Mason et al. 2005).

18.6 COCOON STRUCTURE: SURFACE TOPOLOGY AND ULTRASTRUCTURAL PROPERTIES

Several structural similarities are shared between different cocoon types (primarily hard-shelled and membranous), including a strong, impervious membrane, and one or more glue-like opercula (Fig. 18.3). Nevertheless, distinct structural characteristics occur

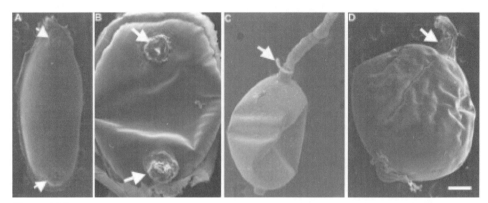

Figure 18.3 Morphology of various clitellate cocoons. (A) *Erpodbella punctata*: arrows identify opercula. (B) *Theromyzon tessulatum*: upper arrow—inward operculum, lower arrow—outward operculum. (C) *Tubifex tubifex*, with juvenile exiting: arrow identifies operculum. (D) *Eisenia fetida*: arrow identifies operculum. Scale bar = 100 mm.

between cocoon types (e.g. variation in step angles between successive layers, fibrillar organization) that may account for differences in the mechanical strength, rigidity and resilience of the membrane types (Dimitriu and Shain 2004; Marotta and Shain 2007). While the majority of clitellate cocoons comprise a tough, protective membrane, those of most aquatic leeches are soft and flexible (with exceptions including the erpobdellid leeches).

The membranous *T. tessulatum* cocoon (ovoid, ~3 mm length; Dimitriu and Shain 2004) comprises a flattened base that is cemented to a substrate and an upper curved portion containing two asymmetrically positioned opercula. These upper and lower regions are divided by a circuitous bulge likely formed by the solidification of a cementitious material that functions to "glue" cocoons to the substrate, as well as to adjacent cocoons. It remains unclear whether this cementitious material shares similar adhesive/structural properties with opercula. The initial secretion that encompasses the clitellar epidermis during cocoon production corresponds to an amorphous mucous layer (i.e. EML) that covers the outer surface of the cocoon, constituting the material into which the remaining fibrillar portion of the cocoon wall is deposited.

18.6.1 Organization of the *T. tessulatum* Cocoon Membrane

The *T. tessulatum* cocoon comprises a multilayered membrane, with each layer containing an array of unidirectional fibrils (Dimitriu and Shain 2004; Marotta and Shain 2007). Fibrils may be organized as cables (e.g. *T. tessulatum* cocoon) or planes (e.g. *E. punctata* cocoon) (Fig. 18.4; Marotta and Shain 2007). Successive fibrillar layers are stacked upon each other at various angles, with no detectable periodicity (Marotta and Shain 2007). Step-angle values are lower for hard-shelled *E. punctata* cocoons (7.5°–62.5°) than for the membranous cocoons of *T. tessulatum* (12.5°–87°), which may account for differences in their rigidity (Marotta and Shain 2007). A supramolecular organization, characterized by parabolic (i.e. C- and S-like) patterns of bow-shaped lines (Fig. 18.5; Dimitriu and Shain 2004), is typical among fibrous biological tissues displaying a twisted, plywood-like architecture (Bouligand 1965, 1972; Knight and Hunt 1974; Gaill and Bouligand 1987;

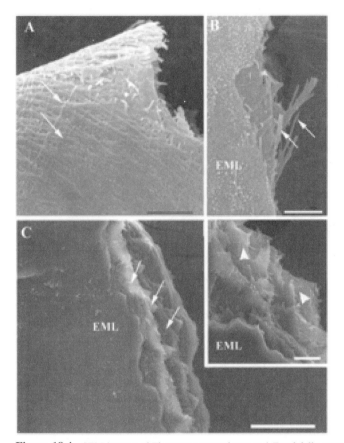

Figure 18.4 SEM images of *Theromyzon tessulatum* and *Erpobdella punctata* cocoon membranes. (A) Interior view of the *T. tessulatum* membrane. Note the apparently random pattern of interwoven cables (arrows). Scale bar = 12 μm. (B) Fracture through the *T. tessulatum* membrane displaying cables of various diameters (arrows). Scale bar = 5 μm. (C) Fracture through the *E. punctata* membrane displaying layers (arrows) stacked on each other. Scale bar = 60 μm. Inset: higher magnification of the layers (arrowheads) constituting the *E. punctata* cocoon membrane. Scale bar = 5 μm.
Source: Modified from Marotta and Shain (2007), with permission from Elsevier Limited.

Neville 1993; Giraud-Guille et al. 2003) and originates from the passage of the protein through a cholesteric liquid mesophase during polymerization (Friedel 1922; Bouligand 1972; Feher and Kam 1985). Bouligand (1972) predicted that the fibrous component of twisted tissue is initially released in the form of liquid crystals; once nucleated, these may transform into solidified fibrils (Friedel 1922; Bouligand 1972; Feher and Kam 1985).

TEM analysis of *Tubifex hattai* cocoons reveal the presence of vesicles within the clitellum containing polymerized fibrils prior to their release (Suzutani-Shiota 1980), suggesting that fibril polymerization and cocoon hardening may be independent processes. In *T. tessulatum*, cocoon hardening is a gradual process, as evidenced by depressions on the ventral and dorsolateral surfaces of the inner cocoon membrane, which conform to the shape of embryos (Coleman et al. 2008).

The *T. tessulatum* EML is disrupted by small, exteriorly projecting protuberances (height = 0.5 μm, semi-width = 0.3 μm), i.e. triangular prisms arranged parallel to each other, and separated by a distance of ~1.6 μm (Dimitriu and Shain 2004; Dimitriu et al.

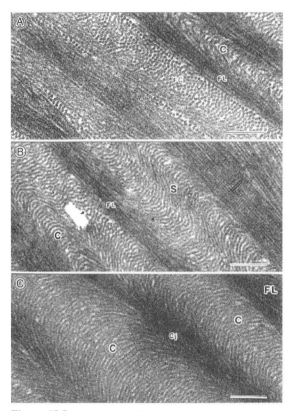

Figure 18.5 Transmission electron microscopy sections through the *Theromyzon tessulatum* cocoon membrane. (A) Section perpendicular to the cocoon membrane; layers of cross-sectioned fibrils (FC) alternate with layers of fibrils sectioned longitudinally (FL). A limited zone shows a C-like pattern of bow-shaped lines (C). (B) Section oblique through the cocoon membrane. Layers of bow-shaped lines, in C-like patterns (C) and S-like patterns (S), alternate with layers of fibrils sectioned longitudinally (FL). An 11-nm banding pattern (lines) is superimposed over the fibrils. The arrowhead shows a space having a polygonal shape. (C) The C-like patterns (C) of bow-shaped lines predominate. C_j represents the junction between two C-like patterns. FL represents fibrils sectioned longitudinally. Scale bars = 200 nm.

Source: Modified from Dimitriu and Shain (2004), with permission from Elsevier Limited.

2006). Protuberances share some elements of the twisted arrangement and are likely formed by tension in the cocoon membrane between opercula (Dimitriu et al. 2006; Coleman et al. 2008). They are localized to the exterior surface of the cocoon membrane and can be confused with relief folds, which are found in regions surrounding opercula on the upper aspect of the *T. tessulatum* cocoon membrane. Relief folds likely result from the constriction of cocoon ends as opercula form (Coleman et al. 2008).

18.6.2 Opercula

Opercula (each ~200 μm in diameter) are the most prominent features in the *T. tessulatum* cocoon, accounting for ~40% of the cocoon's total volume (Coleman et al. 2008). Opercula of an individual cocoon appear to differ in their spatial orientation, protruding either into the cocoon lumen or toward the cocoon's exterior (presumably the anterior and posterior

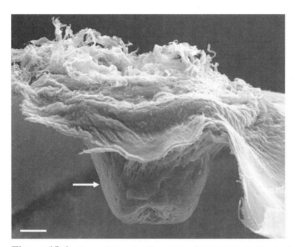

Figure 18.6 Inwardly protruding operculum. Arrow indicates the portion of the operculum that projects into the cocoon lumen. Scale bar = 50 μm.

Source: Reprinted from Coleman et al. (2008), with permission from John Wiley & Sons, Inc.

opercula, respectively; Nagao 1957; Coleman et al. 2008). These differences likely result from the mechanical process by which most clitellates remove cocoons (i.e. over the anterior end). Inwardly protruding opercula extend ~200–300 μm into the cocoon lumen (Fig. 18.6).

In contrast to the fibrous architecture of the membrane, opercula appear to be secreted in an unpolymerized state that is morphologically similar (i.e. microporous arrays of closed cells ranging in size from 1.0 to 7.5 μm) to the tube cement secreted by some marine polychaetes (e.g. *Pharmacopia californica*; Stewart et al. 2004). TEM analysis of *T. tessulatum* opercula has revealed at least four organized regions of variable electron density that may correlate with different phases of polymerization (R. Marotta, pers. comm.). Interestingly, some opercula appear to be deposited in a spiral-like pattern consistent with the 360° rotation of the worm's anterior end during cocoon production and egg laying (Wilkialis and Davies 1980; Coleman et al. 2008).

18.6.3 *Theromzyon* Cocoon Protein (Tcp)

T. tessulatum cocoons are resilient to heat and denaturing reagents (e.g. autoclaving in chaotropic salts and reducing agents; Mason et al. 2004). Upon incubating cocoons with 10% acetic acid at 50 °C for several hours, however, Mason et al. (2004) successfully isolated two protein fragments (~35 and 40 kDa) that represent a major structural protein from the *T. tessulatum* cocoon, designated Tcp (Mason et al. 2004). Tcp contains an unusually high cysteine content (~17.8%), which results from six repeats of a conserved series of 12 ordered Cys residues (Mason et al. 2004, 2006). The first six Cys residues in each repeat (Cys 1–6) share strong sequence similarity with antistasin (a leech anticoagulant; Han et al. 1989), and the region from Cys 7 to Cys 12 within each repeat is similar to notch, a key regulator of cell fate throughout the Animalia (Vardar et al. 2003; Karon 2003). Interestingly, Tcp has several hypothetical homologues in metazoan phyla including Nematoda and Arthropoda, which contain a single Cys 1–12 repeat flanked by a putative membrane-spanning domain. A Tcp homologue from the hard-shelled cocoon secreted by

Table 18.1   Solubility the *Theromyzon tessulatum* cocoon

Denaturants	Membrane	Opercula	EML
Distilled water (1 h at 121 °C)	−	−	−
1% SDS (1 h at 100 °C)	−	−	+
6 M GIT (1 h at 121 °C)	−	−	−
1% β-ME (1 h at 121 °C)	−	−	−
6 M GIT/1% β-ME (1 h at 121 °C)	−	+	+
Tween 20 (1 h at 100 °C)	−	−	−
Formamide* (1 h at 121 °C)	−	+/−	−
10% acetic acid (4 h at 50 °C)	−	+	+
1 N NaOH (20 min at 50 °C)	+	+	+
1 N HCl (72 h at 22 °C)	−	+	−
6 N HCl (1 h at 22 °C)	+	+	+
Proteinase K (1 h at 50 °C)	+	+	+

*Formamide only solubilized relatively old opercula (e.g. ~6 months after secretion).
Categories: − = insoluble; + = soluble.

the distantly related leech *E. punctata* was found to share ~60% sequence identity with Tcp (Lopez Coral 2007), suggesting that Tcp-like proteins may occur throughout the Clitellata. Utilizing crystal structures and known Cys cross-linking patterns from each domain (i.e. antistasin and notch), Mason et al. (2006) proposed a protein model for Tcp that appears to be stabilized predominantly by hydrophobic associations (Mason et al. 2004, 2006), consistent with the thermostabile properties of the *T. tessulatum* cocoon.

Scanning electron microscopy (SEM) was employed to assess the cocoon's structural integrity upon exposure to various denaturing conditions (Table 18.1). Interestingly, opercula are more susceptible to denaturing conditions (e.g. autoclaving in chaotropic salt) than the cocoon membrane, consistent with their role as an escape route for hatching juveniles (Mason et al. 2004; Coleman et al. 2008). Opercula were solubilized by several treatments (e.g. autoclaving in guanidinium isothiocyanate containing beta-mercaptoethanol—a chaotropic salt and a reducing agent, respectively), while the *T. tessulatum* cocoon membrane was resilient to these treatments. A 10% acetic acid solution resulted in the selective solubilization of the opercula and caused dissociation of the EML; thus, the original source of the Tcp protein described by Mason et al. (2004) is likely the EML and/or the opercula. The latter is favored, however, since the opercula comprise ~40% of the cocoon's total volume (compared with ~5% for the EML), and thus more likely accounts for the robust proteins bands generated in the original solubilization experiments (Mason et al. 2004).

18.6.4 Composition of Clitellate Cocoons

The *T. tessulatum* cocoon membrane and opercula have essentially identical amino acid profiles (Table 18.2; Mason et al. 2005), suggesting that these two structures are built from the same protein (i.e. Tcp). Their very different morphologies, however, raise the possibility that the cocoon membrane and opercula represent different polymerization states and/or are differentially posttranslationally modified. A variety of molecules (e.g. mucopolysaccharides, glycoproteins) and inorganic elements (e.g. phosphorous, magnesium, calcium) are associated with extramural invertebrate proteins (Defretin 1971; Truchet and

Table 18.2 Amino acid composition of *Theromyzon tessulatum* cocoon components (Approximate molar %)

Amino Acid Residue	Predicted Tcp Protein[a]	Membrane	Membrane—Less EML	Opercula
Asp + Asn	7.4	10.5	10.3	11.3
Thr	4.4	7.8	7.6	8.3
Ser	4.2	10.4	12.0	10.2
Glu + Gln	13.0	8.3	8.1	9.3
Pro	8.8	10.0	10.8	8.6
Gly	5.9	11.2	12.7	11.4
Ala	5.6	5.4	5.9	5.3
Cys	17.9	6.5	4.2	2.5
Val	10.3	4.5	4.0	4.3
Met	0.2	0.3	0.7	1.9
Ile	3.7	3.6	3.8	4.4
Leu	4.4	4.5	4.2	5.6
Tyr	3.4	2.4	1.9	1.8
Phe	2.0	2.2	2.3	2.8
Lys	5.1	6.7	5.9	6.4
His	0.7	1.2	1.1	1.3
Arg	2.9	4.5	4.7	4.6
Total	100	100	100	100

[a]Mason et al. (2004).

Vovelle 1977; Vovelle 1979; Gaill and Hunt 1986; Gaill and Bouligand 1987; Jensen and Morse 1988; Talmont and Fournet 1991; Stewart et al. 2004). For example, tube cements secreted by marine polychaetes (e.g. *Alvinella pompejana*) contain methylated glycoproteins (Talmont and Fournet 1991). Similarly, differential staining of the *T. tessulatum* cocoon reveals that opercula and the EML stain with aniline blue suggesting polysaccharide content, while the membrane is proteinaceous (i.e. staining red with azocarmine; Sayers et al. submitted). Likewise, Mason et al. (2004) determined that the cocoon is proteinaceous, as both the cocoon membrane and opercula were solubilized with proteinase K and had similar amino acid composition profiles. Taken together, it appears that the opercula may be heavily glycosylated (consistent with aniline blue staining) and its "sugar coat" likely interferes with the azocarmine's ability to penetrate into the proteinaceous core. This data suggest that differential posttranslational modification (i.e. glycosylation) of the Tcp protein within CGCs (i.e. type II cells; see above) may be associated with the operculum's unpolymerized state; specifically, the opercula may represent a glycosylated, monomeric form of the polymerized fibrils that constitute the *T. tessulatum* cocoon membrane.

18.7 EVOLUTION OF CLITELLATE COCOONS AND THEIR SECRETION

The construction of distinct cocoon types (i.e. hard-shelled, membranous and gelatinous) in different habitats adds an evolutionary context to cocoon production. Whether these cocoons evolved independently (i.e. convergent evolution), or which—if either—is plesiomorphic among clitellates (or annelids in general) remains a debated topic.

All annelid cocoons, regardless of ultimate form, are initially secreted as an unmodified, transparent egg case that resembles a simple mucous capsule. This structure may have been gradually modified (e.g. perhaps being embedded in a hardened proteinaceous matrix), forming the elaborate structural designs of membranous and hard-shelled cocoons (Dimitriu and Shain 2004; Marotta and Shain 2007). Mann (1962) suggests that membranous cocoons are plesiomorphic to hard-shelled cocoons, but this notion has been disputed by Apakupakul et al. (1999), who argue that more primitive annelids (e.g. *Acanthobdella peledina*; Siddall and Burreson 1995) deposit hard-shelled cocoons. Apakupakul et al. (1999) instead suggest that hard-shelled cocoons and membranous cocoons evolved independently, perhaps as parallel attempts to facilitate adaptations to two very diverse habitats (terrestrial versus aquatic). Thus, no clear evolutionary order can be resolved.

Cocoon production among annelids is not exclusive to clitellates. Several marine polychaetes, from which clitellates are believed to have evolved, also secrete cocoons (Schroeder and Hermans 1975; Giere and Riser 1981; Westheide 1997; Westheide et al. 1999; McHugh 2000; Purschke et al. 2000). Polychaetes, which typically disperse their gametes into the open ocean, may have developed cocoons as a means of ensuring reproductive success upon inhabiting swamps and lake margins where adverse conditions (e.g. low light levels, scarce food supplies) are encountered (Swedmark 1959). Although clitellates have diverged dramatically from polychaetes and also from each other (e.g. compare leeches and earthworms), Clitellata is a monophyletic group that shares one unique feature, the clitellum, and thus the origin of this structure is of considerable evolutionary interest. Interestingly, the major difference between clitellar versus nonclitellar segments in the leech *T. tessulatum* is the presence of a single cell type, the type II/III PPC, which proliferates to form PC clusters (Sayers et al. submitted). The default state of the PC appears to be an alcian blue-staining type II cell, which generates the glue-like opercula. Since disparate clitellate cocoons contain opercula, it seems likely that type II-like cells are present throughout Clitellata, and probably functioned in an ancestral clitellate, as well as the basal polychaete from which clitellates arose.

Many extant polychaetes (and oligochaetes) secrete tubes into which they burrow, and some are constructed by secreting an adhesive that mixes with available sand and debris to form the tube (Jensen and Morse 1988; Stewart et al. 2004; Zhao et al. 2005). We propose that the glandular cell type responsible for secreting cementitious material in polychaetes may be evolutionarily linked to the cell type (i.e. type I-like) that gave rise to the clitellate type II/III PC. In principle, a localized signal leading to deglycosylation could then generate the two major cell types necessary for constructing a clitellate cocoon: one that produces glycosylated protein granules for the opercula (type II), and another that produces deglycosylated, proteinaceous granules for the cocoon membrane (type III).

18.8 BIOMATERIALS APPLICATIONS

Fibrous structural proteins self-assemble into complex material architectures (e.g. silks, collagens, elastins), often functioning as supportive or protective structures in nature. Included among these are the fibrous clitellate cocoons that comprise relatively thin, flexible membranes with extraordinary biophysical properties (e.g. thermostability, chemical/proteolytic resilience) (Mason et al. 2004; Coleman et al. 2008). The controlled deposition of fibrous layers that characterizes cocoon assembly makes this structure appealing as a biomembrane or as a scaffold for use in tissue engineering.

The bioglue, or opercula, that seals the cocoon ends is a naturally produced, and in many cases, underwater adhesive (i.e. in aquatic oligochaetes and leeches). Biological adhesion is generally mediated by insoluble multiprotein complexes (Kamino et al. 2000; Zhao et al. 2005; Kamino 2008), but clitellate opercula may be generated by only one primary protein (i.e. Tcp; Mason et al. 2004). Posttranslational modifications (e.g. glycosylation) appear to modulate the biophysical properties of opercula (e.g. structure, function) by preventing polymerization and may also facilitate adhesion, as does modifications such as 3, 4 dihydroxyphenylalanine (DOPA) in marine invertebrates (e.g. mussels, tubeworms; Jensen and Morse 1988; Anderson and Waite 1998; Zhao and Waite 2006). Recent advances in adhesive technology rely on the use of organic solvents in the development of synthetic adhesives, and therefore these adhesives are generally incompatible with water. Not only are clitellate opercula impervious to water, but they are also ecologically safe and biodegradable; thus, a more comprehensive understanding of the molecular system of clitellate opercula may provide insight into one mechanism of underwater adhesion.

ACKNOWLEDGMENT

This work was supported by NSF grants IBN-0417081000 and DBI-0216233 to D.H.S.

REFERENCES

ALBERT, R. 1975. Zum Lebenszyklus von *Enchytraeus coronatus* Nielsen and Christensen, 1959 (Oligochaeta). Mitt. Hamb. Zool. Mus. Inst. 72, 79–90.

ANDERSON, D.T. 1973. *Embryology and Phylogeny in Annelids and Arthropods, International Series of Pure and Applied Biology*. Oxford: Zoological Division Pergamon Press.

ANDERSON, K.E., WAITE, J.H. 1998. A major protein precursor of zebra mussel (*Dreissena polymorpha*) byssus: deduced sequence and significance. Biol. Bull. 194, 150–160.

APAKUPAKUL, K., SIDDALL, M.E., BURRESON, E.M. 1999. Higher level relationships of leeches (Annelida: Clitellata: Euhirudinea) based on morphology and gene sequences. Mol. Phylogenet. Evol. 12, 350–359.

AUTRUM, H. 1939. Hirudineen. Geographische Verbreitung, Stellung im system und literatur. In *Klassen und ordnungen des tierreichs*, Band 4, Abt. III, Buch 4, Teil 2, edited by H.G. Bronn. Leipzig: Akademische Verlagsgesellschaft, pp. 497–662.

BHATIA, M.L. 1939. The prostomial glands of the Indian leech, *Hirudinaria granulosa*. J. Morphol. 64, 37–46.

BHATTACHARJEE, G., CHAUDHURI, P.S. 2002. Cocoon production, morphology, hatching pattern and fecundity in seven tropical earthworm species—a laboratory-based investigation. J. Biosci. 27, 283–294.

BOULIGAND, Y. 1965. Sur une architecture torsadke repandue dans de nombreuses cuticules d'Arthropodes. C. r. hebd. Sbanc. Acad. Sci., Paris 261, 3665–3668.

BOULIGAND, Y. 1972. Twisted fibrous arrangements in biological materials and cholesteric mesophases. Tissue Cell 4, 189–217.

BRANDES, G. 1901. Die Begattung der Hirudineen. Abh. Naturf. Ges. Halle 22, 1–22.

BRUMPT, E. 1900. Reproduction des Hirudinées. Mem. Soc. Zool. Fr. 13, 286–430.

CALOW, P., RILEY, H. 1982. Observations on reproductive effort in British erpobdellid and glossiphoniid leeches with different life cycles. J. Anim. Ecol. 50, 697–712.

COLEMAN, J., MAROTTA, R., SHAIN, D.H. 2008. Surface topology and structural integrity of the *Theromyzon tessulatum* (Hirudinea: Glossiphoniidae) cocoon. J. Morphol. 269, 812–819.

DALES, R.P. 1970. *Annelids*. London: Hutchinson University Library.

DASH, M.C., SENAPATI, B.K. 1980. Cocoon morphology, hatching and emergence pattern in tropical earthworms. Pedobiologia 20, 317–324.

DAVIES, R.W., GOVEDICH, F.R. 2001. Annelida: Euhirudinea and Acanthobdellidae. In *Ecology and Classification of North American Freshwater Invertebrates*, 2nd edn., edited by J.H. Thorp and A.P. Covich. San Diego: Academic Press, pp. 465–504.

DAVIES, R.W., OOSTHUIZEN, J.H. 1992. A new species of duck leech from North America formerly confused with *Theromyzon rude* (Rhynchobdellida: Glossiphoniidae). Can. J. Zool. 71, 770–775.

DAVIES, R.W., REYNOLDSON, T.B. 1976. A comparison of the life-cycle of *Helobdella stagnalis* (Li. 1758) (Hirudinoidea) in two different geographic areas in Canada. J. Anim. Ecol. 45, 457–470.

DAVIES, R.W., SINGHAL, R.N. 1988. Cosexuality in the leech, *Nephelopsis obscura* (Erpobdellidae). Int. J. Invertebr. Reprod. Dev. 13, 55–64.

DAVIES, R.W., WILKIALIS, J. 1982. Observations on the ecology and morphology of *Placobdella papillifera* (Verrill) (Hirudinoidea: Glossiphoniidae) in Alberta, Canada. Am. Midl. Nat. 107, 316–324.

DEFRETIN, R. 1971. The tubes of polychaete annelids. In *Comprehensive Biochemistry*, edited by M. Florkin and E.H. Stotz. Amsterdam: Elsevier, pp. 713–747.

DIMITRIU, C., SHAIN, D.H. 2004. Ultrastructural properties of the *Theromyzon* (Annelida: Hirudinae) cocoon membrane. Micron. 35, 281–285.

DIMITRIU, C., SAYERS, C.W., COLEMAN, J., SHAIN, D.H. 2006. Two-dimensional ultrastructural elements lead to three-dimensional reconstruction of protuberances on the cocoon membrane of the leech, *Theromyzon tessulatum*. Tissue Cell 38, 35–41.

DITLEVSEN, A. 1904. Studien an Oligochaten. Z. Wiss. Zool. 77, 389–480.

EDWARDS, C.A., BOHLEN, P.J. 1996. *Biology and Ecology of Earthworms*, 3rd edn. London: Chapman and Hall.

ERSÉUS, C. 2005. Phylogeny of oligochaetous Clitellata. Hydrobiologia 535/536, 357–372.

FARNESI, R.M., TEI, S. 1975. Histochemical and ultrastructural studies of the cocoon of *Branchiobdella pentodonta* with (Annelida: Oligochaeta). Boll. Soc. Ital. Biol. Sper. 51, 1184–1189.

FEHER, G., KAM, Z. 1985. Nucleation and growth of protein crystals: general principles and assays. Methods Enzymol. 114, 77–111.

FLEMING, T.P., BARON, P.J. 1982. The histochemistry of the clitellum of *Tubifex tubifex* (Annelida: Oligochaeta). Folia Histochem. Cytochem. (Krakow) 20, 109–128.

FRAGOSO, C., KANYONYO, J., MORENO, A., SENAPATI, B.K., BLANCHART, E., RODRIGUEZ. C. 1999. A survey of tropical earthworms: taxonomy, biogeography and environmental plasticity. In *Earthworm Management in Tropical Agroecosystems*, edited by P. Lavelle, L. Brussaard and P. Hendrix. New York: CABI Publishing, pp. 1–26.

FRIEDEL, G. 1922. Les etats mesomorphes de la mattiere. Ann. Phys. (Paris) 19, 273–274.

GAILL, F., HUNT, S. 1986. Tubes of deep sea hydrothermal vent worms *Riftia pachyptila* (Vestimentifera) and *Alvinella pompejana* (Annelida). Mar. Ecol. Prog. Ser. 34, 267–274.

GAILL, F., BOULIGAND, Y. 1987. Alternating positive and negative twist of polymers in an invertebrate integument. Mol. Cryst. Liq. Cryst. 153, 31–41.

GIERE, O.W., RISER, N.W. 1981. Questidae—polychaetes with oligochaetoid morphology and development. Zool. Scr. 10, 95–103.

GIRAUD-GUILLE, M.M., BESSEAU, L., MARTIN, R. 2003. Liquid crystalline assemblies of collagen in bone and in vitro systems. J. Biomech. 36, 1571–1579.

GOVEDICH, F.R. 2004. Tender loving leeches. Australian Sci. 25, 16–22.

GOVEDICH, F.R., DAVIES, R.W. 1998. The first record of the genus *Helobdella* (Hirudinoidea:Glossiphoniidae) from Australia, with a description of a new species, *Helobdella papillornata*. Hydrobiologia 389, 45–49.

GRASSE, P.P. 1959. Classes des Oligochètes et Hirudinées. In *Traité de Zoologie*, vol. 5 (Part 1). Paris: Masson et Cie, 224–713.

GROVE, A.J. 1928. The passage of the spermatozoa into the cocoon in the brandling worm (*Eisenia fetida*, Sav.). Q. J. Microsc. Sci. 71, 283.

GROVE, A.J., COWLEY, L.M. 1927. The relation of the glandular elements of the clitellum of the branding worm (*Esenia fetida*, Sav.) to the secretion of the cocoon. Q. J. Microsc. Sci. 71, 31–46.

HAN, J.H., LAW, S.W., KELLER, P.M., KNISKERN, P.J., SILBERKLANG, M., TUNG, J.S., GASIC, T.B., GASIC, G.J., FRIEDMAN, P.A., ELLIS, R.W. 1989. Cloning and expression of cDNA encoding antistasin, a leech-derived protein having anti-coagulant and anti-metastatic properties. Gene 30, 47–57.

HERTER, K. 1937. Die Okologie der Hirudineen. In *Klassen und Ordnungen des Tierreichs*, Band 4, Abt. III, Buch 4, Teil 2, edited by H.G. Bronn. Leipzig: Akademische Verlagsgesellschaft, pp. 321–496.

HESS, R.T., VENA, J.A. 1974. Fine structure of the clitellum of the annelid *Enchytraeus fragmentosus*. Tissue Cell 6, 503–514.

HIRAO, Y. 1965. Cocoon formation in *Tubifex*, with its relation to the activity of the clitellar epithelium. J. Fac. Sci. Hokkaido Univ. Ser. VI Zool. 15, 625–632.

JAMIESON, B.G.M. 1988. On the phylogeny and higher classification of the Oligochaeta. Cladistics 4, 367–402.

JENSEN, R.A., MORSE, D.E. 1988. The bioadhesive of *Phragmatopoma californica* tubes: a silk-like cement containing L-DOPA. J. Comp. Physiol. [B] 158, 317–324.

KAMINO, K. 2008. Underwater adhesive of marine organisms as the vital link between biological science and material science. Mar. Biotechnol. 10, 111–121.

KAMINO, K., INOUE, K., MARUYAMA, T., TAKAMATSU, N., HARAYAMA, S., SHIZURI, Y. 2000. Barnacle cement proteins: Importance of disulfide bonds in their insolubility. J. Biol. Chem. 275, 27360–27365.

KARON, M. 2003. An overview of the Notch signaling pathway. Semin. Cell Dev. Biol. 14, 113–119.

KASTER, J.L. 1980. The reproductive biology of *Tubifex tubifex* Muller (Annelida: Tubificidae). Am. Midl. Nat. 104, 364–366.

KNIGHT, D.P., HUNT, S. 1974. Mollecular and ultrastructural characterization of the egg capsule of the leech *Erpobdella octoculata*. Comp. Biochem. Physiol. A Comp. Physiol. 47, 871–880.

KOJIMA, S. 1998. Paraphyletic status of Polychaeta suggested by phylogenetic analysis based on the amino acid sequences of elongation factor-1 α. Mol. Phylogenet. Evol. 9, 255–261.

KUTSCHERA, U. 1984. Untersuchungen zur Brutpflege und Fortpflanzungsbiologie beim Egel *Glossiphonia complanata* L. (Hirudinea: Glossiphoniidae). Zool. Jb. Syst. 111, 427–438.

KUTSCHERA, U., WIRTZ, P. 1986. Reproductive behaviour and parental care of *Helobdella striata* (Hirudinea: Glossiphoniidae): a leech that feeds its young. Ethology 72,132–142.

KUTSCHERA, U., WIRTZ, P. 2001. The evolution of parental care in freshwater leeches. Theory Biosci. 120, 115–137.

LEARNER, M.A., LOCHHEAD, G., HUGHES, B.D. 1978. A review of the biology of British Naididae (Oligochaeta) with emphasis on the lotic environment. Freshw. Biol. 8, 357–375.

LEE, K.E. 1985. *Earthworms, Their Ecology and Relationships with Soils and Land Use*. New York: Academic Press.

LODEN, M.S. 1981. Reproductive ecology of Naididae (Oligochaeta). Hydrobiologia 83, 115–123.

LOPEZ CORAL, A. 2007. Characterization of cocoon-associated proteins from arthropods and the aquatic leech Erpobdella punctata (Annelida: Hirudinea). MS Thesis, Rutgers University.

MCHUGH, D. 1997. Molecular evidence that echiurans and pogonophorans are derived annelids. Proc. Nat. Acad. Sci. U.S.A. 94, 8006–8009.

MCHUGH, D. 2000. Molecular phylogeny of the Annelida. Can. J. Zool. 78, 1873–1884.

MAITLAND, P.S., PHILLIPS, D.S., GAYWOOD, M.J. 2000. Notes on distinguishing the cocoons and the juveniles of *Hirudo medicinalis* and *Haemopis sanguisuga* (Hirudinea). J. Nat. Hist. 34, 685–692.

MALECHA, J., PRENSIER, G. 1974. Les glandes clitelliennes de Piscicola geometra L.: structure et cycle annuel. Bull. Soc. Zool. Fr. 99, 433–440.

MALECHA, J., VINCKIER, D. 1983. Formation du cocoon chez l'hirudinee Rhynchobdelle *Piscicola geometra* L. Arch. Biol. 94, 183–205.

MANN, K.H. 1957. A study of a population of the leech *Glossiphonia complanata* (L.). J. Anim. Ecol. 26, 99–111.

MANN, K.H. 1962. *Leeches (Hirudinea). Their structure, Physiology, Ecology and Embryology*. New York: Pergamon Press, Inc.

MAROTTA, R., SHAIN, D.H. 2007. Irregular helicoids in leech cocoon membranes. J. Struct. Biol. 158, 336–343.

MATHER, C.K. 1954. *Haemopis kingi*, new species (Annelida: Hirudinea). Am. Midl. Nat. 52, 460–468.

MASON, T.A., MCILROY, P.J., SHAIN, D.H. 2004. A cysteine-rich protein in the *Theromyzon* (Annelida: Hirudinea) cocoon membrane. FEBS Lett. 561, 167–172.

MASON, T.A., SAYERS, C.W., PAULSON, T.L., COLEMAN, J., SHAIN, D.H. 2005. Cocoon deposition and hatching in the aquatic leech, *Theromyzon tessulatum* (Annelida, Hirudinea, Glossiphoniidae). Am. Midl. Nat. 154, 78–87.

MASON, T.A., MCILROY, P.J., SHAIN, D.H. 2006. Structural model of an antistasin/notch-like fusion protein from the cocoon wall of the aquatic leech, *Theromyzon tessulatum*. J. Mol. Model. 12, 829–834.

MATTHAI, G. 1921. Preliminary observations on cocoon-formation by the common Lahore leech, *Limnatis (Poe-*

cilobdella) granulosa (Sav.). J. Asiat. Soc. Bengal. 16, 341–346.

MEHRA, H.R. 1920. On the sexual phase in certain Indian Naididae. Proc. Zool. Soc. Lond. 31, 457–465.

MICHAELSEN, W. 1926. Oligochaten aus dem Ryck bei Greifswald und von benachbarten Meeresgebieten. Mitt. Hamb. Zool. Mus. Inst. 42, 21–29.

MILNE, I.S., CALOW, P. 1990. Costs and benefits of brooding in glossiphoniid leeches with special reference to hypoxia as a selection pressure. J. Anim. Ecol. 59, 41–56.

MOORE, J.P. 1959. Hirudinea. In *Freshwater Biology*, 2nd edn., edited by W.T. Edmonson. New York: Wiley, pp. 542–557.

NAGAO, Z. 1957. Observations on the breeding habits in a freshwater leech, *Herpobdella lineate* O.F. Muller. J. Fac. Sci. 13, 192–196.

NAGAO, Z. 1958. Some observations on the breeding habits in a freshwater leech, *Glossiphonia lata* Oka. Jap. J. Zool. 12, 219–228.

NEVILLE, A.C. 1993. *Biology of fibrous composites: development beyond the cell membrane*. Cambridge, UK: University Press.

OISHI, M. 1930. On the reproductive process of the earthworm *Pheretima communissima*. Sci. Rep. Tohoku Imp. Univ. 4, 509–524.

PAWLOWSKI, L.K. 1955. Observations biologiques sur les sangues. Bull. Soc. Sci. Lett., Lodz 6, 1–21.

PURSCHKE, G. 1999. Terrestrial polychaetes—models for the evolution of the Clitellata (Annelida). Hydrobiologia 406, 87–99.

PURSCHKE, G., HESSLING, R., WESTHEIDE, W. 2000. The phylogenetic position of the Clitellata and Echiura—on the problematic assessment of absent characters. J. Zoolog. Syst. Evol. Res. 38, 165–173.

REINECKE, A.J., VILJOEN, S.A. 1988. Reproduction of the African earthworm, *Eudrilus eugeniae* (Oligochaeta)—cocoons. Biol. Fertil. Soils 7, 23–27.

REYNOLDS, J.W., COOK, D.G. 1993. Nomenclatura oligochaetologica. Supplementum tertium. A catalogue of names, descriptions and type specimens of the Oligochaeta. New Bruns. Mus. Monograph. Ser. (Nat. Sci.) 9, 1–33.

RICHARDS, K.S. 1977. The histochemistry and ultrastructure of the clitellum of the enchytraeid *Lumbricillus rivalis* (Oligochaeta: Annelida). J. Zool. (Lond.) 183, 161–176.

ROUSE, G.W., PLEIJEL, F. 2001. *Polychaetes*. Oxford: Oxford University Press.

SAWYER, R.T. 1970. Observations on the natural history and behavior of *Erpobdella punctata* (Leidy) (Annelida: Hirudinea). Am. Midl. Nat. 83, 65–80.

SAWYER, R.T. 1971. The phylogenetic development of brooding behaviour in the Hirudinea. Hydrobiologia 37, 197–204.

SAWYER, R.T. 1972. *North American Freshwater Leeches*. Illinois Biological Monographs 46, 1–154.

SAWYER, R.T. 1986. *Leech Biology and Behavior*, Vol I–III. Oxford, UK: Clarendon Press.

SAYERS, C.W., COLEMAN, J., SHAIN, D.H. 2008. Cell dynamics during cocoon secretion in the aquatic leech, *Theromyzon tessulatum* (Annelida: Clitellata: Glossiphoniidae). Tissue & Cell (in press).

SCHROEDER, P.C., HERMANS, C.O. 1975. Annelida: Polychaeta. In *Reproduction of Marine Invertebrates*, vol. 3, edited by A.C. Geise and J.S. Pearse. New York: Academic Press, pp. 1–213.

SHANKLAND, M., SAVAGE, R.M. 1997. Annelids, the segmented worms. In *Embryology: Constructing the Organism*, edited by S.F. Gilbert and A.M. Raunio. Sunderland, MA: Sinauer, pp. 219–235.

SIDDALL, M.E., BURRESON, E.M. 1995. Phylogeny of the Euhirudinea: independent evolution of blood feeding by leeches? Can. J. Zool. 73, 1048–1064.

SIDDALL, M.E., BURRESON, E.M. 1996. Leeches (Oligochaeta?: Euhirudinea), their phylogeny and the evolution of life history strategies. Hydrobiologia 334, 227–285.

SIDDALL, M.E., BUDINOFF, R.B., BORDA, E. 2005. Phylogenetic evaluation of systematics and biogeography of the leech family Glossiphoniidae. Invertebr. Syst. 19,105–112.

STEPHENSON, J. 1930. *The Oligochaeta*. Oxford: Oxford University Press.

STEWART, R.J., WEAVER, J.C., MORSE, D.E., WAITE, J.H. 2004. The tube cement of *Phragmatopoma californica*: a solid foam. J. Exp. Biol. 207, 4727–4734.

SUZUTANI, C. 1977. Light and electron microscopical observations on the clitellar epithelium of *Tubifex*. J. Fac. Sci. Hokkaido Univ. Ser. VI, Zool. 21, 1–11.

SUZUTANI-SHIOTA, C. 1980. Ultrastructural study of cocoon formation in the freshwater Oligochaete, *Tubifex hattai*. J. Morphol. 164, 25–38.

SWEDMARK, B. 1959. On the biology of sexual reproduction of the interstitial fauna of marine sands. Proc. 15th Int. Congr. Zool., Lond. 1958, 327–329.

TALMONT, F., FOURNET, B. 1991. Isolation and characterization of methylated sugars from the tube of the hydrothermal vent tubiculous annelid worm *Alvinella pompejana*. FEBS Lett. 281, 55–58.

TRUCHET, M., VOVELLE, J. 1977. Study of the cement glands of a tubicolous polychaete [*Pectinaria* (=Lagis) *koreni*] with the help of electron microprobe and ion microanalyzer. Calcif. Tissue Res. 24, 231–236.

VAN DER LANDE, V.M., TINSLEY, R.C. 1979. Studies on the anatomy, life history and behaviour of *Marsupiobdella africana* (Hirudinea: Glossiphoniidae). J. Zool. (Lond.) 180, 537–563.

VAN PRAAGH, B.D., HINKLEY, S.D. 2002. Survey of the Giant Gippsland Earthworm, *Megascolides australis* in areas potentially affected by a realignment of the South Gippsland Highway—Bena to Korumburra. Mus. Vic. Sci. Rep. 3, 1–5.

VARDAR, D., NORTH, C.L., SANCHEZ-IRIZARRY, C., ASTER, J.C., BLACKLOW, S.C. 2003. Nuclear magnetic resonance structure of a prototype Lin12-Notch repeat module from human Notch1. Biochemistry 42, 7061–7067.

VOVELLE, J. 1979. Cement glands of *Petta pusilla Malmgren*, an Amphictenidae Tubicole Polychaete, and their organomineral secretion. Arch. Zool. Exp. Gen. 120, 219–246.

WESTHEIDE, W. 1997. The direction of evolution within the Polychaeta. J. Nat. Hist. 31, 1–15.

WESTHEIDE, W., MULLER, M.C. 1996. Cinematographic documentation of enchytraeid morphology and reproductive biology. Hydrobiologia 334, 263–267.

WESTHEIDE, W., MCHUGH, D., PURSCHKE, G., ROUSE, G. 1999. Systematization of the Annelida: different approaches. Hydrobiologia 402, 291–307.

WILKIALIS, J., DAVIES, R.W. 1980. The reproductive biology of *Theromyzon tessulatum* (Glossiphoniidae: Hirudinea) with comments on *Theromyzon rude*. J. Zool. (Lond.) 192, 421–429.

ZHAO, H., WAITE, J.H. 2006. Linking adhesive and structural proteins in the attachment plaque of *Mytilus californianus*. J. Biol. Chem. 281, 26150–26158.

ZHAO, H., SUN, C., STEWART, R.J., WAITE, H.J. 2005. Cement proteins of the tube-building polychaete *Phragmatopoma californica*. J. Biol. Chem. 280, 42938–42944.

Index

α chains, 292–293
Acanthobdella, 23, 117, 317
Acanthobdella peledina, 320–321, 342
Acanthobdellida peledina, 320
Acanthodrilidae, 258, 272
acetylated tubulin (ACT), 159–160
Aciculata, 14, 16
Acoelomorpha, 4
Acrocirridae, 22
acrosome tube, 314, 318, 324
ACT. *See* acetylated tubulin
actin, 172
actin microfilaments, 121–122
adenosine triphosphate (ATP), 310–311
    citrate lyase, 289
afterhyperpolarization (AHP), 143
agriculture, 274
AHP. *See* afterhyperpolarization
albumenotrophy, 333
Alciopidae, 212
algal "gardening" behavior, 214
Alitta virens, 206
Allolobophora celtica, 274
Allolobophora chlorotica, 39–40
alpha-tubulin, 94
Alvinella pompejana, 211, 287–290, 341
    adaptation studies of, 297–298
    amino acid composition of, 293
    cellular temperature adaptation, 293–297
    chemical and thermal stability of, 291–299
    temperature comfort of, 290–291
    tubes, 291–292
Alvinella spp., integument types in, 291
Ampharetidae, 211
Amphinomidae, 22, 50, 211
Amphitrite spp., 213
Amynthas, 275
Amynthas corticis, 273
Amynthas diffringens, 273
Amynthas gracilis, 273
Amynthas hawayanus, 273
Amynthas hendersonianus, 276
Amynthas minimus, 269

Amynthas spp., 269
Amynthas tokioensis, 273
Anaitides, 219
Andiorrhinus amazonicus, 232
Anisochaeta spp., 268
Annelida, 4. *See also* Lophotrochozoa
    and brachiopod/phoronid/nemertean clade, 5
    cellular and behavorial properties of learning in, 135–150
    culture techniques for, 48–50
    developmental events, heterochronic shift in, 79
    diversity in cleavage of, 78–79
    in evo-devo biology, 65–82
    genetic and developmental tools for, 5–6
    genomics, 108–109
    groups within, 78
    hydrothermal vent, 287–298
    life cycle cultures of, 47–59
    major lineages of, 47
    and metazoan phylogeny, 67–68
    as model organisms, 6–9, 23–24
        criteria for selection of, 48
    molecular analysis of, 13–26
    molecular relationships with allies, 16–22
    phylogeny of, 4–5, 13–26, 24–25
        molecular, 16–22
        problems inferring, 23–24
    radiation, 25
    segmentation, molecular mechanism of, 73–74
    studies of subtaxa relationships among, 16
    subtaxa, 23
        summary of representative studies of, 15
    temperature adaption in, 290–291
anoxia, 207
anterior-posterior (AP) axis, 70, 72, 186
    polarity, 106–108
Aphelochaeta sp., 212
Aplysia, 139, 143
Aplysia californica, 5
Aporrectodea caliginosa, 39–40, 237
Aporrectodea rosea, 39–40
Aporrectodea turgida, 273
Archiannelida, 16

Annelids in Modern Biology, Edited by Daniel H. Shain
Copyright © 2009 John Wiley & Sons, Inc.

Arenicola marina, 207, 213–214, 221, 293
Arenicolidae, 16
Arthopoda, 4, 68, 81, 339
astral microtubules, 122
ATP. *See* adenosine triphosphate
Aulodrilus, 322
Autolytinae, 49
Autolytus, 49
Autolytus brachycephalus, 56
Autolytus prolifera, 56
axial patterning, 4, 66, 77

BAFs. *See* bioaccumulation factors
"bandlet," 71, 121, 124–127
bar coding, 34, 110
basal taxon, 66
Begemius, 258, 272, 275
behavioral neurobiology, 82
bending, local, 166
benthos, states of change in, 218–219
BI. *See* biotic index
bilateral organisms, clades of, 4–5
Bimastos parvus, 273
Bimastos spp., 268, 273–274
bioaccumulation, 220
    factors, 234
    test
        Enchytraeid, 233–235
        oligochaete, 233–235
        sediment, 235–236
biodiversity comparisons, requirements of,
    269–271
biomonitoring, 217–219
biotic index (BI), 217
bioturbation, 214, 216
blast cells, 71, 72, 126–127
blastomeres, 74–78, 94, 117–118, 120
Boccardia hamata, 210
Boccardia proboscidea, 54
body shortening, 169–170
Brachiopoda, 4–5
branch length, 24
branch point reflection, 149
Branchiobdella italica, 321
Branchiobdella pentodonta, 334
Branchiobdellida, 320–321
Branchiomaldane vincentii, 208
Branchiomma luctuosum, 215, 219
Branchiomma vesiculosum, 147
Branchiura sowerbyi, 235
BrdU, 73, 91, 95
brooding behavior, 334–335
Bryozoa, 4

Caenorhabditis elegans, 3, 13, 23–24, 65, 68, 76,
    81, 99, 108, 124, 150, 169, 179
calcium sensors/signaling, 173, 174
Californian blackworm. *See Lumbriculus variegatus*
Calopteryx, 35
cAMP/PKA messenger pathway, 141
Canalipalpata, 14, 16
Capilloventridae, 23
Capitella capitata, 7, 16, 23–24, 50, 76, 89, 208,
    216, 218
    chromosome numbers in, 90
    life history traits in, 90–91
    mesoderm and nervous system of, 95–97
Capitella sp., 207, 209, 219
Capitella sp. I, 5–8, 13, 23–24, 81, 104, 220
    advantages of, for evo-devo studies, 89
    allozyme patterns and species designations,
        89–90
    colonizing polluted areas, 109
    developmental stages in, 92, 94
    evolution, development and ecology of origin of,
        88–111
    gene expression during gut development of, 98
    metabolizing polycyclic aromatic hydrocarbons,
        109
    metamorphosis in, 93
    morphology, 91–92
    phylogeny of, 110–111
    vasa expression in, 101
Capitella sp. II, 7
Capitella sp. IIIa, 90
Capitellidae, 211
carabid beetles, 234
carbendazim, 232
cathepsin L (*Tt-catl*), 177
cell division, asymmetric, 123–124
centipedes, 234
central arbors, 163–164
central nervous system (CNS)
    development of the, 157–159
    injury, 172–173
    leech, 156–180
        neuromeres in, 156
        and regeneration, 169–171
    oligochaete, 187
central pattern generator (CPG), 144
cephalochordate amphioxus, 107
CGCs. *See* clitellar gland cells
chaetal structures, 5
Chaetopteridae, 16, 22–23, 211
Chaetopterus, 24, 76, 81, 104, 107
Chaetopterus variopedatus, 8–9, 13, 24, 207
Chaetozone setosa, 216

Chironomidae, 230
Chironomus riparius, 230
chitin, 291
Chlamydomonas nivalis, 305
Chordata, 4–5, 96
CI. *See* consistency index
ciliary band markers, 94
Ciona intestinalis, 5
circumferential indentation, 167
Cirratulidae, 16, 50–51, 215–216
Cirratulus cirratus, 209
Cirriformia, 219
Cirriformia spirobrancha, 50–51
cleavage, 120, 123–124
    equal, 75–76, 94
    polar lobe-dependent, 76
    spiral, 74–77, 117–121
    unequal, 76, 91, 117
clitellar gland cells (CGCs), 329–332
Clitellata, 5, 14, 16, 22, 25, 48, 71–73, 78, 305
    assessing phylogenetic relationships in, 314–325
    cocoons, 328–342
        biomaterials applications for, 342
    cryptic speciation in, 31–42
    ectoparasitic, 320
    examples of model organisms of, 34–36
    hermaphroditic reproduction of, 314, 329–330
    as model organisms, 34–40
    morphology, 32–34
    patterns of spermatological characters in, 324–325
    phylogeny of, 23, 318–319
    polyploidy of, 33–34
    popularity of, for research, 31–32
    reproductive biology in, 330
    spermatozoal characters of, 315
    spermatozoon, 324
    variation, sources and kinds of, 32–34
clitellum, 314, 328–331
Clymenella torquata, 14, 206, 213
cnidarians, 4
CNQX-sensitive synapse, 142
CNS. *See* central nervous system
cocoons
    Clitellata, 331–336
    T. tessulatum, 336–338
    types of, 329
coelom, 25, 95, 176–177
Cognettia sphagnetorum, 228
COI. *See* genes, cytochrome oxidase I
collagen, 291–295
Collembola, 304
conduction, 149–150

consistency index (CI), 324
Cossura sp., 207
cox1, 8, 14, 22–23
cox3, 14
C-propeptide, 294
crack propagation, 215
Crassiclitellata, 318
Crassostrea virginica, 210
Ctenodrilidae, 51
Ctenodrilus, 16
Ctenodrilus serratus, 51, 220
ctenograths, 243
ctenophores, 4
culture techniques, 48–50
Cupiennius salei, 105
Cupressus sempervirens, 276
cystatin B *(Tt-cysb)*, 177
cysteine-rich intestinal protein (CRIP), 172
cytaster, 122
cytb, 14
cytokinesis, 76
    proinflammatory, 178
cytoplasm, 76
cytotoxic lymphocytes, 177

Danio rerio, 3
DD-PCR. *See* differential display-PCR
dendritic arbor, 156–158
Dendrobaena attemsi, 274
Dendrobaena octaedra, 39–40, 274
Dendrobaena spp., 269
Dendrodrilus rubidus, 39–40
deoxyribonucleic acid (DNA), 34
    RAPD, 40
destabilase, 177–178
Deuterostoma, 3–4, 67–70
developmental system drift (DSD), 126–127
5,7-DHT, 138–139
Diachaeta, 258
Dichogaster curgensis, 277
Dichogaster spp., 258, 269, 275
differential display-PCR (DD-PCR), 128
differentiation, 95, 96
5,7-dihydroxytryptamine (5,7-DHT), 138–139
Dinophilidae, 16
Dinophilus gyrociliatus, 51–52, 220
Diopatra, 206, 210
dipteran insects, 70
DM. *See* proteloblast, mesodermal DM
D-NAME, 171
DNOPQ. *See* proteloblast, ectodermal DNOPQ
Dodecaceria sp., 210

Doliodrilus, 322

"dorsal pharynx," 97

dorsal-posterior (DP) nerve, 157, 165

Dorvillea, 219, 243–244

Dorvillea articulata, 51

Dorvillea (Schistomeringos) longicornis, 51

Dorvilleidae, 49, 51, 211, 243–253

double-P triplets (GPP), 293

DP. *See* dorsal-posterior nerve

D-quadrant, 75–77

draft genome sequence, 108–109

Drawida barwelli, 273

Drawida japonica, 275

Drawida spp., 269

Drosophila melanogaster, 3, 13, 23–24, 65, 68, 70,
 73–74, 81, 96–97, 99, 101–103, 105–106,
 108, 125, 150, 169, 177, 179

DSCAM, 161

DSD. *See* developmental system drift

Dunaliella sp., 49

earthworm, 7, 230–232
 bioaccumulation test, 232
 common, *see* megadrile oligochaete
 cosmopolitan, 257–278
 field test, 231
 megascolecid, 319
 in soil ecotoxicology, 237
 standardized tests for, 231
 "true," 267

Ecdysozoa, 3–4, 13, 67–70, 96

Echinodermata, 4, 68

Echiura, 5, 14, 16, 22, 25, 74, 78

Eclysippe vanelli, 14

"ecosystem engineers," 228

ectodermal ablations, 159

EF-1α. *See* elongation factor-1α

eggs
 hermaphroditic, 91, 100
 lecithotrophic, 91

Eirene, 49

Eisenia albidus, 41, 234, 237

Eisenia andrei, 39–40, 230, 232, 237

Eisenia fetida, 37, 39–41, 57, 71, 186–187, 228,
 230, 232, 237, 273, 330, 336

Eisenia fetida Eisenia andrei, 39–40

Eisenia japonica, 272, 275

Eisenia koreana, 272

Eisenoides, 273

El Niño-Southern Oscillation (ENSO) index, 311

electrophoresis, 34, 35

embryogenesis, 156, 167
 leech, 118–120

Embryology of Clepsine, The (Whitman), 116

EML. *See* external mucous layer

Enchytraeid, 233

enchytraeid reproduction test (ERT), 233

Enchytraeidae, 32–33, 37–39, 56–57, 301–311,
 317–318, 330
 + Crassiclitellata, 23
 in soil ecotoxicology, 237–238
 soil tests with, 232–235

Enchytraeus, 37–38

Enchytraeus albidus, 33, 56–57, 233–234
 genotypes for, 38–39

Enchytraeus buchholzi, 38–39

Enchytraeus bulbosus, 38–39

Enchytraeus christenseni, 38–39

Enchytraeus coronatus, 95

Enchytraeus crypticus, 38, 56, 233, 237

Enchytraeus japonensis, 186

Enchytraeus luxuriosus, 56, 234

Enchytraeus variatus, 38

endoderm formation, 4

ENSO. *See* El Niño-Southern Oscillation index

Enteromorpha intestinalis, 214

Enteromorpha spp., 49

environmental risk assessment (ERA), 229–230

epsilon proteobacteria, 289

EPSP. *See* excitatory postsynaptic potential

ERA. *See* environmental risk assessment

Erpobdella, 333

Erpobdella lineata, 334

Erpobdella loctoculata, 321

Erpobdellidae, 330, 334

Erpodbella punctata, 334, 336–337, 340

Errantia. *See* Aciculata

EST. *See* express sequence tag

estuaries, 206–207

Eteone, 219

Eucalyptus marginata, 269

Euclymene sp., 207

Eukerria, 258

Eukerria saltensis, 269, 276–277

Eumida, 219

Eunicidae, 16, 211, 244–253

Eurythoe complanata, 50

eusperm, 322

evolutionary development (evo-devo), 66–67, 79–80

Exallopus, 244, 246

excitatory motor neurons (eMNs), 166–167

excitatory postsynaptic potential (EPSP), 142

Exogone lourei, 208

express sequence tag (EST), 4–5, 22, 26, 66, 298

external mucous layer (EML), 332, 336

extracellular matrices (ECMs), 290–291

Fabricia berkeleyi, 208
Faivre's nerve, 140
feeding, 213, 216
Ficopomatus enigmaticus, 209
Flabelligeridae, 16, 22
flatworms, 47, 68, 74
foldases, 173
Fridericia, 39

Galathealinum brachiosum, 14
Galeolaria caespitosa, 48
gamasid mites, 234
gamma-tubulin, 123–124
gastrulation, 78
genes
    18S rRNA (*18S*), 14, 18–20, 22–23, 25, 34,
        318–319
        + 28S rRNA + EF-1α, 21, 25
        analysis of, 17–20
    28S rRNA (*28S*), 22, 25, 34
    brachyury, 108
    "candidate" approach, 73–75, 128
    CapI-Cdx, 100, 107
    CapI-Delta, 105
    CapI-en, 103
    CapI-foxA, 99
    CapI-gataA1, -gataB1, -gataB2, -gataB3, 99
    CapI-Gsx, 107
    CapI-hes1, -hes2, -hes3, -hesr1, -hesr2, 105
    CapI-hh, 100, 103
    CapI-nanos, 100
    CapI-Notch, 105
    CapI-sna1, 100
    CapI-twt1 and *twt2*, 100
    CapI-vasa, 100
    CapI–wnt1, 99–100, 103
    CapI-Xlox, 100, 107
    Cdx, 107
    CH-Hox2, 107
    claudal, 108
    CYP4AT, 109
    CYP331, 109
    cytochrome b (*cytb*), 14
    cytochrome c oxidase subunit I (*cox1*), 8, 14,
        22–23
    cytochrome c oxidase subunit III (*cox3*), 14
    cytochrome oxidase I (COI), 31, 110, 127–128,
        244
    Delta, 105–106
    developmental regulatory, 66, 75
    elongation factor-1α (EF-1α), 14, 22, 25
    engrailed (*en*), 73–74, 102–105, 125, 128
    even-skipped, 103, 108, 125

    expressing, 128–130
    foxA, 99
    "gap," 73–74, 102–103
    GATA1/2/3, 97–99
    GATA4/5/6, 97–98
    GATAb, 98
    Gsx, 107
    hairy, 105–106
    Hau-Pax3/7A, 125
    hedgehog (*hh*), 102–105
    homeobox, 107
    "housekeeping," 66
    Hox, 66, 101–102, 106–107, 124–125, 128
    Hox1-Hox5, 107
    Hro-nos, 125
    hunchback (*hb*), 102–103
    K110, 130
    mitochondrial (mtDNA), 14–16, 31, 40
    mitochondrial COI, 34, 37
    mt16S rRNA (*16S*), 8, 14, 22–23, 34–35, 37
    nad1, nad6, 14
    nanos, 125
    Notch, 105–106
    nuclear, 14, 40
    P450 (CYP), 109
    pair-rule, 102–103
    ParaHox, 107–108
    pax3/7, 103
    Pdu-en, 104
    Pdu-vasa, 100
    Pdu-wnt1, 104
    ProtoHox, 107
    runt, 103
    segment polarity, 102–104
    snail, 96
    translocations of, 15–16
    wingless (*wg*), 102–105
    wnt, 125
    Xlox, 107
genetics, 5–6
germinal band, 71
germinal plate, 71
GFs. *See* giant fibers
giant fibers (GFs), 187
glacier ice worms, 301–311. *See also*
        Mesenchytraeus solifugus; *Mesenchytraeus
        solifugus rainierensis*; *Sinenchytraeus
        glacialis*
    adenylate profiles of, 310
    clades, 307–309
    classification and phylogenetic relationships of,
        305–306
    conservation status of, 311

glacier ice worms (*cont'd*)
  density of, 302–305
  natural history of, 303–305
  origins of, 306–307
  phylogeny of, 307
  physiology of, 309–311
Glossiphonia, 335
Glossiphonia complanata, 118
Glossiphoniidae, 6, 58–59, 71, 117, 330, 334
  brooding behavior in, 334–335
Glossoscolecidae, 232
glutamine (*trnQ*), 14
Glycera dibranchiata, 48, 209–210, 221
glycosylation, 243, 341
  mannosidic, 163–164
Glyphidrilus, 275
Glyphidrilus kuekenthali, 275
Glyphidrilus stuhlmanni, 275
Gly-X-Y amino acid triplets, 292–294
Goniada, 219
Gordiodrilus elegans, 269
Grania, 32
growth zones, 95
guanylate cyclase inhibitor, 171

Haementeria, 334
Haementeria ghilianii, 6, 139, 333
Haemopis spp., 117
Haliotis rufescens, 210
Halosydna, 208
Halosydna brevisetosa, 53
Halosydna johnsoni, 53, 208
Haplotaxidae, 318–319
HARAP. *See* Higher-Tier Aquatic Risk Assessment
  for Pesticides
Harmothoë imbricata, 48, 206
HCB. *See* hexachlorobenzene
heat shock proteins (HSPs), 297–298
Helobdella, 122, 335
Helobdella europaea, 128
Helobdella papillornata, 329
Helobdella robusta, 5–7, 13–14, 40, 72, 77, 82,
  101, 104, 109–110, 122–123, 126–128, 169,
  179
Helobdella spp., 7, 23–24, 40–41, 50, 58, 65, 71,
  76, 81–82, 117–118, 125, 334
Helobdella stagnalis, 127, 330, 335
Helobdella triserialis, 6–7, 58, 107, 126–128
Helodrilus oculatus, 269, 274
hemerythrin message, 175
Hemichordata, 4–5
hermaphrodites, 90–91, 118
Hesiocaeca methanicola, 212

Hesiolyra bergi, 297
Hesionidae, 211
Heteromastus, 219
Heteromastus filiformis, 213
Heteroporodrilus, 258
hexachlorobenzene (HCB), 234–235
Higher-Tier Aquatic Risk Assessment for Pesticides
  (HARAP), 238
Hirudinidae, 6–7, 59, 117, 317, 320
  leeches, 156
  life cycle cultures of, 58–59
Hirudo medicinalis, 6–7, 13, 23, 40, 50, 59, 82,
  107, 117, 135–137, 139–140, 160, 163,
  165–166, 169, 186, 330, 333
  segmental nerves in, 158–159
Hirudo spp., 23–24, 333
Hirudo verbana, 7, 40
histone H3, 14, 22
Hm-inx1, *-inx2*, *-inx5*, *-inx6*, *-inx8*, 169
*Hm*LAR1, -2, 161–163
HNK-1, 94
Hodgkin, Alan, 117
Hrabeiella periglandulata, 206
HSPs. *See* heat shock proteins
5-HT. *See* serotonin
5-HT$_7$, 141
Hudineans, sperm ultrastructure in, 320–321
human migration routes, 275
Huxley, Andrew, 117
HXXE divalent ion, 177
Hydra, 185
Hydra magnipapillata, 97
Hydra vulgaris, 97
Hydroides dianthus, 209
Hydroides elegans, 8–9, 53, 81, 94,
  103–104
Hydroides norvegica, 208
hydrothermal vent, 211
  communities, 287–298
hymenopteran insects, 70

Ilyanassa, 6
Ilyanassa obsoleta, 75–76, 125
Ilyodrilus, 33
immune responses, in leech nervous system,
  156–180
immunity, 176–178
immunoreactivity, cGMP, 171
iMN. *See* inhibitory motor neuron
infaunal macrobenthos, 216
inhibitory motor neuron (iMN), 165–167
Insulodrilus bifidus, 314
  spermatozoa, 316

intercellular communication, 169
interleukin-1β, 178
interleukin-6, 178
internal transcriber spacers (ITS1–2), 34
International Code of Zoological Nomenclature
        (ICZN 1999), 278
interneuron S, 170
intertides, 207
introns, 26
introsperm, 314
Iospilidae, 212
Iphitime, 244
Iphitime paguri, 244–246
Iphitimidae, 244
Isochrysis sp., 49
isolates, tools to compare, 42
"ITS region," 34
ITS1–2. *See* internal transcriber spacers

Janua (Dexiospira) brasiliensis, 54

karyotypes, 90
kinases, 173
Kinorhyncha, 4

lambda-cyhalothrin, 232
Lamellibranchia cf. *luymesi*, 211
large fat cells (LFCs), 177
large ribosomal subunit (LSU), 4
learning, nonassociative, 146–147
leech, 6–7, 40, 48, 78
    cellular and behavorial properties of learning in,
        135–150
    differentiation in, 116–130
    immune responses of nervous system of, 156–180
    innexins, 169
    sanguivorous, 332
    swim circuit, 144–146
    touch-based swim response, 145
    whole-body shortening and the s interneuron,
        137–139
leucopoiesis, 177
Leucosticte tephrocitis, 306–307
leukemia, 6
Limnodriloides monothecus, 322–323
Limnodriloidinae, 317, 321–323, 322
Limnodrilus hoffmeisteri, 35–36, 41
Limnodrilus sp., 33, 95
lindane, 234
lineage tracer, intracellular, 73
linkage, M, N, O, P, Q, 125–128
litterbag test, 231
L-NAME, 171

local bend interneurons (LBIs), 166–168
Loligo pealei, 6
long-branch attraction (LBA), 15, 22, 24
long-term depression (LTD), 148–149
long-term potentiation (LTP), 148–149
Lopadorhynchidae, 212
Lophotrochozoa, 3–9, 67–70, 74
    considerations for a model system for, 4–5
    early diversification of, 4
    genetic and developmental tools for, 5–6
    model taxa, 13
    phylogenetic consderations for, 4–5
Lottia gigantea, 5
LSU. *See* large ribosomal subunit
LTD. *See* long-term depression
LTP. *See* long-term potentiation
Lumbricidae, 39–40, 57, 186, 230, 267, 271–274
    soil tests with, 230–232
Lumbriculidae, 23, 37, 186, 317, 320
    bioaccumulation test, 235–236
Lumbriculus spp., 37
Lumbriculus variegatus, 33, 37, 41, 58, 140, 186,
        228, 236, 310
    asexual reproduction of, 191–192
    life cycle cultures of, 192
    as a model, 185–199
    and PCP toxicity, 235–236
    respiratory apparatus of, 190
Lumbricus rubellus, 7, 232
Lumbricus spp., 24
Lumbricus terrestris, 7, 13–14, 228–229, 237,
        272
Lumbrineridae, 16
Lumbrineris, 208, 219
"lumping," 32
Lymnaea, 140
Lymnaea stagnalis, 75–76, 174
Lysidice collaris, 208
Lysidice ninetta, 208

mAbs. *See* monoclonal antibodies
macromeres, 77–78, 121, 122
magelonids, 215
malate dehydrogenase-1, 90
Maldanidae, 211, 216
Manayunkia speciosa, 206
Marphysa leidi, 221
Marphysa sanguinea, 221
Marsupiobdella africana, 335
matrix-assisted laser desorption/ionization
        (MALDI), 179–180
mechanoreceptor, T, 146
Mediomastus sp., 207–208, 212, 219

megadrile oligochaete, 330
   in archeology and human history, 273–277
   biodiversity of, 268–273
   characteristics of, 268
   cosmopolitan, 258–278
   exotic, 268–273
   global and historical perspective on, 257–278
   introduced exotic, 258
   nonendemic species of, 258–266
   number of species of, 267–273
   population, components of, 258
   resident native, 258
   translocated native, 258
   transportations, 273–279
Megascolecidae, 258, 267, 271–272
Megascolides australis, 331
Melanenchytraeus, 305
Mesenchytraeus altus, 301
Mesenchytraeus armatus, 306
Mesenchytraeus flavus, 306
Mesenchytraeus gelidus, 301, 305–307
Mesenchytraeus hydrius, 301
Mesenchytraeus niveus, 305
Mesenchytraeus pedatus, 306–307
Mesenchytraeus solifugus, 39, 55, 305, 307–310
Mesenchytraeus solifugus rainierensis, 305,
      309
mesodermal presumptive growth zone (MPGZ),
      100–101
metamerism, 70
metamorphosis, 94
Metaphire, 275
Metaphire hilgendorfi, 273
metatrochophore, 93
   gut formation in, 97–100
methylene blue, 171
methysergide, 146
Michaelsen, W., 267
Microdorvillea, 244
microdriles, 330
microglial cells, 171–172
micromeres, 75, 95
Microscolex dubius, 258
Microscolex kerguelarum, 258
Microscolex macquariensis, 258
Microscolex spp., 258
microtubules, 122
mitochondria, 314–317, 324
model organisms, 24, 31
model system, choosing, 3–9
model waxing, for Polychaete study,
      88–111
Mollusca, 4–5, 47, 68, 74
Moniligastridae, 272

monoclonal antibodies (mAbs), 157, 160
Monticellina spp., 206, 217
Mooreonuphis stigmatis, 208
morphallaxis, 186–187, 192–198
morphology
   and clades, 13–14
   plasticity of, 25, 35
MPGZ. *See* mesodermal presumptive growth zone
mRNA, 123
Murchieona minuscula, 269
Mus musculus, 3
mussel beds, 207–208
Mya arenaria, 6
myohemerythrin, 172
myomodulin, 141, 143
Mytilus edulis, 208
Myxobolus cerebralis, 35
Myzostoma seymourcollegiorum, 14
myzostomids, 5, 14

nad1, *nad6*, 14
Naididae, 23, 57, 186, 330
Naidinae, 33, 41, 317
Neanthes acuminata, 52
Neanthes arenaceodentata, 52, 220–221
Neanthes caudata, 52
Neanthes lighti, 207
Neanthes limnicola, 52
Neanthes spp., 219
Neanthes succinea, 52
Nematoda, 4, 68, 81, 339
Nematonereis unicornis, 208
Nematostella vectensis, 97, 99
   CapI-Post1, 106
Nemertea, 4–5, 74
Nephasoma minuta, 108
Nepthys spp., 14, 219
Nereididae, 8, 23, 52–53, 206, 211–212, 214
nereids, 208
Nereis crigognatha, 52
Nereis diversicolor, 214, 221
Nereis grubei, 52
Nereis spp., 8, 76, 213, 219
Nereis succinea, 206
Nereis vexillosa, 214
Nereis virens, 107, 210, 214
Nerillidae spp., 16, 211, 212
nerve injury, response to, 171–172
netrin, 162
   receptor unc-5, 163
   siRNA, 163
neurohemerythrin, 173–174
neuroimmunity, 173–176
neuronal circuits, 165–169

neuronal formation, 4
neurons
    AE and AP, 165
    CD8+ cells, 176
    L, 137, 143
    mechanosensory, 137
    NK-like, 176
    nociceptive (N), 137, 144
    P_D, 160–161
    pressure (P) sensory, 137, 144, 147, 163,
        166–168
    P_V, 161–163
    S, 147
    touch (T) sensory, 137, 144, 163
neuropil, 163–164
nitric oxide (NO), 171–172, 175
NMDA-R, 148–149
notch signaling pathway, 105–106
nuclear factor κB, 178

OBT. *See* oligochaete bioaccumulation test
Ocnerodrilus occidentalis, 269
Octochaetidae, 258, 272
Octochaetus ambrosensis, 258
Octodrilus spp., 276
octopamine, 149
Oligochaeta, 48, 71–72, 78, 97, 330
    central nervous system (CNS) of, 187–189
    for ecotoxicological assessment of soil and
        sediment, 228–238
    escape reflex anatomy, 188–189
    life cycle cultures of, 56–58
    neural regeneration in, 187
    in sediment ecotoxicology, 238
Onuphidae, 214
Onuphis teres, 221
operational taxonomic unit (OTU), 24
opercula, 329, 334, 338–339, 342–343
Ophryotrocha, 243–253
Ophryotrocha adherens, 251–252
Ophryotrocha akessoni, 252
Ophryotrocha alborana, 246
Ophryotrocha cf. *vivipara*, 251
Ophryotrocha claparedii, 251
Ophryotrocha diadema, 8, 51, 95, 246, 250
Ophryotrocha gracilis, 246–248, 250
Ophryotrocha hartmanni, 244–248, 251
Ophryotrocha labidion, 243
Ophryotrocha labronica, 8, 24, 246, 248, 252
Ophryotrocha mandibulata, 243
Ophryotrocha platykephale, 252
Ophryotrocha puerilis, 8, 250, 252
Ophryotrocha sp. *japonica*, 252
Ophryotrocha spatula, 252

Ophryotrocha spp., 8, 13, 23, 49, 211–212, 220
    characteristics of, 243–244
    described, 245
    ecology of, 250–252
    evolution and ecology of, 243–253
    phylogeny of, 247
    reproductive biology of, 247–250
    taxonomy of, 243–247
    undescribed, 246
Orbinia latreillii, 14
Orbiniidae, 14, 16, 22
Osedax spp., 210–211
OTU. *See* operational taxonomic unit
Ougia spp., 244
Oweniidae, 22
2-oxoglutarate:acceptor oxidoreductase, 289
oxygen minimum zones (OMZs), 206

PAHs. *See* polycyclic aromatic hydrocarbons
Palola viridis, 221
Palpata, 14
Paralvinella sulfincola, 287, 297
Paranais litoralis, 9, 57
Paraonidae, 211
Paraonidae sp., 212
Parapionosyllis sp., 208
Parapodrilus, 244
parasperm, 322
Parergodrillidae, 22
Parergodrilus heideri, 206
Parophryotrocha, 244
Parougia, 244
parthenogenesis, 268
Patella vulgate, 75–76
Pax-6, 66
PCP. *See* pentachlorophenol
PCs. *See* precursor cells
PEA. *See* proenkephalin A
PEC. *See* predicted environmental concentration
Pectinaria koreni, 214
Penaeus kerathurus, 221
Penaeus vannani, 221
pentachlorophenol, 232
peptide B, 177
perforatorium, 314
Perinereis nuntia, 221
Perionyx excavatus, 14
peripheral nervous system (PNS), 156
    and DP nerve formation, 160
    and ectopic ganglia, 160–161
    genesis of, 159–160
    P-neuron arbor of, 161–163
    target, disruption of, and abnormal central
        projections, 165

PGCc. *See* primordial germ cells

Pharmacopia californica, 339

Phascolion strombus, 107

Phascolopsis gouldii, 14

phenotype, and developmental change, 66

Pheretima, 275, 276

phoronids, 5

phosphoglucomutase, 89

phosphohexose isomerase, 90

Phreodrilidae, 314, 318–319

Phyllodocidae, 211

phylogenetic analysis, characters for, 15

phylogeny

   metazoan, typical, 68–69

   molecular, 67

      Annelida, 16–22

Piscicola geometra, CGC types in, 331

Piscicolidae, 330

Pisione, 16

Pista cristata, 14

Placobdella papillifera, 331–332

Placopeclen magellanicits, 210

plankton, 212

platyhelminthes, 5

Platynereis, 8, 24, 219

Platynereis bicanaliculata, 208, 214

Platynereis dumerilii, 5, 8, 13–14, 24, 53, 75–76,
     81, 97, 99–100, 102, 104–105, 107–109

   germ cell specification and differentiation in,
     100–101

   vasa expression in, 101

Platynereis massiliensis, 53

Platyzoa, 4

Playthelminthes, 4

Poebius, 16

Poeobiidae, 212

Pogonophora, 5, 22, 25, 78

pollution, 215, 218, 219, 222

Polychaeta, 5, 14, 22, 25, 47–48, 71–72, 78, 267

   and cold methane environments, 211–212

   economic importance of, 221

   as environmental indicators and mediators,
     214–217

   in environmental studies, 205–222

   feeding guilds, 213

   gonochoristic, 212

   habitats and demographics of, 205–206

   holopelagic, 212

   importance of, 205–206, 222

   life cycle cultures of, 50–56

   metals toxic to, 219–220

   model, waxing, 88–111

   selected metals and, 220

   shell-boring, 210

   skeletons, 210–211

   and toxicological testing, 219–221

Polychaete, 291, 342

polycyclic aromatic hydrocarbons (PAHs), 216

Polydora, 219

Polydora ciliata, 55, 209

Polydora commensalis, 210

Polydora concharum, 210

Polydora cornuta, 55

Polydora ligni, 55, 208, 210

Polydora socialis, 210

Polydora spp., 210

Polydora websleri, 210

Polygordiidae, 16

Polynoidae, 53, 206, 208, 211

Polypheretima, 275

Polypheretima brevis, 275

Polypheretima elongata, 276, 277

Polypheretima pentacystis, 275

Polypheretima voeltzkowi, 276

polyploidy, 33–34, 39–41, 268

Pomatoceros triqueter, 54

Pompeii worm. *See Annelida pompejana*

Pontodoridae, 212

Pontodrilus, 258

Pontodrilus litoralis, 258

Pontoscolex, 258

Pontoscolex corethrurus, 258, 272, 277

Potamothrix, 33

PPCs. *See* pro-precursor cells

precursor cells (PCs), 331

predicted environmental concentration (PEC),
     229–230

Priapulida, 4

primordial germ cells (PGCs), 125

Prionospio, 219

Pristina leidyi, 57, 104, 186

proenkephalin A (PEA), 177

proline hydroxylation, 293

Propappidae, 23

Propappus, 317

Propappus glandulosus, 317–319

Propappus Volki, spermatozoon of, 317–319

pro-precursor cells (PPCs), 331

proteins, 291

   cytoskeletal and metabolic, 173

   gliarin, 173

   immunity and regeneration, 176

   modulated after bacterial challenge, 174

   ReN3, 172–173

proteloblast, 122–124

Protodrilidae, 16

Psammoryctides, 33
Pseudonereis, 208
Pseudophryotrocha, 244
Pseudopolamilla reniformis, 210
Pygospio, 219
Pygospio elegans, 55, 209, 214

Questa, 16

receptor protein tyrosine phosphatase (RPTP),
    161
redox potential discontinuity (RPD), 219
regeneration, 71–72, 140, 169–170
    annelid and leech, 185–186
    in *Capitella* spp., 92–93
    epimorphic, 186
    in *Lumbriculus veriegatus*, 185–199
    morphallaxis, 186–187
    neuronal, 169–170, 172
    segmental, 186
reproduction, 41, 48
    in Clitellata, 330
    fission, 330
    fragmentation, 330
    gonochoristic, 248, 250
    hermaphrodic, 330
        sequential, 248, 250
        simultaneous, 248–250
    in *Lumbriculus veriegatus*, 185–199
    parthenogenesis, 330
    self-fertilizating, 118, 330
retention index (RI), 324
Retzius cells, 141–143, 146–147
reverse tricarboxylic acid (rTCA) cycle, 289
Rhododrilus, 258, 272
Rhododrilus edulis, 276
Rhododrilus kermadecensis, 276
Rhododrilus queenslandicus, 276
Rhodomonas sp., 49
Rhyacodrilus, 33
RI. *See* retention index (RI)
Riftia pachyptila, 14, 287, 289, 293–295, 296
rough endoplasmic reticulum protein 1 (RER-1),
    172
RPD. *See* redox potential discontinuity

16S, 8, 14, 22–23, 34–35, 37
18S rRNA (*18S*), 14, 18–20, 22–23, 25, 34,
    318–319
S cell. *See* S interneuron
S interneuron, 137–144
Sabella pavonina, 186
Sabella spallanzanii, 215, 219–220

Sabellaria cementarium, 76
Sabellidae, 16, 22, 209, 211
Saccocirridae, 16
Salmacina (or *Filograna*) spp., 54
Satchellius mammalis, 274
scanning electron microscopy (SEM), 340
Schistomeringos rudolphii, 252
Schistosoma mansoni, 6
Schmidtea mediterranea, 5
Scolecida, 14, 16
Scolelepis, 219
Scoloplos, 208
Scoloplos armiger, 14
Scoloplos robustus, 210
sea grasses, 208
Seepiophyla jonesi, 211
segmental circuitry, and local bending, 166
segmental nerves, 159–160
segmentation, 24–25, 68–74, 88, 97
    Annelida, 101–102
        cellular mechanism of, 71–73
        molecular mechanism of, 73–74
    CapI-hbnl, 97
    CapI-hes2, 97
    Capl-hes3, 97
    de novo, 72
    engrailed, 97
    evolution origin of, 70–71
    homologous, 70
    hunchback , 97
    HyAlx, 97
    interpreting conflicting comparative data about,
        104–106
    leech, ganglion, 156
    muscle, 95
    periodicity in, 71
    vertebrate, 70
sensitization, 141, 146–147, 170
"sensory tile," 161–163
serotonin (5-HT), 138, 141–144, 146–147, 149
Serpula narconensis, 209
Serpulidae, 16, 53–54, 209, 211
shortening, whole-body, 138
Siboglinidae, 5, 14, 16, 22–23, 25, 211
Sipuncula, 5, 14, 16, 22, 25, 74
siRNA, 173
small subunit (SSU), 4
Smithsonidrilus spp., 322
snowfleas, 304
Solea senegalensis, 221
Solea vulgaris, 221
Solenopsis richteri, 207
solifugus, 304

speciation, in Clitellate model organisms, 31–42
species
  novel, using as models, 47
  number of, in animal kingdom, 47
  opportunistic, characteristics of, 218
specification
  axis, 74–77
  gene expression during, 95–100
spermathecae, 330
spermine NONOate (SPNO), 171
Spio spp., 209
Spionidae, 49, 54–56, 210–211, 219
Spiralia, 68, 73–77
Spirobranchus corniculatus, 48
Spirorbis spirorbis, 206
"splitting," 32
springtails, 304
SSU. *See* small subunit
Stauronereis ruldolphi, 51
stem cell
  cytoplasmic rearrangement of, 121
  development, 117–121
  differentiation, 124–125
  genesis, 116–130
    and development, 117–120
    factors affecting, 121–124
  research, tenets of, 116
Sternaspidae, 215
Sternaspis, 216
Sternaspis scutata, 207
strand asymmetry, 15
Streblospio, 219
Streblospio benedicti, 55–56, 206, 210
Stygocapitella subterranean, 206
subtidal benthic communities, 209–212
"subzero effect," 144
sulfated glycosaminoglycans, 291
swim circuit, 168–169
Syllidae, 16, 56, 212
synapsin, 172
synaptic plasticity, in leech CNS, 147–150

Taenia solium, 5
tandem cystatin B (*Tt-cysb*)–cathepsin L (*Tt-catl*), 177
TC. *See* toxic concentration
Tcp. *See Theromyzon tessulatum*, cocoon protein (Tcp)
teloblasts, 71–73, 95, 120–121, 126, 156
teloplasm, 121–123
TER. *See* toxic exposure ratio
Terebellida, 14, 215–217

Terebellides stroemi, 14
Terebrasabella heterouncinata, 210
terrestrial model ecosystems (TMEs), 237
Thalassia testudinum, 208
Thalassodrilides, 322
theromacin, 177
Theromyzon rude, 127
Theromyzon spp., 58–59, 117–118, 122
Theromyzon tessulatum, 6, 58, 81, 104, 126, 128–130, 177, 334–340
  cocoon, 338–341
  CGC types in, 330–333
  segmental nerves in, 159
Theromyzon trizonare, 329
Tierreich, Das: Vermes (Michaelsen), 267
thioredoxin (TRX), 172
TMEs. *See* terrestrial model ecosystems
Tomopteridae, 212
toxic concentration (TC), 229
toxic exposure ratio (TER), 229
transposable elements, 26
trichobranchids, 215–216
Trilobodrilus axi, 206
triploblasty, 95
Tritonia diomedea, 6
trochophore larva, 88, 93
trotifers, 47
Tt-catl. *See* cathepsin L
Tt-cysb. *See* cystatin B
Tubifex hattai, 57–58, 126, 337
Tubifex spp., 117
Tubifex tubifex, 8, 9, 32–33, 35, 41, 57, 65, 71, 76–77, 81, 107, 124, 218, 235, 330, 336
  neighbor-joining tree for, 36
Tubificidae, 23, 34–35, 41, 57–58, 318–319, 330
  bioaccumulation test, 235–236
  physiology of, 320–323
  sperm ultrastructure in, 320–322
Tubificinae, 317, 322
tubulin, 172
tumor necrosis factor-α, 178
28S, 22, 25, 34
Typhloscolecidae, 212
Typosyllis pulchra, 56
Typosyllis spp., 208

U2 snRNA, 22
Ulva sp., 49
Urechis caupo, 14

vent shrimps, 298
ventral nerve cord (VNC), 187

vermes, sex among, 100–101
Vibrio alginolyticus, 215
VNC. *See* ventral nerve cord

Whitman, Charles O., 116, 121, 124

Xenopus laevis, 96
Xenoturbella, 4

zebra fish, 179
Zostera noltii, 214